STUDIES IN INTERFACE SCIENCE

Complex Wave Dynamics on Thin Films

STUDIES IN INTERFACE SCIENCE

SERIES EDITORS
D. Möbius and R. Miller

Vol. 1 Dynamics of Adsorption at Liquid Interfaces. *Theory, Experiment, Application.* By S.S. Dukhin, G. Kretzschmar and R. Miller

Vol. 2 An Introduction to Dynamics of Colloids. By J.K.G. Dhont

Vol. 3 Interfacial Tensiometry. By A.I. Rusanov and v.A. Prokhorov

Vol. 4 New Developments in Construction and Functions of Organic Thin Films. Edited by T. Kajiyama and M. Aizawa

Vol. 5 Foam and Foam Films. By D. Exerowa and P.M. Kruglyakov

Vol. 6 Drops and Bubbles in Interfacial Research. Edited by D. Möbius and R. Miller

Vol. 7 Proteins at Liquid Interfaces. Edited by D. Möbius and R. Miller

Vol. 8 Dynamic Surface Tensiometry in Medicine. By V.M. Kazakov, O.V. Sinyachenko, V.B. Fainerman, U. Pison and R. Miller

Vol. 9 Hydrophile-Lipophile Balance of Surfactants and Solid Particles. *Physicochemical Aspects and Applications.* By P.M. Kruglyakov

Vol. 10 Particles at Fluid Interfaces and Membranes. *Attachment of Colloid Particles and Proteins to Interfaces and Formation of Two-Dimensional Arrays.* By P.A. Kralchevsky and K. Nagayama

Vol. 11 Novel Methods to Study Interfacial Layers. By D. Möbius and R. Miller

Vol. 12 Colloid and Surface Chemistry. By E.D. Shchukin, A.V. Pertsov, E.A. Amelina and A.S. Zelenev

Vol. 13 Surfactants: Chemistry, Interfacial Properties, Applications. Edited by V.B. Fainerman, D. Möbius and R. Miller

Vol. 14 Complex Wave Dynamics on Thin Films. By H.-C. Chang and E.A. Demekhin

Complex Wave Dynamics on Thin Films

Hsueh-Chia Chang

University of Notre Dame, IL, USA

Evgeny A. Demekhin

Kuban State University, Russia

2002
ELSEVIER
Amsterdam – London – New York – Oxford – Paris – Tokyo – Boston – San Diego
San Francisco – Singapore – Sydney

ELSEVIER SCIENCE B.V.
Sara Burgerhartstraat 25
P.O. Box 211, 1000 AE Amsterdam, The Netherlands

© 2002 Elsevier Science B.V. All rights reserved.

This work is protected under copyright by Elsevier Science, and the following terms and conditions apply to its use:

Photocopying
Single photocopies of single chapters may be made for personal use as allowed by national copyright laws. Permission of the Publisher and payment of a fee is required for all other photocopying, including multiple or systematic copying, copying for advertising or promotional purposes, resale, and all forms of document delivery. Special rates are available for educational institutions that wish to make photocopies for non-profit educational classroom use.

Permissions may be sought directly from Elsevier Science Global Rights Department, PO Box 800, Oxford OX5 1DX, UK; phone: (+44) 1865 843830, fax: (+44) 1865 853333, e-mail: permissions@elsevier.co.uk. You may also contact Global Rights directly through Elsevier's home page (http://www.elsevier.com), by selecting 'Obtaining Permissions'.

In the USA, users may clear permissions and make payments through the Copyright Clearance Center, Inc., 222 Rosewood Drive, Danvers, MA 01923, USA; phone: (+1) (978) 7508400, fax: (+1) (978) 7504744, and in the UK through the Copyright Licensing Agency Rapid Clearance Service (CLARCS), 90 Tottenham Court Road, London W1P 0LP, UK; phone: (+44) 207 631 5555; fax: (+44) 207 631 5500. Other countries may have a local reprographic rights agency for payments.

Derivative Works
Tables of contents may be reproduced for internal circulation, but permission of Elsevier Science is required for external resale or distribution of such material.
Permission of the Publisher is required for all other derivative works, including compilations and translations.

Electronic Storage or Usage
Permission of the Publisher is required to store or use electronically any material contained in this work, including any chapter or part of a chapter.

Except as outlined above, no part of this work may be reproduced, stored in a retrieval system or transmitted in any form or by any means, electronic, mechanical, photocopying, recording or otherwise, without prior written permission of the Publisher.
Address permissions requests to: Elsevier Science Global Rights Department, at the mail, fax and e-mail addresses noted above.

Notice
No responsibility is assumed by the Publisher for any injury and/or damage to persons or property as a matter of products liability, negligence or otherwise, or from any use or operation of any methods, products, instructions or ideas contained in the material herein. Because of rapid advances in the medical sciences, in particular, independent verification of diagnoses and drug dosages should be made.

First edition 2002

Library of Congress Cataloging in Publication Data
A catalog record from the Library of Congress has been applied for.

British Library Cataloguing in Publication Data
A catalogue record from the British Library has been applied for.

ISBN: 0 444 50970 4
ISSN: 1383 7303

♾ The paper used in this publication meets the requirements of ANSI/NISO Z39.48-1992 (Permanence of Paper).
Printed in The Netherlands.

Preface

Waves dynamics have fascinated us through out the history of civilization. Ocean waves like tsunamis, tidal swells, ship bow waves, breaking beach waves etc. have been studied for centuries. They have even inspired new fields of mathematics. There is, however, a different type of waves that have only been scrutinized in the last 50 years. Yet their dynamics are even richer than deep-water (ocean) waves. In fact, the system that exhibits these waves is a simple prototype hydrodynamic instability that demonstrates many features of other flow instabilities. As a result, it can become a testing ground for many new hydrodynamic concepts/theories.

We are referring to interfacial wave dynamics on a thin film that flows down an inclined plane, the subject of this book. These waves appear on the windshields of our cars, in the evaporator tubes of our refrigerators, within industrial/commercial cooling towers and mass-transfer units, during coating processes etc. However, we shall not focus on the engineering aspects of wave dynamics but rather on a mathematical theory for its intriguing spatio- temporal dynamics. This description represents a significant extension of classical Orr-Sommerfeld type linear hydrodynamic theory and offers, for the first time, quantitative delineation of the complex wave dynamics.

Our ability to quantitatively describe complex wave dynamics on thin films is due to a fundamental physical fact - the waves are localized as solitary waves and shocks for most,but not all, conditions. Such localized "coherent structures" are also observed in many other extended-domain dynamics but its mathematical analysis is most developed for thin films. Mass conservation and viscous effects render the dynamics of these structures very different from energy-conserving deep-water solitons. Consequently, a new nonlinear mathematical approach, different from inverse scattering and transformation theories, must be formulated

Our effort is also simplified by certain physical symmetries of the coherent structures that allow their dynamics to be describe by a few discrete dynamic zero modes. As such, the complex spatio-temporal dynamics can be captured by low-dimensional dynamical systems. We present these new tools and concepts for thin-film wave dynamics , as well as some classical ones, in this book.

Our new approach hence combines concepts from Dynamical Systems Theory, Soliton Spectral Theory and Stochastic Methods with advanced computational methods to explore a classical hydrodynamic instability. It contains results from a six-year collaboration at Notre Dame between the authors, after working independently on the same subject for an earlier six years. During both periods, our colleagues, collaborators and students contributed to the results we report here. They include Y.Ben, M.Cheng, E. Kalaidin, S. Kalliadasis, D. Kopelevich, M. J. McCready, M. Sangalli, S. Saprikin, V. Shkadov and Y.Ye.

It is our hope that this monograph will trigger similar approaches to other flow instabilities.

Contents

Preface		v
1	**Introduction and History**	1
2	**Formulation and Linear Orr-Sommerfeld Theory**	5
	2.1 Navier-Stokes Equation with interfacial conditions	5
	2.2 Linear stability of the trivial solution to two- and three-dimensional pertrubations.	11
	2.3 Longwave expansion for surface waves.	14
	2.4 Unusual case of zero surface tension.	20
	2.5 Surface waves: The limit of $R \to \infty$.	22
	2.6 Numerical solution of the Orr-Sommerfeld equations.	25
3	**Hierarchy of Model Equations**	32
	3.1 Kuramoto-Sivashinsky(KS), KdV and related weakly nonlinear equations	33
	3.2 lubrication theory to derive Benney's longwave equation	46
	3.3 Depth-averaged integral equations	50
	3.4 Combination of Galerkin-Petrov method with weighted residuals.	57
	3.5 Validity of the equations	60
	3.6 Spatial and temporal primary instability of the Skadov model	61
4	**Experiments and Numerical Simulation**	69
	4.1 Experiments on falling-film wave dynamics	70
	4.2 Numerical formulation	91
	4.3 Numerical simulation of noise-driven wave transitions	97
	4.4 Pulse formation and coarsening	103
5	**Periodic and Solitary Wave Families**	111
	5.1 Main properties of weakly nonlinear waves in an active/dissipative medium.	111
	5.2 Phase space of stationary KS equation.	115
	5.3 solitary waves and Shilnikov theorem	120

- 5.4 Bifurcations of spatially periodic travelling waves and their stability 132
- 5.5 Normal Form analysis for the Kawahara equation. 145
- 5.6 Nonlinear waves far from criticality – the Shkadov model. 151
- 5.7 Stationary waves of the boundary layer equation and Shkadov model 160
- 5.8 Navier-Stokes equation of motion – the effects of surface tension 174

6 Floquet Theory and Selection of Periodic Waves 179
- 6.1 Stability and selection of stationary waves. 180
- 6.2 Stable intervals from a Coherent Structure Theory 187
- 6.3 Evolution towards solitary waves 192

7 Spectral Theory for gKS Solitary Pulses 198
- 7.1 Pulse spectra 199
- 7.2 Some numerical recipes to construct eigenfunctions and obtain spectra. 202
- 7.3 Stability of gKS pulses. 205
- 7.4 Attenuation of radiation wave packet by stable pulses. 215
- 7.5 resonance pole–a discrete culmination of the continuous spectrum. 218
- 7.6 resonance pole description of mass drainage 228
- 7.7 Suppression of wave packets by a periodic train of pulses. 239

8 Spectral Theory and Drainage Dynamics of Realistic Pulses 243
- 8.1 Role of drainage in pulse coalescence. 243
- 8.2 Spectrum of the solitary pulse. 250
- 8.3 Quasi-jump decay dynamics 257
- 8.4 Essential and resonance pole spectra of the pulses 262

9 Pulse Interaction Theory 271
- 9.1 Coherent Structure theory due to translational zero mode 271
- 9.2 Repulsive pulse interaction of the gKS pulses 274
- 9.3 Coupled drainage and binary interaction dynamics of pulses for the Shkadov model 287

10 Coarsening Theory for Naturally Excited Waves 293
- 10.1 Spatial evolution, linear filtering and excitation of low-frequency band 294
- 10.2 A theory for the characteristic modulation frequency 299
- 10.3 Universal coarsening rate based on Δ. 310
- 10.4 Noise-driven wave dynamics 313

11 Transverse Instability 316
- 11.1 Coupled oblique waves and triad resonance. 317
- 11.2 Transverse breakup of equilibrium 2D-waves 321
- 11.3 scallop waves 325

11.4 Stability of nonlinear localized patterns. 327

12 Hydraulic Shocks   **340**
12.1 Governing equations 342
12.2 Numerical simulation 345
12.3 Coherent wave structures
     and self-similarity 349
12.4 Self-similar coarsening dynamics 354
12.5 Summary and discussion 362

13 Drop Formation on a Coated Vertical Fiber   **363**
13.1 Pulse coalescence dynamics 364
13.2 Equilibrium subcritical pulses and stability 368
13.3 Growth dynamics of supercritical pulses 374
13.4 Discussion 381

CHAPTER 1

Introduction and History

The scientific community was first introduced to wave dynamics on a falling film when the father-son team of the Kapitza family (Kapitza & Kapitza, 1949) took the elegant photographs shown in Figure 4.3. Under house arrest, they were compelled to make do with the simplest experimental apparati in their kitchen . Yet, a rich variety of wave forms were elicited by their humble experiments – sinusoidal waves with rounded crests and troughs, solitary waves of lonely drops, trains of pearly capillary drops that repel, coalesce and otherwise exhibit fascinating motions. They can be boringly rythmatic and also be unpredictably irregular – traits that have been exploited in fountains and table-top sculptures and extolled by poets. In short, they represent the simplest hydrodynamic instability that can produce the wonderfully rich spatio-temporal patterns of more elaborate instabilities. Their simplicity lies in the fact that, although it is an instability driven by inertia, it can occur at extremely low Reynolds numbers under very common conditions. In fact, the most practical application of falling film waves is at a Reynolds number of about 100. Hence, simple tap water flow is sufficient to produce a wavy film– a fact important to the Kapitzas. Car windshields and window panes on a rainy day are other common examples. Their simplicity also lies in the fact that, for a vertical column and plane, they can be parameterized by a single dimensionless parameter with some mild approximations. All the rich wavy dynamics hence depend only on a single dimensionless number.

Aesthetics aside, interfacial waves on a falling film are also of the utmost practical importance. Interfacial heat and mass transfer is known to increase by an order of magnitude when the waves are present (Dukler, 1977, Nakoryakov et al., 1987). Cooling towers, condenser tubes, compact multi-phase heat exchangers, scrubbers etc all benefit from the interfacial waves. As a direct result, while empirical correlations for wave effects have existed for over a century, earnest engineering research on film waves has also been around since mid twentieth century. Engineering researchers like Dukler of Houston, who first introduced one of the authors to the subject, have devoted their careers to the subject. Dukler, Alekseenko in Novosibirsk, Whitaker at UC Davis, Stainthrop and Allen at Manchester, Portalski and Clegg at Surrey, Goren at Berkeley, Brauner and Maron at Tel-Aviv etc. carried out the definitive experimental characterization of wave statistics under natural conditions. Their data are still among the best in the literature.

Kapitzas' report arrived when the applied mathematics community was actively formulating the basic linear Orr-Sommerfeld theories for the now-classical flow instabilities – Kelvin-Helmholtz, Tollmein-Schlicting, Couette, Poiseuille, Rayleigh etc. Falling-film instability has the added complexity of involving a free surface and hence became even more appealing. In the late fifties and early sixties, some of the top hydrodynamicists and applied mathematicians pursued this problem with zest. In Cambridge, Benjamin first understood the origin of instability with a long-wave expansion. His theory was later improved by Yih at Michigan who explained the all-important role of surface tension. In an attempt to extend the theory to include nonlinear effects, as the Ginzburg-Landau theory has done for short-wave instabilities, Benney at MIT pioneered the lubrication approach, Shkadov of Moscow State began the averaging method (culminating in the Shkadov model) and Sivashinsky, first in Russia and then in Haifa, derived the famed Kuramoto-Sivashinsky equation for the falling film instability.

With such intense scrutiny by both the scientific and engineering communities (as ably documented by Lin in 1983), our understanding of falling-film waves was still woefully lacking by the late seventies. Like all classical instability theories, only the primary instability of the flat-film basic state was fully explored by the linear theories available at that point. Nonlinear theories, began by Benney, Shkadov and Sivashinsky, were still at their infancy and their application still uncertain. This implies that all interesting wave dynamics, those that first appealed to the Kapitzas and to the romantics, remain a mystery. Equivalently, primary instability theory can only describe the wave dynamics of the first 10 cm of a vertical column or plane. Waves continue to evolve for typically another meter with much richer dynamics that escape the linear theory (Figure 1.1). With all that fanfare, only the most boring 10% had been captured by the best theoretical hydrodynamicists. The same assessment applied to almost all other classical instabilities. The sequence of rich evolution depicted in Figure 1.1 is generic to many open-flow instabilities. Yet, theoretical descriptions were, and still are, only available for the first primary instability region.

Drastic improvements began in the eighties, particularly for the falling-film instability. Dramatically improved computational power allows simulation of wave dynamics within a one-meter channel to fully capture the entire sequence of wave evolution. Although only model equations and not the full equations of motion can be simulated, such numerical efforts by one of the authors at Moscow State, Davis and Bankoff at Northwestern, Villadsen at Lyngby, Scriven at Minnesota, Patera and Brown at MIT etc. began to produce wave dynamics and statistics unavailable from empirical measurements. More importantly, the fledgling field of nonlinear dynamics made a fundamental impact. A pioneering numerical paper by Pumir, Manneville and Pomeau (1983) pointed out that the complex interfacial dynamics on a falling film often consist of quasi-stationary travelling waves that propagate for long distances without changing their speed or shape. In particular, such travelling waves can be associated with limit -cycles, homoclinic orbits and heteroclinic orbits of dynamical systems theory. Several efforts in the late eighties, by both authors independently and by Kevrekidis at Princeton, then raced to classify all traveling-wave solutions both theoretically and numerically. The complete zoo was only recently completed by the collaborative efforts of both authors.

However, quasi-stationary travelling waves must still evolve on the falling film to give rise to the rich dynamics in Figure 1.1. Such evolution is the target of our collaboration in the last six years. Our approach is analogous to that of coherent structure theory in shear turbulence – the spatio-temporal dynamics are driven by the relatively slow and long-range interaction of several localized traveling waves. However, although the theory remains controversial for shear turbulence, our effort has established its validity for the falling film. We are helped in our pursuit by the parallel experimental efforts of Gollub at Haverford and Alekseenko at Novosibirsk.

The dominant travelling waves are shown to be solitary waves. Their formation, mutual attraction, repulsion, coalescence and interaction with the substrate essentially contribute to all the spatio-temporal dynamics beyond the first primary instability region in Figure 1.1. To quantify such dynamics, we have invoked discrete and essential spectral theory for solitary waves, a new weighted resonance pole spatial theory, Center Manifold Theory and a host of new mathematical theories. These new weapons complement the existing arsenals of matched asymptotics, two-scale averaging, bifurcation theory and Shilnikov homoclinic theory. In essence, a new library of mathematical tools have been assembled to study the dynamics of these solitary waves that do not conserve energy. They are the counterparts of the inverse scattering theory and transformation methods for intergrable solitons that do.

We summarize most known results on falling film waves, including those from the new approach of our collaboration, in this book. We shall begin with a complete linear Orr-Sommerfeld analysis of the flat-film basic state. This is followed by derivation of simple model equations from the Navier-Stokes equation, numerical solution of which is still impossible in an extended domain. We then carry out numerical simulation and approximate and exact construction of the localized structures by asymptotic and numerical methods. The solution branches of these structures will be classified by dynamic singularity theories (double-zero, steady-Hopf and Shilnikov homoclinity) and normal form theories. The stability of such structures will then be analyzed by a spectral analysis. Their spectra include a finite number of dominant discrete modes and a continuum of essential modes. A weighted spectral theory is used to collpse the latter continuum into a single discrete resonance pole. The existence and degeneracy of the resonance poles and discrete spectra will be shown to be related to mass conservation and translational symmetries of the structures. They also capture several unique hydrodynamic features like receptivity, susceptibility, transient growth etc. of localized structures. Exploiting the low number of discrete modes responsible for the dynamics of a single structure, we extend the analysis of single-structure dynamics to an interaction theory that determines the long-range forces between structures. This binary theory is then incorporated within a statistical theory for many structures. The final front instability that triggers the formation of three-dimensional waves is then studied to complete our analysis of the entire wave evolution sequence. The book finishes with two other examples, hydraulic shocks and fiber coating, where our approach to wave dynamics has been successful applied.

This systematic approach quantifies several outstanding hydrodynamic features beyond the Orr-Sommerfeld theories - formation dynamics of localized waves due to convective amplification of inlet noise, coarsening dynamics of localized waves, breakup of two-dimensional structures into three-dimensional ones, interaction between structures through long-range interfacial forces and "radiative" wavepackets etc. Such delineation from basic principles should be possible for the myriad of spatio-temporal dynamics where localized structures exist.

We shall demonstrate this new approach with a variety of governing equations. The concepts are introduced via the simplified model equations valid only for very restrictive conditions, like the various generalized Kuramoto-Sivashinsky equations and depth-averaged equations, for clarity. However, the full Navier-Stokes equation will also be tackled, whenever possible, with all existing numerical arsenal to obtain more accurate quantification.

The net result of our six-year collaboration is that the entire sequence in Figure 1.1 is now fully understood by fundamental theories. Such theories allow us to quantitatively predict the wave pattern, the wave dimensions and its evolution time scale for the entire meter-long channel. The falling-film instability has become the first open-flow hydrodynamic instability to be fully deciphered from inception to fully developed spatio-temporal dynamics. We hope that it is a comprehensive examination of this intriguing instability and that the book will inspire similar approaches to other hydrodynamic instabilities.

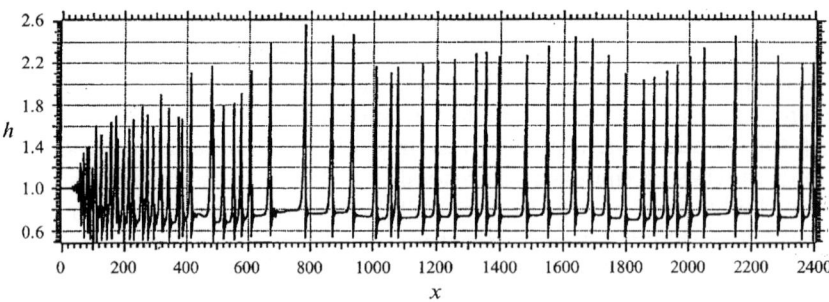

Figure 1.1: Snapshot of the evolving waves from a numerical simulation of the Shkadov model at $\delta = 0.217$. Random forcing is placed at the inlet and a 'soft' boundary condition is placed at the exit to minimize upstream feedback from there. One unit of the scaled x coordinate is roughly 1 mm.

CHAPTER 2

Formulation and Linear Orr-Sommerfeld Theory

We begin with a full statement of the pertinent mathematical equations–the Navier-Stokes equation with in Figure 2.1 free-surface conditions. The usual dimensional analysis would then show that the inclined film problem is specified by three parameters. Several possibilities are possible but we prefer to use the Reynolds number, the inclination angle and a unique Kapitza number that depends only on the physical properties of the fluid and not on the flow rate. The more conventional Weber number can be related to these three parameters. However, with some weak longwave approximations, we shall show that only two parameters are necessary and one of them vanishes for a vertical film. This implies that the rich wave dynamics on a vertical film is only parameterized by a single parameter. The numerical effort to quantify all wave dynamics is considerably reduced by this parameter reduction. The important generic parameter is a normalized Reynolds number with particular scalings with respect to the Reynolds, Weber and Kapitza numbers. The classical linear theory for the flat-film and some recent extensions are then presented. In particular, we present exceptional conditions where the classical longwave instability is replaced by interfacial and internal shortwave instabilities. The theories contain almost all the perturbation methods that have been developed to solve the Orr-Sommerfeld equation, including a large Reynolds number expansion for shear waves. We conclude with a short summary of the latest numerical schemes and results for the spectrum of the flat falling film.

2.1 Navier-Stokes Equation with interfacial conditions

We consider a fluid layer freely falling by gravity down an inclined surface, y=0, at an angle θ with respect to the horizontal plane and having a free interface, $y =$

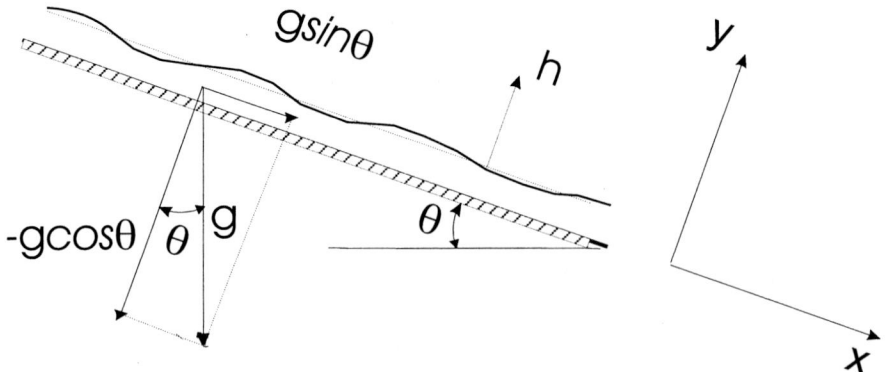

Figure 2.1: Physical picture of the flow

$h(x, z, t)$. Normal stress on the gas side of this interface is just the atmospheric pressure, $P_0 = const$. The normal stress undergoes a jump across the interface with curvature $K(h)$ due to capillary forces. Tangential stresses at the interface are equal to zero. The pertinent variables are reduced into dimensionless form by some characteristic quantities, length l_0, velocity u_0 and the fluid density ρ. We then obtain the properly scaled Navier-Stokes equations and the continuity equation:

$$\frac{\partial u}{\partial t} + u\frac{\partial u}{\partial x} + v\frac{\partial u}{\partial y} + w\frac{\partial u}{\partial z} = -\frac{\partial p}{\partial x} + \frac{\partial \tau_{xx}}{\partial x} + \frac{\partial \tau_{xy}}{\partial y} + \frac{\partial \tau_{xz}}{\partial z} + G\sin\theta \quad (2.1)$$

$$\frac{\partial v}{\partial t} + u\frac{\partial v}{\partial x} + v\frac{\partial v}{\partial y} + w\frac{\partial v}{\partial z} = -\frac{\partial p}{\partial y} + \frac{\partial \tau_{yx}}{\partial x} + \frac{\partial \tau_{yy}}{\partial y} + \frac{\partial \tau_{yz}}{\partial z} - G\cos\theta \quad (2.2)$$

$$\frac{\partial w}{\partial t} + u\frac{\partial w}{\partial x} + v\frac{\partial w}{\partial y} + w\frac{\partial w}{\partial z} = -\frac{\partial p}{\partial z} + \frac{\partial \tau_{zx}}{\partial x} + \frac{\partial \tau_{zy}}{\partial y} + \frac{\partial \tau_{zz}}{\partial z} \quad (2.3)$$

$$\frac{\partial u}{\partial x} + \frac{\partial v}{\partial y} + \frac{\partial w}{\partial z} = 0 \quad (2.4)$$

with boundary conditions at the interface $y = h$ and at the wall $y = 0$,

$$y = h(x, z, t):$$
$$-p - WK(h) + \tau_{nx}n_x + \tau_{ny}n_y + \tau_{nz}n_z = 0 \quad (2.5)$$
$$\tau_{nx}n_y - \tau_{ny}n_x = 0 \quad (2.6)$$
$$\tau_{nz}n_y - \tau_{ny}n_z = 0 \quad (2.7)$$
$$v = h_t + uh_x + wh_z \quad (2.8)$$
$$y = 0:$$
$$u = v = w = 0 \quad (2.9)$$

$$\begin{cases} \tau_{nx} = \tau_{xx}n_x + \tau_{xy}n_y + \tau_{xz}n_z \\ \tau_{ny} = \tau_{xy}n_x + \tau_{yy}n_y + \tau_{yz}n_z \\ \tau_{nz} = \tau_{xz}n_x + \tau_{yz}n_y + \tau_{zz}n_z \end{cases} \tag{2.10}$$

$$\begin{cases} n_x = -\dfrac{h_x}{(1+h_x^2+h_z^2)^{\frac{1}{2}}} \\ n_y = \dfrac{1}{(1+h_x^2+h_z^2)^{\frac{1}{2}}} \\ n_x = -\dfrac{h_z}{(1+h_x^2+h_z^2)^{\frac{1}{2}}} \end{cases} \tag{2.11}$$

Curvature of the surface K is described by

$$K = \left(\frac{h_x}{\sqrt{1+h_x^2+h_z^2}}\right)_x + \left(\frac{h_z}{\sqrt{1+h_x^2+h_z^2}}\right)_z \tag{2.12}$$

For Newtonian fluids:

$$\tau_{xx} = \frac{2}{R}\frac{\partial u}{\partial x}, \quad \tau_{yy} = \frac{2}{R}\frac{\partial v}{\partial y}, \quad \tau_{zz} = \frac{2}{R}\frac{\partial w}{\partial z} \tag{2.13}$$

$$\tau_{xy} = \frac{1}{R}\left(\frac{\partial u}{\partial y} + \frac{\partial v}{\partial x}\right) \tag{2.14}$$

$$\tau_{yz} = \frac{1}{R}\left(\frac{\partial w}{\partial y} + \frac{\partial v}{\partial z}\right) \tag{2.15}$$

$$\tau_{xz} = \frac{1}{R}\left(\frac{\partial u}{\partial z} + \frac{\partial w}{\partial x}\right) \tag{2.16}$$

For any fluid parameters, the equations (2.1)-(2.16) have a trivial solution describing the Nusselt base state:

$$\frac{1}{R}\frac{\partial^2 U}{\partial y^2} + G\sin\theta = 0 \tag{2.17}$$

$$\frac{\partial P}{\partial y} = -G\cos\theta \tag{2.18}$$

$$y = 1: \quad \frac{\partial U}{\partial y} = 0 \tag{2.19}$$

$$y = 0: \quad U = 0 \tag{2.20}$$

This trivial solution, known as the Nusselt flat-film basic state, has a constant thickness which we shall take as a basic length

$$h \equiv 1 \tag{2.21}$$

and a semi-parabolic velocity profile U in the direction of the gravity is

$$U = GR\sin\theta\left(y - \frac{1}{2}y^2\right) \tag{2.22}$$

while the other components are zero.

Choosing characteristic parameters, the waveless Nusselt thickness l_0 or h_0 and the average velocity, $u_0 = \frac{1}{3}\frac{gl_0^2}{\nu}\sin\theta$ and density ρ, these parameters can be related to

$$R = \frac{u_0 l_0}{\nu} = \frac{1}{3}\frac{gh_0^3}{\nu^2}\sin\theta \qquad \text{— Reynolds number;}$$

$$G = \frac{gl_0}{u_0^2} = 3^2 \frac{\nu^2}{gh_0^3 \sin^2\theta} = \frac{3}{R\sin\theta} \qquad \text{— Froude number;}$$

$$W = \frac{\sigma}{\rho l_0 u_0^2} = \frac{3^2 \sigma \nu^2}{\rho g^2 h_0^5 \sin^2\theta} \qquad \text{— Weber number;}$$

and velocity distribution now is

$$U = 3(y - \frac{1}{2}y^2) \tag{2.23}$$

Here, ν is the kinematic viscosity coefficient and g is the gravitational acceleration. Equations (2.1) through (2.16) contain three parameters: R, W and θ. Instead of these parameters, it is often more convenient to take R, γ and θ, where

$$\gamma = \frac{\sigma}{\rho}\nu^{-\frac{4}{3}}g^{-\frac{1}{3}} \tag{2.24}$$

is the Kapitza number which is a dimensionless parameter characterizing only fluid properties and not the flow rate. The replaced W can now be related to these three new parameters

$$W = \frac{3^{\frac{1}{3}}\gamma}{R^{\frac{5}{3}}(\sin\theta)^{\frac{1}{3}}} \tag{2.25}$$

At moderate Reynolds numbers, we can greatly simplify the equations by taking advantage of the fact that the waves are long, i.e. $\frac{\partial}{\partial x}, \frac{\partial}{\partial z} \ll \frac{\partial}{\partial y}$, under most conditions. (For Reynolds numbers up to several hundreds, the water film thickness is on the order of fractions of a millimeter to several millimeters, where as the wavelength of the waves on the surface is as long as several centimeters.)

We first transform the velocity v, coordinate x and the time t:

$$v \to v\varkappa, \quad x \to x/\varkappa, \quad t \to t/\varkappa,$$

where the stretch parameter $\varkappa$ is yet undetermined. This transforms the equations (2.1)-(2.4) for the two-dimensional case with $\theta = \frac{\pi}{2}$ to the form:

$$\frac{\partial u}{\partial t} + u\frac{\partial u}{\partial x} + v\frac{\partial u}{\partial y} = -\frac{\partial p}{\partial x} + \frac{\varkappa}{R}\left(\frac{\partial^2 u}{\partial y^2} + 3 + \frac{1}{\varkappa^2}\frac{\partial^2 u}{\partial x^2}\right) \tag{2.26}$$

$$\frac{1}{\varkappa^2}\left(\frac{\partial v}{\partial t} + u\frac{\partial v}{\partial x} + v\frac{\partial v}{\partial y}\right) = -\frac{\partial p}{\partial y} + \frac{1}{\varkappa R}\left(\frac{\partial^2 v}{\partial y^2} + \frac{1}{\varkappa^2}\frac{\partial^2 v}{\partial x^2}\right) \tag{2.27}$$

$$\frac{\partial u}{\partial x} + \frac{\partial v}{\partial y} = 0 \tag{2.28}$$

$$y = h(x,t): \quad v = h_t + uh_x \tag{2.29}$$

$$p + \frac{W}{\varkappa^2} \frac{h_{xx}}{\left(1 + \frac{h_x^2}{\varkappa^2}\right)^{3/2}} - \frac{2}{\varkappa R} \frac{\partial v}{\partial y} \frac{1 + \frac{h_x^2}{\varkappa^2}}{1 - \frac{h_x^2}{\varkappa^2}} = p_0 \tag{2.30}$$

$$\frac{\partial u}{\partial y} + \frac{1}{\varkappa^2} \frac{\partial v}{\partial x} + \frac{4h_x}{\varkappa^2 \left(1 + \frac{h_x^2}{\varkappa^2}\right)} \frac{\partial v}{\partial y} = 0 \tag{2.31}$$

$$y = 0: \quad u = v = 0 \tag{2.32}$$

We choose $\varkappa$ on the basis of the relation $WR/(3\varkappa^3) = 1$. In such a case, the problem is described by two parameters, conveniently taken to be:

$$\delta = \frac{R}{15\varkappa}, \quad \varepsilon = \frac{1}{\varkappa R}. \tag{2.33}$$

where ε will be shown to be small for most conditions. If we retain terms upto $O(\varepsilon)$ and introduce $p \to p/\delta$, the equations become:

$$\delta\left(\frac{\partial u}{\partial t} + u\frac{\partial u}{\partial x} + v\frac{\partial u}{\partial y}\right) + \frac{\partial p}{\partial x} - \frac{1}{15}\left(\frac{\partial^2 u}{\partial y^2} + 3\right) = \varepsilon\delta\frac{\partial^2 u}{\partial x^2} \tag{2.34}$$

$$\frac{\partial p}{\partial y} = \varepsilon\delta\left[\frac{\partial^2 u}{\partial y^2} - 15\delta\left(u\frac{\partial v}{\partial x} + v\frac{\partial v}{\partial y} + \frac{\partial v}{\partial t}\right)\right], \quad \frac{\partial u}{\partial x} + \frac{\partial v}{\partial y} = 0 \tag{2.35}$$

$$y = h(x,t): \quad v = h_t + uh_x$$

$$p + \frac{1}{5}h_{xx} - p_0 = 2\varepsilon\delta\left[\frac{9}{4}h_{xx}h_x^2 + \frac{\partial v}{\partial y} - h_x\frac{\partial u}{\partial y}\right] \tag{2.36}$$

$$\frac{\partial u}{\partial y} = 15\varepsilon\delta\left[h_x^2\frac{\partial u}{\partial y} - \frac{\partial v}{\partial y} - 4h_x\frac{\partial v}{\partial y}\right] \tag{2.37}$$

$$y = 0: \quad u = v = 0$$

One can then rewrite (2.33) as:

$$\delta = 3^{-7/9} 5^{-1} \gamma^{-1/3} R^{11/9} \tag{2.38}$$

$$\varepsilon = 3^{2/9} \gamma^{-1/3} R^{-7/9} \tag{2.39}$$

$$\varkappa = 3^{-2/9} \gamma^{1/3} R^{-2/9} \tag{2.40}$$

We shall omit ε and $\varkappa$ in subsequence approximations. The lone key parameter for the vertical case is then the generic δ, a normalized Reynolds number that replaces R.

For most liquids under normal conditions and at Reynolds number from zero to a few hundreds, $\varepsilon \ll 1$ and $\varepsilon\delta \ll 1$. We give below ε and δ as functions of the Reynolds number for mercury $\gamma = 28000$, for water $\gamma = 2850$ and glycerin $\gamma = 0.18$. We find ε and $\varepsilon\delta$ to be small for mercury up to $R = 1000$, for water up to 200-300. On the other hand, ε is large when $R = 1$ for glycerin.

	Mercury (20°C) $\gamma = 28000$		H_2O (15°C) $\gamma = 2850$		Glycerin (20°C) $\gamma = 0.18$	
R	5δ	$\epsilon \times 10^{-4}$	5δ	$\epsilon \times 10^{-3}$	5δ	ϵ
1	0.014	1.2	0.030	0.54	0.755	0.341
5	0.100	2.4	0.214	1.10	5.390	0.697
10	0.233	3.3	0.501	1.51	12.57	0.948
50	1.670	6.7	3.580	3.12	89.85	1.939
100	3.900	9.1	8.350	4.21	209.7	2.638
500	27.881	18.1	59.72	8.53	1499.3	5.395
1000	65.04	25.0	139.3	12.0	3498.0	7.341

Table 2.1. Typical values of δ and ϵ for $\theta = \frac{1}{2}\pi$.

We can see the same trend for silicon oil which is absent in the table. However, it is obvious from (2.39) that for any fixed γ, $\varepsilon \to 0$ when $R \to 0$.

If $u, v, \frac{\partial}{\partial x}, \frac{\partial}{\partial y}$ are of unity order, terms on the right-hand side of (2.34) can be ignored when $\varepsilon \to 0$. The equations then become:

$$\frac{\partial u}{\partial t} + u\frac{\partial u}{\partial x} + v\frac{\partial u}{\partial y} = \frac{1}{5\delta}\left(\frac{\partial^3 h}{\partial x^3} + \frac{1}{3}\frac{\partial^2 u}{\partial y^2} + 1\right), \quad \frac{\partial u}{\partial x} + \frac{\partial v}{\partial y} = 0 \quad (2.41)$$

$$y = h(x,t): \quad v = h_t + uh_x, \quad \frac{\partial u}{\partial y} = 0 \quad (2.42)$$

$$y = 0: \quad u = v = 0 \quad (2.43)$$

It is sometimes convenient to replace the kinematic condition (2.29) by its equivalent:

$$\frac{\partial h}{\partial t} + \frac{\partial}{\partial x}\int_0^h u\,dy = 0 \quad (2.44)$$

All the transformations can be done in the same way, with $z \to \varkappa z$, for the three-dimensional case on a general inclined plane $\theta \neq \frac{\pi}{2}$.

The equations are then the boundary layer (BL) equation,

$$\frac{\partial u}{\partial t} + u\frac{\partial u}{\partial x} + v\frac{\partial u}{\partial y} + w\frac{\partial u}{\partial z} = \frac{1}{5\delta}\left(h_{xxx} + h_{xzz} + \frac{1}{3}\frac{\partial^2 u}{\partial y^2} + 1\right) - 3\chi\frac{\partial h}{\partial x} \quad (2.45)$$

$$\frac{\partial w}{\partial t} + u\frac{\partial w}{\partial x} + v\frac{\partial w}{\partial y} + w\frac{\partial w}{\partial z} = \frac{1}{5\delta}\left(h_{xxz} + h_{zzz} + \frac{1}{3}\frac{\partial^2 w}{\partial y^2}\right) - 3\chi\frac{\partial h}{\partial z} \quad (2.46)$$

$$\frac{\partial u}{\partial x} + \frac{\partial v}{\partial y} + \frac{\partial w}{\partial z} = 0 \quad (2.47)$$

$$y = h(x,y,t): \quad v = h_t + uh_x + wh_z, \quad \frac{\partial u}{\partial y} = \frac{\partial w}{\partial y} = 0 \quad (2.48)$$

$$y = 0: \quad u = v = w = 0 \quad (2.49)$$

Instead of (2.44), we can take:

$$\frac{\partial h}{\partial t} + \frac{\partial}{\partial x}\int_0^h u\,dy + \frac{\partial}{\partial z}\int_0^h w\,dy = 0 \qquad (2.50)$$

Here, $\chi = \frac{\cot\theta}{R}, \delta, \varkappa$ and ε are determined by (2.38)-(2.40) and W depends on γ, R, θ as in (2.25). The equations (2.45)-(2.50) are valid for not very large deviations of the inclination angle from $\pi/2$. Near $\theta = 0$, χ approaches infinity and the scaling breaks down.

The above boundary layer (BL) equation offers several advantages. Pressure is explicity eliminated and the original three parameters are reduced to two — χ and δ. In fact, for the vertical film ($\chi = 0$), a single normalized Reynolds number δ parameterizes the entire system. Nevertheless, the boundary layer equations (2.45)-(2.50) are still extremely complicated to study.

2.2 Linear stability of the trivial solution to two- and three-dimensional pertrubations.

Nearly forty years have passed since the survey paper by Yih (1963). In his paper, the known linear stability results of thin layers up to then were summarized, beginning with the work of Yih (1955) and Benjamin (1957). However, several issues remained unclear and some conclusions incorrect. Dozens of papers appeared after Yih's work to address these issues and are summarized in this chapter. Some new results are also reported.

Consider the stability of the Nusselt flat film to infinitesimal disturbances of the form:

$$u = U + \varepsilon\hat{u}e^{i(\mathbf{kx}-ct)}, \qquad v = \varepsilon\hat{v}e^{i(\mathbf{kx}-ct)},$$
$$w = \varepsilon\hat{w}e^{i(\mathbf{kx}-ct)}, \qquad h = 1 + \varepsilon e^{i(\mathbf{kx}-ct)}, \qquad \mathbf{k} = (\alpha, \beta) \qquad (2.51)$$

By substituting these expressions into the Navier-Stokes equation (2.1)-(2.16) and taking the limit $\varepsilon \to 0$, we obtain the linearized boundary value problem:

$$i\alpha(U-c)\hat{u} + U'\hat{v} = -i\alpha\hat{p} + \frac{1}{R}(\hat{u}'' - (\alpha^2+\beta^2)\hat{u}) \qquad (2.52)$$

$$i\alpha(U-c)\hat{v} = -\hat{p}' + \frac{1}{R}(\hat{v}'' - (\alpha^2+\beta^2)\hat{v}) \qquad (2.53)$$

$$i\alpha(U-c)\hat{w} = -i\beta\hat{p} + \frac{1}{R}(\hat{w}'' - (\alpha^2+\beta^2)\hat{w}) \qquad (2.54)$$

$$\hat{v}' + i(\alpha\hat{u} + \beta\hat{w}) = 0 \qquad (2.55)$$

$y = 1$:

$$-\hat{p} + G\cos\theta\tilde{h} + (\alpha^2 + \beta^2)W\hat{h} + \frac{2}{R}\frac{\partial \hat{v}}{\partial y} = 0 \qquad (2.56)$$

$$\hat{u}' + i\alpha\hat{v} = -U''\hat{h} \qquad (2.57)$$

$$\hat{w}' + i\beta\hat{v} = 0 \qquad (2.58)$$

$$\hat{v} = i\alpha(U - c)\hat{h} \qquad (2.59)$$

$y = 0$: $\qquad \hat{u} = \hat{v} = \hat{w} = 0 \qquad (2.60)$

For temporal instability, we impose real wavenumber components (α, β) and solve for the complex eigenvalues $\omega = \alpha c$. If for any (α, β), $\omega_i = Im(\omega) > 0$, the basic state is unstable. For spatial instability, we consider (ω, β) as real and seek a complex α which is now an eigenvalue.

We define the complex stream function:

$$\alpha \hat{u} + \beta \hat{w} = \alpha \phi' \qquad (2.61)$$

$$\hat{v} = -i\alpha\phi \qquad (2.62)$$

After differentiating (2.52) by y, multiplying (2.53) by $-i(\alpha^2 + \beta^2)$ and summing these equations, we obtain the Orr-Sommerfeld equation. After substituting (2.61) into (2.52)-(2.60), we obtain

$$\phi'''' - 2(\alpha^2 + \beta^2)\phi'' + (\alpha^4 + 2\alpha^2\beta^2 + \beta^4)\phi =$$
$$i\alpha R[(U - c)(\phi'' - (\alpha^2 + \beta^2)\phi) - U''\phi] \qquad (2.63)$$

$y = 1$:

$$\alpha\phi''' - 3\alpha(\alpha^2 + \beta^2)\phi' + i\alpha^2 R(c - \frac{3}{2})\phi' -$$
$$3i(\alpha^2 + \beta^2)\cot\theta - i(\alpha^2 + \beta^2)^2 RW \qquad (2.64)$$

$$\phi'' + (\alpha^2 + \beta^2)\phi = 3 \qquad (2.65)$$

$$\phi = c - \frac{3}{2} \qquad (2.66)$$

$y = 0$: $\quad \phi = \phi' = 0 \qquad (2.67)$

After transformation

$$\alpha R = \tilde{\alpha}\tilde{R}, \qquad \frac{\cot\theta}{\alpha} = \frac{\cot\tilde{\theta}}{\tilde{\alpha}}, \qquad \frac{W}{\alpha^2} = \frac{\tilde{W}}{\tilde{\alpha}^2} \qquad (2.68)$$

where $\tilde{\alpha}^2 = \alpha^2 + \beta^2$, our system becomes:

$$\phi'''' - 2\tilde{\alpha}^2\phi'' + \tilde{\alpha}^4\phi = i\tilde{\alpha}\tilde{R}[(U-c)(\phi'' - \tilde{\alpha}^2\phi) - U''\phi] \tag{2.69}$$

$y = 1$:

$$\phi''' - 3\tilde{\alpha}^2\phi' + i\tilde{\alpha}\tilde{R}(c - \frac{3}{2})\phi' - 3i\tilde{\alpha}\cot\tilde{\theta} - i\tilde{\alpha}^3\tilde{R}\tilde{W} = 0 \tag{2.70}$$

$$\phi'' + \tilde{\alpha}^2\phi = 3 \tag{2.71}$$

$$\phi = c - \frac{3}{2} \tag{2.72}$$

$y = 0$: $\quad \phi = \phi' = 0 \tag{2.73}$

If we let $\beta = 0$ and $\hat{w} = 0$ in (2.52)-(2.60), which means the disturbances are two-dimensional, then the system (2.69)-(2.73) coincides with this system. As a result, a Squire theorem can be formulated for the falling film in the following way: the stability of a flat falling film with Reynolds number R, inclination angle θ and dimensionless surface tension W to three-dimensional disturbances with wave-number $\mathbf{k} = (\alpha, \beta)$ is equivalent to its stability with Reynolds number $\tilde{R} = \frac{\alpha R}{(\alpha^2+\beta^2)^{\frac{1}{2}}}$, a new inclination angle determined by $\cot\tilde{\theta} = (\alpha^2+\beta^2)^{\frac{1}{2}}/\alpha \cdot \cot\theta$ and Weber number $\tilde{W} = (\alpha^2+\beta^2)\frac{W}{\alpha^2}$ to two-dimensional disturbances with normalized wave-number $((\alpha^2+\beta^2)^{\frac{1}{2}}, 0)$. We have hence reproduced the result of Yih (1955). Another more physical conclusion from the Squire theorem is possible, as C.-C. Lin (1967) pointed out (chapter 3). Let us turn our coordinate system in the $x-z$ plane so that the x-axis is in the direction of wave propogation and rewrite our equations in this coordinate system. It is easy to see that the projection of the mean flow velocity would only be in the direction of the wave propogation in the governing equations. So the new Reynolds number, with the mean velocity in its numerator, is reduced by $\alpha/\sqrt{\alpha^2+\beta^2}$ times and W, with square of this velocity is in the denominator, accordingly increases by $(\alpha^2+\beta^2)/\alpha^2$ times and inclination cotangent increases by $\sqrt{\alpha^2+\beta^2}/\alpha$ times. Hence, the Nusselt flat-film parallel flow base state is more stable to infinitesimal three-dimensional disturbances than two-dimensional ones. This may not be true for finite amplitude perturbations and for secondary instability.

Sometimes, it is more convenient to solve the Orr-Sommerfeld equation with an equivalent set of boundary conditions:

$$\phi'''' - 2\alpha^2\phi'' + \alpha^4\phi = i\alpha R[(U-c)(\phi'' - \alpha^2\phi) - U''\phi] \tag{2.74}$$

$y = 1$: $\quad \phi'' + \alpha^2\phi\left[1 - \frac{3}{\alpha^2(c-\frac{3}{2})}\right] = 0 \tag{2.75}$

$$\phi''' - 3\alpha^2\phi' + i\alpha R(c - \frac{3}{2})\phi' - \frac{3i\alpha\cot\theta}{c - \frac{3}{2}}\phi = 0 \tag{2.76}$$

$y = 0$: $\quad \phi = \phi' = 0 \tag{2.77}$

In an experiment, the perturbations grow in space and are periodic in time t at each point in space. It is hence more realistic to take some real $\omega = \alpha c$ and complex α seek as an eigenvalue of the problem. In the vicinity of the neutral values, $\omega_i = 0$ or $\alpha_i = 0$, the rates of spatial and temporal growth are connected by Gaster's relation (Gaster, 1963):

$$\omega_i = \alpha_r c_i = -\alpha_i \frac{\partial \omega_r}{\partial \alpha}$$

where subscripts i and r denote imaginary and real parts as usual.

Hence, in the vicinity of the neutral curve, both approaches give the same results. Far from the neutral curve, it is necessary to solve the eigenvalue problem numerically: we seek a complex α for a given ω, as we shall pursue in the next chapter.

2.3 Longwave expansion for surface waves.

At low Reynolds numbers, the dominant instability of the Nusselt flat-film base state is known to involve pronounced interfacial distortion and to have wavelengths much longer than l_0, the thickness of the waveless state. Such waves are also known to propagte at a speed about three times u_0, the average flat film fluid velocity. These characteristic behaviors of the long interfacial waves are best captured with a long-wave expansim of (2.63). Let $\alpha \to 0$ and $\beta \to 0$:

$$\omega \sim \alpha \omega_{10} + i\alpha^2 \omega_{20} + \alpha^3 \omega_{30} + i\alpha^4 \omega_{40} + \alpha \beta^2 \omega_{12}$$
$$+ i\beta^2 \omega_{02} + i\alpha^2 \beta^2 \omega_{22} + i\beta^4 \omega_{04} \qquad (2.78)$$

$$\psi \sim \alpha \psi_{10} + i\alpha^2 \psi_{20} + \alpha^3 \psi_{30} + i\alpha^4 \psi_{40} + \alpha \beta^2 \psi_{12} + i\beta^2 \psi_{02}$$
$$+ i\alpha^2 \beta^2 \psi_{22} + i\beta^4 \psi_{04} \qquad (2.79)$$

where $\psi = \alpha \phi$ and $\omega = \alpha c$. To each order, we have

$$O(\alpha):$$

$$\psi_{10}'''' = 0;$$
$$y = 1: \quad \psi_{10}''' = 0; \qquad (2.80)$$
$$\phi_{10}'' = 3$$
$$\omega_{10} = \psi_{10} + \frac{3}{2}$$
$$y = 0: \quad \psi_{10} = \psi_{10}' = 0$$

$O(\alpha^2)$:

$$\psi_{20}'''' = R\{U\psi_{10}'' - \omega_{10}\psi_{10}'' + 3\psi_{10}\}$$

$y = 1:$ $\psi_{20}''' + R\{\omega_{10}\psi_{10}' - \frac{3}{2}\psi_{10}'\} - 3\cot\theta = 0$

$\psi_{20}'' = 0$ \hfill (2.81)

$\omega_{20} = \psi_{20}$

$y = 0:$ $\psi_{20} = \psi_{20}' = 0$

$O(\alpha^3)$:

$$\psi_{30}'''' - 2\psi_{10}'' = R\{-U\psi_{20}'' + \omega_{10}\psi_{20}'' + \omega_{20}\psi_{10}'' - 3\psi_{20}\}$$

$y = 1:$ $\psi_{30}''' - 3\psi_{10}' + R\{-\omega_{10}\psi_{20}' - \omega_{20}\psi_{10}' + \frac{3}{2}\psi_{20}'\} = 0$

$\psi_{30}'' + \psi_{10} = 0$ \hfill (2.82)

$\omega_{30} = \psi_{30}$

$y = 0:$ $\psi_{30} = \psi_{30}' = 0$

$O(\alpha^4)$:

$$\psi_{40}'''' - 2\psi_{20}'' = R\{U(\psi_{30}'' - \psi_{10}) - \omega_{10}\psi_{30}'' +$$
$$+ \omega_{20}\psi_{20}'' - \omega_{30}\psi_{10}'' + \omega_{10}\psi_{10} + 3\psi_{30}\}$$

$y = 1:$ $\psi_{40}''' - 3\psi_{20}' +$ \hfill (2.83)

$+ R\{\omega_{10}\psi_{30}' - \omega_{20}\psi_{20}' + \omega_{30}\psi_{10}' - \frac{3}{2}\psi_{30}'\} - RW = 0$

$\psi_{40}'' + \psi_{20} = 0$

$\omega_{40} = \psi_{40}$

$y = 0:$ $\psi_{40} = \psi_{40}' = 0$

$O(\alpha\beta)$:

$$\psi_{12}'''' - 2\psi_{10}'' = R\{-U\psi_{02}'' + \omega_{10}\psi_{02}'' + \omega_{02}\psi_{10}'' - 3\psi_{02}\}$$

$y = 1:$ $\psi_{12}''' - 3\psi_{10}' + R\{-\omega_{10}\psi_{02}' - \omega_{02}\psi_{10}' + \frac{3}{2}\psi_{02}'\} = 0$

$\psi_{12}'' + \psi_{10} = 0$ \hfill (2.84)

$\omega_{12} = \psi_{12}$

$y = 0:$ $\psi_{12} = \psi_{12}' = 0$

$O(\beta^2)$:

$$\psi_{02}'''' = 0$$

$y = 1:$ $\psi_{02}''' - 3\cot\theta = 0$

$\psi_{02}'' = 0$ \hfill (2.85)

$\omega_{02} = \psi_{02}$

$y = 0:$ $\psi_{02} = \psi_{02}' = 0$

$O(\alpha^2\beta^2)$:

$$\psi_{22}'''' - 2\psi_{02}'' - 2\psi_{20}'' = R\{U(\psi_{12}'' - \psi_{10}) - \omega_{10}\psi_{12}'' +$$
$$+ \omega_{20}\psi_{02}'' - \omega_{12}\psi_{10}'' + \omega_{02}\psi_{20}'' + \omega_{10}\psi_{10} + 3\psi_{12}\}$$

$y = 1:\quad \psi_{22}''' - 3\psi_{02}' - 3\psi_{20}' +$ \hfill (2.86)

$$+ R\{\omega_{10}\psi_{12}' - \omega_{20}\psi_{02}' + \omega_{12}\psi_{10}' - \omega_{02}\psi_{20}' - \frac{3}{2}\psi_{12}'\} - 2RW = 0$$

$\psi_{22}'' + \psi_{02} + \psi_{20} = 0$

$\omega_{22} = \psi_{22}$

$y = 0: \psi_{22} = \psi_{22}' = 0$

$O(\beta^4)$:

$$\psi_{04}'''' - 2\psi_{02}'' = R\{\omega_{02}\psi_{02}''\}$$

$y = 1:\quad \psi_{04}''' - 3\psi_{02}' + R\{-\omega_{02}\psi_{02}'\} - RW = 0$ \hfill (2.87)

$\omega_{04} = \psi_{04}$

$y = 0:\quad \psi_{04} = \psi_{04}' = 0$

Let us consider the physical mechanism behind the instability which becomes obvious from our expansions.

The $O(\alpha)$ solution (2.80) immediately captures the phase speed 3 and the interfacial character of the waves. Because of the kinematic condition [(2.57), (2.65) and (2.71) in its different guises], the normal velocity is exactly out of phase with the interfacial perturbation, as represented by the factor $i\alpha$ in front of $\hat{v}$ in (2.57). As a result, in this leading-order aproximation without inertia, the velocity field responds instantaneously to any sinusoidal interfacial perturbation. The response produces a disturbance velocity that does not amplify the interfacial amplitude — its normal velocity is zero at the wave troughs and crests. Instead, the finite normal velocities at the nodes cause the wave to translate at a phase velocity of 3. This phase velocity is numerically shown as the quantity ψ_{10}'' at $y = 1$ in (2.80). Physically, it arises because of the quadratic velocity profile (2.22) of the Nusselt base state. Interfacial perturbation produces normal and tangential disturbance velocities of different magnitude across the film because of this non-uniform base state profile.

The $O(\alpha^2)$ solution (2.81) attributes the instability mechanism to an inertial delay of the velocity disturbance introduced by the interfacial waves. This delay is proportional to R and is in the inhomogeneous part of (2.81). Due to this delay, the normal disturbance velocity is now positive at the crest of the initial interfacial perturbation and negative at the trough. The instability is hence due to an inertial delay and the resulting positive feedback of the disturbance velocities crested by a sinusoidal interfacial perturbation.

The remaining higher order terms then capture the effects of inertia, surface tension and viscosity on shorter waves. One expects the destabilizing effect of inertia to diminish and the stabilizing effect of the latter two forces to increase for shorter waves. The shorter waves cannot generate disturbances deeper into the film and hence cannot induce a larger normal velocity. Their smaller vortices should induce more viscous dissipation and their higher curvature capillary smoothing of the interfacial waves. These expectations are borne out, for the most part, in the higher order expansions. The result is a classical longwave instability with a band of unstable wave numbers extending from zero to the neutral wavenumber α_0. This longwave instability will be shown in later sections to be responsible for the creation of pulse-like wave structures observed downstream. It is hence an important feature of the falling-film instability. There are, however, some singular regions where the expected physics go astray and a small short-wave instability region exists. Disturbances with wavenumbers close to zero are stable in such regions.

Systematic solution of (2.80) to (2.87) yields the following expansions of the complex frequency

$$\omega_{10} = 3 \tag{2.88}$$

$$\omega_{20} = \frac{6}{5}R - \cot\theta \tag{2.89}$$

$$\omega_{30} = -3 - \frac{12}{7}R^2 + \frac{10}{7}R\cot\theta = -3 - \frac{12}{7}R(R - \frac{5}{6}\cot\theta) \tag{2.90}$$

$$\omega_{40} = -\frac{WR}{3} - \frac{75872}{25025}R^3 + \frac{17363}{5775}R^2\cot\theta - \frac{2}{5}R\cot^2\theta$$
$$- \frac{1413}{224}R + \frac{9}{5}\cot\theta \tag{2.91}$$

$$\omega_{50} = \frac{10}{21}R^2 W + \{51429178368R^4 - 60566751360R^3 \cot\theta$$
$$+ 14757142400R^2 \cot^2\theta + 116025 \cdot 1093417R^2$$
$$- 63201138000R\cot\theta + 47171124000\} \frac{1}{8576568000} \tag{2.92}$$

$$\omega_{60} = -(-132463235196 5184R^5 + 1817901884743680R^4 \cot\theta$$
$$- 635313045667840R^3 \cot^2\theta + 20995R^3(-4978319360W - 181226024757)$$
$$+ 33571791052800R^2 \cot^3\theta + 607756800R^2 \cot\theta(45760W + 4431679)$$
$$- 219607189401600R\cot^2\theta + 148987238400R(-420W - 15671)$$
$$+ 342670648320000\cot\theta)\frac{1}{104291066880000} \tag{2.93}$$

$$\omega_{02} = -\cot\theta \tag{2.94}$$

$$\omega_{12} = -3 + \frac{10}{7}R\cot\theta \tag{2.95}$$

$$\omega_{22} = -\frac{2RW}{3} + \frac{17363}{5775}R^2\cot\theta - \frac{4}{5}R\cot^2\theta - \frac{1413}{224}R + \frac{18}{5}\cot\theta \tag{2.96}$$

$$\omega_{04} = -\frac{1}{3}RW - \frac{2}{5}R\cot\theta + \frac{9}{5}\cot\theta \tag{2.97}$$

Let us now examine relations (2.88) to (2.97) and obtain the well-known results from Kapitza and Kapitza (1949) and Benjamin (1957).

The first coefficient again yields the classical phase speed of interfacial waves on a film — it is three times the waveless mean velocity u_0 and twice the interfacial velocity of a flat film.

From (2.88), we have Benjamin's (1957) result for the critical Reynolds number R_*:

$$R_* = \frac{6}{5}\cot\theta \qquad (2.98)$$

The coefficient ω_{30} characterizes dispersion of the waves,

$$\frac{1}{2}\frac{\partial^2 c_r}{\partial\alpha^2} = \omega_{30} \qquad (2.99)$$

At the critical Reynolds number $R = R_*$

$$\frac{1}{2}\frac{\partial^2 c_r}{\partial\alpha^2} = 3 > 0 \qquad (2.100)$$

Beyond R_*, ω_{20} is positive and the growth rate is positive for small wave numbers. If ω_{40} is negative in such region, a classical longwave instability then results with an unstable band from $\alpha = 0$ to the neutral wavenumber α_0.

There is another critical Reynolds number R_{**} beyond which the dispersion ω_{30} changes sign. For angles θ close to $\frac{\pi}{2}$, this critical R is too large to justify a longwave expansion but for sufficiently small angles of inclination, its value is meaningful:

$$\frac{R}{R_{**}} = 1 + \frac{126}{25}\tan^2\theta \qquad (2.101)$$

For $R < R_{**}$, the dispersion is positive, while for larger Reynolds numbers dispersion becomes negative, $\frac{\partial^2 c_r}{\partial\alpha^2} < 0$. The growth rate of disturbances has the form:

$$\alpha c_i = \alpha^2\omega_{20} + \alpha^4\omega_{40} \qquad (2.102)$$

Hence, if $R > R_*$ and $\omega_{40} < 0$, the neutral wave number is

$$\alpha_0 = \sqrt{-\frac{\omega_{20}}{\omega_{40}}} \qquad (2.103)$$

and the wave number with the maximum growth rate:

$$\frac{\alpha_m}{\alpha_0} = \frac{\sqrt{2}}{2} \sim 0.707 \qquad (2.104)$$

and

$$\alpha c_i^m = -\frac{1}{4}\frac{\omega_{20}^2}{\omega_{40}} \qquad (2.105)$$

It looks as if for large surface tension, the first term in the relation for ω_{40} is the largest and we can neglect all other terms. This is not always true. This approximation is not valid at sufficiently large Reynolds numbers and sufficiently small angles. Let us take (2.101) at $R = R_* = \frac{5}{6}\cot\theta$ and let us fix the physical properties of the liquid. Instead of W, we shall consider the Kapitza number γ,

$$\omega_{40} = -\gamma(\frac{2}{5})^{2/3}\frac{(\sin\theta)^{1/3}}{(\cos\theta)^{2/3}} + \frac{1}{36036}(\cot\theta)^3 - \frac{7743}{2240}\cot\theta \quad (2.106)$$

At small angles, the first capillary term can be comparable in magnitude to the largest viscous term, the third one. It will happen at

$$\frac{(\cos\theta)^{5/3}}{(\sin\theta)^{4/5}} = \gamma\frac{2240}{7743}(\frac{2}{5})^{2/3} \approx 0.1571\gamma \quad (2.107)$$

For water at $\gamma = 2850$, this angle occurs $\theta \approx 0.56°$. For typical water-glycerin solutions with $\gamma = 300$, this angle is $\theta \approx 3.18°$. Below these angles, surface tension can be neglected in the two-dimensional wave dynamics.

If we want to know when viscosity is as important as capillary forces for 3D-waves, we should consider

$$\frac{(\cos\theta)^{5/3}}{(\sin\theta)^{4/5}} = \gamma\frac{9}{5}(\frac{2}{5})^{2/3} \approx 0.2147\gamma \quad (2.108)$$

Hence, 3D-effects are slightly less sensitive to θ variations. For water, the critical angle is

$$\theta \approx 1'35'' \quad (2.109)$$

and for typical glycerin-water solutions

$$\theta \approx 18'50'' \quad (2.110)$$

There is another interesting behavior–the classical longwave instability of a falling film can become a shortwave instability under extreme conditions. If we decrease the inclination angle, there are certain critical angles when the stabilizing members ω_{40}, ω_{22} and ω_{04} will first vanish and then become positive! If ω_{40} becomes positive and R is less than the critical Reynolds number R_*, such that ω_{20} is negative, the growth rate first decreases from zero wavenumber and then increases–a shortwave instability. If we neglect surface tension, (2.106) stipulates that this occurs when the inclination angle is less than

$$\theta = \arctan\sqrt{\frac{80}{9965241}} = 10' \approx 0.1623° \quad (2.111)$$

There is little sensivity of this critical angle to surface tension. For water, $\gamma = 2850$ and $\theta \approx 9'$, its value is practically the same as the above limit. This implies that the neutral curve has a nose at small θ — the instability changes from a longwave variety to a shortwave instability!

We have checked these facts numerically (see Figure 2.2 and Figure 2.3) and the calculations confirm the existence of a nose in the neutral curve. Moreover, the results for water-glycerine mixture with $\gamma = 300$ coincide with the ones for

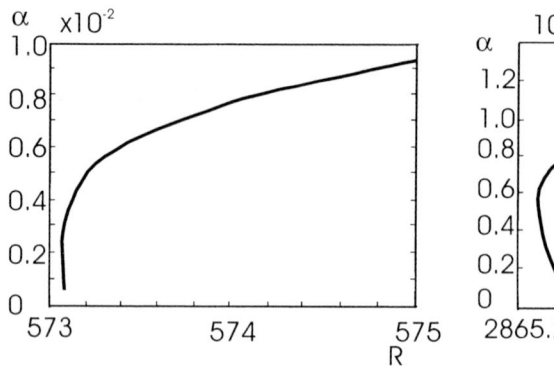 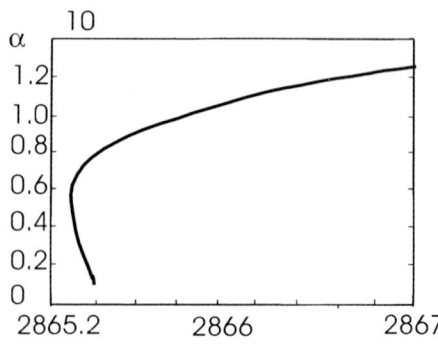

Figure 2.2: Vicinity of the critical Reynolds number, $\theta = 5'$, water, $\gamma = 3000$

Figure 2.3: Vicinity of the critical Reynolds number, $\theta = 1'$, water, $\gamma = 3000$

water within graphical accurcy. Hence, for sufficiently small angles, we should include ω_{50} and ω_{60} terms for the two-dimensional expansion of the dispersion relationship with respect to α. Moreover, the dispersion coefficient ω_{12} changes its sign at

$$\theta = \arctan\sqrt{\frac{25}{63}} \approx 21.6^o \tag{2.112}$$

Thus, for a severely inclined plane, the nature of instability and the wave dynamics should change drastically !

2.4 Unusual case of zero surface tension.

The case $\gamma = 0$ is a unique one. We return to (2.69)- (2.73) and set $\gamma = W = 0$. The form $\frac{c_i}{R}$ cancels at the very point $\alpha = \alpha_0 \approx 1.8017$, but not at $\alpha = 0$. Thus, the neutral stability curve has a gap at the point $R = 0$: $\alpha_0|_{R=0} = 0, \alpha_0|_{R=+0} = 1.8017$. Benjamin(1957) avoided the study of $\gamma = 0$ and Yih (1963) made several mistakes during the investigation of this case. Whitaker and Jones (1966) avoided the case $\gamma = 0$ and $R \to 0$ in their numerical examination. Pierson and Whitaker (1977) carried out a numerical investigation without surface tension in the vicinity of $R = 0$ and found that $\alpha \to 0$ when $R \to 0$. In the last work, the dependence $\alpha_0(R)$ is depicted and data of the graph show that at $0.1 \leq R \leq 5$, $\alpha_0 = const = 1.8$, this coincides with our computed results in Figure 2.7. However, when $R \leq 0.1$ (the calculations are presented up to $R = 10^{-9}$) one can see from their graph that α_0 begins to reduce and tends to zero when $R \to 0$. This is, in fact, wrong and it is easy to explain the error. When $R \to 0$, one can see from (2.102) that the value of c_i reduces proportionally to R. However, it is difficult to catch the point $c_i = 0$ numerically because of computer errors. Alas, in the work of Goncharenko and Urintsev(1975) the same erroneous conclusion was made...

The correct answer was found recently by Shkadov, Belikov and Epikhin (1980). We present their results by correcting minor mistakes in the derivations. Let us consider an expansion of ϕ in (2.69)- (2.73) in R without any restriction on α, as was first carried out by Yih. The expansion parameter is the Reynolds number R, which is assumed small. Hence, we are restricted to the case of near-vertical planes with small critical Reynolds number. We consider only the vertical case here. To leading order,

$$\phi \sim \varphi_0 + iR\varphi_1 \tag{2.113}$$

$$c \sim c_0 + iRc_1 \tag{2.114}$$

and the zeroth order approximation of (2.63)- (2.66) becomes

$$\varphi_0^{IV} - 2\alpha^2 \varphi_0'' + \alpha^4 \varphi_0 = 0 \tag{2.115}$$

$$y = 0: \quad \varphi_0 = \varphi_0' = 0 \tag{2.116}$$

$$y = 1: \quad \varphi_0''' - 3\alpha^2 \varphi_0' = 0 \tag{2.117}$$

$$\varphi_0'' + \alpha^2 \varphi_0 = 3 \tag{2.118}$$

$$c_0 = \varphi_0 + \frac{3}{2} \tag{2.119}$$

At the next approximation,

$$\varphi_1^{IV} - 2\alpha^2 \varphi_1'' + \alpha^4 \varphi_1 = \alpha[(U - c_0)(\varphi_0'' - \alpha^2 \varphi_0) - U'' \varphi_0] \tag{2.120}$$

$$y = 0: \quad \varphi_1 = \varphi_1' = 0 \tag{2.121}$$

$$y = 1: \quad \varphi_1''' - 3\alpha^2 \varphi_1' = \alpha^3 W - \alpha(c_0 - \frac{3}{2})\varphi_0' \tag{2.122}$$

$$\varphi_1'' + \alpha^2 \varphi_1 = 0 \tag{2.123}$$

$$c_1 = \varphi_1 \tag{2.124}$$

From (2.115) to (2.119):

$$\varphi_0 = A(\sinh \alpha y - \alpha y \cosh \alpha y) + By \sinh \alpha y \tag{2.125}$$

$$A = -\frac{3 \cosh \alpha}{\alpha(1 + 2\alpha^2 + \cosh 2\alpha)} \tag{2.126}$$

$$B = \frac{3(\cosh \alpha - \alpha \sinh \alpha)}{\alpha(1 + 2\alpha^2 + \cosh 2\alpha)} \tag{2.127}$$

$$c_0 = \frac{3}{2} + \frac{3}{1 + 2\alpha^2 + \cosh 2\alpha} = \frac{3(3 + 2\alpha^2 + \cosh 2\alpha)}{2(1 + 2\alpha^2 + \cosh 2\alpha)} \tag{2.128}$$

which coincides with Yih's result (1963). This first approximation gives us only the real part of the wave velocity. In order to find the imaginary, let us consider the next approximation. Substituting φ_0 into (2.120) with the following solution for the nonhomogeneous ODE gives

$$\varphi_1 = M(\sinh\alpha y - \alpha y \cosh\alpha y) + Ny \sinh\alpha y + \varphi_* \qquad (2.129)$$

$$\varphi_* = (F_1 y^2 + F_2 y^3 + F_3 y^4)\sinh\alpha y + (D_1 y^2 + D_2 y^3 + D_3 y^4)\cosh\alpha y \qquad (2.130)$$

where φ_1 satisfies the boundary conditions on the wall. From the boundary conditions for stresses, one finds F_k, D_k, N and M. We omit the tedious expressions for them,

After substituting φ_1 into the kinematic boundary conditon, we get the imaginary part of the phase velocity,

$$c_1 = \frac{D}{2\alpha^2 E} + \frac{1}{32\alpha^2 E}[-36\alpha\sinh 2\alpha + (-45 + 9\alpha^2 - 12\alpha^2 c_0)\cosh 2\alpha + (-45 - 27\alpha^2 - 12\alpha^2 c_0)] \qquad (2.131)$$

where

$$D = -\alpha^3 W(\sinh 2\alpha - 2\alpha) + \frac{1}{32\alpha E}(-81 + 63\alpha^2 - 9\alpha^4 - 12\alpha^2 c_0 + 12\alpha^4 c_0)\cosh 4\alpha + \frac{1}{16 E}(180\alpha^2 + 24\alpha^4 + 48\alpha^4 c_0)\sinh 2\alpha \qquad (2.132)$$

$$+ \frac{\alpha}{8E}(45 + 45\alpha^2 - 9\alpha^4 + 12\alpha^2 c_0 + 12\alpha^4 c_0)\cosh 2\alpha$$

$$+ \frac{1}{32\alpha E}(81 + 765\alpha^2 + 261\alpha^4 - 12\alpha^6 + 12\alpha^2 c_0 + 132\alpha^4 c_0 + 144\alpha^6 c_0) \qquad (2.133)$$

$$E = \alpha(1 + 2\alpha^2 + \cosh 2\alpha)$$

If we suppose $W = 0$ or $\gamma = 0$, we can obtain the zero of c_1 at $\gamma = 0$. This zero is located at the wave number $\alpha = 1.8017$ instead of $\alpha = 0$, see Figure 2.4. This immediately suggests that the wavenumbers near $\alpha = 0$ are stable when surface tension is negligible. A shortwave instability is possible even for vertical films at vanishingly small surface tensions !

2.5   Surface waves: The limit of $R \to \infty$.

From the above analyses, it is clear that the classical longwave interfacial instability does not occur for severely inclined planes or for fluids with low surface tension. Another shortwave instability also occurs at high R that does not involve interfacial disturbances at all. The parablic velocity profile of the Nusselt base state suggests that it admits a viscous Tollmien-Schlicting shear instability like those of Couette and Poisuelle flows. This viscous instability is expected to occur at high R with only disturbances internal to the film layer.

Let us begin with the inviscid limit

$$(U - c)(\varphi'' - \alpha^2 \varphi) - U''\varphi = 0 \qquad (2.134)$$

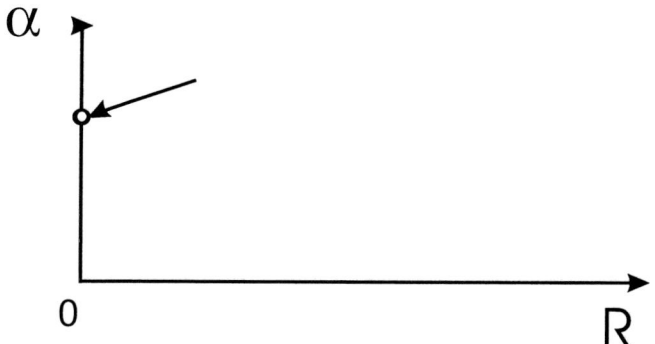

Figure 2.4: Vicinity of the critical Reynolds number for zero surface tension

$$y = 0: \quad \varphi = 0, \qquad (2.135)$$

$$y = 1: \quad (c - \frac{3}{2})^2 \varphi' - \alpha^2 W \varphi = 0 \qquad (2.136)$$

where we have dropped the no-slip condition and the zero shear stress condition.

There is no analytical solution to (2.134) to (2.136) for a parabolic profile U and we need some assumptions to evaluate its behavior. In our case, the singular point of (2.134) $U = c$ lies out of the region $y \in [0, 1]$, because the phase velocity is larger than liquid velocity on the surface. For an approximate solution of (2.134), let us replace it by

$$(\frac{3}{2} - c)(\varphi'' - \alpha^2 \varphi) + 3\varphi = 0 \qquad (2.137)$$

Hence, we suppose $U(y)$ is equal to the surface velocity but $U'' = -3$ as for the real parabolic profile. It is then easy to find the dispersion equation

$$\kappa(c - \frac{3}{2})^2 - \alpha^2 W \tanh \kappa = 0 \qquad (2.138)$$

$$\kappa^2 = \alpha^2 [1 \pm \frac{3}{\alpha^2(c - 3/2)}] \qquad (2.139)$$

At $|\frac{3}{\alpha^2(c-3/2)}| \ll 1$, this means the waves are short or have a velocity much larger than the surface velocity, then $\kappa = \alpha$ and

$$c - \frac{3}{2} = \pm\sqrt{W\alpha \tanh \alpha} \qquad (2.140)$$

We have hence reproduced the solutions which coincide with the disturbances for inviscid irrotational channel flow. One of the waves propagates downstream and the other upstream.

An inviscid analysis, however, does not allow us to get the imaginary part of c and to study the stability. For this, we need to study the Orr-Sommerfeld equation. In particular, we shall determine the neutral curve α_0 at $R \to \infty$

and the phase speed. The neutral wave number is expected to increase with the Reynolds number. Let us suppose that this growth obeys the power law:

$$\alpha_0 = kR^m \tag{2.141}$$

where m is unknown for now. Substituting this α_0 into the Orr-Sommerfeld equation (2.74), we obtain

$$y = 1 + \eta R^{-m} \tag{2.142}$$

where $\eta = O(1)$. This yields the well-known result that a wave can introduce velocity disturbance within a distance below the surface of the order of the wave-length. Besides that, we get

$$c = \frac{3}{2} + \Lambda R^{-2m} \tag{2.143}$$

If we substitute these relations for α_0, y and c into the Orr-Sommerfeld equation (2.74) and boundary conditions, we obtain

$$W = 3^{\frac{1}{3}}\gamma R^{-\frac{5}{3}} \tag{2.144}$$

and from the condition for normal stress

$$m = \frac{1}{3} \tag{2.145}$$

Notice that if the dependence $W = W(R)$ has a power not equal to $-5/3$, then m would have a different power.

After substituting,

$$\alpha_0 = kR^{\frac{1}{3}}, \quad y = 1 + R^{-\frac{1}{3}}\eta, \quad c = \frac{3}{2} + \Lambda R^{-\frac{2}{3}} \tag{2.146}$$

in (2.74) as the first approximation, we obtain:

$$\phi'''' - 2k^2\phi'' + k^4\phi = ik(-(\Lambda + \frac{3}{2}\eta^2)(\phi'' - k^2\phi) + 3\phi) \tag{2.147}$$

$$\eta = 0: \quad \phi'' + k^2\phi(1 - \frac{3}{k^2\Lambda}) = 0 \tag{2.148}$$

$$\phi''' + (ik\Lambda - 3k^2)\phi' + \frac{ik^3 3^{\frac{1}{3}}\gamma}{\Lambda}\phi = 0 \tag{2.149}$$

$$\eta \to -\infty: \quad \phi \to 0, \quad \phi' \to 0 \tag{2.150}$$

There is not analytical solution for (2.147). The numerical solution gives a universal dependence of k and Λ on the Kapitza number γ. The results of the solution are depicted in Figure 2.5 and Figure 2.6.

At constant γ and when $R \to \infty$, the upper neutral curve hence tends to infinity as $R^{1/3}$ while lower one still remains $\alpha_0 = 0$. The phase speed of the neutral modes tends to the liquid surface velocity, which is a well-know fact from experiments: for small and moderate Reynolds numbers the characteristic wave velocity is twice that of the surface, while for large R it is roughly the same as the surface velocity.

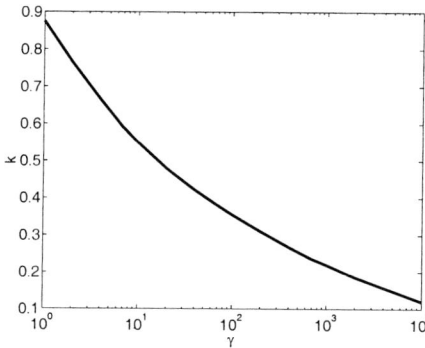
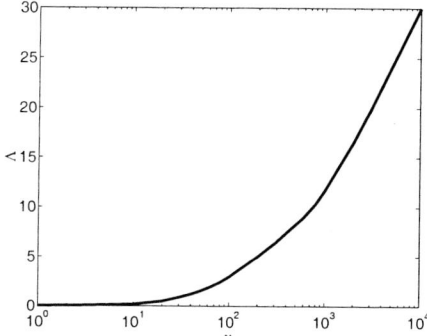

Figure 2.5: Normalize wavenumber k as a function of Kapitza number γ

Figure 2.6: Normalize velocity Λ as a function of Kapitza number γ

2.6 Numerical solution of the Orr-Sommerfeld equations.

We have considered mainly analytical methods and we now know the characteristic features of the stability behavior. Nevertheless, analytical methods can not replace numerical ones. The first numerical effort for the linear stability problem was by Sterling and Tauer in a private publication. We then point to the papers of Whitaker (1964) and Pierson and Whitaker (1977). In these works, finite difference method with uniform grid was applied and the problem reduces to an eigenvalue problem for five - diagonl matrix. Iterations was used and so only one eigenvalue was sought. (The iteration method was an approximate version of Thomas (1953)).

Eigenfunctions of the problem are smooth in some rigion of $y \in [0,1]$ and changes rapidly in some (for large αR). It is hence numerically convenient to reduce it to a Cauchy problem with the follwing numerical integration with adaptive steps. These methods are excellent one-page home work problems (Betchov and Criminale (1967), Goldshtik, Sapozhnikov and Stern (1975)). At large Reynolds numbers, these and related modifications (Kaplan (1964)) allow us to overcome the stiffness difficulties connected with the small parameter $1/(\alpha R)$ at the largest derivative.

These methods, however, have some significant disadvantages: *First*, it is necessary to utilize an initial guess to start the procedure; *Second*, only one eigenvalue can be obtained. Galerkin methods allow us to overcome these disadvantages. For one-phase flow, these methods were applied by Orzag (1971), for two-phase flow by Shkadov, Belikov and Epikhin (1986). Galerkin algorithms are very popular now, because the method can be automatically transformed from one problem to another by changing the boundary conditions of the Orr-Sommerfeld equation.

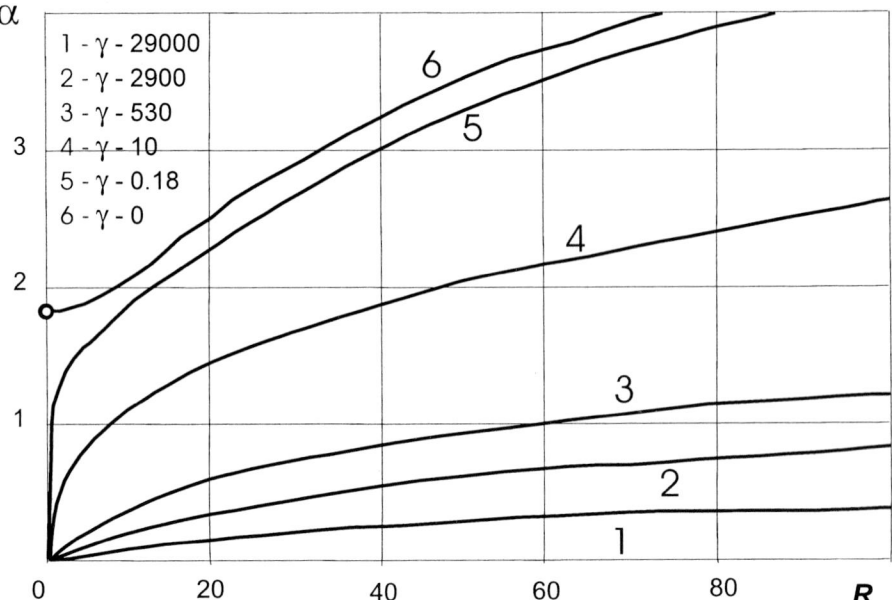

Figure 2.7: Neutral stability curves for flowing down film on a vertical wall for various Kapitza number γ

It is convenient to use Chebysev polynomials to approximate the solution,

$$\phi = \sum_{k=0}^{N} a_n T_n(y) \qquad (2.151)$$

Chebysev polynomials $T_k(y) = \cos(k \arccos y)$, $k = 0, 1, \ldots$, form a complete set of functions, which are orthogonal with a weight $(1 - y^2)^{-1/2}$. This eigenvalue problem can be solved by standard methods.

Orr-Sommerfeld equation after substitution of (2.151) transforms into a linear system

$$\sum_{k=0}^{N} \sum_{m=0}^{N} A_{km} a_m T_k(y) = \varphi_{km}$$

where φ_{km} are small residuals which we take orthogonal to $T_p(y)$ in order to obtain a linear system

$$\sum_{m=0}^{N} A_{km} a_i = 0$$

where A_{km} are functions of $\alpha, Re,$ and c. The functions $T_k(y)$ do not satisfy the boundary conditions (2.148), (2.149) and (2.150), so after substituting the expansion into them, we get four additional equations. Following Orszag (1971), let us replace the last four equations by these conditions in a Galerkin-Tau formulation. Eventually, we arrive at the eigenvalue problem for the matrix $A(c) = A_0 + A_1 c + A_2 c^2$.

$$det|A_0 + A_1 c + A_2 c^2| = 0$$

where A_k do not depend on c.

Let us consider the results of our calculations and relate them to our analytical results. We always have two eigenvalues that corespond to surface waves: one of these modes is stable for all parameters values and propogates upstream in the frame moving with the surface velocity. The second mode propogate downstream and can grow. This is the mode we investigate in our analytical expansion. The other eigenvalues correspond to internal shear waves.

Let us first consider the surface waves. In Figure 2.7, the neutral stability curves are shown γ vs Reynolds number for various. Our calculation is in good agreement with ones by Pierson and Whitaker(1977), exept for the case $\gamma = 0$, as was mentioned in section 2.4. For $\alpha R > 100$, our stability curves approach the analytic solution (2.146). Numerical analysis confirms the specific analitycal predictions: as $R \to 0$ $(\gamma \neq 0)$, $c_0 \to 3$; as $R \to \infty$, $c_0 \to \frac{3}{2}$ and this transition is monotonic. The growth the αc_i has a maximum with respect to R which depends on γ. As $R \to 0$, $c_i \to 0$ as $R^{\frac{10}{3}}$ (at $\gamma \neq 0$) and as $R \to \infty$, $c_i \to 0$ as $R^{-\frac{1}{3}}$.

At small R and W, solution c_r is described by (2.128)

$$c_r - 3/2 = \frac{3}{1 + 2\alpha^2 + \cosh 2\alpha}$$

and at large W there is a transition to

$$c_r - 3/2 = \sqrt{W \alpha \tanh \alpha}$$

for the invicid irrotational liquid. This transition is faster at larger Reynolds number R. At small Reynolds numbers, this transition occurs later but with a very sharp jump from the first asymptote to the next. The second interfacial mode has a similar behavior, but with the asymptote

$$c_r - 3/2 = -\sqrt{W \alpha \tanh \alpha}$$

Let us now consider shear instability. This problem has been studied in the papers of Goncharenko and Uritsev (1975), Bruin (1974), Lin (1967), Shkadov, Belikov and Epikhin (1980) and Floryan, Davis and Kelly (1987).

The shear eigenvalue spectrum is analogous to the spectrum of plane Poiseuille flow. The phase velocities are in the interval of (0,3/2). However, a unique characteristic feature of shear Tollmein-Shlichting waves in liquid layer with free boundary is its dependence on the Weber number W. The influence of surface tension upon the internal shear waves is destabilizing, while the surface waves are stabilized by surface tension. This was first mentioned by Goncharenko and Uritsev(1975), but they report that destabilization increased with W. As was later shown by Belikov and Shkadov there are certain values of

W at which the corresponding wave is stable for any given pair of α and R. Let us present the normal and tangential stresses (??)-(??) boundary conditions at $y = 1$ in the form

$$\phi = \frac{1}{i\alpha^3 RW}\left[\left(c - \frac{3}{2}\right)\phi''' - 3\alpha^2\left(c - \frac{3}{2}\right)\phi' + i\alpha R\left(c - \frac{3}{2}\right)^2 \varphi'\right]$$

$$\phi'' = -\alpha^2 \phi\left[1 - \frac{3}{\alpha^2\left(c - \frac{3}{2}\right)}\right]$$

At fixed α and $W \to \infty$, these conditions become

$$\phi = \phi'' = 0$$

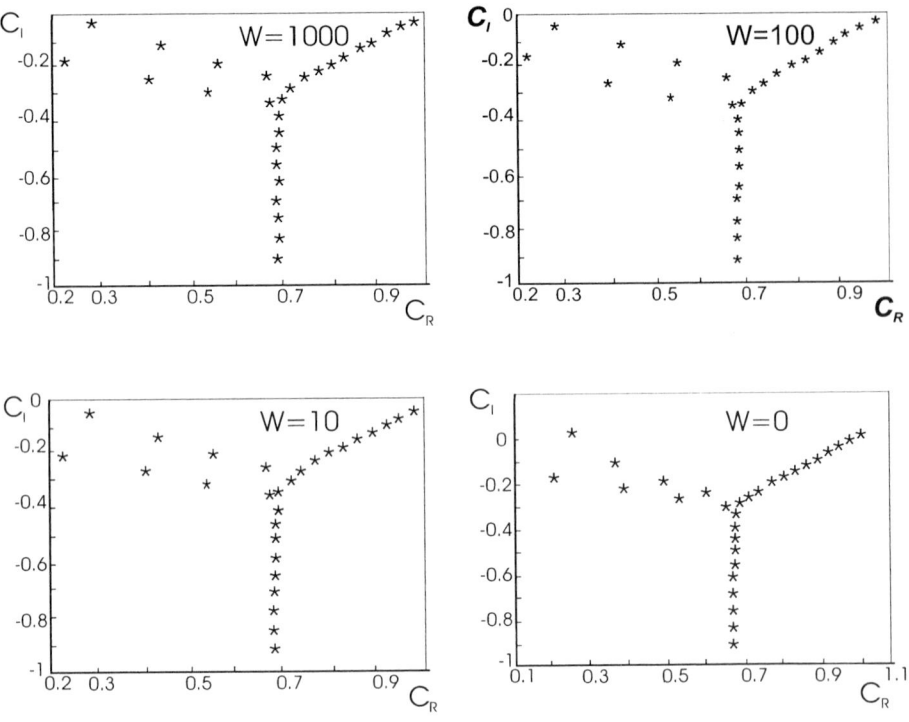

Figure 2.8: Spectra of surface and shear modes for different Weber number, W

The second analog corresponds to a similar condition for anti-symmetric disturbanes for plane Poiseuille flow. Consequently, at $W \to \infty$, the spectrum approaches the anti-symmetric part of plane Poiseuille spectrum. These reasonings are confirmed by calculations. In Figure 2.8 the spectrum without interfacial modes is shown for $\alpha = 1$ and $R = 6667$ for several values of $W = 0$, 1, 10, 100 and 1000. They resemble the anti-symmetric modes in Poiseuill flow (Orsag, 1971). From Figure 2.8, the eigenvalues are seen to be discrete and infinite in number. At $W = 0$, there is an unstable surface mode and one unstable shear mode. At larger W, they both become stable. As in Poiseuille flow(Mack,1976), the shear eigenvalues are divided into three families. The A-family (to the left hand-side) is finite in number; P-family (to the right hand-side) and S-family (at the bottom of the pictures) are infinite in number.

At low Reynolds numbers, c_i is of the order of $-1/(\alpha R)$ and at $R \to 0$ the shear part of spectrum goes to infinity.

As shear wave stability depends on W, one can plot a neutral curve in the $W - R$ plane at any α and find critical parameter values. These values define a relation $R_*(\alpha)$ in the plane α of R. At the nose of the neutral curve, these is another set of critical parameter values: $R_* = 2507, \alpha_* = 1.80$ and $\gamma_* = 127036(W_* = 0.396)$, as seen in Figure 2.9

Hence, liquid with $\gamma = \gamma_*$ first loses stability to viscous shear waves. In this case the critical Reynolds number is three times less than that for Poiseuille flow!

In Figure 2.9, the neutral stability curve for shear instability for water at the critical Kapitza number $\gamma = 2850$ and for vertical flow is shown. One can see that this curve lies inside the instability region of surface waves, see Figure 2.10 for comparison. The dominant instability for vertical falling films is the interfacial instability.

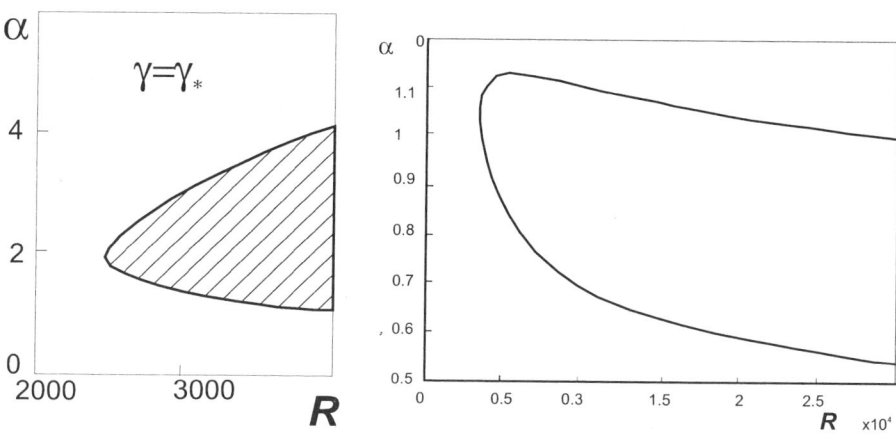

Figure 2.9: Neutral stability curve for the critical Kapitza number, shear waves

Figure 2.10: Neutral stability curve for water, $\gamma = 2850$, shear waves

For vertical flow, both instabilities depend on surface tension. The situation changes for very small incline angles θ. We have already shown in section 2.3 that for a small inclination angles θ, the influence of surface tension can be negligible. Let us consider shear waves for small θ.In Figure 2.11 and 2.12 the neutral stability curves taken from the paper of Floryan, Davis and Kelly(1987) are shown and recalculated for our parameters. It is easy to see that for angles less than $\theta < 1'$, surface instability is independent of surface tension as we have already shown analytically. Hence, for sufficiently small angles, the stability curve of shear waves is outside the instability region of the surface mode. In other words, for sufficiently small angles, the viscous Tollmein-Shlichting instability dominates the interfacial instability for all surface tensions.

Another interesting case is $\pi \geq \theta > \pi/2$, where one can have a third kind of instability - Taylor's instability due to the destabilizing gravitational force. In Figure 2.13, the neutral stability curve α_0 is shown for R=50 and γ=2850 vs the angle θ. One can see

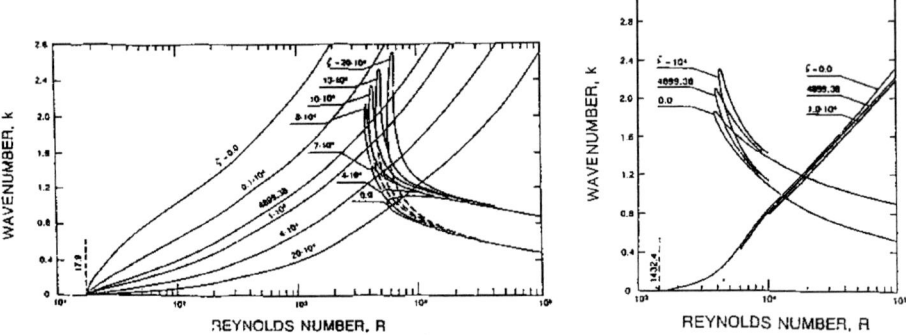

Figure 2.11: Neutral curves for $\theta = 4°$ and various Kapitza number γ for the shear and surface modes

Figure 2.12: Neutral curves for $\theta = 3'$ and various Kapitza number γ for the shear and surface modes

that the instability is in fact nearly independent of θ and only for small angles $\theta \approx 0$ and $\theta = \pi$, does α_0 change sharply. The case $\theta = \pi$ is singlular because the basic state no longer has a mean flow. Let us change the basic values: waveless thickness, l_0; velocity, $u_0 = \frac{1}{3}\frac{gl_0^2}{\nu}$; density, ρ_0. Without mean flow in this horizontal case,

$$U = U'' = 0$$

and the system (49)- (53) becomes:

$$\phi^{IV} - 2\alpha^2\phi'' + \alpha^4\phi = -\frac{3i\alpha c}{G}(\phi'' - \alpha^2\phi) \tag{2.152}$$

$$y = 1:$$

$$\phi''' - 3\alpha^2\phi' + \frac{3i\alpha c}{G}\phi' + 3i\alpha - \frac{3i\alpha^3 W}{G} = 0 \tag{2.153}$$

$$\phi'' + \alpha^2\phi = 0 \tag{2.154}$$

$$\phi = c \tag{2.155}$$

$$y = 0: \quad \phi = \phi' = 0 \tag{2.156}$$

where instead of Reynolds number as a control parameter, we have the Froude number G. Hence, if the Kapitza number is given,

$$W = \frac{\gamma}{3^{4/3}} G^{5/3}; \quad G = \frac{9\nu^2}{gl_0^3}$$

In Figure 2.14, we show the neutral stability curve $\alpha_0 = \alpha_0(G)$ for water.

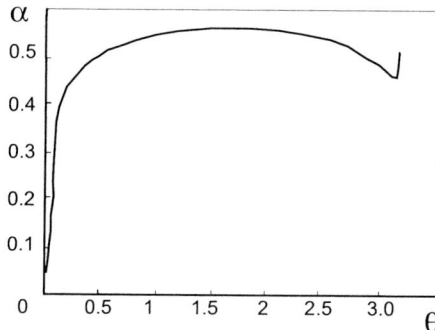
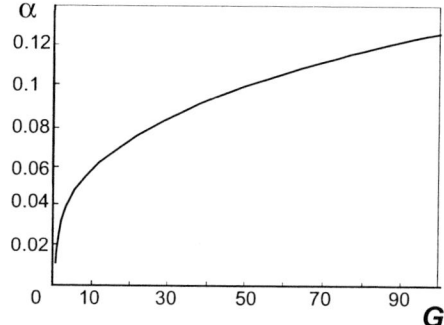

Figure 2.13: Neutral stability curve for water $\gamma = 2850, R = 50$ as a function of inclination angle, θ

Figure 2.14: Neutral stability curve for water $\gamma = 2850$ on the "ceiling", $\theta = \pi$, as a function of the Froude number G

CHAPTER 3

Hierarchy of Model Equations

The waves triggered by the linear instability in the previous chapter evolve rapidly downstream due to nonlinear effects. Such downstream nonlinear wave dynamics change continuously over more than one meter into the channel–none of which can be described by the linear theory. The waves travel much faster than the phase velocity predicted by linear theory . They assume shapes very different from the sinusoidal normal modes of the linear theory. They even have wavelengths as much as one order of magnitude larger than the linear waves. Moreover, all these wave characteristics evolve continuously downstream for one meter or longer (see Figure 1.1 of Chapter 1.)In fact, only waves within a mere 10 cm from the inlet can be described by the linear theory. Even with modern computers, such downstream nonlinear wave dynamics on a falling film still cannot be simulated with the complete Navier-Stokes equation. As a result, simpler model equations are welcomed substitutes provided they exhibit robustness and fidelity. If so, they also lend themselves to analysis to produce compact and useful closed-form correlations. Two approaches exist for such modeling attempts — exact weakly nonlinear theory under stringent limiting conditions and empirical with larger regions of validity. Both suffer from distinct weaknesses. The latter, being an empirical formulation, must be validated against experimental data. The former "rigorous" approach relies on a prori order assignments on the wave dimensions, paticularly the wave amplitude, and the system parameters. While the parameters correspond to physical conditions that can be controlled, the wave dimensions are unknown. Weakly nonlinear theory hence impose a priori assumptions about the wave dynamics. As such, the resulting equations are only appropriate if the true wave dynamics have length and time scales consistent with the imposed assumptions. Such a luxury is often unavailable in real life. Moreover, under the same physical conditions, waves of different dimensions and time scales can coexist and they are described by different model equations. A single equation is hence inadequate to capture the full dynamics.

Nevertheless, bits of the true dynamics can be captured by each. Moreover, they are the longwave analogs of the complex Ginzburg-Landau equation for short-wave instability, which was also derived with similar assumptions about wave dimensions.As such, they represent an important development in bifurcation theory, pattern formation and nonlinear dynamics. What they lack in physical relevance, they more than make up in mathematical rigor and elegance.

3.1 Kuramoto-Sivashinsky(KS), KdV and related weakly nonlinear equations

All weakly nonlinear equations assume small amplitude and longwaves. As such, the neutral wave number α_0 from the linear versions of the equations must be small. However, this α_0 can be small with different parameter conditions. Many weakly nonlinear equations then differ because they choose different parameter conditions to achieve a small α_0. As a result, some have dominant dispersion (KdV) and some have negligible dispersion (KS). Their derivations are also quite different. Some preassume the parameter conditions and carry out the longwave expansion in α_0 at each order by retaining only the relevant terms. One unique Whitham approach actually assigns the parameter orders at the very end. In all cases, however, the dominant linear terms are balanced by the dominant nonlinear term $H\frac{\partial H}{\partial \xi}$ and no others. This dominant nonlinear term arises from the nonlinear correction to the phase speed–a nonlinear kinematic effect that captures how larger waves move faster than smaller ones. The dominant balance between the dominant nonlinear term and the dominant linear terms than specify the order of the amplitude.

We begin such derivations with one approach that preassumes near-critical conditions, $R = R_*$. We shall use the method of multiple scales, see Van Dyke (1975) and Nayfeh (1973). As $R \to R_*$ the neutral wavenumber α_0 tends to zero for a longwave instability, regardless of the order of all other system parameters. It is convenient to take $\alpha_0 = \varepsilon$ as a small parameter in the system where α_0 is defined by the equation

$$\omega_{20} + \alpha_0^2 \omega_{40} = 0 \tag{3.1}$$

where $\omega_{40} \neq 0$ and the two frequencies are the second and fourth order expansions of the imaginary part of the complex frequency in (2.78).

We shall also consider the equations in a frame moving with the longwave linear phase velocity 3 , see (2.88), $\frac{\partial}{\partial t} \to \frac{\partial}{\partial t} - 3\frac{\partial}{\partial x}$ and introduce slow variables $\xi = \varepsilon^n x$, $\zeta = \varepsilon^n z$ and $\tau = \varepsilon^m t$, so that

$$\frac{\partial}{\partial x} = \varepsilon^n \frac{\partial}{\partial \xi}, \quad \frac{\partial}{\partial z} = \varepsilon^n \frac{\partial}{\partial \zeta}, \quad \frac{\partial}{\partial t} = \varepsilon^m \frac{\partial}{\partial \tau} \tag{3.2}$$

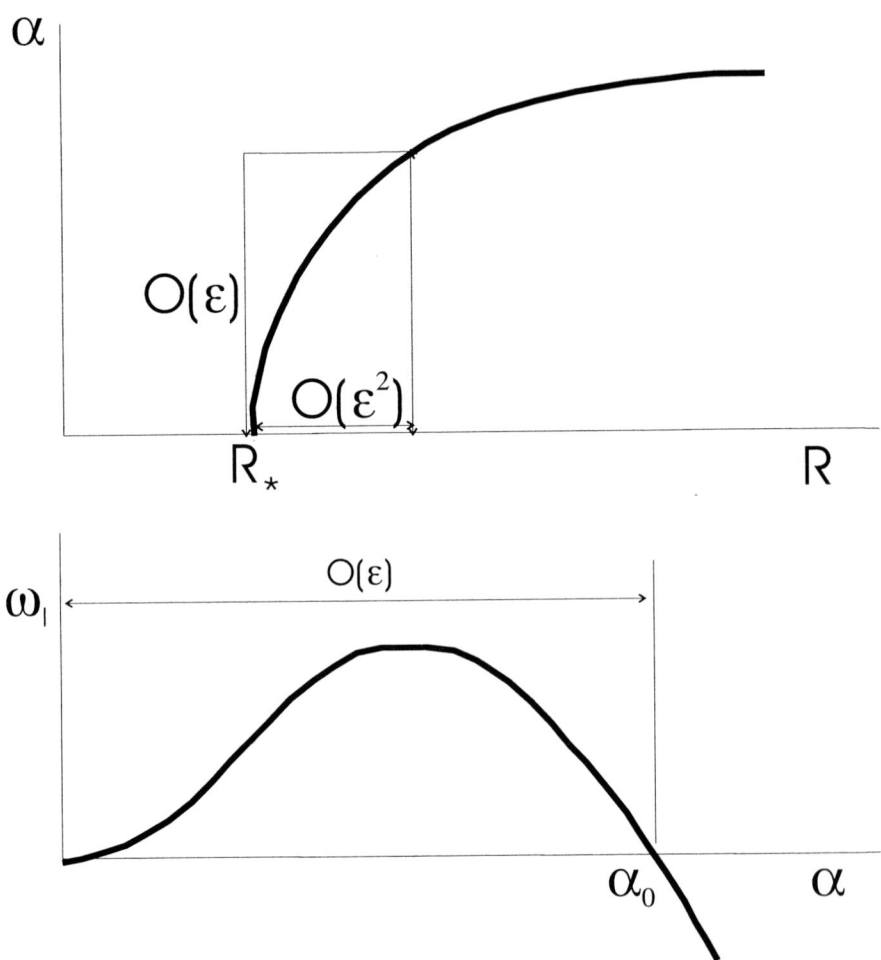

Figure 3.1: longwave instability

We expand the unknowns

$$u \sim 3\left(y - \frac{1}{2}y^2\right) + \varepsilon^p u_1 \tag{3.3}$$

$$v \sim \varepsilon^{p+1} v_1 \tag{3.4}$$

$$w \sim \varepsilon^p w_1 \tag{3.5}$$

$$h \sim 1 + \varepsilon^p H \tag{3.6}$$

The actual order assignment of each relative to ε determines the dominant terms in the final weakly nonlinear equation. For the linear terms, one can prechoose the dominant dispersion terms in (2.88)-(2.97) and then specify all order assignments above except p. Balancing the dominant nonlinear term $H\frac{\partial H}{\partial \xi}$ then completes the derivation. Such algebraic exercise is rather tedious and we present only the final result here. The calculations are similar to the linear analysis at $\alpha, \beta \to 0$ in the pervious chapter, we obtain that the proper choice for the expansion ansatz is $p = 2$ and $m = 3$ and in the 2D-case the KdV equation if ω_{40} is $O(1)$,

$$\frac{\partial H}{\partial \tau} + 6H\frac{\partial H}{\partial \xi} - \omega_{30}\frac{\partial^3 H}{\partial \xi^3} = 0 \tag{3.7}$$

instead of usual Kuramoto-Sivashinsky (KS) equation. The dispersion term originates from the third order expansion of the real part of the complex frequency in (2.78). If next-order terms are included, we obtain

$$\frac{\partial H}{\partial \tau} + 6H\frac{\partial H}{\partial \xi} - \varepsilon\omega_{40}\left(\frac{\partial^2 H}{\partial \xi^2} + \frac{\partial^4 H}{\partial \xi^4}\right) - \omega_{30}\frac{\partial^3 H}{\partial \xi^3} = 0 \tag{3.8}$$

with $\omega_{40} < 0$, so that the coefficient before the second and fourth derivatives is positive.

A key order assignment in the derivation of the KdV equation is that ω_{40} has order $O(1)$. This is true when capillary forces are not large. At the critical point $R = R_*$, (2.93) yields

$$\omega_{40} = -\gamma(\frac{2}{5})^{2/3}\frac{(\sin\theta)^{1/3}}{(\cos\theta)^{2/3}} + \frac{1}{36036}(\cot\theta)^3 - \frac{7743}{2240}\cot\theta \tag{3.9}$$

and at any $\gamma \neq 0$, $\omega_{40} \to \infty$ at $\theta \to \frac{\pi}{2}$. Hence, the KdV equation is inappropriate for the vertical film. For large surface tension, one must also severely incline the plane to reach the KdV limit.

Consider now the other limit, when the inclination angle θ is fixed, $\theta \neq 0$, and $R - R_* = O(1)$, but the Kapitza number γ tends to infinity. This approach obtains longwaves with large surface tension instead of the near-critical condition of the KdV equation. Let $\varepsilon = 1/\gamma$. The proper expansion ansatz now has exponents $n = p = \frac{1}{2}$ and $m = 1$ we get

$$\frac{\partial H}{\partial \tau} + 6H\frac{\partial H}{\partial \xi} + \varepsilon\omega_{20}\frac{\partial^2 H}{\partial \xi^2} + a_4\frac{\partial^4 H}{\partial \xi^4} - \varepsilon^{1/2}\omega_{30}\frac{\partial^3 H}{\partial \xi^3} = 0 \tag{3.10}$$

$$\omega_{20} = O(1), \quad \omega_{30} = O(1)$$
$$a_4 = \frac{1}{3^{2/2} R^{2/3} (\sin\theta)^{1/3}} = O(1)$$

At $\varepsilon \to 0$, the last equation transforms into the Kuramoto-Sivashinsky (KS) equation, if the inclination angle θ is not close to zero and the Reynolds number is not excessively large. Near-critical assumption is unnecessary, however.

The near-critical vertical film can stll yield the KdV equation but with a different length scaling. Let $\theta = \frac{\pi}{2}$ and γ has any finite value, $\gamma \neq 0$. Let $\varepsilon = R \to 0$ at $n = \frac{5}{6}$, $m = \frac{5}{2}$ and $p = \frac{5}{3}$, the weakly nonlinear equation then has the form

$$\frac{\partial H}{\partial \tau} + 6H \frac{\partial H}{\partial \xi} - 3 \frac{\partial^3 H}{\partial \xi^3} + \varepsilon^{1/6} \left(\frac{5}{6} \frac{\partial^2 H}{\partial \xi^2} + a_4 \frac{\partial^4 H}{\partial \xi^4} \right) = 0 \quad (3.11)$$

where
$$a_4 = \frac{\gamma}{3^{2/3}}$$

which at $\varepsilon = R \to 0$ tends to the KdV equation. Hence, for vertical films with large γ, one evolves from the KdV equation into the the KS equation as $R - R_*$ increases from order ε to unit order.

When ω_{40} is small, a short-wave instability is possible, see Figure 3.2 The expansion strategy is entirely different for longwave and shortwave instabilities. To differentiate the two, we need to keep three terms in our near-critical longwave expansion. More specifically, the neutral wave numbers is determined by

$$\omega_{20} + \alpha_0^2 \omega_{40} + \alpha_0^4 \omega_{60} = 0 \quad (3.12)$$

In order to keep both roots $O(\varepsilon^2)$, the dispersion coefficients of (2.90) to (2.93) must have the following order assignments:

$$\omega_{20} = O(\varepsilon^4) \quad (3.13)$$
$$\omega_{40} = O(\varepsilon^2) \quad (3.14)$$
$$\omega_{60} = O(1) \quad (3.15)$$

Let
$$\omega_{20} = \Omega_{20} \varepsilon^4 \quad (3.16)$$
$$\omega_{40} = \Omega_{40} \varepsilon^2 \quad (3.17)$$

and for $n = 1$, $p = 2$ and $m = 3$, the model equation will be

$$\frac{\partial H}{\partial \tau} + 6H \frac{\partial H}{\partial \xi} - \omega_{30} \frac{\partial H^3}{\partial \xi^3} + \varepsilon^2 \left(\Omega_{20} \frac{\partial^2 H}{\partial \xi^2} - \Omega_{40} \frac{\partial^4 H}{\partial \xi^4} + \Omega_{60} \frac{\partial^6 H}{\partial \xi^6} \right) = 0 \quad (3.18)$$

and at $\varepsilon \to 0$ we obtain the same KdV equation.

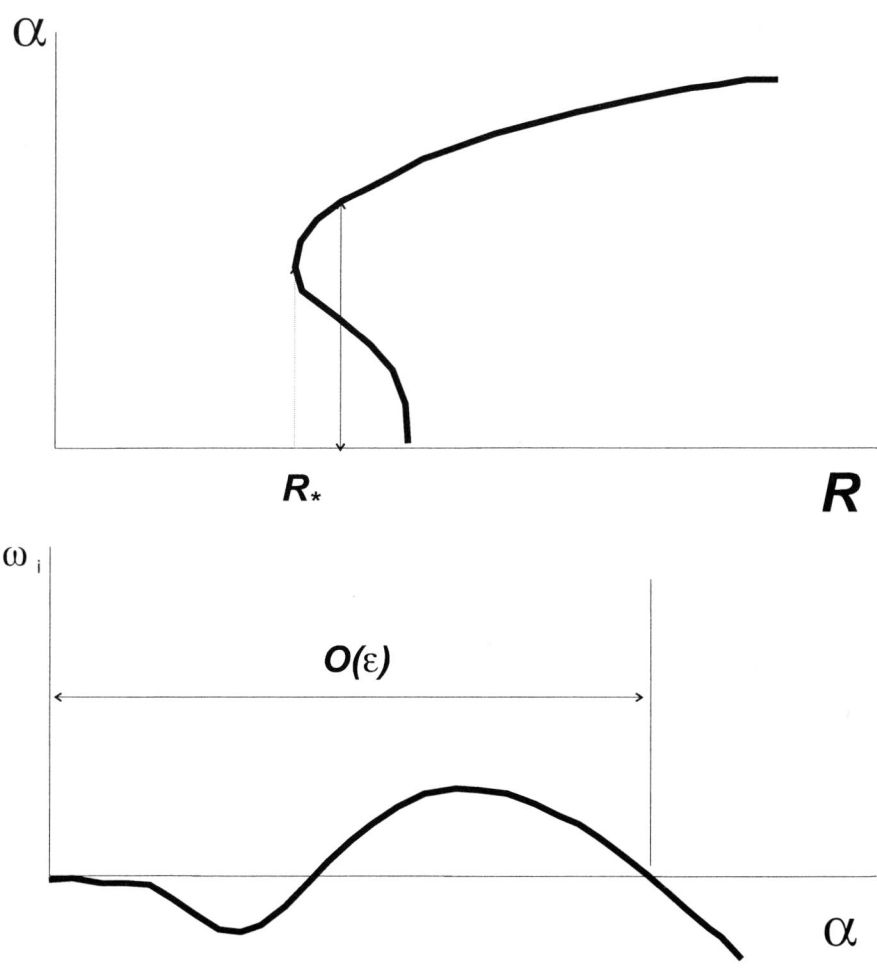

Figure 3.2: Short-wave instability

The model equations (3.8), (3.10) and (3.11) have the generic form

$$\frac{\partial H}{\partial \tau} + 6H\frac{\partial H}{\partial \xi} + \alpha\frac{\partial^2 H}{\partial \xi^2} + \beta\frac{\partial^3 H}{\partial \xi^3} + \gamma\frac{\partial^4 H}{\partial \xi^4} = 0 \qquad (3.19)$$

and by the transformation

$$\tau \to \sqrt{\frac{2}{3}}\alpha^{-5/4}\gamma^{3/4}\tau, \quad \xi \to \sqrt{\frac{\gamma}{\alpha}}\xi, \quad H \to \sqrt{\frac{2}{3}}\alpha^{3/4}\gamma^{-1/4}H$$

it is transformed into

$$\frac{\partial H}{\partial \tau} + 4H\frac{\partial H}{\partial \xi} + \frac{\partial^2 H}{\partial \xi^2} + \delta\frac{\partial^3 H}{\partial \xi^3} + \frac{\partial^4 H}{\partial \xi^4} = 0 \qquad (3.20)$$

where $\delta = \sqrt{\frac{2}{3}}\beta\alpha^{1/4}\gamma^{-1/4}$, or by the transformation

$$\tau \to \gamma^{3/2}\alpha^{-3/2}\beta^{-1}\tau, \quad \xi \to \sqrt{\frac{\gamma}{\alpha}}\xi, \quad H \to \frac{2}{3}\frac{\alpha\beta}{\gamma}H$$

to another form

$$\frac{\partial H}{\partial \tau} + 4H\frac{\partial H}{\partial \xi} + \frac{\partial^3 H}{\partial \xi^3} + \varepsilon\left(\frac{\partial^2 H}{\partial \xi^2} + \frac{\partial^4 H}{\partial \xi^4}\right) = 0 \qquad (3.21)$$

where $\varepsilon = \frac{\sqrt{\alpha\gamma}}{\beta}$ (Topper and Kawahara, 1978).

Both forms have only one parameter; the first equation is valid near the KS limit for small δ while the second equation is applicable near KdV limit for small ϵ. We shall call both equations generalized Kuramoto-Sivashinsky (gKS) equation or the Kawahara equation. (Kawahara and his coworkers first investigated this equation for problems unrelated to the falling film (1983, 1987).)

There is a very important difference between the KdV and gKS equations for $\varepsilon \to 0$. Stationary travelling waves (waves that propagate at constant speed without changing their shape) of the KdV equation are possible due to a balance between nonlinearity and dispersion and we have a two-parametric family of periodic solutions. The gKS equation has an energy source term with a second derivative and a fourth-derivative dissipation term acting as a energy sink. For stationary running waves, even for small ε, a balance between energy supply and sink must be obeyed. Indeed, let us multiply (3.20) by H and integrate over period L

$$\frac{\partial}{\partial t}\int_0^L H^2 d\xi = \int_0^L \left[\left(\frac{\partial H}{\partial \xi}\right)^2 - \left(\frac{\partial^2 H}{\partial \xi^2}\right)^2\right] d\xi \qquad (3.22)$$

The first positive term in the right-hand side originates from the destabilizing $\partial^2 H/\partial \xi^2$ in the equation, while the second negative term is connected with the fourth-order dissipation derivative.

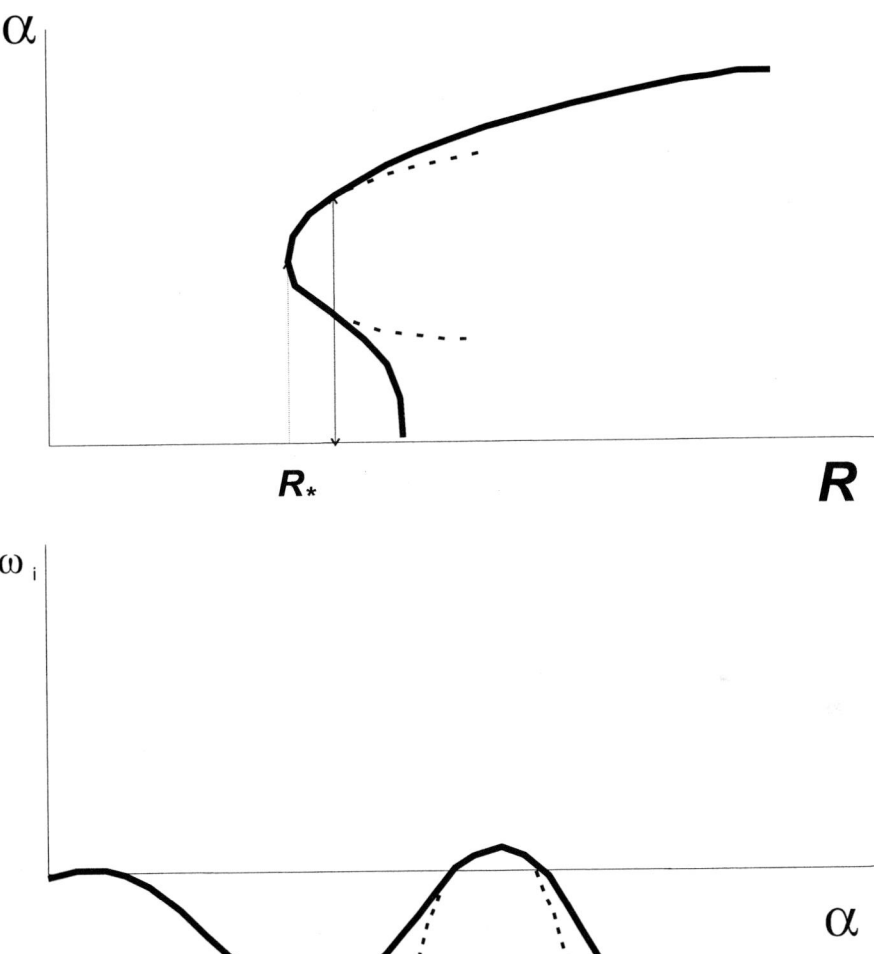

Figure 3.3: Expansion near α_* and ω_*

For a stationary travelling wave in the moving frame, the integral $\int_0^L H^2 d\xi$ is independent of time and we have an energy balance

$$\int_0^L \left(\frac{\partial H}{\partial \xi}\right)^2 d\xi = \int_0^L \left(\frac{\partial^2 H}{\partial \xi^2}\right)^2 d\xi \qquad (3.23)$$

This energy balance involves energy input at long waves and energy dissipation at shorter waves. As a result of this additional energy balance, the two-parameter family of KdV periodic solutions becomes a one-parameter KS family with the wave number as a parameter. This also means that we cannot neglect small terms with second and fourth derivative.

For small angles θ it is possible that the coefficient for the second-derivate term is negative and this term becomes stabilizing. Alternatively, the coefficient of the fourth derivative term can become negative such that it becomes destabilizing. This high order destabilizing term must be balanced by an even higher order stabilizing term to produce travelling waves — energy transfer must be from long to short waves. It is then necessary to include a stabilizing six-derivative term. That is the case for equation (3.18). By transforming its variables

$$\tau \to -\frac{\Omega_{60}^{3/2}}{\omega_{30}\Omega_{40}^{3/2}}\tau, \quad \xi \to \sqrt{\frac{\Omega_{60}}{\Omega_{40}}}\xi, \quad H \to -\frac{2}{3}\omega_{30}\frac{\Omega_{40}}{\Omega_{60}}H$$

we obtain the two-parameter equation

$$\frac{\partial H}{\partial \tau} + 4H\frac{\partial H}{\partial \xi} + \frac{\partial^3 H}{\partial \xi^3} + \varepsilon\left(-r\frac{\partial^2 H}{\partial \xi^2} - \frac{\partial^4 H}{\partial \xi^4} + \frac{\partial^6 H}{\partial \xi^6}\right) = 0 \qquad (3.24)$$

with

$$\varepsilon = -\frac{\omega_{30}}{\Omega_{40}}\sqrt{\frac{\Omega_{60}}{\Omega_{40}}}$$

$$r = -\frac{\Omega_{20}\Omega_{60}}{\Omega_{40}^2}$$

Energy balance now is

$$\frac{\partial}{\partial t}\int_0^L H^2 d\xi = \varepsilon\left[-r\int_0^L \left(\frac{\partial H}{\partial \xi}\right)^2 d\xi + \int_0^L \left(\frac{\partial^2 H}{\partial \xi^2}\right)^2 d\xi - \int_0^L \left(\frac{\partial^3 H}{\partial \xi^3}\right)^2 d\xi\right] \qquad (3.25)$$

The neutral stability curve of the trivial state in (3.24), corresponding to the Nusselt flat film, is shown in Figure 3.4. At $r > r_* = \frac{1}{4}$ the trivial solution is stable, while ta $0 < r < r_*$ a short-wave instability takes place. For $r < 0$ the instability is the longwave type as the gKS-equation.

The shortwave instability requires a different expansion analogous to the classical derivation of the complex Ginzburg-Landau (CGL) equation. At the critical point $r_* = \frac{1}{4}$ and $\alpha_* = \frac{\sqrt{2}}{2}$, we have a sinusoidal solution

$$H = ae^{i(\alpha_* x - \omega_* t)}$$

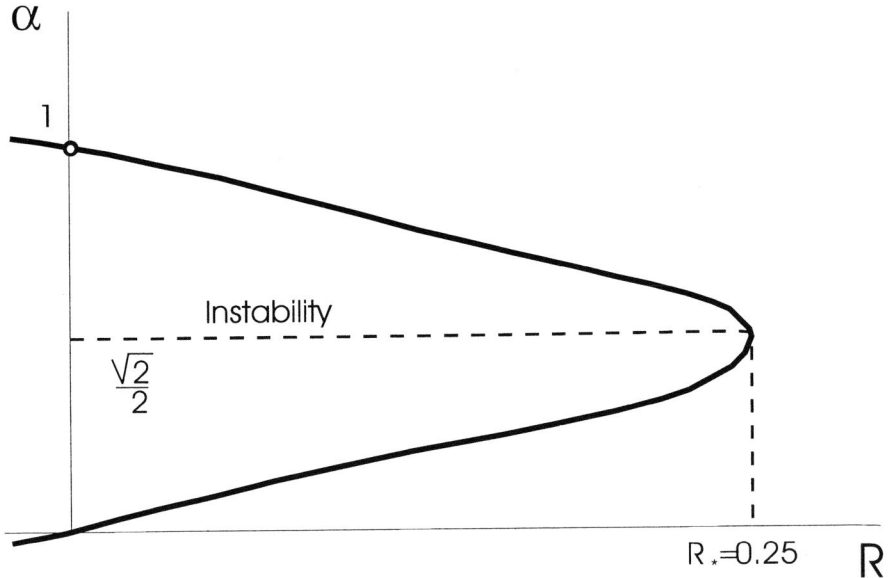

Figure 3.4: Instability region for the equation (3.24)

with infinitesimal amplitude A and real α_* and ω_*. Let us consider a small vicinity of the critical point,
$$r = r_* - \mu^2$$
with complex amplitude
$$H = \mu A e^{i(\alpha_* x - \omega_* t)}$$
and also introduce slow time and space variables,
$$T = \mu\tau$$
$$X = \mu\xi$$

Eventually, we get the complex Ginzburg-Landau (CGL) equation
$$\frac{\partial A}{\partial T} = \gamma_1 A + \gamma_2 \frac{\partial^2 A}{\partial X^2} + \gamma_3 |A|^2 A \qquad (3.26)$$
with real γ_1, positive for supercritical case, and complex γ_2 and γ_3
$$\gamma_1 = 2r_*\varepsilon^2 \quad \gamma_2 = 3r_*\varepsilon\sqrt{2} + 2i \quad \gamma_3 = 3 + r_*\varepsilon^2 i$$
or, in a standart form, after transformation,
$$A \to \chi A, \quad \frac{\partial}{\partial X} \to \varkappa \frac{\partial}{\partial X}, \quad \frac{\partial}{\partial T} \to \gamma_1 \frac{\partial}{\partial T}$$
$$\chi^2 = \frac{\gamma_1}{Real(\gamma_3)}, \quad \varkappa^2 = \frac{\gamma_1}{Real(\gamma_2)}$$

$$\frac{\partial A}{\partial T} = A + (1+ic_1)\frac{\partial^2 A}{\partial X^2} - (1+ic_2)|A|^2 A \qquad (3.27)$$

All above derivations presuppose the orders of the parameters and retain only the relevant terms in each order of the longwave expansion. A different strategy is offered by Whitham (1974). He uses the longwave dispersion relationship (2.90)-(2.99) which contains a minimum of order assignment and then impose further order assignments to reach the various equations. This approach is best used for three-dimensional wave dynamics where the order expansion in both dimensions of the free surface is difficult. It begins with a simple correspondence between the dispersion relation and derivatives

$$\frac{\partial}{\partial \tau} \to -i\omega, \quad \frac{\partial}{\partial \xi} \to i\alpha, \quad \frac{\partial}{\partial \zeta} \to i\beta, \qquad (3.28)$$

to relate the linear dispersion relationship of the previous chapter to the weakly nonlinear equations here.

From the dispersion relation (2.88) to (2.97), we can recover the linearized version of the model equation.

$$\frac{\partial \hat{h}}{\partial t} - 3\frac{\partial \hat{h}}{\partial x} + \omega_{20}\frac{\partial^2 \hat{h}}{\partial x^2} - \omega_{30}\frac{\partial^3 \hat{h}}{\partial x^3} - \omega_{40}\frac{\partial^4 \hat{h}}{\partial x^4} + \omega_{02}\frac{\partial^2 \hat{h}}{\partial z^2} - \omega_{12}\frac{\partial^3 \hat{h}}{\partial x \partial z^2} - \omega_{22}\frac{\partial^4 \hat{h}}{\partial x^2 \partial z^2} - \omega_{04}\frac{\partial^4 \hat{h}}{\partial z^4} = 0 \qquad (3.29)$$

where $\hat{h} = h - 1$. In order to obtain a weakly nonlinear extension of this equation we have to know only main nonlinear term, $6\hat{h}\frac{\partial \hat{h}}{\partial x}$, adding which to (3.29) we obtain the weakly nonlinear 3D equation:

$$\frac{\partial \hat{h}}{\partial t} + 6\hat{h}\frac{\partial \hat{h}}{\partial x} + a_{20}\frac{\partial^2 \hat{h}}{\partial x^2} + a_{30}\frac{\partial^3 \hat{h}}{\partial x^3} + a_{40}\frac{\partial^4 \hat{h}}{\partial x^4} + a_{02}\frac{\partial^2 \hat{h}}{\partial z^2} + a_{12}\frac{\partial^3 \hat{h}}{\partial x \partial z^2} + a_{22}\frac{\partial^4 \hat{h}}{\partial x^2 \partial z^2} + a_{04}\frac{\partial^4 \hat{h}}{\partial z^4} = 0 \qquad (3.30)$$

where $h = 1 + \hat{h}$, we consider equation in a frame moving with the linear phase speed 3. Relating the coefficients a_{ij} to ω_{ij} of (2.88)-(2.97) and their three-dimensional extensions, we get

$$a_{20} = \frac{6}{5}R - \cot\theta \qquad (3.31)$$

$$a_{30} = 3 + \frac{12}{7}R^2 - \frac{10}{7}R\cot\theta \qquad (3.32)$$

$$a_{40} = \frac{1}{3}WR + \frac{75872}{25025}R^3 - \frac{17363}{5775}R^2\cot\theta + \frac{2}{5}R\cot^2\theta +$$
$$+ \frac{1413}{224}R - \frac{9}{5}\cot\theta \tag{3.33}$$
$$a_{02} = -\cot\theta \tag{3.34}$$
$$a_{12} = 3 - \frac{10}{7}R\cot\theta \tag{3.35}$$
$$a_{22} = \frac{2}{3}RW - \frac{17262}{5775}R^2\cot\theta + \frac{4}{5}R\cot^2\theta + \frac{1413}{224}R - \frac{18}{5}\cot\theta \tag{3.36}$$
$$a_{04} = \frac{1}{3}RW + \frac{2}{5}R\cot^2\theta - \frac{9}{5}\cot\theta \tag{3.37}$$

For a vertical plane, $\theta = \pi/2$

$$a_{20} = \frac{6}{5}R; \quad a_{30} = -3 + \frac{12}{7}R^2 \tag{3.38}$$
$$a_{40} = \frac{1}{3}WR + \frac{75872}{25025}R^3 + \frac{1413}{224}R \tag{3.39}$$
$$a_{02} = 0; \quad a_{12} = 3 \tag{3.40}$$
$$a_{22} = \frac{2}{3}RW + \frac{1413}{224}R \tag{3.41}$$
$$a_{04} = \frac{1}{3}RW \tag{3.42}$$

Let us consider the case of large surface tension, $W \to \infty$ (or $\gamma \to \infty$):

$$\alpha_0^2 = \frac{18}{5W} \to 0: \tag{3.43}$$

$$\xi = \alpha_0 x, \quad \zeta = \alpha_0 z, \quad \tau = \frac{6}{5}R\alpha_0 t, \quad \hat{h} = \frac{6}{5}R\alpha_0 H \tag{3.44}$$

and we obtain the 3D gKS equation, see Roskes (1970), Lin and Krishna (1977) and Nepomnyashchy (1974),

$$\frac{\partial H}{\partial \tau} + 6H\frac{\partial H}{\partial \xi} + \frac{\partial^2 H}{\partial \xi^2} + \nabla^4 H + \delta\frac{\partial}{\partial \xi}\nabla^2 H = 0 \tag{3.45}$$

where

$$\nabla^2 \equiv \left(\frac{\partial^2}{\partial \xi^2} + \frac{\partial^2}{\partial \zeta^2}\right)$$

$$\delta^2 = \frac{9}{2R^2 W} = \frac{3^{5/3}}{2\gamma R^{\frac{1}{3}}}$$

At $\gamma \to \infty$ and fixed R, $\delta \to 0$, but at γ large but fixed and $R \to 0$, $\delta \to \infty$. These two limits have very different behavior due to the role of dispersion. It is hence important to know which limit is realized. For water, $\gamma = 3000, \delta = 2.5 * 10^{-2}$ for $R = 5$, and the influence of dispersion is negligible. On the other hand, for oils in Krantz and Goren's experiment (1970, 1971), $\alpha = 0.3$ is small as our longwave expansion requires but $\gamma = 1.92$ and $R = 0.29$ to yield a finite $\delta = 1.58$. The extra three-dimensional dispersion term can hence be important even at large surface tension. This is distinct from the two-dimension KS limit at large surface tension.

For an inclined plane, $\theta < \frac{\pi}{2}$, there are three parameters to the system and a different equation can be obtained for three dimensional waves with judicious combinations of these parameters. Instead of the original three parameters in the last chapter, it is convenient here to choose γ, $\chi = \frac{R}{R_*}$ and the inclination angle θ. The dispersion coefficients now become

$$a_{20} = \cot(\theta)(\chi - 1)$$

$$a_{30} = 3 - \frac{25}{21}\chi(\chi - 1)\cot^2(\theta)$$

$$a_{40} = \left(\frac{2}{5}\right)^{\frac{2}{5}} \frac{\gamma \sin^{\frac{1}{3}}(\theta)}{\chi^{\frac{2}{5}} \cos^{\frac{2}{5}}(\theta)} + \frac{1}{3}\chi\left(\frac{47420}{9009}\chi^2 - \frac{17363}{2772}\chi + 1\right)\cot^3(\theta) +$$

$$+ 3\left(\frac{785}{448}\chi - \frac{3}{5}\right)\cot(\theta)$$

$$a_{02} = -\cot(\theta)$$

$$a_{12} = 3 + \frac{25}{21}\chi \cot^2(\theta)$$

$$a_{22} = 2\left(\frac{2}{5}\right)^{\frac{2}{5}} \frac{\gamma \sin^{\frac{1}{3}}(\theta)}{\chi^{\frac{2}{5}} \cos^{\frac{2}{5}}(\theta)} - \frac{1}{3}\chi\left(\frac{17363}{2772}\chi - 2\right)\cot^3(\theta) +$$

$$+ 3\left(\frac{785}{448}\chi - \frac{6}{5}\right)\cot(\theta)$$

$$a_{04} = \left(\frac{2}{5}\right)^{\frac{2}{5}} \frac{\gamma \sin^{\frac{1}{3}}(\theta)}{\chi^{\frac{2}{5}} \cos^{\frac{2}{5}}(\theta)} + \frac{1}{3}\chi \cot^3(\theta) - \frac{9}{5} ctg(\theta)$$

Let us consider large surface tension $\gamma \to \infty$ limit. At any angle and χ:

$$a_{40} = \left(\frac{2}{5}\right)^{\frac{2}{5}} \frac{\gamma \sin^{\frac{1}{3}}(\theta)}{\chi^{\frac{2}{5}} \cos^{\frac{2}{5}}(\theta)} = \frac{1}{3} WR \to \infty$$

$$a_{22} = 2a_{40} \to \infty$$

$$a_{04} = a_{40} \to \infty$$

all other a_{ij} are the same. Let us use the two-dimensional neutral wave number

$$\alpha_0^2 = -\frac{\omega_{40}}{\omega_{20}} = \frac{a_{40}}{a_{20}}$$

$$ctg(\theta)(\chi - 1) = \frac{1}{3} W R \alpha_0^2$$

$$\alpha_0^2 = \frac{3(\chi - 1)\cot(\theta)}{WR}$$

to scale the spatial coordinates x and z. We then obtain

$$\frac{\partial \hat{h}}{\partial t} + 6\alpha_0 \hat{h} \frac{\partial \hat{h}}{\partial \xi} + \cot(\theta)(\chi - 1)\alpha_0^2 \left[\frac{\partial^2 \hat{h}}{\partial \xi^2} + \frac{\partial^4 \hat{h}}{\partial \xi^4} + 2 \frac{\partial^4 \hat{h}}{\partial \xi^2 \partial \zeta^2} \right.$$
$$\left. + \frac{\partial^4 \hat{h}}{\partial \zeta^4} - \frac{1}{\chi - 1} \frac{\partial^2 \hat{h}}{\partial \zeta^2} \right] + \alpha_0^3 \left[a_{30} \frac{\partial^3 \hat{h}}{\partial \xi^3} + a_{13} \frac{\partial^3 \hat{h}}{\partial \xi \partial \zeta^2} \right] = 0 \quad (3.46)$$

$$\hat{h} = \cot(\theta)(\chi - 1)\alpha_0 H$$
$$\tau = \cot(\theta)(\chi - 1)\alpha_0^2 t$$

$$\frac{\partial H}{\partial \tau} + 6H \frac{\partial H}{\partial \xi} + \frac{\partial^2 H}{\partial \xi^2} + \left(\frac{\partial^2}{\partial \xi^2} + \frac{\partial^2}{\partial \zeta^2} \right)^2 H - \frac{1}{\chi - 1} \frac{\partial^2 H}{\partial \zeta^2} +$$
$$+ \delta \left(a_{30} \frac{\partial^3 H}{\partial \xi^3} + a_{12} \frac{\partial^3 H}{\partial \xi \partial \zeta^2} \right) = 0$$

$$\delta^2 = \left(\frac{5}{2} \right)^{\frac{2}{3}} \frac{\chi^{\frac{2}{5}}}{\chi - 1} \frac{1}{\gamma} \frac{\cos^{\frac{5}{3}}(\theta)}{\sin^{\frac{4}{3}}(\theta)} \to 0 \text{ at } \gamma \to \infty$$

This finally produces the Nepomnyashchy equation (Nepomnyashchy, 1974),

$$\frac{\partial H}{\partial \tau} + 6H \frac{\partial H}{\partial \xi} + \frac{\partial^2 H}{\partial \xi^2} + \left(\frac{\partial^2}{\partial \xi^2} + \frac{\partial^2}{\partial \zeta^2} \right)^2 H - \frac{1}{\chi - 1} \frac{\partial^2 H}{\partial \zeta^2} = 0 \quad (3.47)$$

Consider the two-dimensional case

$$\xi = \alpha x$$

where α is some small unknown value,

$$\frac{\partial \hat{h}}{\partial t} + 6\alpha \hat{h} \frac{\partial \hat{h}}{\partial \xi} + ctg(\theta)(\chi - 1)\alpha^2 \frac{\partial^2 \hat{h}}{\partial \xi^2} + 3\alpha^3 \frac{\partial^3 \hat{h}}{\partial \xi^3} + a_{40}\alpha^4 \frac{\partial^4 \hat{h}}{\partial \xi^4} = 0$$

$$a_{40} = o(1)$$
$$a_{40}\alpha^2 = (\chi - 1)\cot(\theta)$$

$$\frac{\partial \hat{h}}{\partial t} + 6\alpha \hat{h} \frac{\partial \hat{h}}{\partial \xi} + 3\alpha^3 \frac{\partial^3 \hat{h}}{\partial \xi^3} + a_{40}\alpha^4 \left(\frac{\partial^2 \hat{h}}{\partial \xi^2} + \frac{\partial^4 \hat{h}}{\partial \xi^4} \right) = 0$$

With $\hat{h} = 3\alpha^2 H$ and $\tau = 3\tau^3 t$, we again obtain the gKS equation with this approach

$$\frac{\partial H}{\partial \tau} + 6H\frac{\partial H}{\partial \xi} + \frac{\partial^3 H}{\partial \xi^3} + \delta\left(\frac{\partial^2 H}{\partial \xi^2} + \frac{\partial^4 H}{\partial \xi^4}\right) = 0$$

where the normalized Reynolds number for this scaling is

$$\delta^2 = 1/9 a_{40}(\chi - 1) = 1/9\left[\left(\frac{2}{5}\right)^{\frac{2}{5}} \gamma \frac{\sin^{4/3}(\theta)}{\cos^{5/3}(\theta)} - \frac{\cot^2(\theta)}{36036} + 7743/2240\right](\chi - 1)$$

This parameter δ vanished at the limit of $\chi \to 1$.

Whitham (1974) also proposed an entirely empirical appoach which is applicable to weakly nonlinear waves. Instead of PDE's, Whitham offered an integrodifferential equation

$$\frac{\partial H}{\partial \tau} + 4H\frac{\partial H}{\partial \xi} + \int_{-\infty}^{+\infty} K(\xi - \eta)\frac{\partial H}{\partial \eta}(\eta, \tau)d\xi = 0 \qquad (3.48)$$

This linearized equation has an elementary solutions $H = e^{i\alpha x - i\omega t}$ with complex velocity

$$c(\alpha) = \int_{-\infty}^{+\infty} K(\eta)e^{-i\alpha\eta}d\eta$$

If we already have $c(\alpha)$ from, for example, a numerical solution of the Orr-Sommerfeld equation, we can calculate the kernel $K(\eta)$ by the inverse Fourier transform of the comlex phase speed,

$$K(\xi) = \frac{1}{2\pi}\int_{-\infty}^{+\infty} c(\alpha)e^{i\alpha x}d\alpha \qquad (3.49)$$

The Whitham approach was utilized for falling films by Kliakhander (2000), who took a real dispersion relation from the numerical solution of the Orr-Sommerfeld equation. He found that in a wide range of parameters, *real* dispersion relation exhibits remarkably different behavior than that predicted by conventional theories. Reconstructing the pertinent dispersion relation and combining this with associated nonlinear terms leads to a new type of evolution equation for the dynamics of falling films.

3.2 lubrication theory to derive Benney's long-wave equation

The weakly nonlinear equations of the previous section can only resolve wave dynamics with small amlpitude and long wavelength at lagre time. The small amplitude approximation restricts the validity of the weakly nonlinear equations to near-critical conditions. Lubrication theories allow us to relax the small amplitude approximation to obtain equations valid further from criticality. It provides asymptotic solutions of the full problem for disturbances whose length scales are large compared to the film thickness: for ε small. We should still be near critical but can now be further from R_*. This corresponds to finite Reynolds numbers, and the parameters R and W will be supposed to be of order one in the subsequent analysis. We shall follow the derivation by Chang (1986), different from the classical method by Benney (1966), Gjevik (1970), Lin (1974), Nakaya (1975), Atherton and Homsy (1976) and Nakaya (1989).

Expanding the liquid-phase velocity and pressure in an asymptotic series in ε,

$$u = u_0 + \varepsilon u_1 + \varepsilon^2 u_2 + \ldots \tag{3.50}$$

$$v = v_0 + \varepsilon v_1 + \varepsilon^2 v_2 + \ldots \tag{3.51}$$

$$p = p_0 + \varepsilon p_1 + \varepsilon^2 p_2 + \ldots \tag{3.52}$$

and collecting terms in the dimensionless equations (2.1) to (2.11), one obtains for the following $O(\varepsilon^0)$ zeroth-order equation which contains no inertial terms,

$$\frac{\partial^2 u_0}{\partial y^2} + 3 - \frac{\partial p_0}{\partial x} = 0 \tag{3.53}$$

$$\frac{\partial p_0}{\partial y} = 0 \tag{3.54}$$

$$\frac{\partial u_0}{\partial x} + \frac{\partial v_0}{\partial y} = 0 \tag{3.55}$$

and the boundary conditions

$$y = 0, \quad u_0 = v_0 = 0 \tag{3.56}$$

$$y = h, \quad \frac{\partial u_0}{\partial y} = 0 \tag{3.57}$$

$$p_0 = 0 \tag{3.58}$$

From equations (3.54) and (3.58), p_0 vanishes, reflecting the fact that the pressure gradient is $O(\varepsilon)$. Hence, equations (3.53)-(3.58) are rescaled versions of the flat-film equations. In the present case, however, u_0 and h_0 are functions of x, y and t. Also, v_0 is nonvanishing. Specifically,

$$u_0(x, y, t) = -3\frac{y^2}{2} + 3hy \tag{3.59}$$

$$v_0(x, y, t) = -\int_0^y \frac{\partial u_0}{\partial x} dy \tag{3.60}$$

where the t and x-dependence arises from $h(x,t)$. The resemblance is because, at zeroth order, the long waves appear as flat surfaces locally.

The steady-state flat-film solution can be derived by substituting unity for h in equation (3.59). Hence,

$$u_0(y) = -3\frac{y^2}{2} + 3y \tag{3.61}$$

The $O(\varepsilon)$ first-order equations are

$$\frac{\partial^2 u_1}{\partial y^2} - 3\frac{\partial^2 p_1}{\partial y^2} = R\left(\frac{\partial u_0}{\partial t} + u_0\frac{\partial u_0}{\partial x} + v_0\frac{\partial u_0}{\partial y}\right) \tag{3.62}$$

$$\frac{\partial p_1}{\partial y} = 0 \tag{3.63}$$

$$\frac{\partial u_1}{\partial x} + \frac{\partial v_1}{\partial y} = 0 \tag{3.64}$$

$$y = 0, \quad u_1 = v_1 = 0 \tag{3.65}$$

$$y = h, \quad \frac{\partial u_1}{\partial y} = 0 \tag{3.66}$$

$$p + \varepsilon W h_{xx} = 0. \tag{3.67}$$

Applying the zeroth-order solutions, we can solve for u_1 and v_1,

$$u_1(x,y,t) = 3\varepsilon W h_{xxx}\left(hy - \frac{y^2}{2}\right) + 3Rh_t\left(\frac{y^3}{6} - h^2\frac{y}{2}\right) + 3Rh_x\left(\frac{y^4}{8} - h^3\frac{y}{2}\right) \tag{3.68}$$

and

$$v_1(x,y,t) = -\int_0^y \frac{\partial u_1}{\partial x} dy \tag{3.69}$$

We next invoke the kinematic condition at the interface

$$h_t = uh_x - v = 0 \quad \text{at} \quad y = 1 \tag{3.70}$$

which yields

$$h_t + u_0 h_x - v_0 + \varepsilon[u_1 h_x - v_1] = O(\varepsilon^2) \quad \text{at} \quad y = 1. \tag{3.71}$$

Since

$$u_i h_x - v_i = \frac{\partial}{\partial x}\int_0^{h(x,t)} u_i dy = \frac{\partial}{\partial x}Q_i \tag{3.72}$$

for $i = 0$ and 1, where the integral corresponds to the flow rate Q_i, we substitute equations (3.59)-(3.60) and (3.68) into equations (3.71) and (3.72) to obtain the following Benney equation for h which is correct to $O(\varepsilon)$, including $\alpha^3 W$ for large surface tension

$$h_t + \left[h^3 + \left(\frac{6}{5}Rh^6 - \cot\theta h^3\right)h_x + Wh^3 h_{xxx}\right]_x = 0 \qquad (3.73)$$

The quantity within the square bracket is the flow rate. A good exercise is to carry out a weakly nonlinear expansion, with the proper order assignment for the length scales, to show that the Kuramoto-Sivashinsky (KS) equation is approached by the above equation in the limit of small amplitude.

Equation (3.73) is derived for long waves without restriction on their amplitude. Pumir, Manneville and Pomeau (1983) found that equation (3.73) has a singularity. Stationary travelling solitary wave solutions can be realized only for a limited range of Reynolds numbers. Nonstationary waves during their numerical solution have a tendency to self-focus into solitary waves, followed by a blow-up in its amplitude in finite time.

One may suspect that some defect in the derivation procedure of the long-wave equations is responsible for their failure. We also recall several well-known equations with travelling wave solution that do not blow up, such as the Korteweg-de Vries (KdV) equation, Burgers equation, and so on have been derived in a somewhat different way For example, to derive the KdV equation for inviscid shallow water waves, it is necessary to perform the longwave expansion in combination with the amplitude expansion. On the other hand, the longwave equations of film flows were derived without the amplitude expansion. The expansion procedure for equation (3.73) was based on the smallness of $\mu(\sim \partial_x)$, but it included no assumption like $\eta \sim \mu^p$ (where $h = 1 + \eta$); rather, h was regarded as $O(1)$ and left unexpanded, so that equation (3.73) contains excessively strong nonlinearity. This longwave expansion is valid for waves of any amplitude only if the waves are long enough. The premise of long waves, however, is broken when the wave amplitude is large, because shorter wave components can be produced by nonlinear interactions of long wave.

Ooshida (1999) proposed a method to break through the limitation of the traditional longwave expansion and regularize the equation. The basic idea of the regularization method is to regard the longwave expansion as a power series expansion in terms of the differential operator, which we then replace by the Padé approximation. The regularized longwave expansion method reduces the complicated dynamics of film flows into a single equation of h, that is, a surface equation that we will refer to as the *regularized equation* of film flows. Unlike the traditional longwave equations, the regularized equation is valid not only for $R - R_c \leq 1$ but also for $R - R_c \sim \mu^{-1} \gg 1$. Meanwhile it inherits from the traditional longwave equations two favorable properties, namely the advantage of being a surface equation, and that it includes only a first-order time derivative.

The traditional approach is to truncate the power series (3.50), retaining terms up to the flow rates Q_1 or Q_2, or another particular order of μ, and to close the equation by combining it with the mass conservation equation (3.71).

In particular, equation (3.73) is obtained by retaining terms up to Q_1. Equation (3.73) was numerically tested by Salamon et al. (1994) and, unfortunately, was found to make false predictions. They also found that including more terms (up to Q_2) does not save the truncated equation from the failure. The latter result implies that there is a problem in the convegence property of the power series (3.50).

There is an excellent remedy for a poorly convegent power series, known as the Padé approximation. Suppose that a function $\psi(k)$ is given in terms of a power series, as

$$\psi = c_0 + c_1 k + c_2 k^2 + \cdots . \tag{3.74}$$

This series may be only poorly convergent or even not convergent at all, and in such cases the truncated power series becomes a quite bad approximant to ψ. While the power series tries to express ψ in terms of a single polynomial, the Padé approximation expresses ψ as a ratio of two polynomials involves converting the power series (3.74) into another series,

$$f = g\psi = a_0 + a_1 k + a_2 k^2 + \cdots , \tag{3.75}$$

so that $\psi = f/g$, where $g = 1 + b_1 k + \ldots + b_n k^n$. What we expect is that a suitable choice of g can improve the convergence of the power series, as the singularity of $\psi(k)$ may be canceled (at least approximately) by the zero of $g(k)$. Actually, in most cases the Padé approximant f/g provides with a better approximation than the corresponding power series truncated at a certain order, expecially when $|k|$ is comparable to (or even greater than) the convergence radius of the power series (3.74).

Applying these ideas to our problems, Ooshida (1999) derived the improved or regularized equation

$$\partial_t h - \frac{4}{21} R \partial_x \partial_t (h^5) - \partial_x (h^2 \partial_x \partial_t h) +$$
$$+ \frac{2}{3} \partial_x \left[h^3 - \partial_x \left(\frac{\cot \alpha}{4} h^4 + \frac{72}{245} R h^7 \right) + W h^3 \partial_x^3 h \right] = 0. \tag{3.76}$$

The equation has a cross-derivative with respect to t and x.

3.3 Depth-averaged integral equations

An entirely empirical model, motivated by the Karman-Polhausen integral (see Schlichting, 1968) theory for boundary layers, can describe large amplitude waves far from criticality. The key is to retain the dynamic inertial term (the instantaneous acceleration) in the momentum equation. Such a term is omitted in the lubrication equation of the previous section and in the weakly nonlinear analysis of the section before that. Due to the ad hoc nature of its derivation, its must be validated against experimental data. Such validation is surprisingly robust even at reasonably high R (a few hundred for the vertical film). We begin the derivation of such depth-averaged integral equation with the boundary layer BL equations derived in Chapter 2 :

$$\frac{\partial u}{\partial t} + u\frac{\partial u}{\partial x} + v\frac{\partial u}{\partial y} + w\frac{\partial u}{\partial z} = \frac{1}{5\delta}(h_{xxx} + hxzz + \frac{1}{3}\frac{\partial^2 u}{\partial y^2} + 1) - 3\chi\frac{\partial h}{\partial x} \tag{3.77}$$

$$\frac{\partial w}{\partial t} + u\frac{\partial w}{\partial x} + v\frac{\partial w}{\partial y} + w\frac{\partial w}{\partial z} = \frac{1}{5\delta}(h_{xxz} + hzzz + \frac{1}{3}\frac{\partial^2 w}{\partial y^2}) - 3\chi\frac{\partial h}{\partial z} \tag{3.78}$$

$$\frac{\partial u}{\partial x} + \frac{\partial v}{\partial y} + \frac{\partial w}{\partial z} = 0 \tag{3.79}$$

$$\frac{\partial h}{\partial t} + \frac{\partial}{\partial x}\int_0^h u\,dy + \frac{\partial}{\partial z}\int_0^h w\,dy = 0 \tag{3.80}$$

$$y = h(x,z,t): \qquad \frac{\partial u}{\partial y} = \frac{\partial w}{\partial y} = 0 \tag{3.81}$$

$$y = 0: \qquad u = v = w = 0 \tag{3.82}$$

where $\delta = \frac{R^{11/9}(\sin\theta)^{1/9}}{5\gamma^{1/3}37^{7/9}}$ is modified Reynolds number and $\chi = \frac{\cot\theta}{R}$

Further simplification can be achieved by assuming a velocity profile in the normal y direction and average over this direction. This removes the y dependence and one spatial dimension in our problem — a drastic simplification. Such averaging yields integral model equations that are often ad hoc models. However, many of them can actually be considered as a particular truncation of the exact Petrov-Galerkin method . We introduce this method by first using the curvilinear coordinate

$$\tau = t,\ \xi = x,\ \eta = \frac{y}{h(x,z,t)},\ \zeta = z$$

which maps the film into a strip.

Let us expand u and v with a set of basis functions that obey the boundary conditions (3.81))-(3.82):

$$\psi_k(0) = 0,\ \frac{\partial\psi_k(1)}{\partial\eta} = 0$$

$$\psi_1 = \eta - \frac{1}{2}\eta^2,\ \psi_2 = \eta^2 - \frac{2}{3}\eta^3,\ldots,\psi_k = \eta^k - \frac{k}{(k+1)}\eta^{k+1}$$

We note that ψ_1 is simply the parabolic velocity profile of a flat film adapted to this wavy film. It is hence chosen as the first basis function such that its expansion coefficient is simply the local flow rate. The remaining independent bases in the basis set ψ_k all conveniently consist of two monomials of succesive order. Each basis satisfies the above two boundary conditions at the wall and the interface, such that the coefficents of the two monomials are specified. Other bases are possible, for example,

$$\psi_1 = \eta - \frac{1}{2}\eta^2, \quad \psi_2 = \eta^2 - \eta^3 + \frac{1}{4}\eta^4, \quad \psi_3 = \eta^3 - \eta^4 + \frac{1}{5}\eta^5, \ldots$$

For moderate Reynolds numbers, $R < 100$, three to five Galerkin's functions are enough to have a good accuracy, which practically does not depend on the type of Galerkin projection, see Chapter 4.

We limit ourselves to a finite number of base functions N and derive the vertical velocity component v from the continuity equation,

$$u = \frac{1}{h}\sum_{k=1}^{N} A_k(\xi, \zeta, \tau)\psi_k(\eta)$$

$$w = \frac{1}{h}\sum_{k=1}^{N} B_k(\xi, \zeta, \tau)\psi_k(\eta)$$

$$v = -\sum_{k=1}^{N}\left(\frac{\partial A_k}{\partial \xi} + \frac{\partial B_k}{\partial \zeta}\right)\tilde{\psi}_k + h_\xi \eta u + h_\zeta \eta w$$

$$\tilde{\psi}_k = \frac{1}{(k+1)}\left(\eta^{k+1} - \frac{k}{(k+2)}\eta^{k+2}\right)$$

We project the equation of motion (3.77) to (3.78) with the weighting functions $\varphi_1 = 1$, $\varphi_2 = \eta, \ldots, \varphi_N = \eta^{N-1}$. The unit first weighting function is again to preserve the first expansion coefficent as the flow rate. The others are just convenient weights with no specific meanings. Since they have to be independent, the monomials are convenient choices. Thus we have $2N+1$ unknowns $h, A_1, B_1, \ldots, A_N, B_N$ and $2N+1$ equations. For the planar case, the resulting system is:

$$\frac{\partial}{\partial \tau}\left\{h\int_0^1 \varphi_k u\, d\eta\right\} + \frac{\partial}{\partial \xi}\left\{h\int_0^1 \varphi_k u\, d\eta\right\} - \frac{\partial h}{\partial \tau}\int_0^1 \varphi_k' uv\, d\eta + \frac{\partial h}{\partial \xi}\int_0^1 \varphi_k' u^2\, d\eta =$$

$$\frac{1}{5\delta}\left\{h\left(\frac{\partial^3 h}{\partial \xi^3} - 15\delta\chi\frac{\partial h}{\partial \xi}+\right)\int_0^1 \phi_k\, d\eta + \frac{1}{3h}\int_0^1 \phi_k''\, d\eta - \frac{1}{3h}\left[\phi_k'(1)u(0) + \phi_k(0)\frac{\partial u(0)}{\partial \eta}\right]\right\}$$

(3.83)

$$\frac{\partial h}{\partial \tau} + \frac{\partial}{\partial \xi}\left\{h\int_0^1 u\, d\eta\right\} = 0$$

(3.84)

If we restrict ourselves to $N = 1$, with a parabolic velocity profile, for the three-dimensional case (Demekhin and Shkadov, 1984):

$$\frac{\partial q}{\partial t} + \underbrace{\frac{6}{5}\frac{\partial}{\partial x}\left(\frac{q^2}{h}\right) + \frac{6}{5}\frac{\partial}{\partial z}\left(\frac{qp}{h}\right)}_{I} = \frac{1}{5\delta}\left\{h(\underbrace{h_{xxx} + h_{xzz}}_{II} + \underbrace{1}_{III}) - \underbrace{\frac{q}{h^2}}_{IV}\right\} - \underbrace{3\chi h h_x}_{V}$$
(3.85)

$$\frac{\partial p}{\partial t} + \underbrace{\frac{6}{5}\frac{\partial}{\partial x}\left(\frac{qp}{h}\right) + \frac{6}{5}\frac{\partial}{\partial z}\left(\frac{p^2}{h}\right)}_{I} = \frac{1}{5\delta}\left\{h(\underbrace{h_{xxz} + h_{zzz}}_{II}) - \underbrace{\frac{p}{h^2}}_{IV}\right\} - \underbrace{3\chi h h_z}_{V}$$
(3.86)

$$\frac{\partial u}{\partial t} + \frac{\partial q}{\partial x} + \frac{\partial p}{\partial z} = 0 \qquad (3.87)$$

Here it is convenient to choose as unknown functions:

$$q = \frac{1}{3}A_1 = h\int_0^1 u\,d\eta \quad \text{flow-rate in } x\text{-direction}$$

$$p = \frac{1}{3}B_1 = h\int_0^1 w\,d\eta \quad \text{flow-rate in } z\text{-direction}$$

If we introduce a flow-rate vector $\omega = (q,p)$, the system (3.85) to (3.87) may be represented in an invariant vector from with respect to the coordinates:

$$\frac{\partial \omega}{\partial t} + \frac{6}{5}\left[\omega\nabla\left(\frac{\omega}{h}\right) + \left(\frac{\omega}{h}\nabla\right)\omega\right] = \frac{1}{5\delta}\left[\mathbf{i}h + h\nabla(\nabla^2 h) - \frac{\omega}{h^2}\right] - 3\chi h h_x$$

$$\frac{\partial h}{\partial t} + \nabla\omega = 0$$

Here, $\mathbf{i} = \frac{\mathbf{g}}{|\mathbf{g}|}$ is a unit gravity vector and $\nabla = \mathbf{i}\frac{\partial}{\partial x} + \mathbf{j}\frac{\partial}{\partial z}$ is the gradient operator.

This system is reminiscent of equations for two-dimensional unsteady problem in gas dynamics, where the role of density is played by the layer thickness h. An essential difference from the latter problem and the problem of mathematical hydraulics Dressler (1949), Brock (1970) and Hwang and Chang (1987), is the presence of the term $h\nabla(\nabla^2 h)$, responsible for the action of surface tension.

If we set $p = 0$ and $\frac{\partial}{\partial z} = 0$, then (1.2.33) becomes the Shkadov model for two-dimensional waves (Skadov, 1967 and 1968):

$$\frac{\partial q}{\partial t} + \underbrace{\frac{6}{5}\frac{\partial}{\partial x}\left(\frac{q^2}{h}\right)}_{I} = \frac{1}{5\delta}\left(\underbrace{hh_{xxx}}_{II} + \underbrace{h}_{III} - \underbrace{\frac{q}{h^2}}_{IV}\right) - \underbrace{3\chi hh_x}_{V} \qquad (3.88)$$

$$\frac{\partial h}{\partial t} + \frac{\partial q}{\partial x} = 0 \qquad (3.89)$$

The equations (3.88) and (3.89) are averaged versions the x momentum equation and mass balance equation where the averaing is over the transverse coordinate. Term I is responsible for the conection of acceleration, terms II and III characterize capillary forces and gravitational forces, term IV viscous dissipation in the layer and V the influence of tangential gravitational pull due to the inclination. These equations are generalization of the Karman-Polhausen integral boundary-layer approach for momentum boundary layers to non-stationary flows with free surface. It was first derived by Shkadov (1967) and will be called the Shkadov model, see also Levich (1962) and Lee (1969).

The assumption of an equilibrium parabolic velocity profile ($N = 1$) in deriving the Shkadov model must not significantly influence such quantities as film thickness distribution $h(x, z)$, wave velocity and so on. Physically, the wavy processes in the fluid film are surface phenomena, and thus they are expected to be weakly dependent on the velocity profile. This is one of the differences of instability in the Poiseuille, Couette, Blasius boundary layer plane flows, or in free jet flow, mixing layers, flow behind a solid body from the thin films flows. The correctness of the parabolic velocity profile will be discussed subsequently.

A weakly nonlinear version of the Shkadov model, combining depth averaging and small-amplitude expansion, was derived by Alekseenko, Nakoryakov and Pokusaev (1979) for a vertical film and by Nakoryakov and Alekseenko (1980) for an inclined film. For the case of the combined flow of horizontal film and gas flow it was derived by Jurman and McCready (1989).

Let us consider the case of slightly nonlinear waves assuming

$$q = 1 + Q, \ h = 1 + H, \ Q, H \ll 1,$$

where Q, H are dimensionless perturbations of flow rate and thickness, respectively. From the order estimation of equation (3.88), we get $Q \sim H$. From equation (3.88) taking into account the quadratic nonlinearity alone we get

$$\frac{\partial Q}{\partial t} + \frac{12}{5}\frac{\partial Q}{\partial x} - \frac{6}{5}\frac{\partial H}{\partial x} + 3\frac{Q}{Re} - 9\frac{H}{Re} + 3\frac{\cot\theta}{Re}\frac{\partial H}{\partial x} - \frac{3W}{Re}\frac{\partial^3 H}{\partial x^3} =$$
$$9\frac{H^2}{Re} - 2\frac{\partial Q}{\partial t}H - 2\frac{6}{5}\left(Q\frac{\partial Q}{\partial x} + H\frac{\partial Q}{\partial x} - Q\frac{\partial H}{\partial x}\right) - 9\frac{\cot\theta}{Re}H\frac{\partial H}{\partial x}, \qquad (3.90)$$

$$\frac{\partial H}{\partial t} + \frac{\partial Q}{\partial x} = 0. \qquad (3.91)$$

In the first equation all nonlinear terms are on the right-hand side. To obtain one equation for H from these two equations we differentiate equation (3.90) with respect to x and eliminate $\frac{\partial Q}{\partial x}$ using equation (3.91)

$$\left[\frac{\partial^2 H}{\partial t^2} + \frac{12}{5}\frac{\partial^2 H}{\partial t \partial x} + \left(\frac{6}{5} - \frac{3\cot\theta}{Re}\right)\frac{\partial^2 H}{\partial x^2}\right] + \frac{3}{Re}\frac{\partial H}{\partial t} + \frac{9}{Re}\frac{\partial H}{\partial x} + \frac{3W}{Re}\frac{\partial^4 H}{\partial x^4} =$$
$$\frac{18}{Re}H\frac{\partial H}{\partial x} + 2\left(H\frac{\partial Q}{\partial t}\right)_x + \frac{12}{5}\left(Q\frac{\partial Q}{\partial x} + H\frac{\partial Q}{\partial x} - Q\frac{\partial H}{\partial x}\right)_x + 9\frac{\cot\theta}{Re}\left(H\frac{\partial H}{\partial x}\right)_x. \quad (3.92)$$

Now the value of Q remains only in the nonlinear terms which have a higher infinitesimal order compared to the linear terms therefore one can apply a convenient quasi-stationary approximation. To this end, we write (3.91) in a coordinate moving with the phase velocity c_0 i.e. we transform the coordinates $(t, x) \to (t, \xi = x - c_0 t)$. For falling films, (2.88) shows $c_0 = 3$ for near-critical conditions. We then assume the wave is quasi-stationary in this moving frame — an approximation that is only valid for near-critical conditions. equation (3.91) will then acquire the form

$$\frac{\partial H}{\partial t} - c_0\frac{\partial H}{\partial \xi} + \frac{\partial Q}{\partial \xi} = 0. \quad (3.93)$$

and omitting the time derivative due to the nearly quasi-stationary approximation, we obtain

$$Q = c_0 H. \quad (3.94)$$

Substitute this relation into equation (3.91) and we obtain

$$\frac{\partial}{\partial t} = -c_0\frac{\partial}{\partial x} \quad (3.95)$$

These expressions are exact for stationary travelling waves with speed c_0.

Using (3.94), we can replace Q by $c_0 H$ in the nonlinear terms of equation (3.92). In this case the velocity c appears only in combination with a derivative with respect to x therefore the operator $c_0\frac{\partial}{\partial x}$ can be substituted by $-\frac{\partial}{\partial t}$ by using equation (3.95). Thus both Q and c_0 vanish and the resulting equaton for slightly nonlinear quasistationary waves takes the form

$$\left(\frac{\partial}{\partial t} + c_0\frac{\partial}{\partial x}\right)H + \frac{Re}{3}\left(\frac{\partial}{\partial t} + c_1\frac{\partial}{\partial x}\right)\left(\frac{\partial}{\partial t} + c_2\frac{\partial}{\partial x}\right)H + W\frac{\partial^4 H}{\partial x^4} +$$
$$6H\frac{\partial H}{\partial x} - \frac{2}{15}Re\frac{\partial}{\partial t}\left(H\frac{\partial H}{\partial t}\right) - 3\cot\theta\frac{\partial}{\partial x}\left(H\frac{\partial H}{\partial x}\right) = 0 \quad (3.96)$$

$$c_0 = 3, \quad c_{1,2} = 1.2 \pm \sqrt{0.24 + 3\cot\theta/Re}. \quad (3.97)$$

The two-wave nature of the equation is related to the fact that equation (3.96) contains two types of waves, i.e. the lower order waves described by the first order derivatives and the higher order waves descibed by the second order derivatives.

In the previous chapter we showed from the linear stability analysis that decreasing the inclination angle from $\theta = \pi/2$ will eventual break down the approximations in the BL derivation and, hence, the IBL-model. Prokopiou, Cheng and Chang (1991) offered a second-order integral equation which can adequately describe waves for small inclination.

Integrating the hydrostatic head of (2.1) $y = h$ to y and substituting the value of the liquid pressure at $y = h$ from the normal stress condition (2.5), an expression for the normal variation of the pressure is obtained. Upon substituting this into the x-momentum equation and again integrating over the film, one obtains the averaged x-momentum equations

$$\frac{\partial q}{\partial t} + \frac{\partial}{\partial x}\left(\frac{\Gamma q^2}{h}\right) = \frac{2}{Re}\int_0^h \frac{\partial}{\partial x}\left(\frac{\partial u}{\partial x}(y=h)\right)dy + We(hh_{xxx}-$$

$$-\frac{3}{2}hh_x^2 h_{xxx} - 3hh x h_{xx}^2) + Fr(h\cos\theta - hh_x\sin\theta) + \frac{2}{Re}\int_0^h \frac{\partial^2 u}{\partial x^2}dy-$$

$$-\tau_w + \frac{4}{Re}h_x\frac{\partial u}{\partial x}(y=h), \qquad (3.98)$$

where

$$\Gamma = \frac{h}{q^2}\int_0^h u^2 dy \qquad (3.99)$$

is the shape factor (Hanratty, 1983) and

$$\tau_w = \frac{1}{Re}\left(\frac{\partial u}{\partial y}\right)(y=0) \qquad (3.100)$$

is the wall shear. The continuity equation and the tangential stress condition have been used in deriving (3.98).

A specific profile must now be imposed in this theory. However, for the flow of interest, Alekseenko et al. (1994) have experimentally established that a parabolic profile is more appropriate for moderate Reynolds number. Consequently, we impose the following self-similar profile:

$$u = \frac{3q}{h}\left[\left(\frac{y}{h}\right) - \frac{1}{2}\left(\frac{y}{h}\right)^2\right]. \qquad (3.101)$$

The normal velocity v can also be derived from (3.101) but the only pertinent information is that $\partial v/\partial x$ vanishes at the wall such that the wall shear is simply

$$\tau_w = 3q/Re\, h^2. \qquad (3.102)$$

The parabolic profile also yields a shape factor of 1.2 Introducing these expressions into (3.98) yields the final averaged x-momentum equation and mass balance equation

$$\frac{\partial q}{\partial t} + \frac{\partial}{\partial x}\left(\frac{1.2q^2}{h}\right) = \frac{5}{Re}q_{xx} + We(hh_{xxx} - \frac{3}{2}hh_x^2 h_{xxx} - 3hh x h_{xx}^2)+$$

$$\frac{3}{Re\cos\theta}(h\cos\theta - hh_x\sin\theta) + \frac{3}{Re}\left(\frac{2qh_x^2}{h^2} - \frac{2q_x h_x}{h} - \frac{2qh_{xx}}{h} - \frac{q}{h^2}\right). \qquad (3.103)$$

$$\frac{\partial h}{\partial t} + \frac{\partial q}{\partial x} = 0 \qquad (3.104)$$

Two important terms due to the inclusion of the $O(\varepsilon^2)$ normal shear term are the dissipation terms $(5/Re)q_{xx}$ and $(6/Re)qh_{xx}/h$ which do not appear in leading-order boundary-layer theory. If We is $O(\varepsilon)$ or smaller, which occurs at high Re, the surface tension terms associated with We in (3.103) are negligible in the present resolution. This implies that without the dissipation terms, which contains second derivatives of q and h with respect to x. (3.103) is a nonlinear hyperbolic equation equivalent to the shallow-water equations of Dressler (1949) and Stoker (1957). As Brock (1970) has shown, such a hyperbolic equation does not allow periodic and solitary waveforms.

3.4 Combination of Galerkin-Petrov method with weighted residuals.

In the longwave expansion, the flow variables are supposed to be strictly enslaved to the local thickness h which plays the role of an effective degree of freedom governed by a Benney-like evolution equation. The model developed below is an attempt to implement this general idea in the most "economical" way for the Petrov-Galerkin method of deriving integral equations — we shall extend the one-mode approximation of Shkadov model. This approach was first suggested by Ruyer-Quil and Manneville (1998,2000) and Nquyen and Balakotaiach (2000)

We choose a specific extension of the one-mode approximation in the Petrov-Galerkin expansion

$$u(x,y,t) = b_0(x,t)\left(-\frac{1}{2}\eta^2 + \eta\right) + b_1 \frac{1}{6}\left(\frac{1}{4}\eta^4 - \eta^3 + \eta^2\right) \qquad (3.105)$$

such that the flow rate becomes

$$q = \int_0^h u(y)dy = \frac{1}{3}h\left(b_0 + \frac{1}{15}b_1\right). \qquad (3.106)$$

Accordingly (3.105) can be rewritten as

$$u = \left(\frac{3q}{h} + h\tau\right)\left(-\frac{1}{2}\eta^2 + \eta\right) - 15h\tau\frac{1}{6}\left(\frac{1}{4}\eta^4 - \eta^3 + \eta^2\right). \qquad (3.107)$$

But, h, q and τ still make a set of three variables for which we have only two equations (3.88) and (3.89).

A condition useful at first order is obtained from (3.77) differentiated with respect to y;

$$\partial_{ty}u + u\partial_{xy}u + v\partial_{yy}u - \partial_{y^3}u = 0. \qquad (3.108)$$

If further evaluated at $y = 0$, this equation gives:

$$\partial_t(\partial_y u|_0) - \partial_{y^3}u|_0 = 0. \qquad (3.109)$$

Observing that τ is a first-order correction so that the term $\partial_t \tau$ is of higher order and can be dropped from (3.109), we get from (3.107) and (54) the required supplementary equation

$$\partial_t \left(\frac{3q}{h^2} \right) - 15 \frac{\tau}{h^2} = 0, \qquad (3.110)$$

which relates the correction τ to the time derivative of the other fields. Inserting the resulting evaluation of τ_ω in (3.100), using (3.89) to eliminate $\partial_t h$ to obtain the (here sufficient) zeroth-order estimate $\tau = \frac{6}{5}(q^2/h)$, we obtain

$$\partial_t q = \frac{5}{6}h - \frac{5}{2}\frac{q}{h^2} - \frac{7}{3}\frac{q}{h}\partial_x q + \left(\frac{q^2}{h^2} - \frac{5}{6}Bh \right) \partial_x h + \frac{5}{6}\Gamma h \partial_{x^3} h. \qquad (3.111)$$

When added to (3.89), equation (3.111) completes the model at first order as a system of two partial differential equations for the two unknowns h and q. Having the same structure as Shkadov's model but with slightly different coefficients, it will be called the "modified Shkadov model". The corrections arise from a better account of the fluctuations of τ_ω introduced *via* the third derivative term in (3.109) by the η^3-term in $\frac{1}{6}\left(\frac{1}{4}\eta^4 - \eta^3 + \eta^2\right)$, which can be traced back to the $\partial_t h$ contribution of $u^{(1)}$ in (3.68), which is also where the inertial cause of falling-film instability appears.

Now, let us consider a gradient expansion of (3.89) and (3.111) similar to the previous longwave expansion by assuming $q = q^{(0)} + q^{(1)} + q^{(2)} + \cdots$. We then obtain

$$q^{(1)} = \frac{2}{15}h^6 \partial_x h - \frac{1}{3}Bh^3 \partial_x h + \frac{1}{3}\Gamma h^3 \partial_{x^3} h,$$

which, when replaced in $\partial_t h + \partial_x[q^{(0)} + q^{(1)}] = 0$, gives us the Benney equation (3.73). So, by construction, the near-critical behavior is correctly predicted by our modified Shkadov model, which no longer overestimates the value of the stability threshold.

As a matter of fact, performing the same derivation to second order, one obtains

$$\partial_t h = -\partial_x q \qquad (3.112)$$

$$\partial_t q = h - \frac{3q}{h^2} - \tau + \partial_x \left[\frac{2}{35}h\tau q - \frac{6q^2}{5h} \right] - Bh\partial_x h + \Gamma h \partial_{x^3} h, \qquad (3.113)$$

$$\partial_t \tau = \frac{7}{h} - \frac{21q}{h^4} - \frac{42\tau}{h^2} - \frac{18q^2 \partial_x h}{5h^4} + \frac{6q \partial_x q}{5h^3}$$
$$+ \frac{2q\tau \partial_x h}{5h^2} + \frac{\tau \partial_x q}{15h} - \frac{3q \partial_x \tau}{5h} - \frac{7B\partial_x h}{h} + \frac{7\Gamma \partial_{x^3} h}{h} \qquad (3.114)$$

A tedious computation leads to

$$\partial_t q = \frac{27}{28}h - \frac{81}{28}\frac{q}{h^2} - 33\frac{s_1}{h^2} - \frac{3069}{28}\frac{s_2}{h^2} - \frac{12}{5}\frac{qs_1 \partial_x h}{h^2} - \frac{126}{65}\frac{qs_2 \partial_x h}{h^2} +$$

$$\frac{12}{5}\frac{s_1\partial_x q}{h} + \frac{171}{65}\frac{s_2\partial_x q}{h} + \frac{12}{5}\frac{q\partial_x s_1}{h} + \frac{1017}{455}\frac{q\partial_x s_2}{h} + \frac{6}{5}\frac{q^2\partial_x h}{h^2} -$$
$$\frac{12}{5}\frac{q\partial_x q}{h} + \frac{5025}{896}\frac{q(\partial_x h)^2}{h^2} - \frac{5055}{896}\frac{\partial_x q \partial_x h}{h} - \frac{10851}{1792}\frac{q\partial_{xx} h}{h} +$$
$$\frac{2027}{448}\partial_{xx} q - \frac{27}{28}Bh\partial_x h + \frac{27}{28}\Gamma h\partial_{xxx} h, \qquad (3.115)$$

$$\partial_t s_1 = \frac{1}{10}h - \frac{3}{10}\frac{q}{h^2} - \frac{3}{35}\frac{q^2\partial_x h}{h^2} - \frac{126}{5}\frac{s_1}{h^2} - \frac{126}{5}\frac{s_2}{h^2} + \frac{1}{35}\frac{q\partial_x q}{h} +$$
$$\frac{108}{55}\frac{qs_1\partial_x h}{h^2} - \frac{5022}{5005}\frac{qs_2\partial_x h}{h^2} - \frac{103}{55}\frac{s_1\partial_x q}{h} + \frac{9657}{5005}\frac{s_2\partial_x q}{h} - \frac{39}{55}\frac{q\partial_x s_1}{h} +$$
$$\frac{10557}{10010}\frac{q\partial_x s_2}{h} + \frac{93}{40}\frac{q(\partial_x h)^2}{h^2} - \frac{69}{40}\frac{\partial_x h \partial_x q}{h} + \frac{21}{80}\frac{q\partial_{xx} h}{h} - \frac{9}{40}\partial_{xx} q -$$
$$\frac{1}{10}Bh\partial_x h + \frac{1}{10}\Gamma h\partial_{xxx} h, \qquad (3.116)$$

$$\partial_t s_2 = \frac{13}{420}h - \frac{13}{140}\frac{q}{h^2} - \frac{39}{5}\frac{s_1}{h^2} - \frac{11817}{140}\frac{s_2}{h^2} - \frac{4}{11}\frac{qs_1\partial_x h}{h^2} +$$
$$\frac{18}{11}\frac{qs_2\partial_x h}{h^2} - \frac{2}{33}\frac{s_1\partial_x q}{h} - \frac{19}{11}\frac{s_2\partial_x q}{h} + \frac{6}{55}\frac{q\partial_x s_1}{h} - \frac{288}{385}\frac{q\partial_x s_2}{h} -$$
$$\frac{3211}{4480}\frac{q(\partial_x h)^2}{h^2} + \frac{2613}{4480}\frac{\partial_x q \partial_x h}{h} - \frac{2847}{8960}\frac{q\partial_{xx} h}{h} + \frac{559}{2240}\partial_{xx} q -$$
$$\frac{13}{420}Bh\partial_x h + \frac{13}{420}\Gamma h\partial_{xxx} h, \qquad (3.117)$$

A much simpler model is obtained by assuming s_1 and s_2 to be of higher order than second order. Thus, their derivatives or products with h or q-derivatives can be dropped so that they only enter into the calculation *via* the terms $\frac{1}{h^2}\int_0^h g_j'' u\, dy$ appearing in the evaluation of the residues as noticed previously. Within this crude assumption and because $g_0'' = -1$, s_1 and s_2 do not appear into the evaluation of the first residue. Thus applying the Galerkin method to the second-order problem but with a single function g_0 leads to the consistency condition:

$$\partial_t q = \frac{5}{6}h - \frac{5}{2}\frac{q}{h^2} - \frac{17}{7}\frac{q}{h}\partial_x q + \left(\frac{9}{7}\frac{q^2}{h^2} - \frac{5}{6}Bh\right)\partial_x h + 4\frac{q}{h^2}(\partial_x h)^2 -$$
$$\frac{9}{2h}\partial_x q \partial_x h - 6\frac{q}{h}\partial_{xx} h + \frac{9}{2}\partial_{xx} q + \frac{5}{6}\Gamma h\partial_{xxx} h. \quad (3.118)$$

The new terms are on the second line. They are all generated by the second-order contributions coming from $2\partial_{xx} u + \partial_x[\partial_x u|_h]$ in the momentum equation (3.98) and the boundary conditions (3.81). As such, they include the effect of viscous dispersion that was lacking at first order.

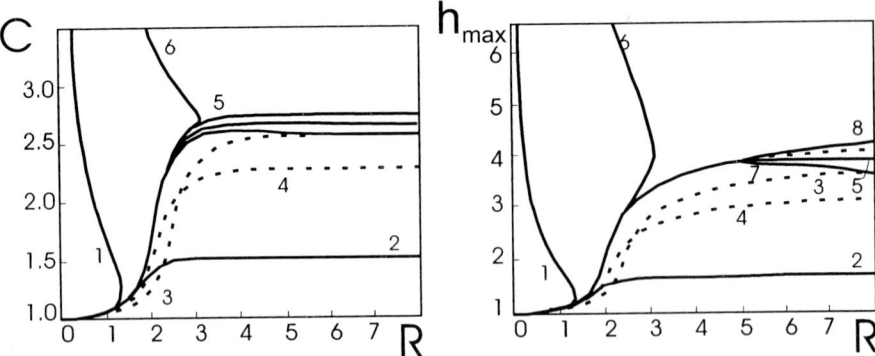

Figure 3.5: Speed (left) and amplitude (right) of one-hump solitary waves as functions of the Reynolds number for the different models considered. The plane is vertical and the Kapitza number is $\gamma = 252$: Curves 0: Benney's equation (3.73) (dashed); Curves 1: second-order Benney's equation (solid); Curves 2: Ooshida's equation (3.76) (solid); Curves 3: Shkadov's model (3.88)-(3.89) (dashed); Cureves 4: first-order model (3.111), dashed; Curves 5: first-order Galerkin model (3.118) (dashed); Curves 6: second-order model (solid); Curves 7: full second-order Galerkin model (solid, thicker) (3.112)-(3.117); Curve 8: Exact solution of the Navier-Stokes equations.

3.5 Validity of the equations

The hierarchy of model equations derived in the previous sections are either rigorously derived with some stringent limiting conditions or empirically formulated without a priori justification. If the former are valid under less stringent conditions and if the latter are accurate at all these simple equations could then be studied or simulated in lieu of the full Navier-Stokes equations. An inadeguacy is first observed by Pumir et al. (1983) who found the rigorously derived Benney equation (3.73) to exhibit local blow-up behavior at sufficiently large R. Since such blow-up behavior produced amplitude and wavelength scalings that are incompatible with the longwave expansion used in deriving the Benney equation, it suggest that the blow-up behavior is unrealistic and, more importantly, the Benney equation breaks down beyond a critical R. Similar blow-up behavior has also been observed for related longwave equations by Joo et al. (1991, 1992).

In Chapter 13, for a different but related problem, we will show that such blow-up behavior does not occur when a stationary solitary wave (pulse) solution exists. Such solitary wave solutions will be constructed in Chapter 5 for the gKS, Shkadov and BL equation of the falling film. However, we have yet to connect the blow-up behavior of the falling-film equations. Nevertheless, experimental measurements discussed in the next chapter will show ample evidence of pulses at R much larger than R_*. In fact, we shall formulate a theory for the falling-film wave dynamics far beyond R_* based on these pulses. It is hence essential that the model equations yield pulse solutions for $R >> R_*$. Similiar blow-up behavior has also been observed for related longwave equation by Joo et al. (1991,1992).

Hence, in anticipation of the construction of the pulse solutions in Chapter 5, we construct here the pulse solution of many of the model equations for vertically falling film in this chapter. As shown in Figure 3.5, the pulse solution branch (branch 1) of the Benney equation exhibits a turning point at $R = 1.2$ and beyond this value of R, no pulse solution exists for this equation. This critical value of R roughly coincides with the value of R beyond which Pumir et al. (1983) observes blow-up behavior. Quite surprisingly, the simple KS equation actually yields a pulse solution for R upto at least 10. (We do not track the branch beyond $R = 10$.) However, it does not yield the correct speed for the solitary pulse. The most complex BL equation produces the most accurate pulse speed and amplitude. For $R > 3.0$, these two values asymptote towards infinity. In comparison , the Shkadov model, despite its simplicity, is within 15% and only slightly less accurate than the BL equation. The Ooshida equation yields a solution that exists till $R = 10$ but is a factor of 2 off. For vertical films, the Shkadov model is the simplest and most accurate model that can produce the all-important pulses.

3.6 Spatial and temporal primary instability of the Skadov model

The simplicity of the Shkadov model allows us to address one important issue — what is the spatial evolution rate of the infinitesimal wavs triggered at the inlet. The linear stability analysis of Chapter 2 is a temporal one. The Gaster transformation of section 3.1 can convert the temporal growth rate to the spatial one, but only for near-critical conditions ($R \sim R_*$ of (2.100)) of periodic forcing with near-neutral wavenumber or wave frequency. We hence need the spatial growth rate for more general conditions. The spatial theory also leads us to another important issue — the convective nature of the instability. This characteristic is responsible for convecting the inlet noise into the channel to trigger the sentive sequence of wave dynamics.

We shall consider the general Shkadov model (3.88) for an inclined film. The basic state is the flat-film solution $(h, q) = (1, 1)$ and it is unstable when Reynolds number R exceeds a critical value $R_c = \cot \theta$. We shall show first that this primary instability is convective and a spatial formulation must be used to study the consequent wave evolution. Denoting $\hat{h} = h - 1$ and $\hat{q} = q - 1$, lenarized equations for $(\hat{h}, \hat{q})$ are

$$\frac{\partial}{\partial t} \begin{bmatrix} \hat{h} \\ \hat{q} \end{bmatrix} = \begin{bmatrix} 0 & -\frac{\partial}{\partial x} \\ \Phi & \Psi \end{bmatrix} \begin{bmatrix} \hat{h} \\ \hat{q} \end{bmatrix}, \qquad (3.119)$$

where

$$\Phi = \frac{3}{5\delta} + (1.2 - 3\chi)\frac{\partial}{\partial x} + \frac{1}{5\delta}\frac{\partial^3}{\partial x^3},$$

$$\Psi = -\frac{1}{5\delta} - 2.4\frac{\partial}{\partial x}.$$

Response of $(\hat{h}, \hat{q})$ to a unit impulse $\delta(x)\delta(t)$ is then, from a Laplace-Fourier transform,

$$\mathbf{G}(x,t) = \left(\frac{1}{2\pi}\right)^2 \int_F d\alpha \int_L \mathbf{D}^{-1}(\alpha, \omega)\mathbf{p} e^{i(\alpha x - \omega t)} d\omega, \tag{3.120}$$

where

$$\mathbf{D}(\alpha, \omega) = \begin{bmatrix} i\omega & -i\alpha \\ \Phi(\alpha) & \Psi(\alpha) + i\omega \end{bmatrix},$$

$$\Phi = \frac{3}{5\delta} + (1.2 - 3\chi)(i\alpha) + \frac{1}{5\delta}(i\alpha)^3,$$

$$\Psi = -\frac{1}{5\delta} - 2.4(i\alpha).$$

and **p** is a constant vector that determines the specific structure of the disturbance. In the above equation, the path F in the complex α plane is initially taken to be the real axis. The contour L in the complex ω plane, however, must be placed high enough so that the causality condition, namely $G(x,t) \equiv 0$ for all $t < 0$, can be satisfied. The determinant of the matrix **D** yields the complex relation $|\mathbf{D}(\alpha, \omega)| = 0$, or equivalently,

$$\omega^2 - i\omega\Psi(\alpha) - i\alpha\Phi(\alpha) = 0, \tag{3.121}$$

which related the complex numbers corresponding to the wave number α and the frequency ω. This relation can also be obtained by substituting the normal mode expansion $\mathbf{v}e^{i(\alpha x - \omega t)}$ for $(\hat{h}, \hat{q})$ into (3.119) and solving the resulting eigenvalue problem. When α is taken to be real in (3.121), it yields two branches of temporal normal modes with complex eigenvalues $\omega_1(\alpha)$ and $\omega_2(\alpha)$.

Simple algebra yields the critical χ to be

$$\chi_* = 1$$

and for $\chi > \chi_*$, the neutral wavenumber of the longwave instability is

$$\alpha_0 = \sqrt{18\delta(1-\chi)}$$

and the mode with the largest growth rate has a wavenumber

$$\alpha_m = \frac{\alpha_0}{\sqrt{2}}.$$

The growth rate at α_m is defined implicitly by (3.121). These are rather simple approximations of the growth rate in (2.90)-(2.99), the critical Reynolds number

(2.100) and the neutral wavenumber (2.105) from a longwave expansion of the full Orr-Sommerfeld equation.

Conversely, when ω is real, one obtains four spatial branches α_j, $j = 1, 2, 3, 4$, as shown in Figure 3.6. We note that except for nearly neutral modes with frequency (wave number) near zero and the neutral frequency ω_0 (wave number α_0), the temporal branches cannot be transformed to the spatial branches by Gaster's transformation. In fact, there are only two temporal modes but four spatial modes. The spatial growth rate should strictly be computed from the dispersion relation (3.121) with a real frequency ω.

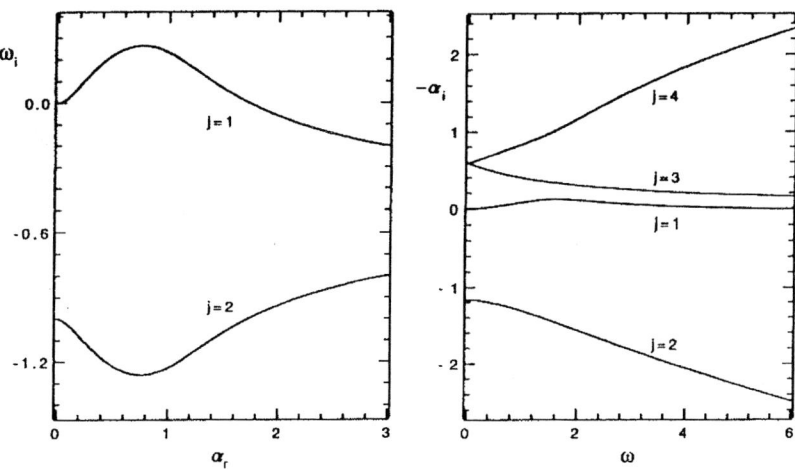

Figure 3.6: Two temporal primary growth rate branches as functions of wave number α and four primary spatial growth rate branches as functions of wave frequency ω for $\delta = 0.2$ and $\chi = 0$. The maximum spatial growth rate of the pertinent downstream-traveling spatial branch α_1 is about 0.1 which is consistent with the inception length in Figure 1. The neutral wave number α_0 is 1.7 while the netral frequency ω_0 is 5.2.

Since $\mathbf{D}^{-1}$ contains a factor $|\mathbf{D}|$ in its denominator, a steepest descent analysis similar to that by Bers (1975) and Huerre and Monkewitz (1990) indicates that, regardless of the structure of the disturbances as represented by $\mathbf{p}$ in (3.120), both components of the vector Green's function $\mathbf{G}$ are dominated at large time for any given position x by the stationary point of the dispersion relation.

$$\mathbf{G}(x, t) \sim \sum_j \exp[-i\omega_j^*(\alpha_j^*)t], \qquad (3.122)$$

where, (ω^*, α^*) are saddle points of the complex dispersion relation and they correspond to a mode with zero group velocity that satisfies the dispersion relation, viz., an order two branch point of (3.121),

$$(\omega^*)^2 - i\omega^* \Psi(\alpha^*) - i\alpha^* \Phi(\alpha^*) = 0, \qquad (3.123)$$

$$\frac{\partial \omega}{\partial \alpha}(\omega^*, \alpha^*) = 0. \tag{3.124}$$

Hence, a positive imaginary part of ω^* signifies that the response to the unit impulse will grow at any position x and the instability is called absolute in nature. However, as Huerre has pointed out, this criterion is not explicit enough as it stands. One has to carefully monitor how the spatial branches $\alpha_j(\omega)$ pinch together and form the saddle points as one moves down the Laplace contour L in the complex ω plane, or equivalenty, decreases the imaginary part of ω. Only those branch points of $\alpha(\omega)$, which pertain to pinching between the upstream and downstream spatial branches $\alpha^+(\omega)$ and $\alpha^-(\omega)$ originating from distinct halves of the α plane, can be used to determine the nature of the instability.

One can show that solution of (3.123) and (3.124) is given by

$$\omega^* = \frac{\Phi(\alpha^*) + \alpha^*(d\Phi/d\alpha)(\alpha^*)}{(d\Psi/d\alpha)(\alpha^*)}, \tag{3.125}$$

where (α^*) is the root of polynomial

$$P(\alpha^*) = \sum_{j=0}^{6} C_j (\alpha^*)^j \tag{3.126}$$

with coefficients

$$C_0 = \left(\frac{3}{5\delta}\right)^2 - 2.4 \left(\frac{3}{5\delta}\right)\left(\frac{1}{5\delta}\right), \quad C_1 = 2.4 \left(\frac{3}{5\delta}\right)(1.2 - 3\chi),$$

$$C_2 = 4(1.2 - 3\chi)^2 - 5.76(1.2 - 3\chi), \quad C_3 = 1.2 \left(\frac{4}{5\delta}\right)\left(\frac{3}{5\delta}\right),$$

$$C_4 = \left(\frac{16}{5\delta}\right)(1.2 - 3\chi) - 5.76 \left(\frac{3}{5\delta}\right), \quad C_5 = 0, \quad C_6 = \left(\frac{4}{5\delta}\right)^2.$$

They are obviously six roots for α^* at any given δ and β. Consequently, there exists six branch points in the complex dispersion relation $\omega = \omega(\alpha)$ defined by (3.121). Numerical analysis of (3.125) further shows that there are only two possible sign patterns associated with the imaginary part of ω^*. These two patterns are + - - - - - - and + + - - - - -. They are separated by the curve $3\chi = \beta_0(\delta)$ in the $\delta - \beta$ plane of Figure 3.7. However, as we see in Figures 3.8 and 3.9, those branch points with positive ω_i^* result from pinching of either $\alpha^+(\omega) - \alpha^+(\omega)$ or $\alpha^-(\omega) - \alpha^-(\omega)$, but not $\alpha^+(\omega) - \alpha^-(\omega)$! This then implies that the Nusselt flat film is convectively unstable to the primary disturbances for all conditions, viz., all values of δ and β.

Convective response of an unit impulse implies that the amplitude of the resulting wave packet does not grow at any given station x but in some comoving frame. Alternatively, when periodic forcing is applied at the inlet, the Green's function (3.120) indicates that the disturbances will propogate and grow in amplitude downstream but not at any given position. The signal at any location

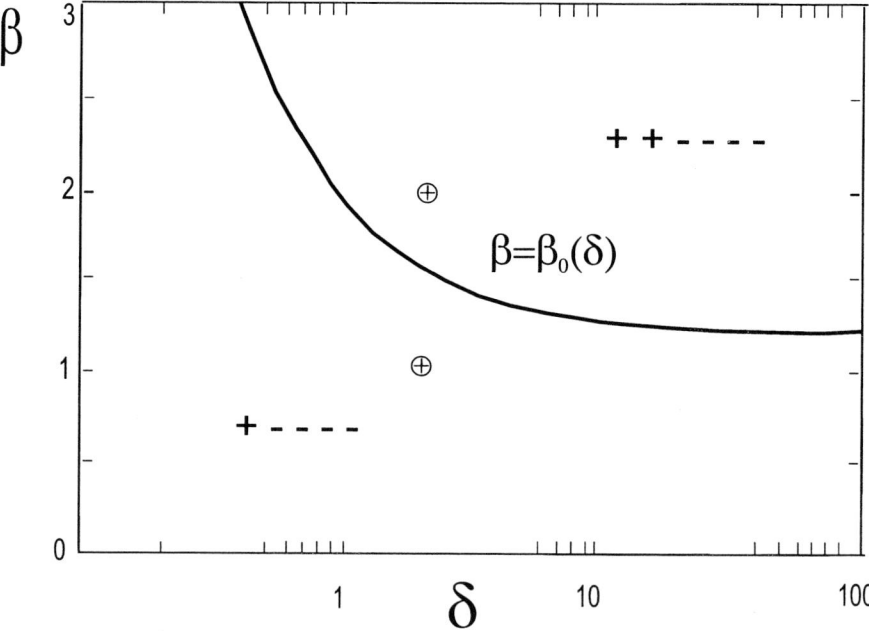

Figure 3.7: The sign of the complex frequency ω_i^* at the six saddle points. Above the curve $3\chi = \beta = \beta_0(\delta)$, there are two saddles with positive signs and below there is just one. However, contour continuation show that the pinching corresponding to the positive saddles lie above the Laplace contour and hence violate causality. This shows the system is convectively unstable and not absolutely unstable.

will be periodic in time after some transient. This implies that the spatial modes in Figure 3.6 of the dispersion relation (3.121) governs the growth in the inception region. However, the α_3 and α_4 branches of Figure 3.6 have positive real wave numbers α_r and they correspond to upstream propogating waves. Branches α_1 and α_2, on the other hand, represent downstream propogating waves. The unpstream propagating waves are not detected in the linear inception region of $x < 20$ in Figure 1.1, although they will contribute to the saturation of the monochromatic fundamental when thay are excited by weakly nonlinear interaction. Since the α_2 mode always decays in the positive x direction, the observed exponential growth in space in the inception region is due to the unstable band for $\omega \in (0, \omega_0)$ of the α_1 mode only.

The α_1 spatial mode exhibits several unique chracteristics that are endowed by the physics of the falling-film instability. A simple algebraic manipulation of the temporal dispersion relation yields an expression for the linear phase speed c,

$$c = 3 + 5\delta\omega_i(2.4 - 2c). \tag{3.127}$$

This indicates that at zero wave number and at the neutral wave number α_0

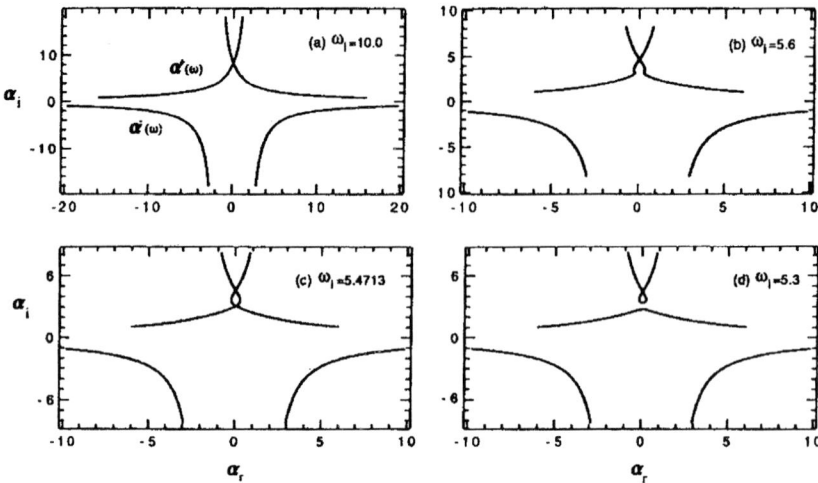

Figure 3.8: The spatial branches for $(\delta, 3\chi) = (2, 1)$ as a function of the complex frequency ω_i. The saddle point with positive $\omega_i^* = 5.4713$ results from a $\alpha^+ - \alpha^+$ pinching.

where ω_i vanishes exactly, the phase speed is exactly three times average velocity u_0. Moreover, within the unstable wave-number band $\alpha \in (0, \alpha_0)$, ω_i is negative and the phase speed is less than 3 since c is always in excess of 1.2. Conversely, within the stable band (α_0, ∞), the phase speed increases monotonically beyond 3. Alekseenko et al. (1994) have shown that, at large α, the phase speed approaches that of the capillary branch of the classical gravity-capillary dispersion relation in a particular moving frame. The monotonic increase in c beyond α_0 is hence due to the usual capillary dispersion mechanism which accelerates shorter waves. The decrease of c below 3 for longer waves, on the other hand, is due to an inertia effect which decelerates growing waves. The neutral points at $\alpha = 0$ and α_0 of the temporal formulation can be mapped exactly to the neutral frequencies of $\omega = 0$ and ω_0 of the α_1 branch. Consequently, the phase speeds are also exactly 3 at these locations in the spatial formulation. With some simple continuation and topological arguments, one can also be convinced that the phase speed is less than 3 in the unstable band of frequencies. In fact, the temporal formulation offers another unique feature of the falling-film dispersion relationship. A simple analysis of the dispersion relation (3.121) indicates that the minimum of the phase speed is located at exactly the maximum-growing wave number α_m where ω_i is at a maximum. This is true for all relatively low Reynolds number films from our earlier analysis of the complete Orr-Sommerfeld equation. It is also approximately true for the spatial formulation. Finally, the stable modes of the α_1 branch have growth rates that are slightly larger than the true values from the Orr-Sommerfeld equation. This is due to the inadequate description of the short waves by the averaged equation and it accounts for most of equantitative errors in our theory.

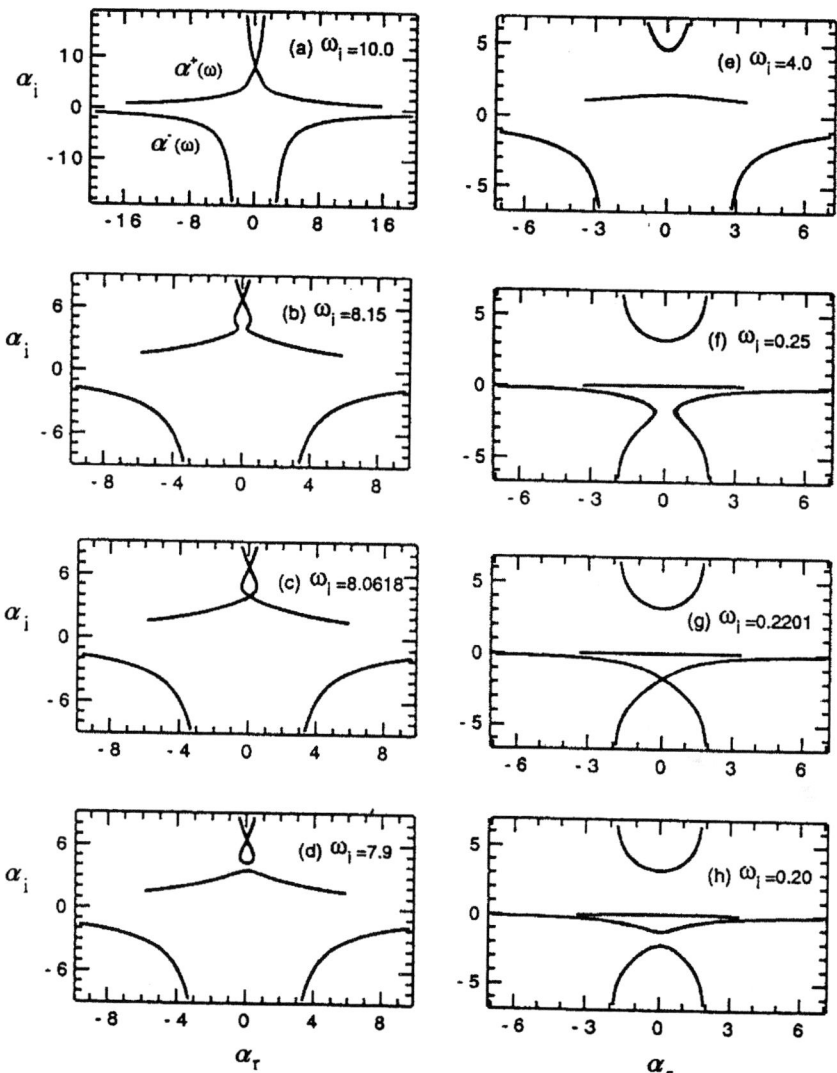

Figure 3.9: Contour continuation for $(\delta, 3\chi) = (2, 2)$. The positive saddles at $\omega_i^* = 8.0618$ and 0.2201 correspond to $\alpha^+ - \alpha^+$ and $\alpha^- - \alpha^-$ pinching, respectively.

CHAPTER 4

Experiments and Numerical Simulation

Wave dynamics on a falling film are triggered by inlet noise and they evolve both temporally and spatially. The spatial evolution continues for a long expanse (more than one meter) before the wave dynamics reach an equilibrium state without further changes in texture, speed and other wave statistics. As such, experiments must be carried out in very long flow channels. Two general types of inlet forcing have been applied in the experiments, periodic forcing and forcing by natural noise (room disturbance). The wave evolution of each is quite distinct. The evolution also changes with the frequency of the periodic forcing. As discussed in Chapter 1, the evolution is often slow at modest Reynolds numbers and the waves seem to evolve from one wave regime to another–an empirical fact that will eventually allow us to construct a theory for the wave evolution. We summarize the wave transitions detected in experiments and by numerical simulations in this chapter.

The extended spatio-temporal wave dynamics require a large computational domain and does not allow the use of periodic conditions in the downstream x direction. If periodicity is imposed, the feedback from the end of the channel would completely destroy the spatial transitions from one wave regime to the next — an important feature of falling-film wave dynamics that we hope to capture. To obtain accurate temporal statistics for each wave regime and yet still capture the spatial connection between them, we typically need to simulate over a channel of length $x = 1000$ (more then 1 meter for a vertical water film) for t in excess of 10000 to 50000 (2 to 10 minutes of physical time for a vertical falling film). Such long computational domains and time rule out a full simulation of the Navier-Stokes equation with white-noise inlet forcing. (Periodic forcing is now possible over a 1-m long channel, see Ramaswamy et al. , 1996). Instead, we shall need to use either the simpler boundary layer equation (2.45) to (2.50) or the Shkadov model (3.88) and (3.89). Whenever possible, comparison to experimental data will be used to verify the validity of these model equations.

4.1 Experiments on falling-film wave dynamics

A comprehensive survey of experimental data for both laminar and turbulent films up to 1964 year was presented by Fulford (1964). He had collected works not only for falling film wave dynamics but also for films flowing in parallel to a gas stream, on the effect of boundary roughness and so on. However, the data are often contradictory and vigorous experimental efforts continued after 1964. We shall discuss only the most important experimental papers on falling films todate and only at moderate Reynolds numbers without flow turbulence.

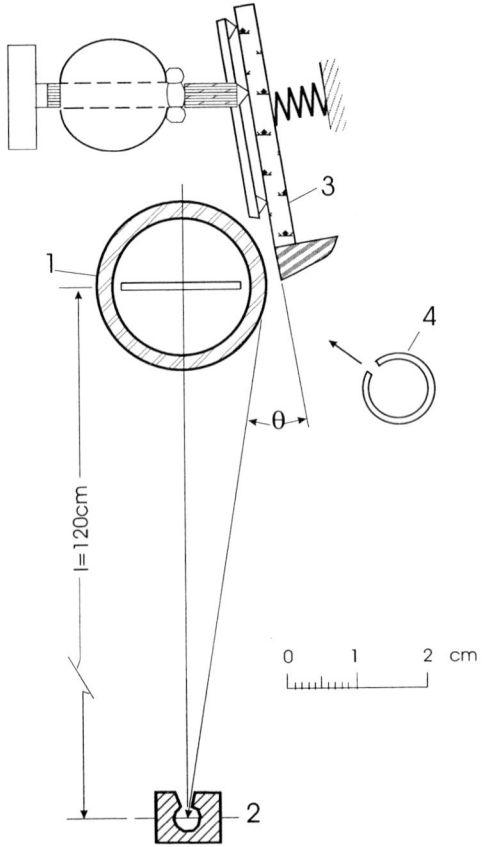

Figure 4.1: Cross-section of the channel. Kapitza (1948,1949).

A number of fundamental results were already obtained by P.L. Kapitza and S.P.Kapitza (1948,1949) in their seminal work. They managed with the most primitive of apparati but with very clever techniques. For historical reference, we shall document their experiments in more detail than some of the later ones.

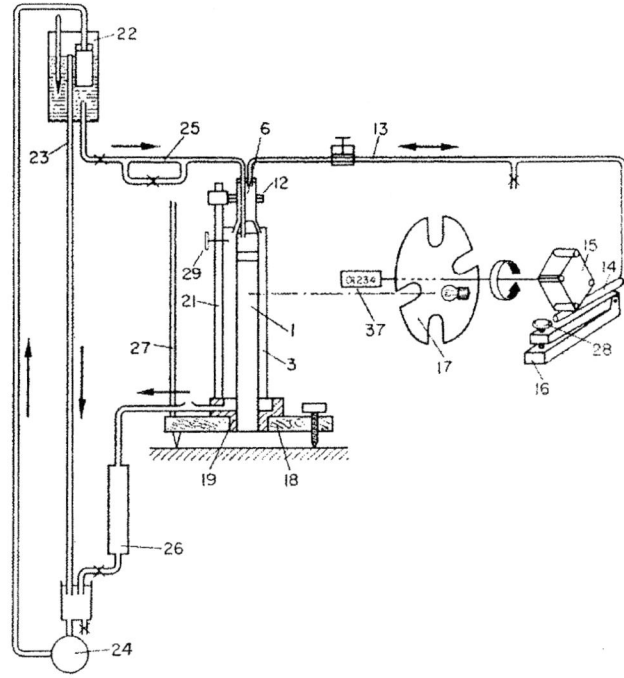

Figure 4.2: The general scheme of the apparatus. Kapitza (1948,1949).

Their experiments were done on a vertical glass tube 25 cm in length with an outside diameter of 35 mm. The length of the test section is a mere 17 cm. The liquids used were water and alcohol. Due to the short length, the wave dynamics were excited by large-amplitude inlet forcing, primarily periodic pulsation of the flow rate with air pressure. A horizontal cross-section of the tube and the optical arrangement are shown in Figures 4.1 and 4.2 to demonstrate their shadowgraph wave visualization technique. This shadowgraph method captured the profile of the flowing layer on the external wall of tube 1. A spark gap serving as a light source for photographs was positioned at point 2. The plane of the plate makes an angle θ with a ray tangential to the edge of the tube, see Figure 4.1. The shadow of the flowing layer fell on the vertical plate 3 of dimension $18*14 cm$ which made an angle θ with a ray tangent to the edge of the tube. The profile shadow of the flowing layer was thus magnified N times in the horizontal direction, $N = \frac{1}{\sin\theta}$, and could then be easily photographed.

The overall flow system that feeds into the above imaging section is shown in Figure 4.2. A well annealed glass tube 20–25 cm long with a diameter of 2–5 cm feeds the fluid through gap 6 between two cones. The fluid was recycled in a circulation loop driven by pump 24 and to prevent liquid contamination, a multilayer filter of textile fabric was used within the loop. Stroboscope 17 excite the flowrate pulses and the subsequent film waves in such a way that they were synchronized with the illumination of the shadowgraph imaging setup.

The average Reynolds number range was $7.6 < R < 23$ in their experiments. A critical Reynolds number R_* for the initiation of ripples was observed and correlated to a semi-empirical relationship: $R_* = 0.61\gamma^{3/11}$, where $\gamma = \sigma\rho^{-1}\nu^{-4/3}g^{-1/3}$ is now known as the Kapitza dimensionless parameter in (2.25) for the physical properties of liquid. For the modified Reynolds number δ of (2.40), this critical value is $\delta = 0.0465$. After searching over a wide range of amplitude and frequency of inlet flow rates, two general stable wave regimes were found: those with approximately sinusoidal waves (the "periodic" regime) and those with isolated (or single) waves (the solitary wave regime), as seen in Figure 4.3. The amplitude, velocity and wavelength λ for the first regime were recorded. For water, $\lambda = 0.85 - 1cm$ and for alcohol, $\lambda = 0.7 - 0.8cm$. In their experiments, quantitative characterization of the solitary wave regime was not offered. We now know that the length of such waves could range from 2.5 to 4 cm, and are hence longer than the working section of the apparati. The Kapitzas did notice that periodic waves tend to coalesce and evolve towards solitary waves at higher Reynolds numbers–indicating a transition from the former to the latter can occur. At low discharges, however, they noted that the periodic waves do not undergo such transitions.

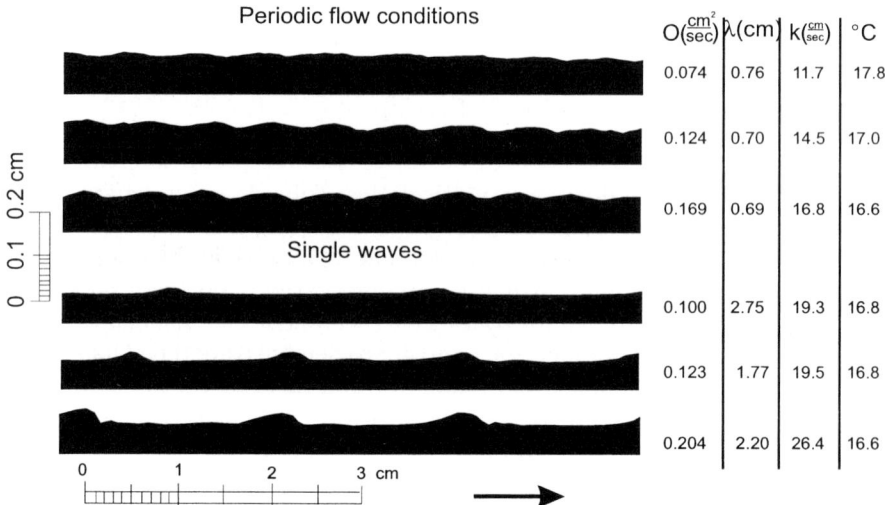

Figure 4.3: Wave profile from Kapitza-Kapitza experiments; liquid-alcohol.

To remedy the strong limitation of a short tube, which obviously suppressed wave transitions, Binnie (1957) carried out new experiments on a vertical glass pipe 1.5 meters long with an outside diameter of 26.5 mm. The Reynolds number range for the water flow he used was rather small: $R = 4 - 5$. One main purpose of his investigation was to find the minimum Reynolds number R_* at which waves can be observed when waves are naturally excited (the first wave transition) rather than periodically forced. Binnie found that for such naturally excited waves $R_* = 4.4$ and the wavelength at inception is $\lambda \simeq 1.14 cm$. A second Binnie study (1959) was made on a flat glass channel 4.9 m long and 8.4 mm wide which can now be inclined at $\theta = 1° - 2.75°$ relative to the horizontal. Experiments were also carried out at the inclination angle $\theta = 45°$ but no data were reported. The critical Reynolds number R_* was measured and the theoretical prediction (2.100) from linear theory, $R_* = \frac{5}{6}\cot\theta$, was confirmed experimentally. The wave velocities were also measured at inception. These velocities were as much as 2.6-2.8 of the mean velocity and they decreased as the angle θ decreased. The inception wave lengths observed were 2.8-2.9 cm.

Tailby and Portalski (1860) carried out their measurements on a vertical plate 53 cm width and 213 cm long with various liquids: water, 27%, 37%, 45% and 82% glycerin solutions in water, iso-propyl alcohol and methyl alcohol. Only naturally excited waves were studied and the Reynolds number R varied

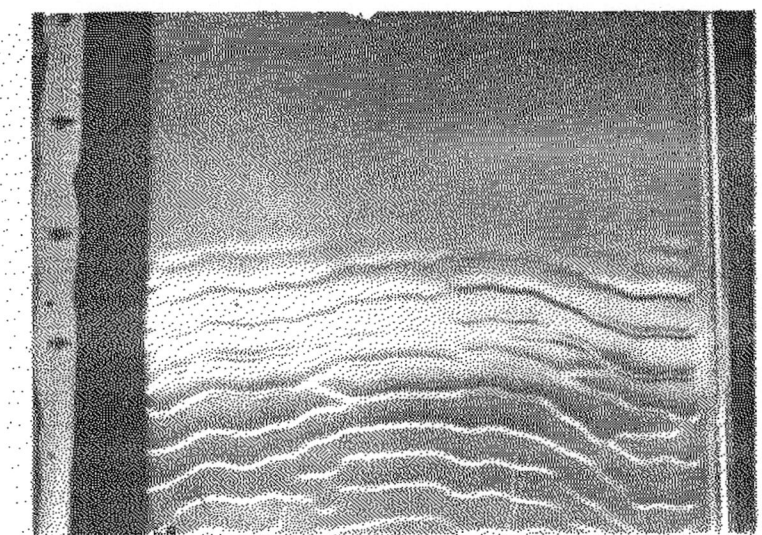

W10: Re = 10·4

Figure 4.4: Two-dimensional waves, 82% (wt) glycerin solutionm Tailby and Portalski (1960).

between 1.2 and 3.4. The authors obverved a definite critical flow rate for the inception of wave motion and generally supported the prediction that a non-zero critical Reynolds number R_* exists for a vertically falling film, in contradiction to the prediction of linear theory in (2.98). For water, the wavelength at the inception position was $\lambda \simeq 0.96 - 1.24 cm$ over the entire range of Reynolds numbers they examined. The authors observed many more wave regimes than the Kapitzas and documented them with photographs (Figures 4.4 and 4.5).

Wilkes and Nedderman (1962) used the tallest vertical tube with a height of 3m and a tube diameter of 25 mm. They also worked with glycerin solutions to allow large variations in the physical parameters. They provided the first measurements of the velocity profiles under thin wavy films. They imaged the velocity field by stereoscopic photography of small air bubbles moving with the liquid. Wilkes and Nedderman concluded that under a wavy film, the average profile is still approximately parabolic, as that in the Nusselt flat film of (2.22), within a maximum deviation of about 20 per cent. The measurements were restricted to small Reynolds numbers, $R \approx R_*$. Entrance effects and wave suppression by surface active agents were examined. Only naturally excited waves were studied. The critical Reynolds number was determined from all their data to be $\delta \simeq 0.044$.

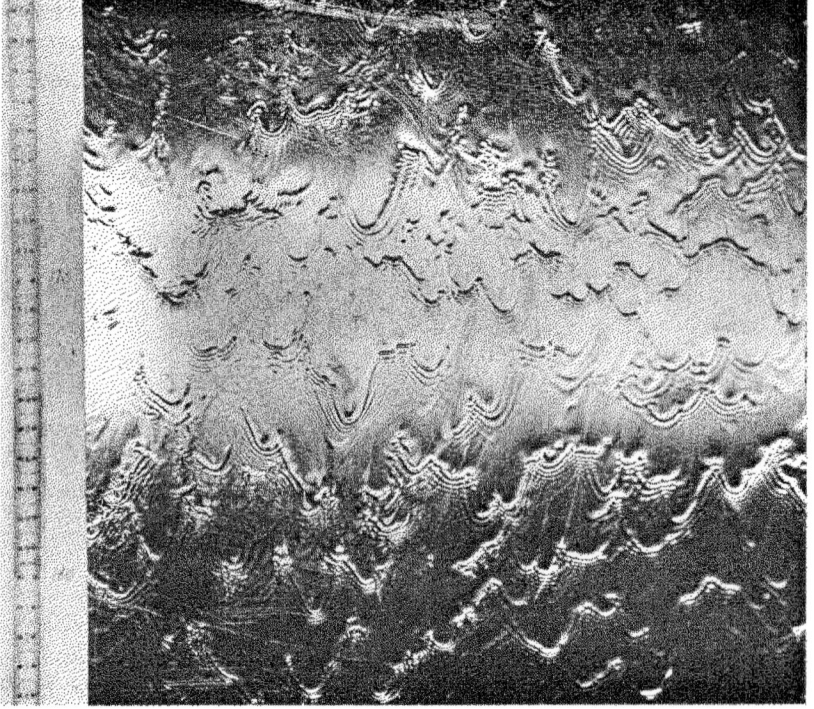

Figure 4.5: Three-dimensional "soliton gas", Tailby and Portalski (1960).

Additional measurements of the velocity field were done by Portalski (1964) with the same apparati as in his previous work. The working fluid was a 71.5% glycerin solution in water and his Reynolds number ranged from 2.5 to 25. It was again found that the average velocity profile differs weakly from the parabolic profile of a flat film.

Then came a important experiment by Stainthorp and Allen (1965). The experiments were carried out with water inside a vertical glass tube 76 cm long and 34.5 cm in inner diameter. The measurements were made for the Reynolds number range $R = 4 - 45$. They studied mostly naturally excited waves except when they determined the critical Reynolds number for wave inception. The first part of the work is devoted to seeking this R_*. For $R < R_* = 4.25$ the authors could not detect any wave anywhere on the tube–small periodic perturbations at the inlet (or anywhere on the tube) were attenuated within a region 30 to 60 cm below the position where the disturbance was introduced. This contradicts the linear stability theory of (2.98) that predicts a zero critical Reynolds number for a vertical film. The authors believe the contradiction is due to the presence of surface active agents which has wave-damping effects. Their measured wave inception distance as a function of R is shown in Figure 4.6. Close to the inception position the wavelength ranges from $\lambda \simeq 1.0 - 1.2 cm$ with an average of $\lambda \simeq 1.13 cm$.

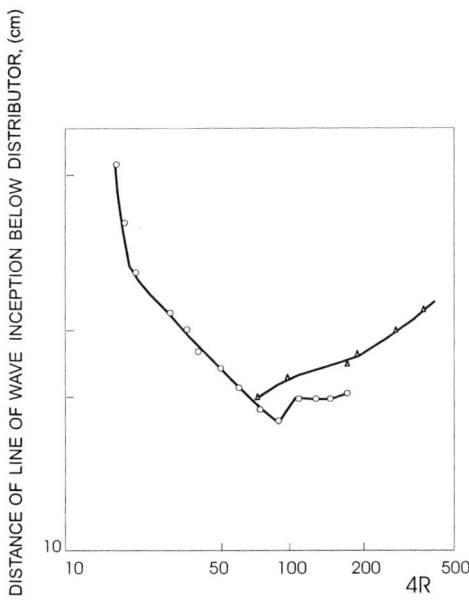

Figure 4.6: The wave inception curve, Stainthorp and Allen (1965), water.

Stainthorp and Allen first found that the wave profile, wavelength and wave velocity changed significantly downstream after wave inception for naturally excited waves–the first documentation of wave transitions. A series of their wave recordings at different downstream stations are shown in Figure 4.7. As seen in Figure 4.8, the wave speed 10 cm beyond inception actually decelerates. Beyond this 10 cm, however, these waves change their shape and evolve into solitary waves with dramatic speed acceleration. In particular, they noticed significant wave steepening downstream and the appearance of small front-running capillary ripples as the waves approach solitary waves, same as those periodically excited in the Kapitzas' first experiments. However, Stainthorp and Allen ob-

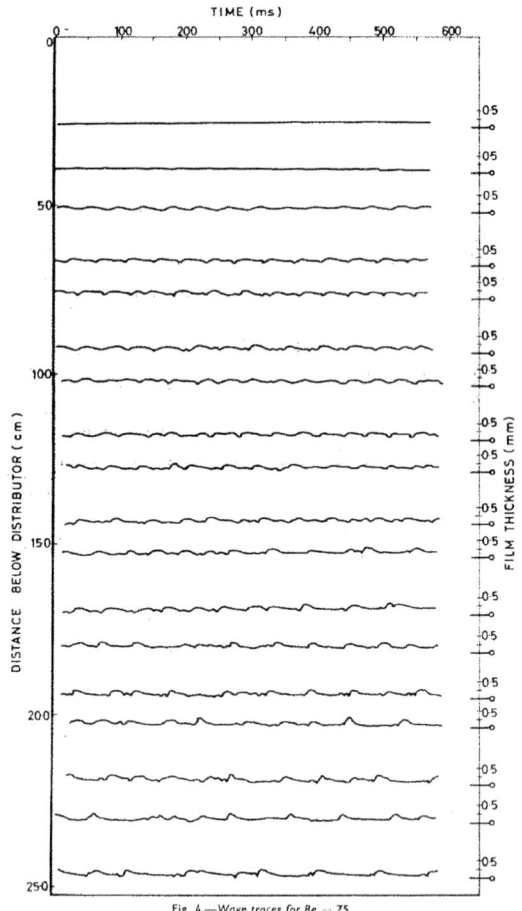

Figure 4.7: Wave traces, Stainthorp and Allen (1965), $4R = 75$, water.

served that all wave transitions eventually lead to these solitary waves. Instead of the large field of periodic waves at inception, these solitary waves are separated by large expanses of flat substrates with a thickness a tenth or smaller than that of the waveless Nusselt flat-film thickness of (2.21). The average separation between these solitary waves is as much as as a factor of 8 larger than the 1 cm wavelength at inception and the wave speed is also larger by as much as a factor of two (Figure 4.8). Although the solitary waves continuously increase their separation, amplitude and speed downstream, they are observed to retain the same self-similar shape regardless of their size and speed. One could almost geometrically scale a big one down to a small one (Figure 4.9). Stainthorp and Allen also noted that their wave speed is proportional to their wave amplitude. Theirs was not only the first experimental documentation of wave transitions to solitary waves but also the first detailed measurement of the solitary wave fea-

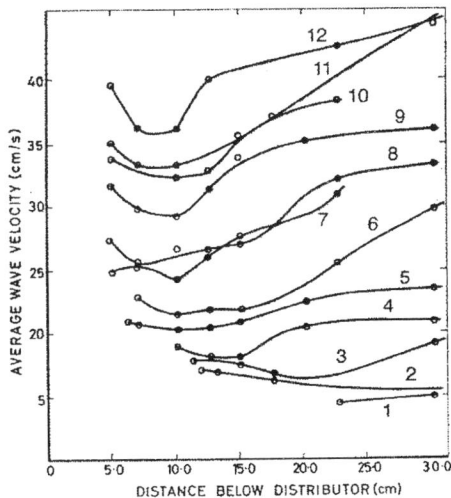

Figure 4.8: Wave velocities, Stainthorp and Allen (1965), water; the waves accelerate downstream.

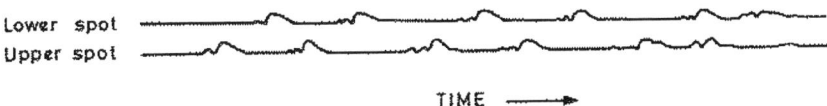

Figure 4.9: Wave profiles at two locations below inlet, $R = 25$, Stainthorp and Allen (1965).

tures. Benjamin first christened these waves as "solitary waves" to distinguish them from solitons in deep-water waves. We shall often refer to them as pulses, a nomenclature from the pattern formation literature.

Jones and Whitaker (1966) investigated natural waves at Reynolds numbers ranging from 7.7 to 124, higher than all earlier studies. They found the average wave length to be 0.9 cm at onset for $27 < R < 33$. Detailed correlations of velocities and wave lengths with respect to R at a distance of 2.5-5 cm from the water distributor (i.e. near the inception line) were documented. solitary waves form from the periodic waves at a distance of 30-60 cm from the inlet for all conditions.

Nemet and Sher (1969) carried out experiments outside vertical copper and glass tubes which were 44.8-69.4 cm long and 18.5-20.2 mm in outer diameter. The authors confirmed earlier observations that the onset wavelength does not change significantly with Reynolds number and is approximately equal to 1 cm. They also noted the wave speed and wavelength increase as the waves evolve, culminating in an average wavelength (solitary wave separation) far from the inlet of $\lambda = 2.6 cm$.

Strobel and Whitaker (1969) added small amounts of surfactant to their water working fluid for Reynolds number in the range of 3.3-67. The stabilizing effects of these surfactant were documented with precise measurements of the critical Reynolds number R_* , wave length, wave velocity and wave inception distance from the distributor .

The experiments of Krantz and Goren (1970, 1971) were done on an inclined flow glass plane 3 mm thick, 1.52 m long and 19 cm wide. The experiments were for two angles: 90° and 74.5° and with Chevron No.5 and No.15 white oils as working fluids. The waves were drivn periodically by vibration at the inlet. Due to the highly viscous fluids, their Reynolds number range was 0.13-1.3, corresponding to $\gamma = 1.1 - 7.8$ and $\delta = 0.0066 - 0.07$. (Thus the wave exists for $\delta > \delta_* = 0.0456$!). The equilibrium amplitudes listed in their tables depended on R and the imposed frequency. Small amplitude waves were most often nearly sinusoid. Larger amplitude waves usually exhibited varying degrees of "teardropping" - i.e. steeper slope on the downstream side of the wave. Large - amplitude long waves on occasion exhibited "reverse teardropping" - i.e., steeper slope on trailing edge of wave - which eventually evolve into to a new type of solitary wave, almost mirror images of those observed in earlier experiments, including those of the Kapitzas. Unfortunately, wave profiles of such negative solitary waves were not presented. Krantz and Goren also observed bimodal waves composing of two waves. Another important observation was that , for two - dimensional waves, the wave length and wave velocity remained constant downstream even when their amplitudes grew towards a final equilibrium value. Numerous correlations of the amplification factor and phase velocity as functions of wave number and Reynolds number , of neutral wave number and wave velocity as functions of Weber number were constructed from their measurements.

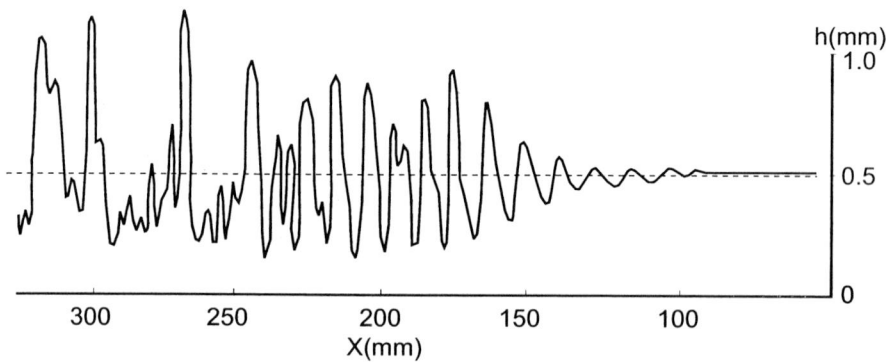

Figure 4.10: Snapshot of a wave profile, 25% water-glycerine solution, $R = 37$, Portalski and Clegg (1972).

The contradiction between the predictions from linear stability theory and earlier experimental observations for the critical Reynolds number and the inception length was finally resolved by a remarkable experiment by by Portalski and Clegg (1972). They also made dramatic discoveries about the wave transition to solitary waves. They used three working liquids (water and 25% and 45% w/w glycerin/water solutions) and their Reynolds number varied between 17.7 - 375 for their naturally excited waves. Their experimental breakthrough involves an imaging technique with a large-plate camera that captures the instantaneous film profile over a 30 cm expanse of the channel and a clever microdensitometer measurement of film height from the the photographic plates. As a result, the instantaneous flowing layer with the overlaying waves can be imaged over a large portion of their channel (Figure 4.10) ! This capability significantly extends previous time series recordings from individual sensors at several points along the column. The authors were of the opinion that the inception of wave motion is a gradual process as the film moves down the column. Most importantly, they concluded that there is no critical "position of inception" and that a non-zero "critical Reynolds number" for the vertical film is fictitious. The waves initiate from the inlet at all non-zero Reynolds numbers, as predicted by the linear theory, but with very small and often undetectable amplitudes. Their detection and the corresponduing R_* are essentially functions of the resolution of the sensors and the length of the apparatus used. Their inception waves appeared to be sinuous and their wavelength is accurately predicted by the maximum-growing mode of (2.104) from the linear theory. They also reported downstream wave steepening of these inception waves. However, unlike any earlier observations, they observed a sudden breakup of these inception waves at a critical downstream distance, accompanied by highly irregular wave patterns . From this patch of random and irregular waves, gradual regularization occured followed by a transition into the solitary wave regime . It was a dramatic first observation of the somewhat violent and abrupt transition from the periodic waves to the solitary waves. As the wave amplitude increases and especially when the solitary wave regime is reached, the average film thickness decreases dramatically and is only a fraction of the Nusselt value, as Kapitza first reported.

Next came the experiments of Telles and Dukler (1970) and Chu and Dukler (1974, 1975) and Dukler (1976). They worked mostly in the high Reynolds number end of the region $R = 225 - 1440$ where turbulent flow is expected in the film. Nevertheless, measurements were also made at moderate Reynolds numbers with laminar films. The authors used a vertical Plexiglass channel 152 mm wide and 5.5 m long ! They studied natural waves generated by room disturbances in water films. The wave structure at these higher Reynolds number is extremely complex even for laminar films which required a statistical analysis. Neverthelss, they reported the coexistence of two distinct wave patterns —"a two-wave system".

Large lumps of liquid which carry a significant portion of all flowing liquid moved down the interface with no change in speed or shape (Figure 4.11). The lumps were asymmetrical with the front steeper than the back and some showed a dip below the substrate level. However, they did not possess the front-running capillary ripples of the solitary waves observed in all earlier experiments at lower Reynolds numbers. These structures were also much faster and larger than the solitary waves. Smaller waves rode on these larger structures and often

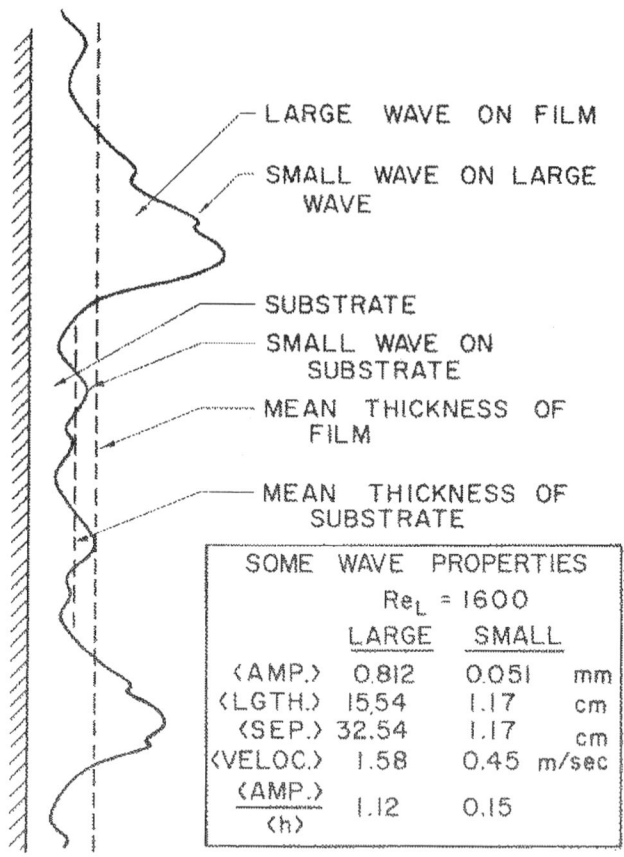

Figure 4.11: System of waves at large Reynolds numbers, Chu and Dukler (1974).

disappeared intermittently. The large waves were separated by very long flat liquid substrates, that were themselves covered by small waves, by distances more then 100 times their amplitudes. It was estimated that the liquid carried by the large waves move 10 to 20 times faster than the liquid flowing in the substrate (Figure 4.12). The localized nature of these lumps was visualized by the following comparison in scale: let the substrate thickness be equal to one unit of length; then the maximum wave height is equal to 5 units, its width at the base is 50 units and the average distance separating the waves is 1000 units! The small waves covering the substrate had amplitudes approximately 5 - 10 times smaller thann the large waves . Even these small substrate waves were highly asymmetrical. It was shown that the substrate waves played an important role in the heat and mass transfer rate across the film. Telles, Chu and Dukler were apparently the first to notice waves that travel three-times

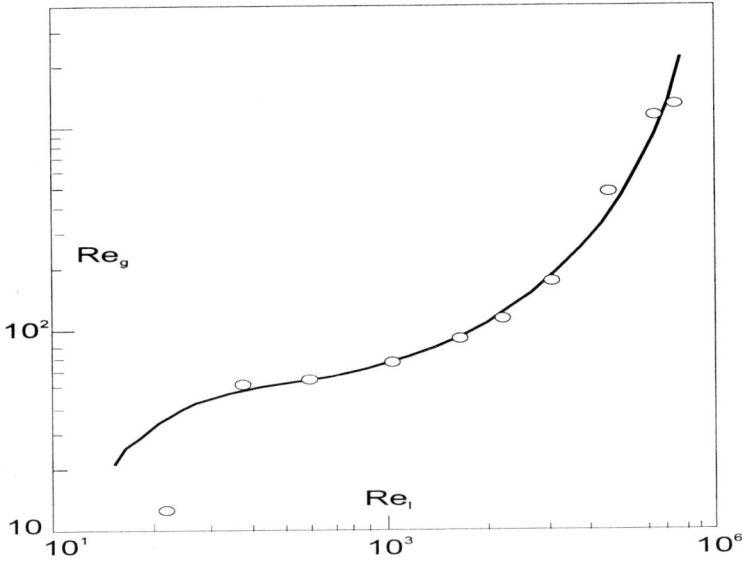

Figure 4.12: Average Reynolds number as a function of substrate Reynolds, Chu and Dukler (1970)

faster than the average velocity in the film, though such waves actually existed in earlier works. In a late chapter, we shall examine hydraulic jumps that are rough approximations of the large localized waves detected by Dukler's group. Nevertheless, wave dynamics for vertical films with Reynolds number in excess of 300 are quite distinct from those at lower Reynolds numbers. Our understanding of the former waves is still woefully lacking.

Brauner and Maron (1983) studied well-developed wavy flow on an inclined glass plate 1m long. The measurements were made for inclination angles θ ranging from $2.2°$ to $90°$ and for the Reynolds number range of $R = 100 - 7000$. They also confirmed the gradual downstream decrease of the wave frequency from inception to an asymptotic equilibrium value downstream. They noticed a specific mechanism for frequency reduction where a subharmonic peak appears repeatedly downstream in their wave spectrum. Every other wave seems to disappear in each downstream wave transition within a cascade of transitions reminiscent of period doubling. At equilibration, the wave shape has been modified from a sinusoid to the steepened solitary wave. They found the equilibrium frequency to decrease with decreasing θ and increasing R. Dependences of frequency, velocity and wave amplitude upon Reynolds number, inclination angle and distance from the inlet were presented and are similar to that of Stainthorp and Allien (1965).

The rich wave data in these experiments suffer from one major drawback. The waves typically suffer pronounced transverse modulation some distance from the inlet but they were still treated as two-dimensional waves. In a number of cases, particularly for small tubes like the Kapitzas' and for inclined films like that of Brauner and Maron (1983), the transverse wavelength was much larger than the downstream wavelength and the two-dimension assumption is justifiable. The transverse modulation is significant, however, under most conditions and this disallowed any meaningful comparison to theory except for the waves near inception.

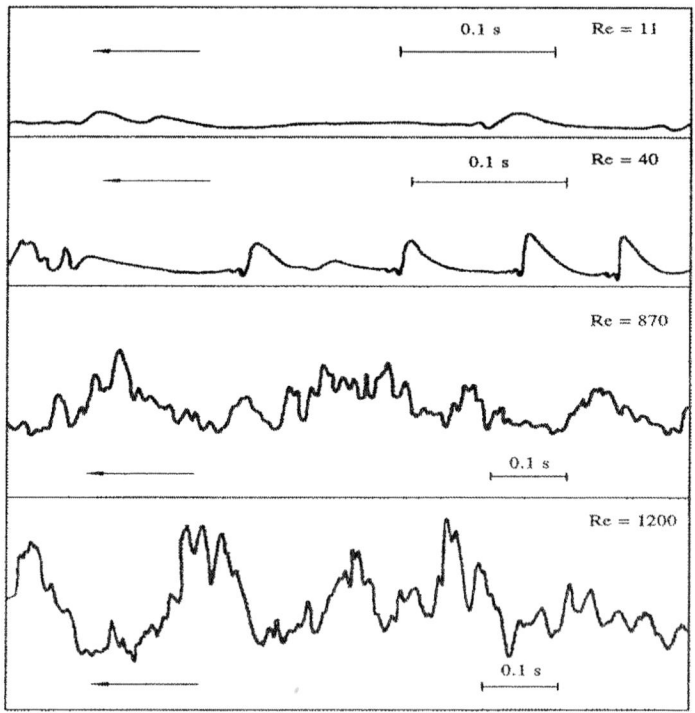

Figure 4.13: Wave profiles at different Reynolds numbers, water, Alekseenko, Nakoryakov and Pokusaev (1994).

A detailed experimental investigation of strictly two-dimensional waves on the vertical plane was carried out only in the late 70's by a Novosibirsk group of scientists: see Alekseenko, Nakoryakov and Pokusaev (1985, 1994), Nakoryakov et. all (1974, 1976, 1987), Pokusaev and Alekseenko (1977). These periodic forcing experiments generate the most accurate two-dimensional wave data at moderate Reynolds numbers (Figure 4.13) and we shall mostly compare our theory to their experimental data.

Nakoryakov, Pokusaev and Alekseenko carried out their experiments with a vertical tube which was 1m long and 60 mm in outside diameter. Water/glycerin and water/alcohol solutions were used as working fluids. Two-dimensional waves were driven by carefully designed axisymmetric inlet disturbances of specific amplitude and frequency . Their excited waves kept their two-dimensional shape and regularity over the entire length of the working tube. The inlet forcing is

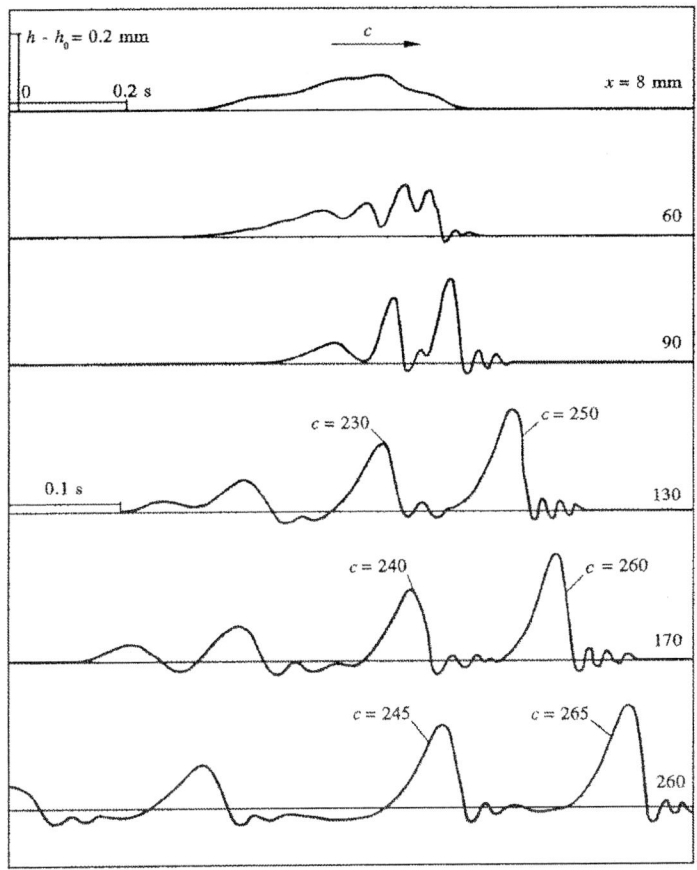

Figure 4.14: Evolution of initial single perturbation, $\nu = 7.5*10^{-6} m^2/s$, $\sigma/\rho = 57*10^{-6} m^3/s^2$, $R = 3.05$, Alekseenko, Nakoryakov and Pokusaev (1994).

so axisymmetric that transverse disturbances were never appreciably excited even after the two-dimensional waves undergo several transitions. They could hence excite and study two-dimensional waves over a wide range of parameters, including a Reynolds number range of $R = 1.5 - 65$. Nakoryakov, Pokusaev and Alekseenko first experimentally showed that the solitary waves were stationary travelling patterns with constant speeds. Abundant data on amplitudes, phase velocities, wave lengths and wave profiles were recorded for both periodic and solitary waves. The evolution of initial disturbances of various forms (periodic, single signals, hydraulic jumps and so on) was examined (Figure 4.14).

These experiments were continued by Nakoryakov, Pokusaev and Radev (1987) and Radev (1989) using the same apparatus. Though the main purpose of their work was to investigate the influence of waves on interfacial mass transfer of weakly soluble gas, spontaneous generation of three-dimensional waves in freely falling films at the Reynolds number range $R = 15 - 400$ was also

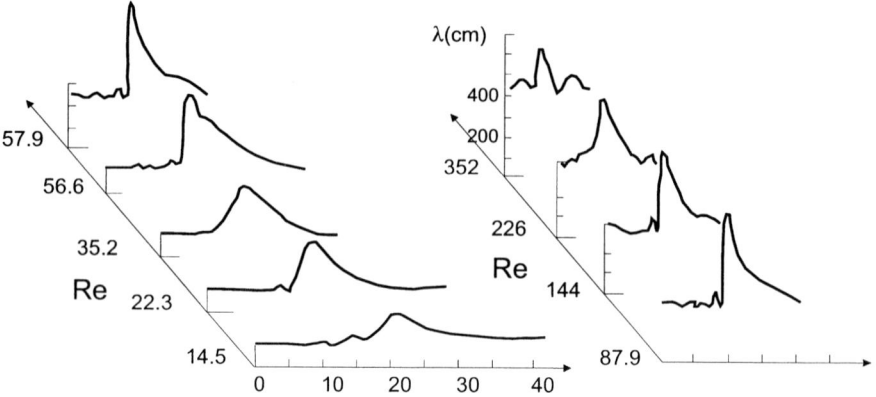

Figure 4.15: "Generalized" wave patterns, water, Nakoryakov, Pokusaev and Radev (1987).

examined . As in Alekseenko (1994), the wave motion was imaged by shadow graph method but now with a more sophisticated statistical averaging softeware to capture the most common three-dimensional wave structure for a seeming stochastic wave field. Up to $R = 400$, they observed a dominant pattern which resembles a three-dimensional solitary wave with a steepened downstream profile and a gently sloping trailing tail (Figure 4.15). The amplitude of such disturbances increased with the Reynolds number up to R = 30-60, where it exhibited a maximum, and decreases beyond that. The wave structure beyond the maximum developed an additional hump at the trailing edge . At even larger R, this second hump fell to the wave base and divided into two waves of smaller amplitude. The amplitude of these secondary humps increased with R while that of the main hump decreased. For turbulent films at $R > 400$, their amplitudes became comparable and their shape became more symmetrical.

In a series of experiment by Gollub's group (Liu and Gollub, 1993, 1994) at Haverford College, two-dimensional wave transition processes were investigated in great detail. Experiments were carried out on an inclined plate $50cm$ wide and $200cm$ long and the inclination angle was changed over the range 0 to 35 degees. Different water-glycerin solutions were used. Their apparati reflected the technological advances over the half century since Kapitzas' first experiment. Nevertheless, one could still detect design innovations that can be traced back to the Kapitzas that , in fact, appear in all the experiments described earlier. We shall hence describe Gollub's apparatus in the same detail as the Kapitzas for historical reference. Their flow and measurement systems are shown schematically in Fig ure 4.16. The fluid was pumped through a filter to limit contamination and through a ballast tank to prevent pump vibrations from reaching the film. It emerged from an input manifold through a narrow but adjustable gap between the film plane and an overlying plate. The dimensions

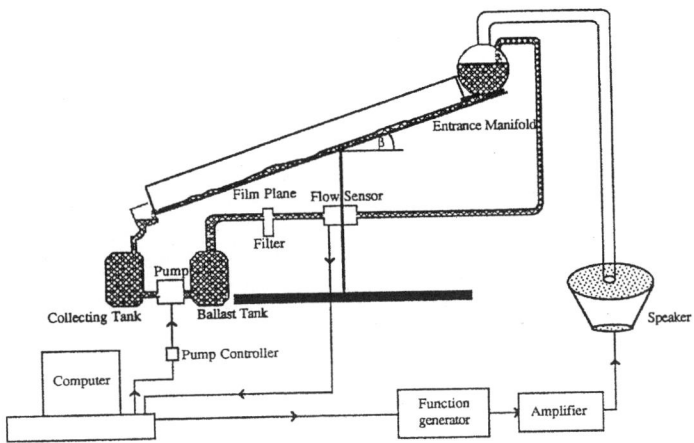

Figure 4.16: The general scheme of the arrangement, Liu, Paul and Gollub, (1993)

of the film plane were 200 cm parallel to the flow by 50 cm transverse to the flow. The supporting framework of the film plane was massive and mounted on rubber feet to reduce the influence of any building vibrations. The input manifold contained a copper mesh as a precaution against fluid oscillations inside the manifold. The system could accommodate fluids with a range of viscosities from 1–10 cs. The angle β could be continuously adjusted over the range 0–35 degrees. This allowed the phase velocity and the amplification rate of waves to be adjusted, so that both the transitional processes and the statistics of the disordered regime could be studied effectively. The flow rate was digitally monitored and computer-controlled.

A system for perturbing the entrance flow rate at frequency ω and amplitude A was based on applying small pressure variations to the entrance manifold. These perturbations could span a wide range in amplitude, waveform, and frequency. Small two-dimensional disturbances could also be generated at downstream position by weak air flow from a tube which had a gap 0.5 mm wide along its length, placed transverse to the flow about 2–3 mm above the liquid film. This "air knife" created waves that are almost perfectly two-dimensional. The flow rate was digitally monitored and computer-controlled, so that the control parameter R could be ramped slowly through secondary bifurcation thresholds. Water and glycerin-water solutions (50% by weight) were used. The latter is less affected by surfactants adsorbed on the film. Also, two-dimensional waves on the surface of glycerin-water films were found to be more stable against three-dimensional disturbances.

Laser beam deflection was used as a local measurement method in the experiments to detect the waves and measure their properties. Position-sensing photodiodes (PSPD's) are used to detect the deflection of normally incident laser beams. This gave a quantitative time series for the local wave slope $h_x(x_0, t)$. The resolution of the wave slope measurement was approximately $5 * 10^{-5}$. For sinusoidal waves with $\lambda = 5cm$, this slope sensitivity corresponded to a wave amplitude of only $0.4 \mu m$.

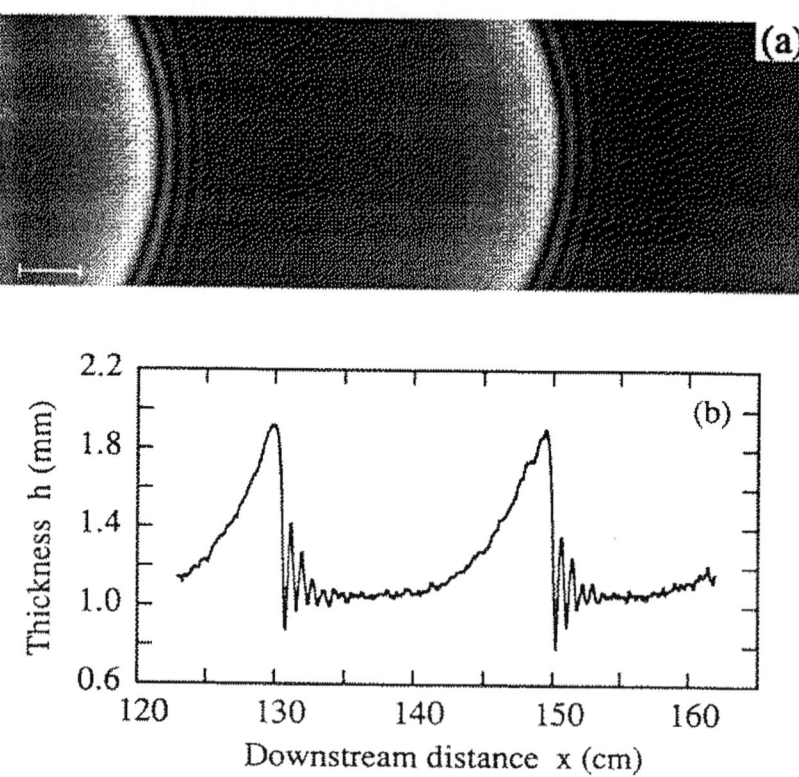

Figure 4.17: solitary waves in film flows. (a) A snapshot of periodic solitary wave train ($\omega = 1.5$Hz, $R = 29$, $W = 35$ and inclination angle $6.4°$). (b) A wave profile read from the center line of the image (a), Liu and Gollub. (1994)

Instead of Kapitzas' shadowgraph technique, "fluorescence imaging" was used to delineate the spatial and temporal dynamics for nonlinear waves and to provide global space-time measurement. Fluorescent dye is dissolved in the liquid and induced to fluoresce by illumination provided by fluorescent "black lights" oriented parallel to the flow direction and located above the lateral edges of the film plane. Calibrations showed that the liquid properties were essentially unaffected by the dye, and that the light intensity in the image plane was linearly correlated to the local film thickness. As a result, the two-dimensional wave field could be imaged with a video camera. An imaging system based on a CCD camera shuttered at 0.001 s allowed low light level quantitative imaging without blurring. The data was recorded on tape and analyzed on an ITI imaging system. Spatial resolution of 512x480 was sufficient to resolve the waveform shape in detail while at the same time viewing many wavelengths (Figure 4.17). In this way, they could study both the nonlinearity within one wavelength and the nonperiodicity over many wavelengths. When the video temporal resolution of 1/30 s was insufficient, a fast-scan one-dimensional CCd camera can be substituted that allowed measurements at 500 points along a line with temporal resolution of 0.001 s.

To avoid edge effects, which were often responsible for pronounced curvatures in wave fields on a plane, two-dimensional waves were imaged near the center line of the film plane. The inlet noise genereated by pressure variations to the entrance manifold were computer-controlled and could span a wide range in amplitude, waveform, and frequency. Thus, the sensitivity of the spatiotemporal patterns to perturbations of various types could be explored precisely , conveniently and without disturbing the flow profile in the entrance region.

Periodically forced waves were studied (as in the work by Kapitza and Alekseenko)with emphasis on transitions from the initial periodic waves. In particular, their sophisticated equipment and detection system allowed the first verification of several important concepts related to falling film waves - convective and absolute instabilities convecting the inlet disturbance (section 3.5); instability of periodical waves (secondary instability) and their transition to solitary pulses, particularly the subharmonic instability observed by Brauner and Maron; pulse-pulse interaction (Figures 4.18 to 4.21) and transverse (tertiary) instability. The pulse-pulse interaction is especially important as it determines how the solitary waves accelerate and coarsen (Figure 4.20) after their inception–the dominant wave dynamics for most of the channel. They imaged how a larger "excited" pulse can capture a smaller front neighbor in an irreversible coalescence event. Both the excited pulse and the large pulse created after the coalescence drained excess liquid from them. As a result, both their amplitude and speed decayed in time. Such drainage dynamics will be shown in our study to determine the rate of coalescence and the coarsening dynamics seen in Stainthorp and Allen's data in Figure 4.8. Our pulse interaction and drainage theory will be verified against Gollub's data.

Quite recently, Bontozoglow's group at the University of Thessaly, Greece, has continued Gollub's efforts in studying solitary wave dynamics.Their experimental apparatus was an inclined plate of plexiglas, 25 cm wide and 80 cm long. The inclination angle was changed from $0°$ to $60°$. Their working liquid was pure water and 28% water/glycerin solutions. The fluorescence imaging method of Gollub was used. Their work concentrated on the existence and properties of the solitary wave (pulse) coherent structures. Solitary-wave interactions, inelastic collisions and other non-adiabatic processes (Figures 4.22 and 4.23) were studied in far greater detail than the pioneering work of Gollub. Theirs represent the latest on-going experimental efforts to understand the complex falling-film wave dynamics.

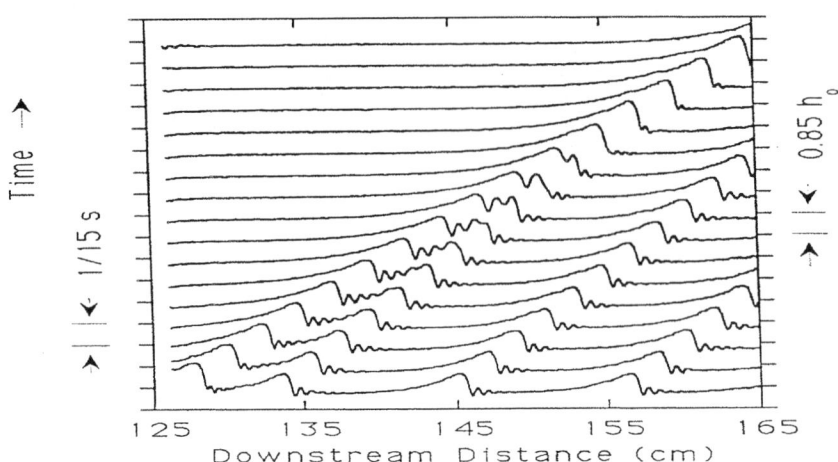

Figure 4.18: Space-time evolution of the film thickness, showing the interaction of solitary waves. Inclination angle $8°$ and $R = 32$, Liu and Gollub (1994)

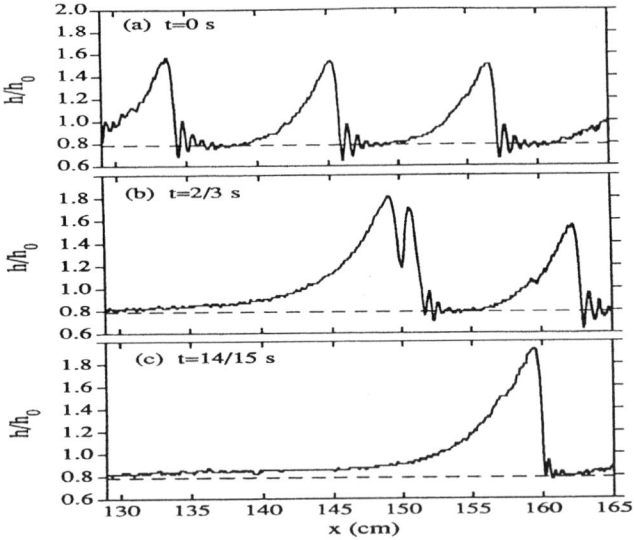

Figure 4.19: Snapshots of solitary waves (pulses) during their coalescence, The dashed line corresponds to substrate thickness. Liu and Gollub (1994).

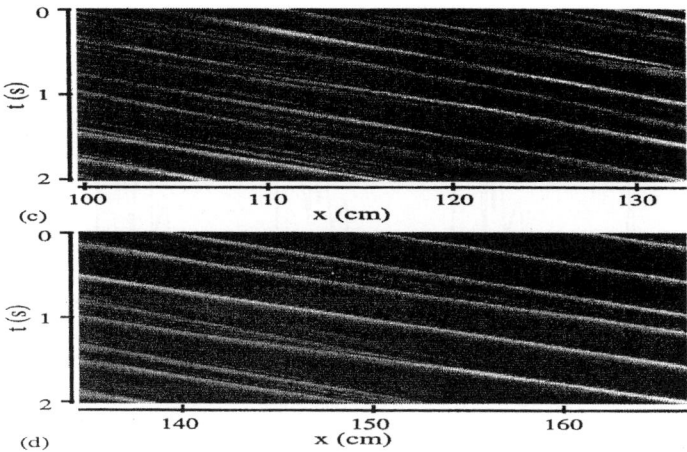

Figure 4.20: World-lines of pulse crests show pulse-pulse interaction and coalescence - the crest density decreases downstream, Liu and Gollub (1994).

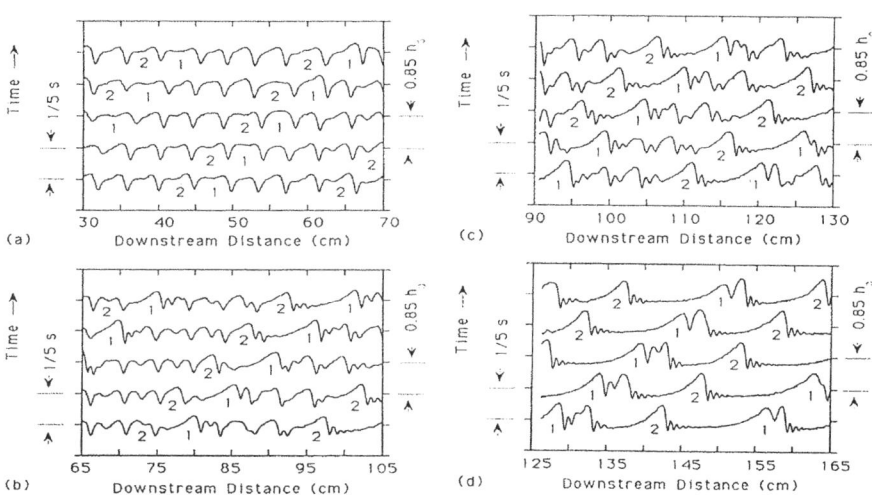

Figure 4.21: Development od solitary wave trains through the interaction of waves forced at both 5 and 6 Hz, ($R = 27$, $W = 33$ and $\beta = 8°$.), Liu and Gollub (1994).

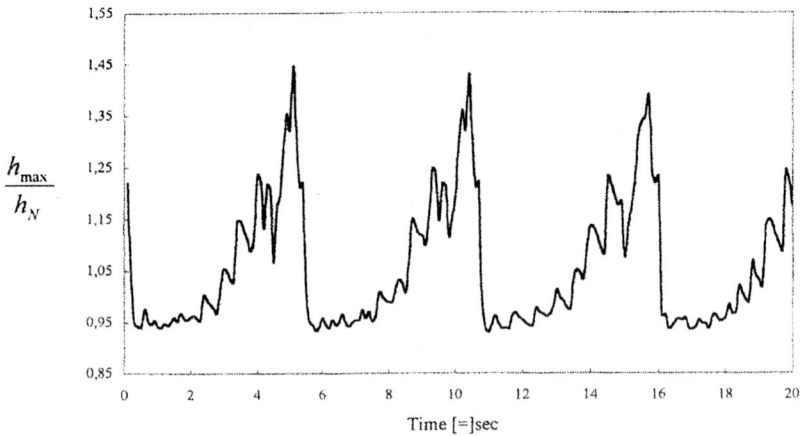

Figure 4.22: Evolution of the crest at fixed station of space, Vlachogiannis and Bontozoglou (private communication).

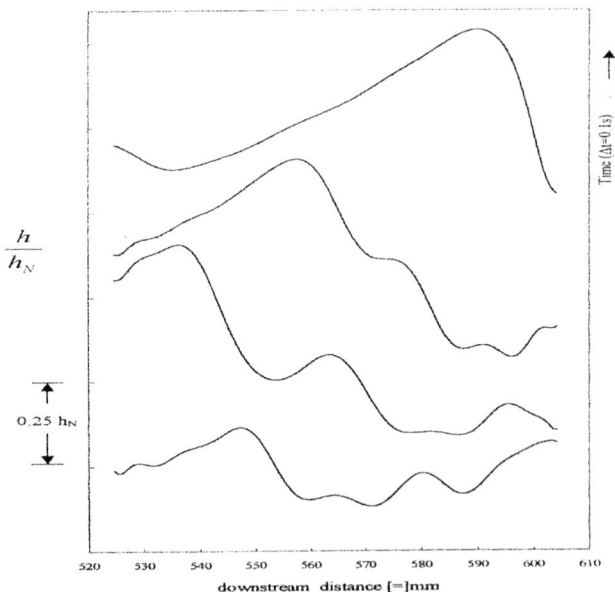

Figure 4.23: Solitary wave coalescence, $Re = 36, \theta = 5°$, Vlachogiannis and Bontozoglow (private communication).

4.2 Numerical formulation

The experimental measurements clearly show a rich variety of wave transitions in falling films. For natural waves excited by room disturbances, Figures 4.7 and 4.10 show evolution from small amplitude waves at inception to solitary waves, with a concomitant rise in the wave velocity in Figure 4.8. From Gollub's experiments in Figures 4.18 to 4.21, we understand such acceleration is because the solitary waves coalesce irreversibly to form larger ones with longer separation. There is also a final transition to three dimensional scallop waves in the "soliton gas" of Figure 4.5 and captured by periodic forcing in Figure 4.15. Although similar transitions from periodic waves to solitary waves are evident in Figures 4.18 to 4.21 for periodic inlet forcing, their transition locations and how the wave speed and wave texture change downstream are different from the naturally excited waves.

Of particular interest is the evolution beyond 10 cm in the naturally excited waves of Figure 4.8 and in the periodically forced experiment of Figure 4.18 to 4.21. The evolution in all these cases involve solitary waves that coalesce irreversibly to increase the average wavelength (pulse separation) and the average wave speed. Since all the solitary waves look self-similar, albeit with widely varying amplitude and speed, it is reasonable to speculate that the coarsening rate during this evolution possesses some generic scalings. For example, the downstream velocity evolution beyond 10 cm in Figure 4.8 seems to be self-similar for all Reynolds numbers and , with proper normalization, can be collapsed into a universal coarsening curve. This will in fact be demonstrated to be true in the later chapters. Since this self-similar stage occurs as soon as the solitary waves are created from the periodic waves, it represents the dominant part of wave evolution on a falling film. A simple universal description of its wave dynamics would hence be very valuable.

However, the waves are created by inlet disturbances and the difference in evolution between periodic forcing and naturally forcing suggest all wave dynamics, including the downstream solitary wave evolution described above, is sensitive to noise. The rate of solitary wave coalescence far downstream must still some how retain some memory of the inlet noise. Although detailed imaging, like that by Gollub's group and by Bontozoglow's group, has revealed the solitary wave coalescence mechanism for the coarsening dynamics, their experiments can only be done on short channels with periodic forcing. Their link to the inlet noise and the first few wave transitions is hence lost. Ideally, one would like to capture such evolution of naturally excited waves with a simple noise model–one that involves as few empirical parameters as possible. However, as is evident from the discussion of the experimental measurements, there is even a controversy on how such noise triggers the first inception of periodic waves on a vertical film. Even with Portalski and Clegg's definitive experiment, it is difficult to use their wave tracings to verify any noise model. One hence needs to develop an accurate simulation package that can faithfully reproduce all measured data and , more importantly, provide sufficient accuracy on the noise-driven wave evolution to allow us to decipher any intrinsic self-similarity and symmetry and derive simple scalings for this complex spatio-temporal wave dynamics.

Numerical scrutiny requires boundary conditions at both ends of the channel. We have examined two kinds of distributors at the inlet: a closed-channel distributor of height H found in many coating processes (Khesghi and Scriven, 1987), corresponding to the boundary condition at $x = 0$

$$u = \frac{6}{H}(\frac{y}{H} - \frac{y^2}{H^2})(1 + F(t)) \quad (4.1a)$$

$$h = H \neq 1 \quad (4.1b)$$

and an open-channel distributor with a flat film

$$u = 3(y - y^2/2)(1 + F(t)) \quad (4.2a)$$

$$h = 1 \quad (4.2b)$$

Since the y-dependent disturbances damp much faster than the mean-flow in a flat film or a closed channel, we have neglected all y-dependent disturbances and lumped the forcing within the flow rate fluctuation $F(t)$ which has a zero mean

$$<F> = \lim_{T \to \infty} \frac{1}{T} \int_0^T F(t)dt = 0 \quad (4.3)$$

We are hence imposing a constant mean flow-rate boundary condition with zero-mean fluctuations. As such, condition (3a) is also augmented by the following constraint that specifies H,

$$\int_0^H \frac{6}{H}(\frac{y}{H} - \frac{y^2}{H^2})dy = 1 \quad (4.4)$$

The boundary conditions at the exit at $x = L$ are more troublesome. As first formulated by Fasel et al. (1987) for other convectively unstable open flows, the best strategy here is to abandon any attempt to model reality but to formulate "soft" boundary conditions that minimize the generation of upstream propagating disturbances to enhance numerical stability. This is permitted since the convective nature of the instability stipulates a predominantly downstream evolution of disturbances and due to the length of the domain, the region affected by upstream propagation of disturbances generated by the exit is only a small fraction of the entire channel. One may then minimize the creation of exit noise to prevent numerical error and reduce computational domain. Roberts (1992) has formulated a systematic derivation of such boundary conditions using Invariant Manifold Theories. We shall, however, adopt only the main idea here which is to induce the film to decay to the flat-film solution at the exit and hence minimize the generation of noise. (This amounts to forcing the system to approach the stable manifold of the flat film in the language of the Invariant Manifold Theory.) Our approach parallels that in section 3.5 to determine the convective instability of the Nusslet flat film and to determine the spatial growth rate. Here, we shall identify and suppress the spatial modes that invade upstream into the channel from the exit.

In a non-stationary problem where the linear problem can be Fourier transformed in time, the spatial eigenvalues that dictate how the film approaches the flat-film solution vary from one frequency mode to another. To select a unique exit condition, we choose the zero frequency mode and forces the steady film to approach the flat-film solution at the exit. Linearizing (2.41) as in Chapter 2 about $h^* = 1$ and $u^* = 3(y - y^2/2)$, one obtains for the linear steady problem

$$\varphi_{yyy} - 15\delta\sigma[u^*\varphi_y - u_y^*\varphi] = 3\sigma^3 \tag{4.5a}$$

$$y = 1 \quad \varphi_{yy} = -3, \quad \varphi = 3/2 \tag{4.5b}$$

$$y = 0 \quad \varphi = \varphi_y = 0 \tag{4.5c}$$

where φ is the disturbed tangential velocity divided by σ, the spatial eigenvalue. The countable infinite number of spatial eigenvalues $\sigma_i(\delta)$ can be obtained with a routine solver and we find the first two to be a complex conjugate pair with positive real parts. These two modes grow in the positive x direction and must hence be suppressed. The other eigenvalues have negative real parts and we choose the largest one $\sigma(\delta)$, which is always real, and stipulate

$$x = L \quad \frac{\partial h}{\partial x} - \sigma(h - 1) = 0 \tag{4.6a}$$

$$\frac{\partial^2 h}{\partial x^2} - \sigma\frac{\partial h}{\partial x} = 0 \tag{4.6b}$$

such that the "unstable" complex pair is always suppressed. Note that when $\sigma = -\infty$, one obtains the usual Danckwerts exit boundary conditions $h = 1$ and $\frac{\partial h}{\partial x} = 0$ at the exit. However, (4.6) is a more precise since σ is typically finite.

We can describe the above approach in more physical terms. An exit boundary condition introduces a perturbation to travelling wave patterns. The disturbances triggered by the perturbed wave pattern can propagate upstream and corrupt the wave dynamics in the computational domain. They can also cause numerical instability. These undesirable effects occur because the true exit condition can never be simulated without expending considerable numerical effort to resolve the exit channel. The solution taken here is to force the time-averaged interface to a flat one at the exit. Since all disturbances triggered by a flat interface travel downstream, upstream disturbances propagation caused by the artificial condition is considerably minimized. It is not entirely removed since we are only dampening the zero-frequency mode. This concept of "soft" boundary conditions was used by Fasel et al. (1987) and Roberts (1992). The artificial flat film is seen at the exit end of the computational domain in Figure 4.24. Consequently, although the computational domain extends to $x = 1300$ in Figure 4.24, wave profiles are presented only upto $x = 1200$. All subsequent figures have been similarly truncated to remove the artificial flat film due to the "soft" exit boundary condition.

94

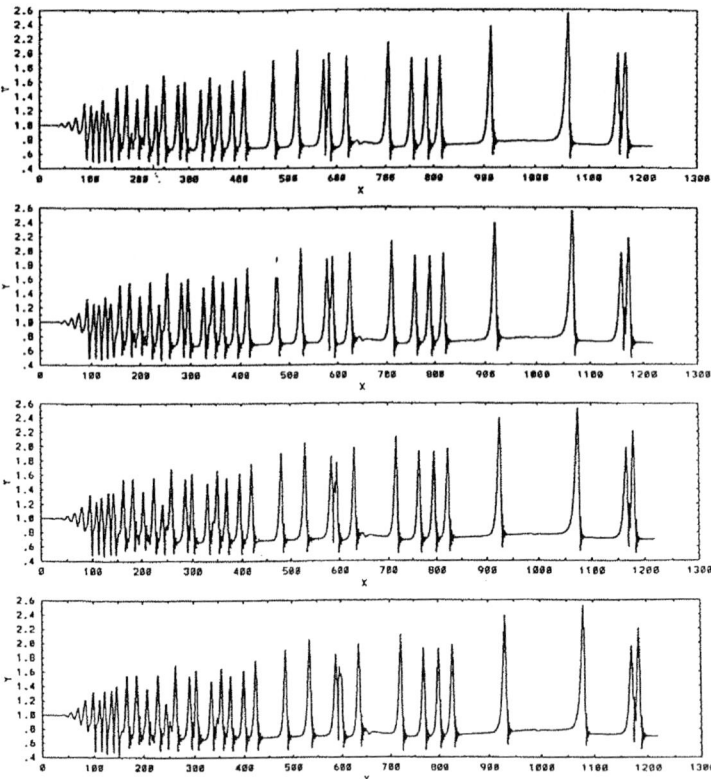

Figure 4.24: Theoretical wave profiles, $\Delta t = 1$, $\delta = 0.216$.

We shall use a domain transformation $\eta = y/h(x,t)$ to transform the film into a long rectangle between $x\epsilon(0,L)$ and $\eta\epsilon(0,1)$, as detailed in section 3.3 where we derived the Shkadov model. However, we shall utilize more modes in the Petrov-Galerkin expansion than the single parabolic basis used in the Shkadov model. For $R < 300$, the film is very thin and the profile is almost parabolic such that the number of modes N is never larger than 7. Nevertheless, convergence with respect to N was carefully verified in our numerical study.

The projected equations from (3.83) and (3.84) can be expressed in the following general form

$$a_{km}\frac{\partial A_m}{\partial t} + b_{kml}\frac{\partial}{\partial x}\frac{A_m A_l}{h} + c_{kml}\frac{A_l}{h}\frac{\partial A_m}{\partial x} = \frac{1}{5\delta}[h(\frac{\partial^3 h}{\partial x^3}+1)\delta_{k1}+d_{km}\frac{A_m}{3h^2}] \quad (4.7a)$$

$$\frac{\partial h}{\partial t} + a_{1m}\frac{\partial A_m}{\partial x} = 0 \quad (4.7b)$$

where

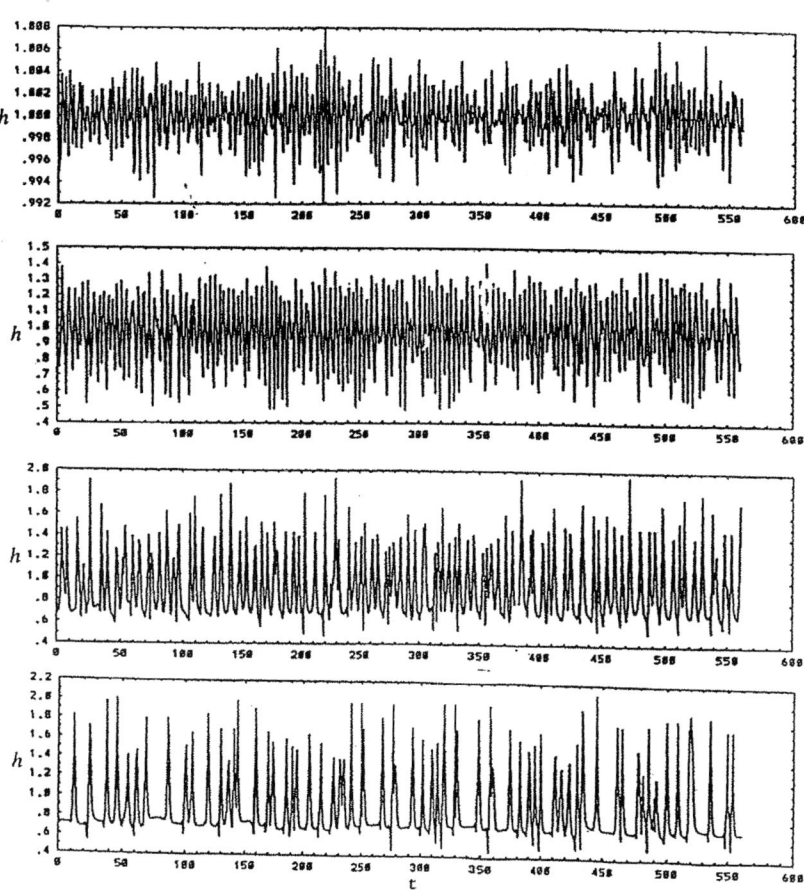

Figure 4.25: Time-records in different stations downstream: $x = 30, 100, 250$ and 500, $\delta = 0.216$.

$$\delta_{k1} = \begin{array}{ll} 0 & k \neq 1 \\ 1 & k = 1 \end{array} \qquad (4.8)$$

and the Einstein summation notation has been used. The tensors **a, b, c** and **d** are determined from the inner products $\int_0^1 \varphi_k \psi_m d\eta$, $\int_0^1 \varphi_k \psi_m \psi_l d\eta$ and $\int_0^1 \varphi_k \psi_m \overline{\psi}_l d\eta$. Because of the simplicity of the bases and weighting functions, they can actually be written explicitly without numerical integration. The next step is then to use a fourth-order finite-difference scheme in x for h in (4.7) due to the $\frac{\partial^3 h}{\partial x^3}$ term and a first order scheme in x for A_m. A fully implicit scheme is then used where the time derivative $\frac{\partial h}{\partial t}$, for example, is approximated by $(h(t_{n+1}) - h(t_n))\Delta t$ but all other quantities are evaluated at t_{n+1}. A set of nonlinear algebraic equations must then be solved at every time step and we use a Newton scheme which usually requires fewer than 4 iterations per time step. Solution of the linear equation by *LU* decomposition during each iteration is found to be unstable. Instead, we use a triangulation formulation by transforming the matrix **A** to a triangular one **T** with the transformation

$$\mathbf{T} = \mathbf{PA} \tag{4.9}$$

where **P** is a product of orthogonal matrices. The linear equation is then solved by backward recurrence during the Newton iteration scheme.

The final information needed is the specification of the zero-mean fluctuation $F(t)$ for the flow rate. We have chosen the following random-phase characterization of noise. Fourier transforming $F(t)$ in time and decomposing the Fourier coefficient into modulus and phase components,

$$\begin{aligned} F(t) &= \int_0^\infty \hat{F}(\omega) e^{-i\omega t} d\omega \\ &= \int_0^\infty |\hat{F}(\omega)| e^{i\theta(\omega) - i\omega t} d\omega \end{aligned} \tag{4.10}$$

where $\theta(\omega)$ is the phase of the complex amplitude $\hat{F}(\omega)$. We approximate (4.10) with M frequency units of width $\triangle\omega = \omega_*/M$ where ω_* is some high frequency cut-off. Hence,

$$F(t) \sim \sum_{k=1}^{M} |\hat{F}(\omega)| e^{-ik\triangle\omega t + i\theta(\omega_k)} \triangle\omega \tag{4.11}$$

where $\omega_k = k\triangle\omega$ and we have omitted the amplitudes of all modes with frequencies larger than ω_*. The phase $\theta_k = \theta(\omega_k)$ is taken from a random number generator from the range $\theta_k \epsilon [0, 2\pi]$ and $|\hat{F}(\omega)|$ can be arbitrarily specified. We have found that most experimental data can be simulated with a cut-off frequency $\omega_* = 2\omega_0$, where ω_0 is the wave-frequency of the neutral mode calculated in chapters 2 and 3. It is approximately $\sqrt{18\delta}$ in dimensionless form which is of the order of $10Hz$ for water under most conditions. We also impose a piece-wise constant forcing spectrum of

$$\begin{aligned} |\hat{F}(\omega)| &= F_0 \quad \omega < \omega_* \\ &= 0 \quad \omega > \omega_* \end{aligned} \tag{4.12}$$

With $M = 1000$ such that the results are independent of M. Hence, the noise is fully characterized by the single index F_0 in this representation if the default cut-off frequency $\omega_* = 2\omega_0$ is used. A program GALA has been prepared in a package form to allow easy usage with arbitrary assignment of ω_* and $|\hat{F}(\omega)|$. Each run of duration $t = M/2\omega_0 = 500/\omega$ and channel length $L = 2000$, corresponding to roughly 2 meters, requires about 400 to 500 hours of CPU time on a Convex computer. Details about the numerical code are presented in the thesis by Kalaidin (1996).

4.3 Numerical simulation of noise-driven wave transitions

A typical run with the open-channel distributor at $\delta = 0.216, \omega_* = 2\omega_0 = 2\times 1.972$ and $F_0 = 1\times 10^{-4}$ is shown in Figure 4.24 which is qualitatively the same an in Portalski and Clegg (1972) experiments, see Figure 4.10. The snapshots in Figure 4.24 show a wave inception point of 50 units, corresponding to 5 cm for water at $R = 18.75$. (A good rule of thumb for the present conditions is that one dimensionless unit in the x-direction corresponds to 0.1 cm.) We shall later examine this inception point more carefully and resolve an earlier controversy–whether the critical Reynolds number is indeed zero for a vertical film, as the linear theory predicts. For now, wave inception is simply when waves are visually detectable at the scales of Figure 4.24. The waves beyond inception quickly develop into pulse-like structures at 100 units and all subsequent pulses have qualitative similar shapes with a steep front preceded by small bow waves (Figure 4.25). The amplitude and spacing between pulses seem to grow slowly within the 1300 unit domain (130 cm for water at $R = 18.75$). As discussed earlier, this gradual increase in the average spacing is caused by an irreversible coalescence of adjacent pulses to from a single pulse. Two coalescence events are seen at $x \sim 600$ and 1200. Figure 4.26 depicts the time-averaged separation, pulse speed, pulse amplitude and substrate thickness as functions of downstream distance. The coarsening evolution of the wave texture is clearly evident. The time-average speed shows a unique behavior that was observed much earlier by Stainthorp and Allen (1965) whose data in Figures 4.7 and 4.8 remain the most accurate recording of how wave velocity and wave profile evolve downstream. It drops precipitously from the onset wave speed of about three times the average film velocity u_0 in the inception region to $2.2u_0$ at the end of the inception region at $x = 50$. A slight decrease in the wavelength λ is also evident here at about 10 cm. This will be shown to be consistent with our result in chapter 5 that the finite-amplitude monochromatic wave selected after the inception region is slower and shorter than its small-amplitude precursor within the inception region.

However, these monochromatic waves quickly lose their periodic structure soon after they exit the inception region and trigger the formation of pulses. We find from our simulations that the only exceptions to this scenario of secondary pulse formation occur for the low flow rate range of δ less than about 0.03, corresponding to R less than 2.0 for water. Under these extreme conditions, the inception region is extremely long and the wave structure of the saturated waves beyond it are of small-amplitude and irregular shape. They resemble localized saturated monochromatic waves. This near-critical exception to the generic scenario of pulse generation lies in the region where the Kuramoto-Sivashinsky (KS) equation or its generalization, the generalized Kuramoto-Sivashinsky equation, is a valid model (chapter 3). In Chapter 7, we shall show that the reason for this deviant behavior is that the pulse travelling wave solution is "convectively" unstable, in a sense analogous to the classical convective stability theory for a

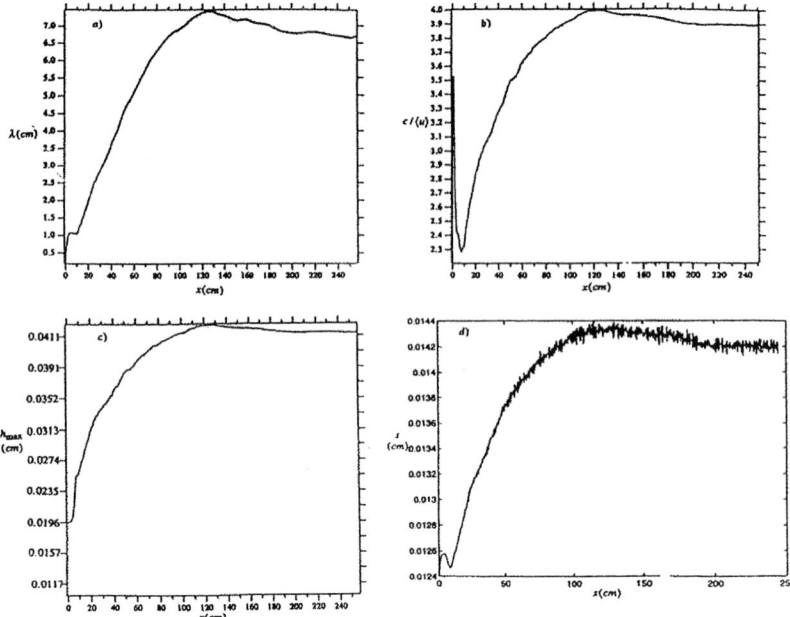

Figure 4.26: Time-averaged 1) separation distance, b) normalized wave velocity, c) thickness maximum and d) substrate thickness for $R = 18.75$, water.

flat film in section 3.5, below a critical δ and hence will never evolve from the inception region. Instead, one sees the highly irregular and small-amplitude KS chaos that has been of some interest recently. However, at higher flow rates when inertia is significant, the distinct large-amplitude pulses are stable. They appear beyond the inception region as seen in Figure 4.24 at higher values of δ and their dynamics is quite generic. The spacing between these pulses is still denoted λ and it is observed to increase downstream in Figure 4.25 and 4.26. As the pulses spread out, their average speed also increases downstream. Both the coarsening and the acceleration seem to evolve in a roughly linear manner with respect to x. At $x \sim 800$ units ($\sim 80cm$), both the spacing and the speed have saturated at their equilibrium values. The average amplitude of the pulses also increases downstream until it also saturates at 800 units. The substrate thickness s on the other hand drops by 50% within the inception region since most of the liquid is drained into the pulses. As a result, a gradual upward drift of the average substrate thickness downstream, in parallel with the coarsening and acceleration, can be detected. The final saturation at 800 units seems to occur because the flat substrates between pulses has increased its width to such an extent that new pulses begin to nucleate on them. When this balances the coalescence rate, the pulse density and hence, pulse speed and substrate thickness, reach statistically stationary values.

Before we explore the statistics of this evolution in more detail, we shall

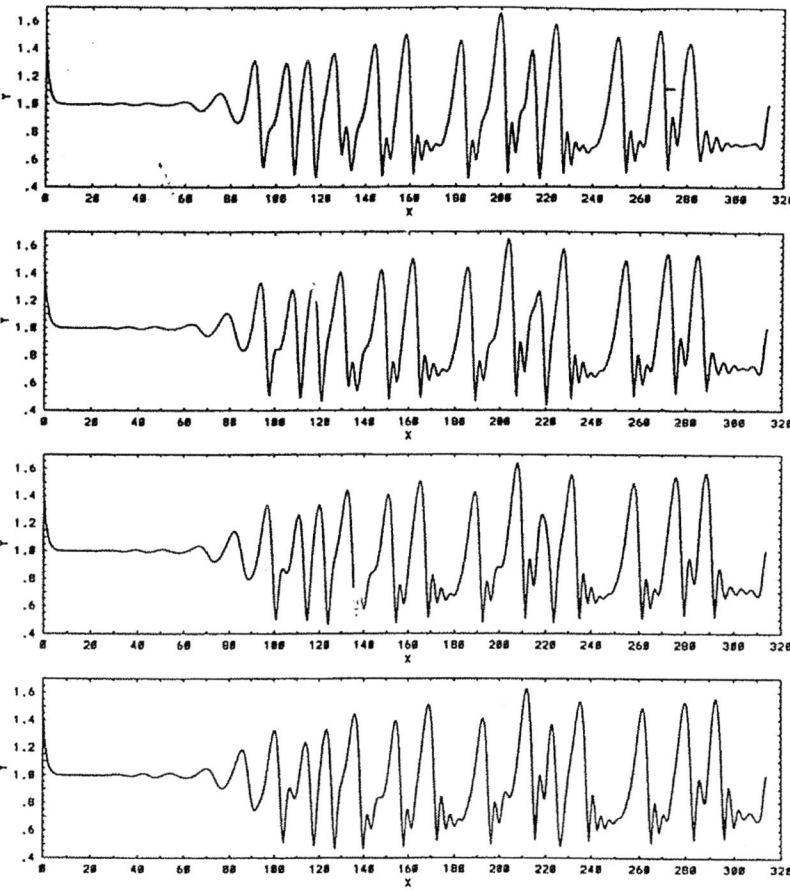

Figure 4.27: Closed distributor, $\Delta t = 1$, $H = 1.5$ and $\delta = 0.216$.

examine several other factors. In Figure 4.27, the closed channel distributor is used for the same conditions as in Figures 4.24 and 4.26 with $H = 1.5$. It is seen that the film height quickly drops from 1.5 of the channel height to 1.0 and since the closed channel flow is more stable, the inception region is lengthened somewhat compared to Figure 4.24. However, subsequent evolution is quite analogous to the open-channel distributor. We hence conclude that, for small-amplitude forcing, the distributor has little effect on the wave dynamics. We also experimented with narrow-banded forcing. As seem in Figure 4.28 for $\omega_* = 0.2\omega_0$ and $0.1\omega_0$, the wave dynamics is now quite different with more regularly spaced pulses due to the narrow-banded noise spectrum. For $\omega_* = 0.1\omega_0$, the spacing is so large that now pulses begin to nucleate between the pulses right after the inception region instead of evolving directly from the inlet disturbance. One nucleating pulse is seen at $x = 120$. Such regular wave dynamics is never observed in realistic noise-driven experiments although Alekseenko, Nakoryakov

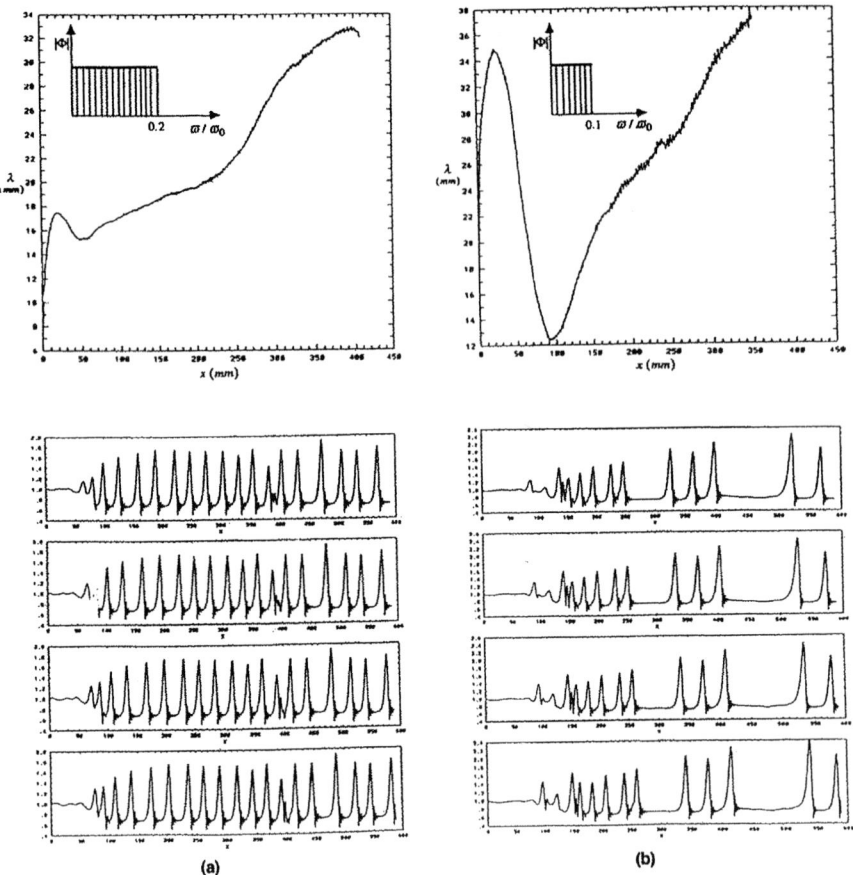

Figure 4.28: Wave behaviour at harrow-banded forcing a) $\omega_* = 0.2\omega_0$ and b) $\omega_* = 0.1\omega_0$.

and Pokusaev (1985), Liu and Gollub (1994) and Liu et al. (1993,1995) have artificially introduced periodic forcing at the inlet to entrain periodically spaced pulses seen in Figure 4.28 (see Figure 4.17 and 4.21). With $\omega_* \sim \omega_0$ or higher, the dynamics is again insensitive to ω_* and is only a function of δ and F_0.

In the case of broadbanded forcing in Figure 4.24, the linear filtering mechanism that selects a single monochromatic wave is very effective eventhough a large band of equal-amplitude disturbance is in play. This is evident in the wave tracing of Figure 4.25 at $x = 100$. Eventhough some envelope modulation is present, the waves are almost monochromatic. Compare this to the irregular tracing at $x = 30$ where the random noise is still felt and to the pulse-like shapes at $x = 500$. To make this selection process more precise, we evaluate the spatial eigenvalues $\alpha_k(\omega)$ at a given frequency with the theory of section 3.5 but now with the full BL equation. There is an infinite number of such spatial eigenvalues from the linearized equation and the film thickness evolves as

$$h - 1 = \int_0^\infty \sum_{k=1}^\infty \phi_k(\omega) e^{i[\alpha_k(\omega)x - \omega t]} d\omega \qquad (4.13)$$

and $\phi_k(\omega)$ is the unspecified Fourier coefficient. Of the infinite number of spatial eigenvalues, only one, with growth rate α_1, is physically appropriate since it propagates downstream with a positive spatial growth rate: $-\alpha_1^i = -Im\{\alpha_1\} > 0$ and $\alpha_1^r = Re\{\alpha_1\} > 0$ for $\omega \epsilon (0, \omega_0)$. For $\delta = 0.216$, this mode exhibits a maximum in $-\alpha_1^i$ at $\omega_m \sim 0.290 \, \omega_0$ where the growth rate is $-\alpha_{1m}^i = 0.120$. Omitting all modes except α_1 in (19), it is clear that the maximum and minimum of $h - 1$ correspond to when the real part of $e^{i[\Theta - \omega t + \alpha_1^r x]}$ is equal to ± 1, respectively, where Θ is the phase of $\phi_1(\omega)$. We hence focus on the behavior of $\int_0^\infty |\phi_1(\omega)| e^{-\alpha_1^i(\omega)x} d\omega$ in the limit of large x which can be estimated from a stationary phase approximation to be

$$\int_0^\infty |\phi_1(\omega)| e^{-\alpha_1^i(\omega)x} d\omega \sim |\phi_1(\omega_m)| \sqrt{\frac{2\pi}{x \frac{d^2\alpha_1^i}{d\omega^2}(\omega_m)}} e^{-\alpha_{1m}^i x} \qquad (4.14)$$

As such, the selected wave mode is $\omega_m \sim 0.290 \, \omega_0$ and its growth rate for both $|h_{max} - 1|$ and $|h_{min} - 1|$ is exponential with an exponent of $-\alpha_{1m}^i = 0.120$. When $|h_{max} - 1|$ and $|h_{min} - 1|$ are tracked downstream in the numerical experiment of Figure 4.24, ω is found to be $0.291 \, \omega_0$ with a spatial growth exponent of 0.108 – very close to the linear theory and fully verifies the filtering mechanism that yields a nearly monochromatic wave within the inception region. The evolution of $|h_{max} - 1|$ is shown in Figure 4.26. These selected monochromated waves are quickly saturated, decelerated and compressed by the weakly nonlinear mechanism just beyond the inception region at $x = 50$.

The length of the inception region is quite sensitive to F_0. In Figure 4.29, we show snapshots of the waves at various values of R for water with $F_0 = 5.0 \times 10^{-5}$. It is clear that, although the critical R is zero for the vertical film, an appreciable inception region is only apparent at R in excess of 1.5, the effective critical Reynolds number. It is also clear from the snapshot of Figure 4.29 for $4R = 21$, corresponding to $\delta = 0.05$, that pulses are not observed within the computation domain. Instead, small-amplitude KS chaos is observed. This is actually slightly beyond the transition δ value and after a long chaotic transient, regular pulses are still generated downstream beyond the computation domain shown in Figure 4.29. At lower δ values ($\delta < 0.03$), the KS chaos persists indefinitely downstream. For higher R values, distinct large-amplitude pulses are generated after the inception region as seen in Figure 4.29. We use Stainthorp and Allen's data in Figure 4.6 and their criterion that the inception line is defined as when the wave amplitude first exceeds 10^{-4} cm to determine the beginning of the inception region for $F_0 = 10^{-4}$ and 10^{-5} for several values R for water. As seen in Figure 4.29, the inception length drops rapidly from extremely high values around 1 meter to about 10cm for realistic values of R at both values of F_0 before it rises slowly again. It is also evident that Stainthorp

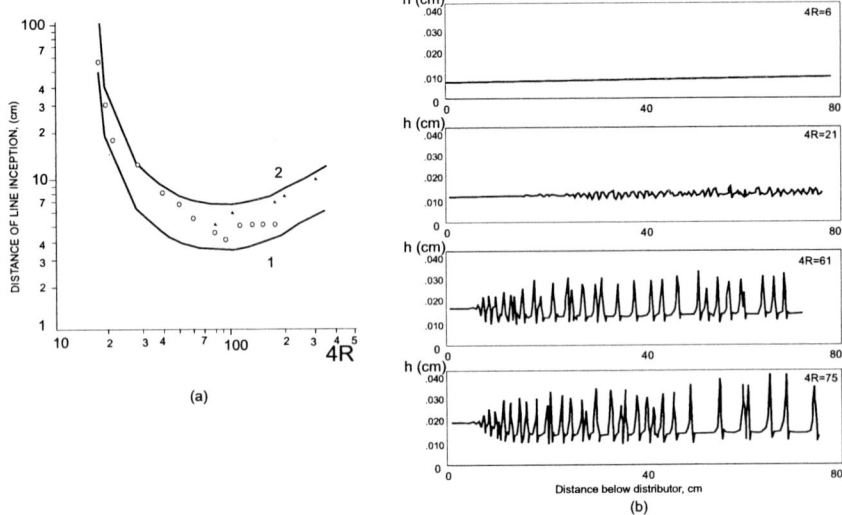

Figure 4.29: (a) Inception line from simulations at $F_0 = 10^{-4}$ (1) and 10^{-5} (2) against experimental data of Stainthorp and Allen (1965) (liquids is water); (b) the snapshots for the indicated R values at $F_0 = 0.5 * 10^{-4}$.

and Allen's inception line can be accurately reproduced with $F_0 = 5 \times 10^{-5}$, which is the value we shall use to reproduce their downstream data.

With F_0 determined from the inception line, the dynamics is fully specified for Stainthorp and Allen's system if our random phase description of the noise is accurate. In Figure 4.30, our simulated wave speeds far beyond the inception region are seen to be in quantitative agreement with their data for water at 3 values of R in a 30 cm long channel. The deceleration from the linear phase speed $3u_0$ by the weakly nonlinear mechanism followed by an acceleration after the pulses are formed are clearly evident. Stainthorp and Allen's channel is too short to reach the saturation region at 80 cm in Figure 4.26. More conclusive demonstration is shown in Figure 4.31, where their measured wave tracings for $4R = 75$ (from the original experimental records of Figure 4.7) are compared directly to the simulated results. The average wave structure and spacing are faithfully reproduced. This fully verifies the accuracy of the BL equation and the ability to capture even some details of the wave dynamics and all the essential statistics with just two parameters - δ and F_0. It should be noted that the waves measured by Stainthorp and Allen in Figure 4.7 are inside a vertical tube and they demonstrate appreciable transverse variation and yet their dynamics and wave shapes are still captured by the two-dimensional simulation. This is consistent with the observation that, until the transverse variation is so pronounced that it begins to pinch off adjacent crests to form "scallop waves", each cross-section across the crest still behaves as a two-dimensional pulse with nearly two-dimensional wave dynamics.

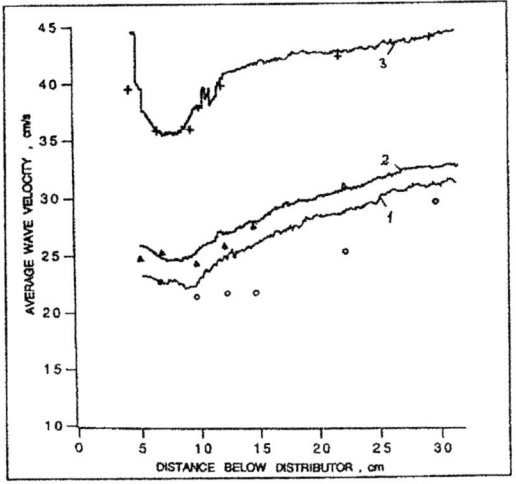

Figure 4.30: Simulated downstream evolution of time-averaged speed for (1) $4R = 180$ with $F_0 = 5*10^{-5}$ against Stainthorp and Allen's data for water represented by the symbols.

4.4 Pulse formation and coarsening

We examine the pulse formation and coarsening dynamics, evident in Figures 4.24 and 4.25, with more numerical resolution here. Wave tracings taken at various downstream stations of a typical run are shown in Figure 4.32 for $\delta = 0.1$ and $F_0 = 7*10^{-5}$. By $x = 200$, about 14 cm for water, the inlet white noise has been filtered into a modulated wave field whose fundamental frequency is limited to a narrow band around $\omega = \omega_m$. This is evident in the Fourier spectrum (square root of the power spectrum) of the corresponding wave tracing in Figure 4.33. Also seen in the Fourier spectrum at $x = 200$ are slight overtone and zero-frequency (zero-mode) peaks. These secondary peaks are absent prior to $x = 200$ and hence result from weakly nonlinear interaction with the primary band.

The modulation frequency is specified by the secondary zero-frequency band. By $x = 280$, this secondary band exhibits a distinct maximum at $\omega = \Delta \sim 0.3$ (see the second frame of Figure 4.33). This is the characteristic modulation frequency that is a key to all subsequent wave dynamics. Within a short interval between $x = 225$ and 300 (a 5 cm interval for water), the sinusoidal waves at inception have evolved into solitary pulses (see Figure 4.32). The narrow band of primary harmonics now expand dramatically, as seen in Figure 4.33, to reflect the large harmonic content of each pulse. A second peak ω_s appears below ω_m within this wide band and becomes the lone maximum beyond $x = 440$. This ω_s represents the average pulse frequency. As the pulse density decreases downstream, the pulse frequency ω_s decreases.

The mechanism behind this downstream decrease in pulse density is apparent

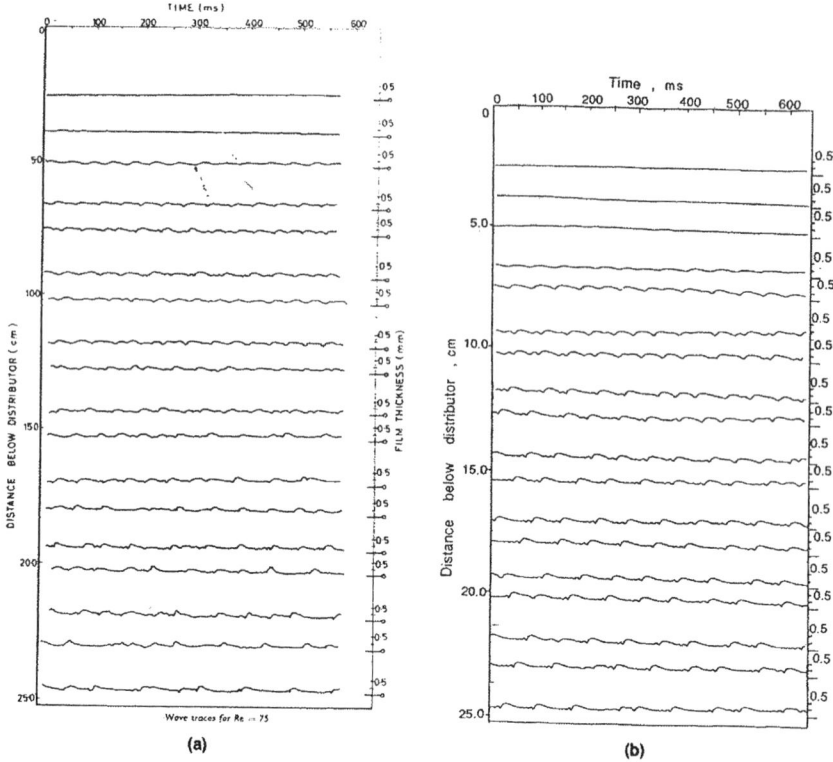

Figure 4.31: (a) Wave tracings of Stainthorp and Allen for $4R = 75$; (b) simulated tracings with $F = 5.0 * 10^{-5}$ and $4R = 75$ for water ($\gamma = 2850$).

from the world lines of Figure 4.34, analogous to the experimental ones measured by Gollub's group for periodically forced waves in Figure 4.20 but with the time axis rotated. These lines track the wave peaks of all waves, sinusoidal waves and pulses, in the $t = x$ plane. Various intersections are evident as steeper world lines terminate at less steep ones. These correspond to coalescence events that occur when slower pulses (steeper world lines) are captured by faster ones (less steep world lines) from behind. Nearly periodic spacing for $x < 200$ corresponds to the filtered band of sinusoidal waves. Some modulation is evident near $x = 200$ and the modulation nodes are seen to produce the "excited" pulses that capture the equilibrium pulses downstream. These excited pulses survive till the exit of the channel, after all the equilibrium pulses between the modulation nodes have been captured.

By determining the average spacing in t of the world lines at a given x, we obtain the average wave period $<t>$ at various stations downstream, as shown in Figure 4.35. Except at the onset of coarsening and at the end when the wave period reaches a final equilibrium value, the wave period is seen to increase downstream, with an almost linear coarsening rate with respect to x.

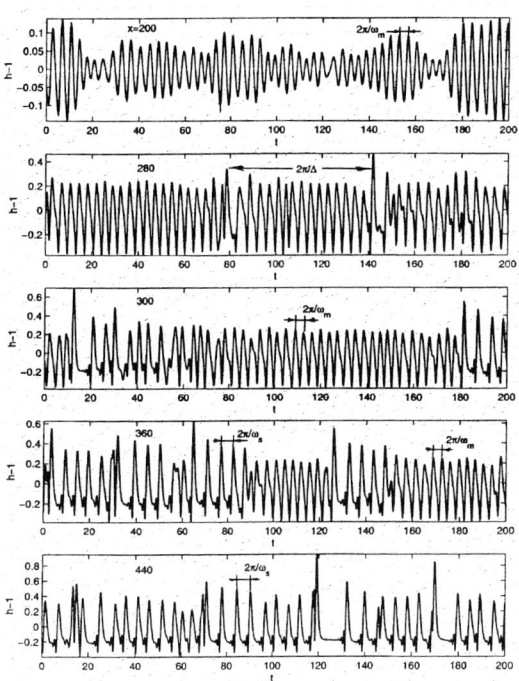

Figure 4.32: Wave tracings measured at stations $x = 200, 280, 300, 360$ and 440 for the noise-driven wave dynamics at $\delta = 0.1$. The initial waves are modulated primary waves of frequency ω_m. Excited pulses form at the dominant modulation period $2\pi/\Delta$ and the pulse separation $2\pi/\omega_s$ increases downstream as the excited pulses eliminate pulses by coalescence. (The specific modulation period shown in the second frame for $x = 280$ is $2\pi/\Delta = 60$ but the average at the location shown in the power spectra of Figure 4.33 is $2\pi/\Delta \sim 20$.)

The noise amplitude ϵ merely determines the onset location of the coarsening but not its rate nor the final wave period $<t>_\infty$. As is consistent with our earlier observations that the excited pulses are created at the modulation nodes and that they are the only surviving pulses at the end of the wave coarsening interval, we find the final wave period to be close to $2\pi/\Delta \sim 20$. (The value from our simulations in Figure 4.35 is 18.)

In Figure 4.36, we present the wave period coarsening dynamics for a range of δ at the same noise amplitude F_0. All coarsening intervals show the same slope. Higher δ values (upto 0.6) are more difficult to simulate and the entire $<t>$ profiles cannot be captured. However, their coarsening interval can be seen to have the same slope. Quite curiously, although δ varies over nearly an order of magnitude (water Reynolds number from 10 to 60), their coarsening rates are

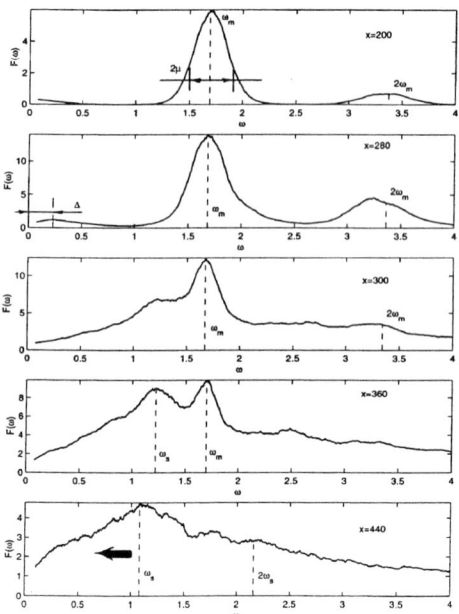

Figure 4.33: Power spectra of the wave tracings in Figure 4.32 with characteristic frequencies ω_m (with standard deviation σ), Δ and ω_s.

nearly constant eventhough their final wave periods $<t>_\infty$ are different. This universal coarsening rate will be a focus of our theoretical and modeling efforts.

This constant coarsening rate breaks down below $\delta = 0.1$ and beyond $\delta = 0.6$. Our visual examination of the simulation displays suggest that, at low δ, the excited pulses do not drain the excess mass after coalescence. The excited pulses hence accelerate downstream, yielding a power-law coarsening range (with an algebraic power larger then unity) instead of a linear one. For δ in excess of 0.6, inertial forces dominate over capillary forces and the excited pulses no only drain their excess mass readily, they also often break up and form pulses smaller than the equilibrium ones. This consumption of excited pulses reduces the coarsening rate below linear with an algebraic power less than unit. It is in the range of 0.1. and 0.6 that the excited pulses remain constant in speed and number, giving rise to universal coarsening dynamics. These invariant properties also render their modeling more feasible. It is fortuitous that this range falls into the most practical conditions.

To further ascertain that the characteristic modulation frequency Δ specifies the density of the excited pulses, the coarsening rate and the final wave period, we introduce periodic forcing with two frequencies - a specific fundamental frequency ω_m and a specific modulation frequency Δ. The primary band and the

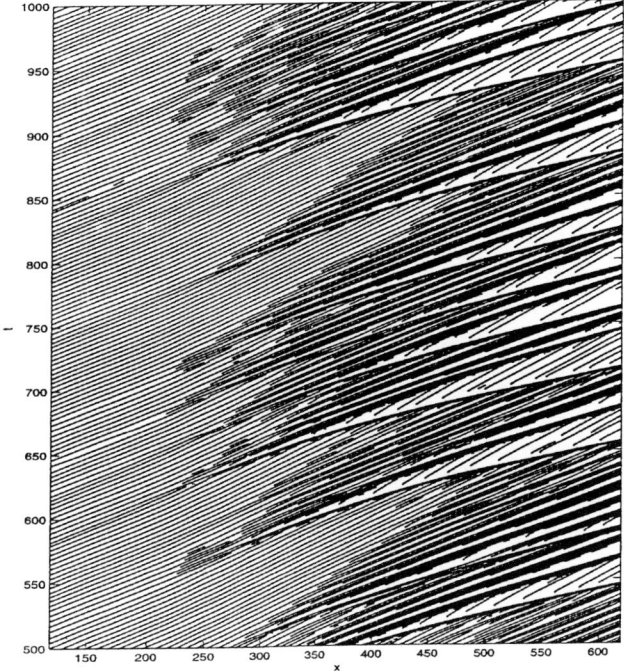

Figure 4.34: World lines tracking the wave crests of noise-driven wave dynamics in the $t - x$ plane for $\delta = 0.1$.

secondary band hence both become delta functions. The world lines shown in Figure 4.37 clearly reflect these periodicities. Eventhough considerable stretching of the steep world lines occur after each coalescence, corresponding to readjustment of the equilibrium pulses between the excited ones, the excited pulses remain oblivious to such adjustment. They propagate at constant speed and eventually capture all equilibrium pulses. Each surviving pulse at the end can be traced back to a node in the periodic modulation.

Finally, it should be emphasized that the rich wave dynamics on the film are driven by the pulse coherent structures. A naive interpretation of the last power spectrum in Figure 4.33, measured at $x = 440$, is that it resembles a "turbulent" spectrum with a power-law "inertial" tail. The implicit assumption is that the phases of the wave Fourier harmonics are random and the power-law decay is due to a weakly nonlinear energy cascade to higher wavenumbers. This interpretation would be incorrect. As seen in the wave tracing of $x = 440$ in Figure 4.32, the wave harmonics phase lock and synchronize to form individual solitary pulses. In fact, the power spectrum of each pulse resembles that of the aggregate in Figure 4.33. The absence of phase information in the wave spectrum can hence be misleading and a turbulent spectral theory for thin-film wave dynamics would be misguided. One needs to capture the pulse

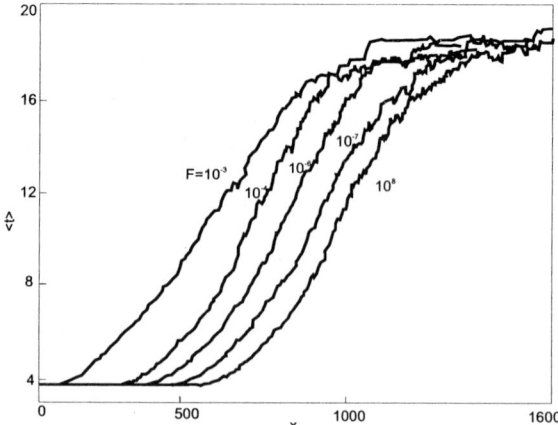

Figure 4.35: The average wave/pulse period at each station x for noise-driven wave dynamics at (a) $\delta = 0.1$ and noise amplitude between 10^{-3} and 10^{-8}.

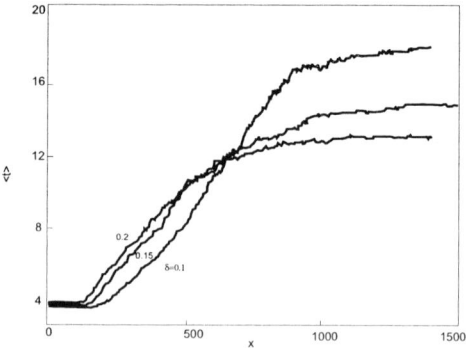

Figure 4.36: The average wave/pulse period at each station x for noise-driven wave dynamics at $\epsilon = 10^{-3}$ and three different values of δ.

dynamics to decipher the rich wave dynamics. This coherent structure approach is fundamental to all our work on falling-film waves.

To ensure that each pulse is a quasi-stationary travelling wave which does not change its speed locally, we normalize the time-averaged local film thickness at every station to unity and rescale the time-averaged pulse speed accordingly by the new Nusselt velocity. The renormalized pulse speed as a function of the renormalized δ at every downstream position is shown in Figure 4.38. We have also plotted the solitary wave speed of the Shkadov model on a unit substrate thicness, which will be constructed in Chapter 5 and is already presented in Figure 3.5. It is seen that, beyond $x = 50$, the average pulse is a quasi-stationary travelling wave. To be sure, there is a wide variance the coalescence dynamics require some larger pulses. This wide variance from the recorded statistics is shown Figure 4.39. The spread is skewed towards faster excited pulses.

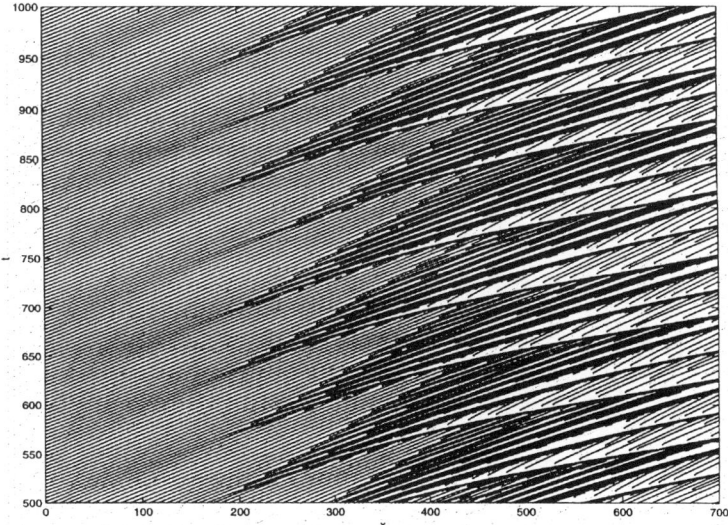

Figure 4.37: World lines with biperiodic forcing at the inlet with a fundamental of $\omega_m = 1.744$ and a modulation frequency Δ of 0.1. The dominant excited pulses are clearly created at the modulation minimum at a frequency of Δ.

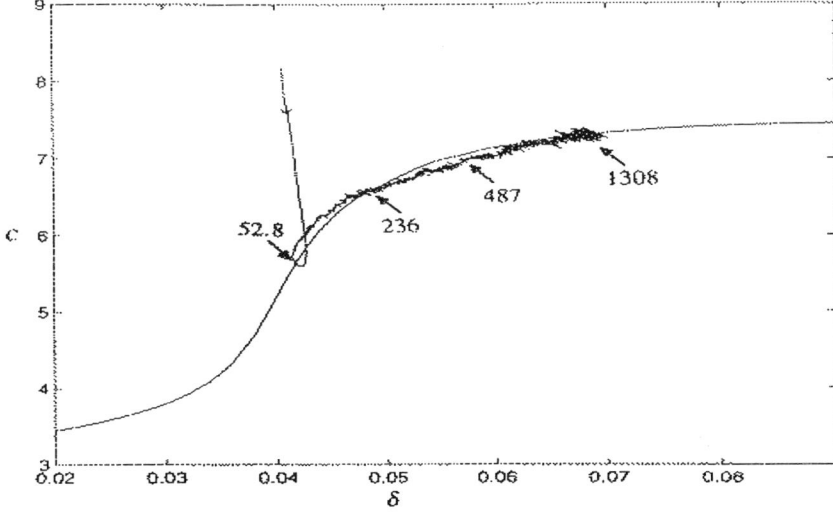

Figure 4.38: Evolution of Figure 4.24 lies along the pulse solution branch $c(\delta)$ after local normalization. The numbers are downstream position x.

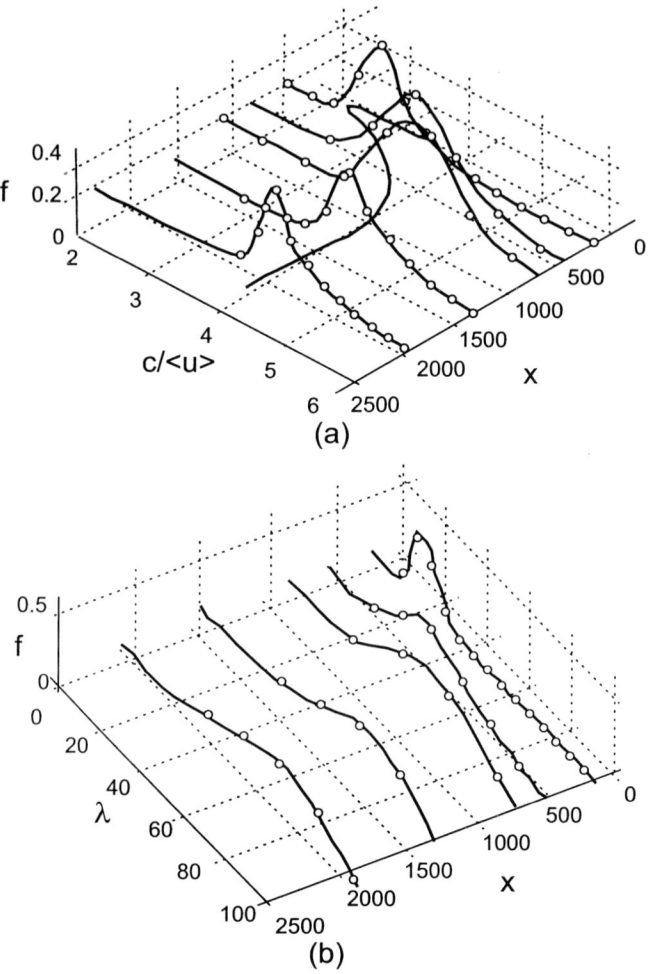

Figure 4.39: Pulse speed and spacing distributions at various spatial stations for conditions of Figure 4.24.

CHAPTER 5

Periodic and Solitary Wave Families

It is clear from the simulations of Chapter 4 that the primary filtering, governed by the linear stability theory of Chapter 2, transforms the inlet white noise into a very narrow band of harmonics. This band of waves resemble a fundamental periodic wave with wavenumber α_m and ω_m of (3.121) with a modulating envelope (see the first frames of Figures 4.32 and 4.33). In some experiments Kapitza and Kapitza (1948, 1949), Alekseenko, Nakoryakov and Pokusaev (1985) and Liu and Gollub (1994), periodic forcing is introduced at the inlet such that the perfectly periodic wave field is observed downstream from the inlet. Consequently, a periodic wave regime exists for both types of inlet forcing. In both cases, these primary periodic waves can destabilize to a modulation instability (Figure 4.33) or to a subharmonic instability (Liu and Gollub, 1994) to form either a solitary wave or another periodic wave which is a subharmonic of the primary. In this chapter, we shall first construct all periodic waves, including the solitary waves with infinite wavelength. Their stability will then be analyzed in the next chapter.

5.1 Main properties of weakly nonlinear waves in an active/dissipative medium.

In Chapter 3 we derived a hierarchy of model equations for different film flows. The simplest equation was the generalized Kuramoto-Sivashinsky (gKS) or Kawahara equation for a medium that both supplies and dissipates energy:

$$\frac{\partial H}{\partial \tau} + 4H\frac{\partial H}{\partial X} + \frac{\partial^2 H}{\partial X^2} + \delta\frac{\partial^3 H}{\partial X^3} + \frac{\partial^4 H}{\partial X^4} = 0 \tag{5.1}$$

Here we shall consider the important KS equation case of $\delta = 0$,

$$\frac{\partial H}{\partial \tau} + 4H\frac{\partial H}{\partial X} + \frac{\partial^2 H}{\partial X^2} + \frac{\partial^4 H}{\partial X^4} = 0 \tag{5.1a}$$

This equation is a generic weakly non-linear model of many physical pattern formation phenomena with a longwave instability. They include our wavy liquid layer falling under gravity down a vertical and inclined planes; hydrodynamic instability of flame fronts,concentration waves of Belousov-Zhabotinsky chemical reactions far from thermodynamic equilibrium , thermocapillary convection in horizontal thin layers, wavy processes in plasma etc. The equation has become the generic equation for active/dissipative systems with a longwave instability as the KdV equation for conservative ones. As a testimony to its generic relevance, it was derived independently for chemical reactions far from thermodynamic equilibrium by Kuramoto and Tsuzuki (1976), for thin liquid films by Nepomnyashy (1974) and for unstable flame fronts by Michelson and Sivashinsky (1977) and Sivashinsky (1977). As the KS equation also captures falling-film waves under some extreme conditions specified in Chapter 3, we shall classify all of its periodic and solitary travelling wave solutions. The relative simplicity of this equation makes it possible to analyze it in detail by using the methods of bifurcation theory and dynamical systems theory. Some of the more important wave families survive outside the regions where the KS equation is valid. In fact, our construction of the wave families of the Shkadov model and the BL equation at the end of this chapter will be guided by the analysis of the KS waves. These latter waves can then be explicitly compared to measured wave profiles.

We consider in this section some general properties of (5.1) that will be relevant to the structure of the wave families and their stability. Solutions of this KS equation are invariant with respect to transformations:

$$H(X, \tau) \to H(X + X_0, \tau + \tau_0) \tag{5.2}$$

$$H(X, \tau) \to -H(-X, \tau) \tag{5.3}$$

$$H(X, \tau) \to H(X - 4\chi\tau, \tau) + \chi \tag{5.4}$$

where X_0, τ_0 and χ are constants. These symmetries are (respectively) translation, reversibility and mass conservation.

Equation (5.1) has a trivial solution $H = const$ which , invoking the mass conservation symmetry (5.4), can be taken as: $H(X) \equiv 0$ without loss of generality. Any non-zero trivial solution can be mapped into this zero trivial solution through (5.4). For small sinusoidal perturbations of the zero trivial solution, $\varepsilon \exp[is(X - c\tau)]$, $\varepsilon \to 0$, a characteristic dispersion equation can be easily derived with the complex phase velocity $c = is(1 - s^2)$.

The trivial solution is stable for $s > 1$ and unstable for $s < 1$. When $s = s^{max} = \sqrt{2}/2 \approx 0.7071$, sc_i the disturbance amplification rate (growth rate) is at its maximum value of $sc_i^{max} = 1/4$. The real part of the phase velocity vanishes, $c_r = 0$, i.e. linear dispersion is absent.

A nonlinear extension of the linear stability theory can be made by multiplying the KS equation by H and integrating over one wavelength to yield the energy balance:

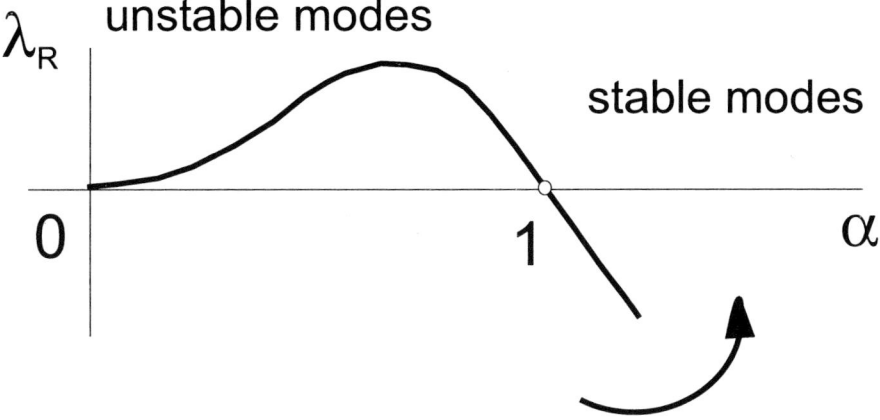

Figure 5.1: Linear stability curve and energy flux for KS equation.

$$\frac{\partial}{\partial t}\int_{-\pi/s}^{\pi/s} H^2 dX = \int_{-\pi/s}^{\pi/s}\left[\left(\frac{\partial H}{\partial X}\right)^2 - \left(\frac{\partial^2 H}{\partial X^2}\right)^2\right]dX \quad (5.5)$$

The left-hand side of (5.5) measures the variation of a total wave energy. The first term on the right originates from $\partial^2 H/\partial X^2$ of the original equation and is responsible for energy supply into the system and the second term corresponds to the fourth derivative term in (5.1) and is responsible for energy dissipation. For a spatially periodic travelling wave, one can always decompose its solution H in a Fourier series,

$$H = \sum_{k=1}^{\infty}(A_k(\tau)\cos skX + B_k(\tau)\sin skX) \quad (5.6)$$

Equation (5.5) then takes the following form:

$$\frac{d}{d\tau}\sum_{k=1}^{\infty}(A_k^2 + B_k^2) = 2\sum_{k=1}^{\infty} s^2 k^2(1 - s^2 k^2)(A_k^2 + B_k^2) \quad (5.7)$$

It follows that if $s > 1$,

$$\frac{d}{d\tau}\sum_{k=1}^{\infty}(A_k^2 + B_k^2) < 0 \quad (5.8)$$

and any perturbation is damped. If $s < 1$, harmonics with wavenumbers $k < 1/s$ are responsible for the energy supply and those with $k > 1/s$ are responsible for energy dissipation.

For a localized solitary wave that vanishes at $X \to \pm\infty$, H $\partial H/\partial X$, $\partial^2 H/\partial X^2 \to 0$, the energy balance equation is transformed into:

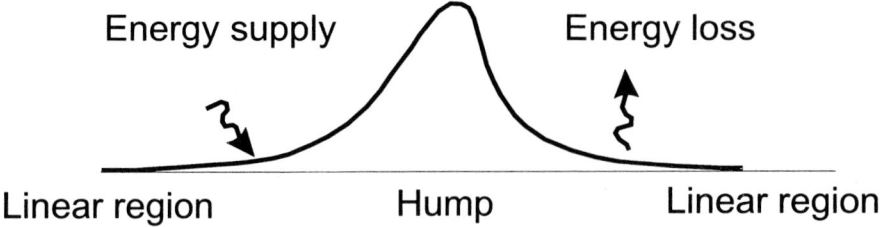

Figure 5.2: Energy distribution in space

$$\frac{d}{d\tau}\int_{-\infty}^{+\infty} H^2 dX = 2\int_{-\infty}^{+\infty}\left[\left(\frac{\partial H}{\partial X}\right)^2 - \left(\frac{\partial^2 H}{\partial X^2}\right)^2\right] dX \qquad (5.9)$$

For such solitary waves, the sum in (5.7) must be converted into an integral. Using the Fourier integral in its complex form, we get:

$$H = \int_{-\infty}^{+\infty} F(s)e^{isX} ds, \quad \frac{d}{d\tau}\int_0^\infty |F|^2 ds = 2\int_0^\infty s^2(1-s^2)|F|^2 ds \qquad (5.10)$$

Here, the third term of (5.1a) is again responsible for the energy supply while the fourth term is responsible for the energy dissipation, see Figures 5.1 and 5.2. The presence of a non-linear term enables the energy supplied at low wavenumbers to propagate down the spectrum to higher wavenumbers to be dissipated. The stationary travelling wave hence exists because the energy supply and dissipation processes are in equilibrium in the wavenumber space due to this "inertial cascade".

For stationary waves that travel at a velocity C, a simple transformation $\partial/\partial\tau = -C\partial/\partial X$ converts (5.1) into an ordinary differential equation which can be integrated once

$$H''' + H' - CH + 2H^2 = Q \qquad (5.11)$$

where Q is an integration constant, corresponding to the deviation liquid flow rate. Since the waveless flat film corresponds to $H = 0$, this flow rate is a deviation flow rate from that of the waveless Nusselt flat film. In such a formulation we assume that average thickness is zero over one wavelength $2\pi/\alpha$

$$<H> = \frac{\alpha}{2\pi}\int_0^{2\pi/\alpha} H dX = 0 \qquad (5.12)$$

However, we can take another assumption that nonlinearity changes the average film thickness but not the flowrate, viz. $Q = 0$,

$$u''' + u' - \lambda u + 2u^2 = 0. \qquad (5.13)$$

The two equations are connected by the relations,

$$\lambda = (8Q + C^2)^{1/2} \tag{5.14}$$

$$u = H - \frac{C}{4} + (\frac{C^2}{16} + Q)^{1/2} \tag{5.15}$$

For solitary pulses, the deviation flow rate Q vanishes exactly and the two formulations are identical. For periodic waves, however, they can either carry no additional flow rate ($Q = 0$) or has the same thickness as the flat film, as required by (5.12), but not both. The symmetry of (5.2) and (5.3) to transformation

$$H \to -H, \ C \to -C, \ X \to -X \tag{5.16}$$

or

$$u \to -u, \ \lambda \to -\lambda, \ X \to -X. \tag{5.17}$$

immediately suggests that for every travelling wave that propagates in one direction, there is another one with an inverted and reflected profile propagating with the same speed but in the opposite direction. It implies the existence of two families of periodic waves with the constant flux condition, one with a negative deviation speed $C_{-1} < 0$ and one with a positive one $C_{+1} > 0$. Moreover, the families are symmetric about C - axis, viz. $C_{-1} = -C_{+1}$, for near-critical conditions. We shall often skip the subscript 1 and refer to these two important wave families as C_+ and C_-. These two symmetric waves persist in the gKS equation and the more sophisticated equations, including the BL and Navier Stokes equations. They are, in fact, the only periodic wave families observed in the experiments among a plethora of periodic wave families generated from bifurcation of the reversible KS equation.

5.2 Phase space of stationary KS equation.

The stationary travelling waves of the KS equation (5.1) are first constructed by Nepomnyashchy (1974), Michelson (1986), Kevrekides et al. (1990), Demekhin et al. (1991), Chang et al. (1992) and Chang et al. (1993). Nonstationary waves were studied in the works, Nyman and Nikolaenko (1988) and Hyman, Nikolaenko and Zalesky (1988). We shall focus on some unreported fine solution structures here. Some of these finer structures will survive at finite δ and become important travelling waves of the boundary layer (BL) equation.

Converted into an autonomous dynamical system, (5.13) produces trajectories in a three-dimensional phase space for each value of λ. Solutions of (5.11) and (5.13) that are bounded at $X \in (-\infty; +\infty)$ will be the travelling wave solutions of interest. Limit-cycle trajectories correspond to periodic travelling waves, homoclinic trajectories to solitary waves, heteroclinic trajectories to shocks, bounded chaotic trajectories to irregular travelling waves etc. Hence, dynamical systems and bifurcation theory tools can come to bear to decipher the kind of travelling wave to expect for each wave speed λ. The symmetries of (5.13) will facilitate the implementation of these theories. Equation(5.13) with symmetry (5.17) is referred to as a reversible system whose general theory has been developed by Arnold (1984), Sevryuk (1986) and Arnold and Sevryuk (1986). They show that reversible systems exhibit dynamics similar to Hamiltonian systems, with a similarly diverse sequence of bifurcations that can be classified by exploiting the reversibility. For example, as in Hamiltonian systems, reversible systems possess a one-parameter continuous family of periodic solutions, corresponding to the fixed points of involution G, for (5.13): $(u, u', u'', u''') \to (-u, u', -u'', u''')$ in the phase space. These solutions are classified as the S-type in Figures 5.18 and 5.19.

Normal form theory in the vicinity of a fixed point or periodic solution of a reversible system almost entirely follows Birkhoff's normal form theory for Hamiltonian systems. For example, we can develop an analogous Kolmogorov - Arnol'd - Moser (KAM) theory similar to Hamiltonian systems. We shall use such a theory on the periodic solution family of S type of equation (5.13) to show that it has five types of travelling waves: periodic, doubly-periodic, solitary, kink, anti-kink and chaotic waves.

For (5.13), it is convenient to introduce a phase space $u_1 = u$, $u_2 = u'$ and $u_3 = u''$ and consider a dynamical system

$$u_1' = u_2 \tag{5.18}$$

$$u_2' = u_3 \tag{5.19}$$

$$u_3' = \lambda u_1 - 2u_1^2 - u_2 \tag{5.20}$$

From the system (5.18)-(5.20) we easily find that divergence of the phase space is zero and, thus, the phase volume of the system remains unchanged as,

$$\text{div}\mathbf{u}' = \frac{\partial u_1'}{\partial u_1} + \frac{\partial u_2'}{\partial u_2} + \frac{\partial u_3'}{\partial u_3} = 0 \tag{5.21}$$

and, hence, the Liouville theorem for phase volume conservation is satisfied as if the system were Hamiltonian. Since (5.13) is invariant with respect to transformation (5.17), we can merely consider $\lambda \geq 0$.

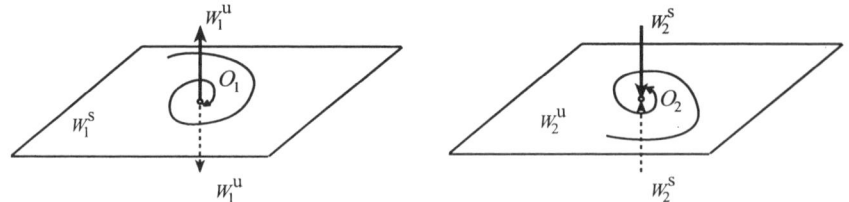

Figure 5.3: Phase portraits near the fixed points of KS equation

Equation (5.13) includes two saddle points O_1 and O_2, $u = u' = u'' = 0$ and $u = c/2$, $u' = u'' = 0$. Their characteristic equations are, respectively:

$$\sigma^3 + \sigma \pm \lambda = 0$$

For O_1, $\sigma_1^{(1)} = 2m > 0$, $\sigma_{2,3}^{(1)} = -m \pm i\beta$, where $\beta = \sqrt{1+3m^2}$;
for O_2, $\sigma_1^{(2)} = -2m < 0$, $\sigma_{2,3}^{(2)} = m \pm i\beta$.

Hence, fixed point O_1 has a two-dimensional stable manifold W_1^S and one-dimensional unstable manifold W_1^u while the second fixed point, O_2, has a two-dimensional unstable manifold W_2^u and a one-dimensional stable manifold, W_2^S, see Figure 5.3.

A solitary wave is a homoclinic trajectory that is doubly asymptotic to O_1. This curve is an intersection of a two-dimensional stable manifold W_1^s and of one-dimensional unstable one W_1^u, $\Gamma = W_1^u \cap W_1^s$. Such intersection between two manifolds of the specified dimensions breaks with a slight perturbation - it is structurally unstable and is eliminated by a small perturbation of the λ-parameter. Thus, solitary waves can exists only for a discrete set of $\{\lambda_k\}$, see Figure 5.4. A key one, corresponding to a one-hump pulse, exists at $\lambda_1 = 1.21615$ and is shown in Figure 5.5.

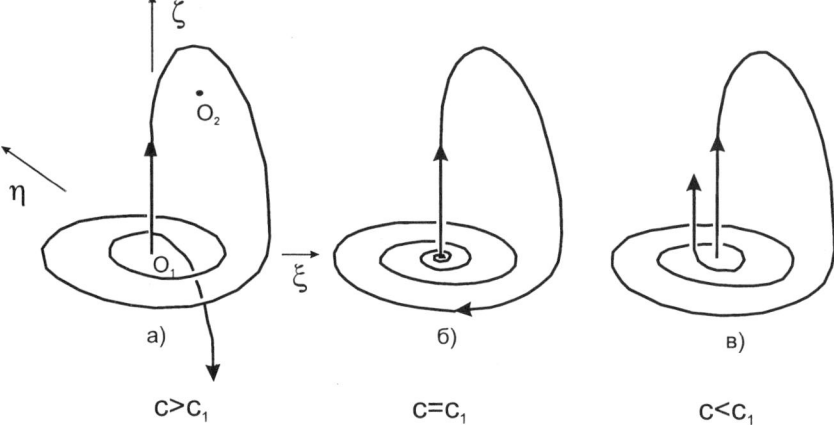

Figure 5.4: Typical phase-space trajectory near pulse eigenvalue $C = C_1$

Equation (5.13) allows shocks corresponding to heteroclinic trajectories. Shocks can appear as kinks and antikinks:

$(1) u \to \dfrac{\lambda}{2}$, $u', u'' \to 0$ at $X \to -\infty$; $u, u', u'' \to 0$ at $X \to +\infty$;

$(2) u, u', u'' \to 0$ at $X \to -\infty$; $u \to \dfrac{\lambda}{2}, u', u'' \to 0$ at $X \to +\infty$;

A heteroclinic trajectory which corresponds to a kink is uncoiling over the unstable two-dimensional manifold W_2^u of the second stationary point O_2 and for $X \to +\infty$, spirals along the stable two-dimensional manifold W_1^S into fixed point O_1. Thus, the heteroclinic trajectory lies on the intersection of two-dimensional

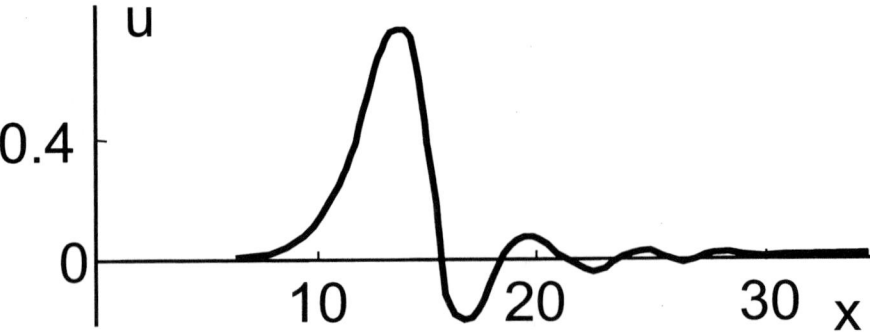

Figure 5.5: One-hump solution, $\lambda = \lambda_1 \simeq 1.21615$

surfaces, $L = W_1^S \cap W_2^u$, which is structurally stable. Therefore, a kink can exist throughout a continuous range of λ values. These theoretical predictions are confirmed by direct numerical calculations that show a one-parameter kink family within the range $\lambda \in (0; \infty)$. For small λ, the kink solution can be constructed analytically, as a limit of doubly periodical solutions, see relation (5.49) in this chapter.

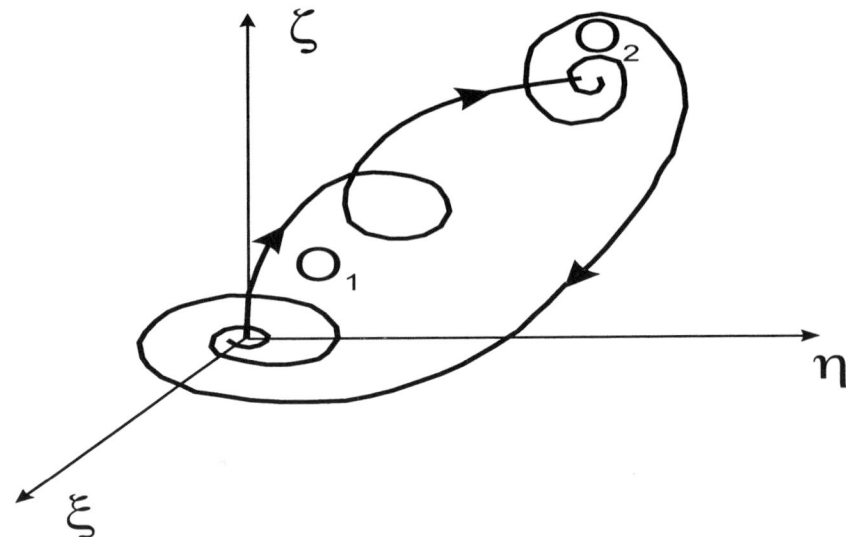

Figure 5.6: Heteroclinic contour of the stationary points O_1 and O_2.

Another case exists where a phase trajectory comes out of O_1 along the one-dimensional unstable manifold W_1^u and connects to the one-dimensional stable manifold of O_2. Such intersections of curves are not structurally stable. Nevertheless, these solutions exists (Kuramoto and Tsuzuki, 1976) and their existence is due to the reversibility symmetry of (5.3). The solution can be

conveniently represented as

$$u = \frac{15}{19}\varkappa(1 - \frac{9}{2}\tan \varkappa X + \frac{11}{2}\tan^3 \varkappa X), \quad \varkappa = \frac{1}{2}\sqrt{\frac{11}{19}}$$
$$\lambda = \frac{60}{19}\varkappa \approx 1.201398 \tag{5.22}$$

which is an exact solution of (5.13).

Taking into consideration the fact that kinks exist for $\lambda \in (0, \infty)$, we obtain at $\lambda = \frac{60}{19}\varkappa$ a structurally unstable heteroclinic contour, Figure 5.6, which consists of two heteroclinic curves: from O_1 to O_2 and from O_2 to O_1.

Now consider homoclinic trajectories in relation to periodic solutions. We shall show during investigation of (5.30) that, at each value of $\lambda \in (0.449; 1.791)$, saddle-type periodic trajectories γ_1 and γ_2 exist. These trajectories generate a countable set of periodic cycles close to the initial γ_1 and γ_2.

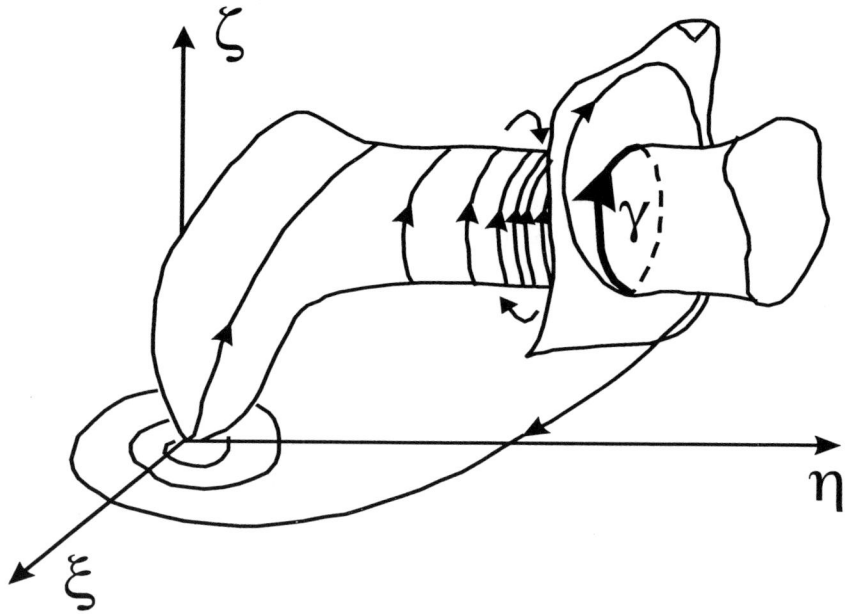

Figure 5.7: Typical heteroclinic contour of the stationary point O_1 and of a saddle-type periodic trajectory.

These limit cycles also possess invariant surfaces, W_γ^s and W_γ^u, which correspond to their stable and unstable manifolds, respectively. The intersection of a two-dimensional unstable manifold W_γ^u and a two-dimensional stable manifold W_1^s of fixed point O_1 is structurally stable. On the other hand, the stable one-dimensional manifold of the origin, W_1^u, and the stable two-dimensional manifold of the cycle can intersect at only some specific values of the λ parameter. At such parameter values, a heteroclinic contour exists which consists of two curves, see Figure 5.7. For $X \to +\infty$, one of the curves "winds" around the γ cycle whereas at $X \to -\infty$, it tends toward O_1. The second of the curves becomes asymptotic to γ, for $X \to -\infty$, and to O_1 if $X \to +\infty$.

Moreover, the intersection of unstable (stable) manifold of one cycle, $W_{\gamma_1}^u$ ($W_{\gamma_1}^s$), with the stable (unstable) manifold of another, $W_{\gamma_2}^s$ ($W_{\gamma_2}^u$) is also structurally stable since both are two-dimensional. Hence, such intersection exists over a continuous range of λ values.

Extending the above arguments, we could expect on a discrete manifold of λ homoclinic contours consisting of two heteroclinic curves. One of the curves comes out of O_1 and for $X \to +\infty$, it winds into a limit cycle γ_1 (or γ_2). The other curve "detaches" from γ_1 (or γ_2) along the unstable manifold $W_{\gamma_1}^u$ and at $X \to +\infty$ it becomes asymptotic to O_1. At similar values of λ, another contour can exists which consists of three heteroclinic trajectories, such as (1) from O to γ_1 (or γ_2); (2) from γ_1 (or γ_2) to γ_2 (or γ_1) and (3) from γ_2 (or γ_1) to O_1. We shall use these qualitative observations and properties of the phase space in the next subsections.

5.3 solitary waves and Shilnikov theorem

In many physical phenomena, initial perturbations eventually evolve into localized nonlinear structures–pulses or solitons. In falling films, as well as in other active-dissipative media, these solitons result from a balance between instability and dissipation. There is a rich variety of these solitary pulses, as we shall demonstrate with the simple KS equation.

Our numerical integration will show that there exists one hump soliton or a primary homoclinic loop Γ_0 in the phase space and, moreover, that the primary homoclinic loop is not unique. Besides a one-hump pulse, we have multi-hump or multiple pulses. It is entirely different from the usual KdV solitons, where we have a one-parameter family of one-hump solitons.

The reason for this complexity is provided by the famous Shilnikov theorem, Shilnikov (1965). There are many versions of this theorem, see Shilnikov (1976), Bykov (1977) and Bykov (1980), and we will present the most convenient one for our purposes.

Assume that we have an n-dimensional dynamical system

$$\frac{dy}{dX} = f(y, \lambda)$$

$$y = (y_1, y_2, ..., y_n)$$

with some parameter λ.

Assume also that, at any λ, the origin is a fixed point, $f(X, \lambda) = 0$, with n simple eigenvalues σ_k. Let us suppose that the smallest eigenvalues by absolute value are:

- $\sigma_1 > 0$ - a positive eigenvalue;

- $\sigma_{2,3} = -m \pm i\beta$ - a pair of complex-conjugate roots with a negative real part, see Figure 5.8. If at some value of the parameter $\lambda = \lambda_0$ the system has

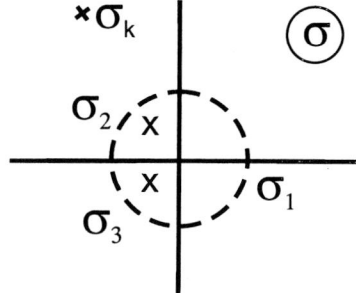

Figure 5.8: Location eigenvalues for Shilnikov's theorem; σ_1 - is a real positive eigenvalue, $\sigma_{2,3}$ is a pair of complex conjugate roots with a negative real parts; $\sigma_1 + Re(\sigma_2) > 0$

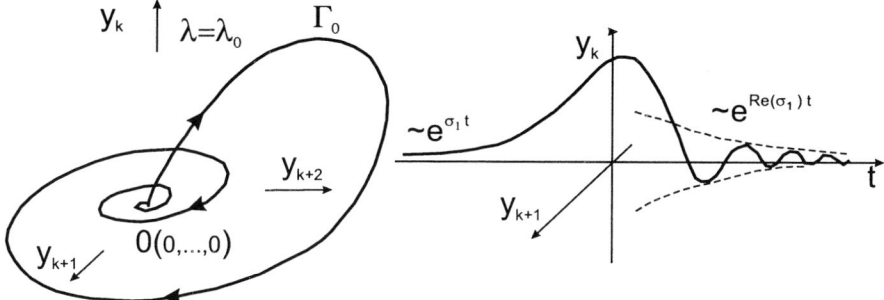

Figure 5.9: Primary homocliinc loop a near saddle-focus fixed point.

a homoclinic orbit Γ_0 see (Figure 5.9) which is doubly asymptotic to the fixed point at $X \to \pm\infty$, and if the saddle value or Shilnikov number

$$S = \sigma_1 + Real(\sigma_2) > 0$$

is positive, then at any vicinity of λ_0 there is a countable number of multi-loop homoclinic orbits.

Physically, it means that if the pulse front has oscillations, and the trailing edge decays monotonically, and if the monotonical decay is faster than the decay with oscillations, there are also multi-hump solitary waves along with the one-hump pulse. A good visualization of Shilnikov theorem is given in Arneodo et al. (1982). We shall follow this paper.

It is clear that the Shilnikov number for the O_1 fixed point of the KS dynamical system is $2m - m > 0$. We hence expect many-hump pulses to also exist for λ values close to $\lambda_1 = 1.21615$ of the one-hump pulse in Figure 5.5. It should be emphasized that the system dimension can be larger than 3 and, hence, the results are applicable not only to the Kuramoto-Sivashinsky equation, but also to much more complex systems.

Let us transform (5.18)-(5.20) to the canonical form near the origin O_1 which has one-dimensional unstable and two-dimensional stable manifolds, W_1^u and W_1^s. We orientate our new axes, ξ tangentially to the unstable manifold and η and ζ tangentially to the stable manifold,

$$\frac{d\xi}{dX} = -m\xi - \beta\eta + P(\xi,\eta,\zeta)$$
$$\frac{d\eta}{dX} = \beta\xi - m\eta + Q(\xi,\eta,\zeta)$$
$$\frac{d\zeta}{dX} = 2m\zeta + R(\xi,\eta,\zeta) \tag{5.23}$$

where

$$P = 2\beta g^2, \quad Q = 2mg^2, \quad R = -4mg^2, \tag{5.24}$$

$$g = \frac{2m\xi + 8m^2\beta\eta + \frac{1}{2}\frac{\lambda}{m}\beta\zeta}{\lambda\beta(9m^2 + \beta^2)} \tag{5.25}$$

The old and the new variables are related by the similarity transform

$$\begin{pmatrix} \xi \\ \eta \\ \zeta \end{pmatrix} = \begin{pmatrix} 0 & 2m\beta & -\beta \\ \lambda & m^2 - \beta^2 & -m \\ \lambda & 4m^2 & 2m \end{pmatrix} \begin{pmatrix} u_1 \\ u_2 \\ u_3 \end{pmatrix}$$

such that the second fixed point in the new variables has coordinates $(0, 1/2\lambda^2, 1/2\lambda^2)$.

The specific form of P, Q and R is not important for the qualitative features

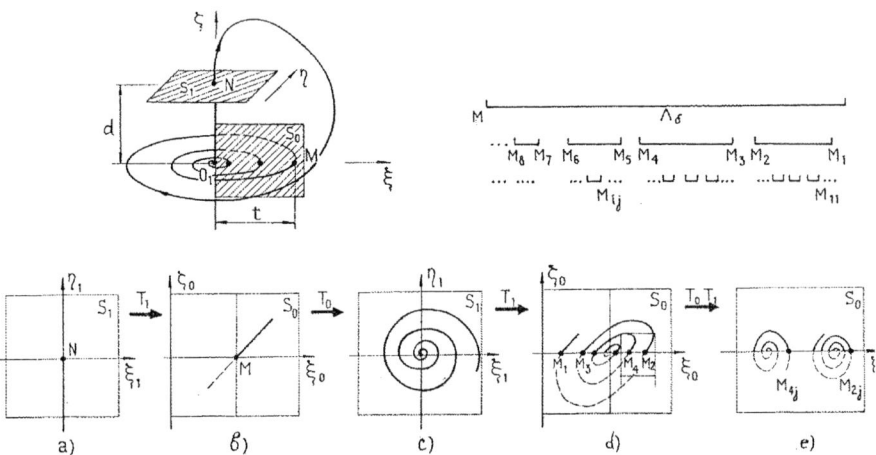

Figure 5.10: Homoclinic orbit Γ_0 and planes S_0 and S_1 which split it into global and local pieces.

of the homoclinic orbits, except that R is negative and that the Shilnikov number S, the sum of the real positive eigenvalue and the real part of the complex eigenvalue, is positive. The surface of the stable manifold separates the phase space into two half-spaces, the upper and lower space, see Figure 5.10,left corner. A tangent plane to the stable manifold of the origin is a horizonal plane $\zeta = 0$. The one-dimensional unstable manifold consists of two branches: the first goes upwards and the second, which is not shown, goes downwards, from the origin, with a tangent perpendicular to the horizontal plane. We find that a point in the lower half-space will never be connected by the flow to the upper half-space. Indeed, for $\zeta < 0$ and taking into account that R is negative, we can see that $\frac{d\zeta}{dX} < 0$.

We assume that, at $\lambda = \lambda_0$, we have a homoclinic orbit Γ_0 that leaves the origin along the unstable one-dimensional manifold and returns along the stable one. We shall construct a Poincare or return map near this orbit. For this purpose we split our phase trajectory into two pieces, in a small vicinity of the origin and outside of this vicinity. This is the main idea of Shilnikov. Thus, we shall study the local, near the fixed point, and global behaviors of the trajectory.

In a small vicinity of the origin, we take two planes, S_0 and S_1, defined by $\eta = 0$ and $\zeta = d$, see Figure 5.10, left corner. We will keep in this vicinity only linear terms and neglect the nonlinear P, Q, and R. Then, in this vicinity, the stable manifold is the horizontal plane and the unstable manifold is a straight line perpendicular to this plane. This line strikes the plane S_1 at a point N(0,0,d).

If we take this point on the unstable manifold as the initial condition, we can determine its entire trajectory in the phase space. If we keep the parameter λ equal to λ_0, then the phase trajectory goes along the homoclinic orbit Γ_0 and intersects the plane S_0 at a countable number of points which converge to the origin. Let us choose one of these points M with coordinates $(t, 0, 0)$.

Let us first consider the map T_1, which transforms the plane S_1 into S_0. The origin of S_1, the point N, with coordinates $(0,0)$, maps into the point M with coordinates (t,0),Figure 5.10, at the bottom, a) into b). It is reasonable to assume that a small vicinity of N will map to some small vicinity of the point M. Hence, taking a Taylor expansion with respect to the space variables and the parameter C and keeping only linear terms, one gets the map T_1:

$$\xi_0 = t + b_1(\lambda - \lambda_0) + a_{11}\xi_1 + a_{12}\eta_1 + ...$$
$$\zeta_0 = b_2(\lambda - \lambda_0) + a_{21}\xi_1 + a_{22}\eta_1 + ...$$

The values of the coefficients b_k and a_{ij} are not important, as long as the map is nonsingular, $\det ||a_{ij}|| \neq 0$, and b_2 is not zero, say negative.

Let us now consider the map T_0 of the plane S_0 into S_1, Figure 5.10 b) into c). In a small vicinity of the origin the system can be linearized :

$$\frac{d\xi}{dX} = -m\xi - \beta\eta$$
$$\frac{d\eta}{dX} = -\beta\xi - m\eta$$

$$\frac{d\zeta}{dX} = 2m\zeta$$

If we divide the first and the second equation by the third one, we shall find the system:

$$\frac{d\xi}{d(ln\zeta)} = -\frac{1}{2}\xi - \frac{\beta}{2m}\eta$$

$$\frac{d\eta}{d(ln\zeta)} = \frac{\beta}{2m}\xi - \frac{1}{2}\eta$$

the solution of which is easy to get by taking $ln(\zeta)$ as an independent variable. We then obtain the map of ξ_0 and ζ_0 on the plane S_0 into the plane S_1 with variables ξ_1 and η_1:

$$\xi_1 = \xi_0\sqrt{\frac{\zeta_0}{d}}\cos[-\frac{\beta}{2m}\ln(\frac{\zeta_0}{d})]$$

$$\eta_1 = \xi_0\sqrt{\frac{\zeta_0}{d}}\sin[-\frac{\beta}{2m}\ln(\frac{\zeta_0}{d})]$$

This is the key map, so let us consider some of its details, see Figure 5.11. We remind ourselves that if a phase point is in the lower half-plane S_0, with

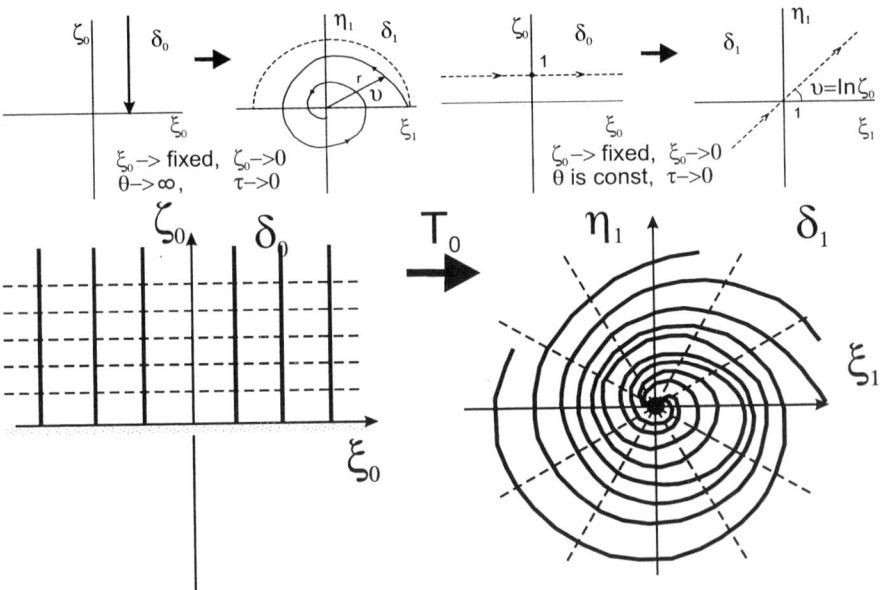

Figure 5.11: Model map T_0, transforms lines parallel to ζ_0 — axis into spirals, and lines parallel to ξ_0 — axis into radial straight lines.

negative ζ_0, it goes to $-\infty$ and so should be excluded from consideration. All the points of the ξ_0-axis shrink into the origin on the S_1-plane. If we fix ξ_0 and change ζ_0 to zero, the radius r will tend to zero, while the angle θ to the infinity, resulting in a spiral. If we fix ζ_0 and moves ξ_0 to zero, the angle will be constant, while the radius will tend to zero. As a result, the map transforms the families of lines parallel to the axes into spirals and radial straight lines.

We have constructed the return map of the plane S_0 and we seek homoclinic orbits which start from the origin, wind along the unstable manifold and cross the plane S_1 at the point N. Thus, we can start with the point N and change the parameter λ. If we change λ, it will create a one-parameter family of trajectories close to Γ_0. This family will cross the plane S_0 along some line, which in a small vicinity of λ_0 is a straight line.

At $\lambda = \lambda_0$ the line crosses the stable manifold at the point M on the ξ_0-axis, which corresponds to the homoclinic curve Γ_0. The lower part of the line, for which ζ_0 is negative (the dashed line in the picture 5.10), corresponds to $\lambda > \lambda_0$, because the coefficient $b_2 < 0$. The points on this part of the curve leave the vicinity of Γ_0 forever and tend to minus infinity as X tends to plus infinity. Thus we should exclude this part of the curve from our consideration, and the corresponding values of $\lambda > \lambda_0$, Figure 5.10. The upper part of the curve, solid line, corresponding to $\lambda < \lambda_0$, will be mapped into a spiral on the plane S_1 with the center in the point N, Figure 5.10 b) to c). In turn, the map T_1 by virtue of its linearity and nonsingularity, transforms the spiral into a deformed spiral at the plane S_0 with the center in the point M, Figure 5.10, c) to d) (linear map transforms circles into ellipses). This provides the answer we desire - the last spiral crosses the ξ_0- axis in a countable number of points $M_1, M_2, ...$, which converge to M. It happens at the parameter λ-values, $\lambda = \lambda_1, \lambda_2, ...$, see Figure 5.10 right upper corner. Thus, these values of λ are eigenvalues at which new homoclinic orbits $\Gamma_1, \Gamma_2, ...$ appear. These homoclinic orbits make two loops close to each other in the phase space before returning to the origin. The parts of the spiral with negative ζ_0 should be excluded from the consideration, and we mark these parts with a dashed line. For negative ζ_0, we have the corresponding intervals of λ: $(\lambda_1, \lambda_2), (\lambda_3, \lambda_4), ...$, which we have to exclude from further consideration (dashed lines). We can repeat the procedure and find 3-loop homoclinic orbits. At the same time, we have to discard the intervals with negative ζ_0.

It is tempting to say that there is a Cantor set of the homoclinic orbits. However, this is not true but there is a countable number of homoclinic trajectories. Besides homoclinic orbits, the return map provides chaotic trajectories, and the number of these trajectories gives a Cantor set.

Homoclinic orbits Γ and the λ values at which they exist can only be determined numerically. As we shall see from the numerical investigation, the behavior of KS pulses indeed are qualitatively guided by the Shilnikov theorem. It is convenient to use the shooting method. Very similar numerical procedure is applicable for more complex equations or systems.

The equation is a third order ODE. At $X \to \pm\infty$, u decays to 0 and hence one can linearize the equation by neglecting the quadratic nonlinear term:

$$u''' + u' - \lambda u = 0, u = \epsilon e^{\sigma X}$$

with the characteristic equation

$$\sigma^3 + \sigma - \lambda = 0$$

Let us supposed λ is positive and fixed; we have one positive root

$$\sigma_1 = 2m > 0$$

and two complex conjugate with a negative real part

$$\sigma_{2,3} = -m \pm i\beta$$

Sufficiently far to the left hand-side only the real root is pertinent; the complex roots correspond spatial modes which grow at $X = -\infty$. Hence, at $X = -\infty$ $u = \epsilon e^{2mX}$, where ϵ is an arbitrary but small value. It can be arbitrary chosen as it corresponds to a shift in the space variable. We now have the initial conditions to integrate the equation through the nonlinear region for a given λ. Upon integrating through the nonlinear region such that if begins to approach zero, one will be again in the linear region, where the solution has to be a linear combination of three independent solutions:

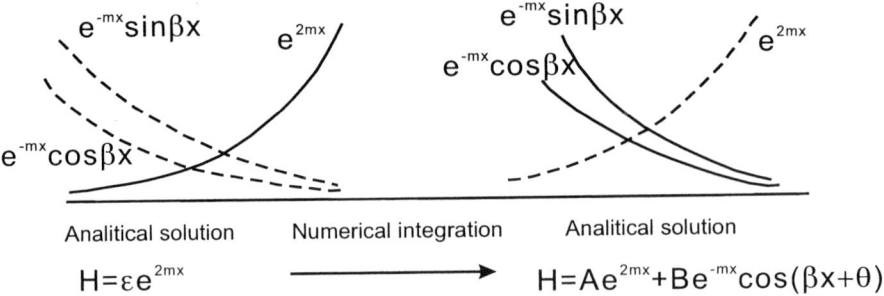

Figure 5.12: Qualitative scheme of numerical solution for the stationary pulse.

$$u = Ae^{2mX} + Be^{-mX}\cos(\beta X + \theta)$$

Now only one of these solutions is unbounded, see Figure 5.12.

The coefficients A, B and, θ are functions of λ. We hence find such λ that the unbounded solution at plus infinity will be suppressed, $A(\lambda) = 0$. Then, the constructed solution will decay both at plus and minus infinities — exactly what we need. The critical value of λ for one hump pulse is 1.21615, and the wave profile is shown in the Figure 5.5.

As it is shown in the calculations by Demekhin and Shkadov (1985, 1986), solitary waves exist in a countable set of parameter λ segments: $C_1 = [\lambda_1, \lambda_1']$, $C_2 = [\lambda_2, \lambda_2']\ldots$. At $n \to \infty$, $\lambda_n, \lambda_n' \to 0$, i.e. point $\lambda = 0$ is the limit to which the segments converge point of the segments convergence. Outside of C_n, homoclinic curves do not exist. In Table 5.1 the values of λ_n, λ_n' for the first eight n are given.

Table 5.1

k	1	2	3	4	5	6	7	8
λ_k	1.2162	0.7046	0.5413	0.4567	0.4031	0.3655	0.3370	0.3146
λ'_k	1.1805	0.6873	0.5269	0.4462	0.3945	0.3579	0.3303	0.3085

To understand how the segments C_n arise, consider the behavior of trajectories in space $\{\xi, \eta, \zeta\}$ and their profiles at various values of obtained numerically

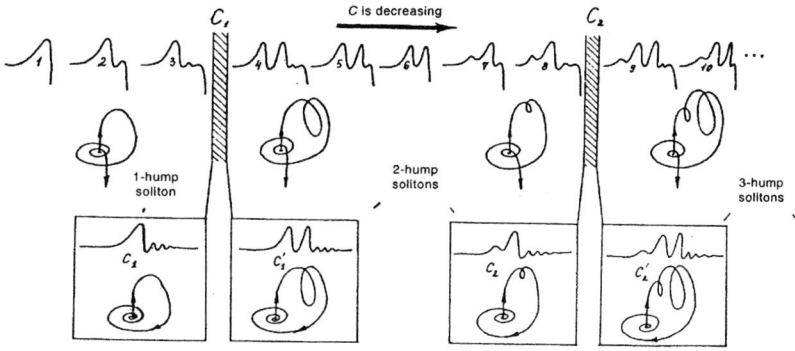

Figure 5.13: Typical profiles and phase-trajectories that initiate from the unstable manifold of the fixed point O_1; homoclinic orbits are absent outside of C_k.

For $\lambda \in (\lambda_{n+1}, \lambda'_n)$, $n = 1, 2, \ldots$ a phase curve makes $n + 1$ turns around O_2, see Figure 5.13. At $\lambda = \lambda_1$ we have a one-hump pulse. At $\lambda = \lambda_2$ a two-hump pulse exists, the second hump of which is apparently larger than the first one. At $\lambda \to \lambda'_2$ the waveform approximates to the three-hump pulse profile, the latter two humps of which have approximately equal amplitudes, and the amplitude of the first hump is significantly smaller. At $\lambda \to \lambda_3 + 0$ we also obtain a three-hump pulse though the first two humps are somewhat smaller and the last a bigger one, etc. The wave speed λ_n corresponds to a homoclinic curve with an $n-1$ smaller loop which gradually "unwinds" close to O_2 and one greater loop which brings the trajectory helically back to O_1. Here, $u(X)$ has n number of humps differing in their forms. At $\lambda = \lambda'_n$ a homoclinic curve has an $n-1$ smaller turn near O_2 and two bigger turns; $u(X)$ has $n-1$ hump. Solitary wave profiles obtained by calculations for $n = 1, 2, 3$ are shown in Figure 5.14 (a),(b),(c). Solitary waves with, $\lambda = \lambda_n$ are illustrated by a continuous line while $\lambda = \lambda'_n$ by a dashed line. In addition, phase projections at $n = 1$ are shown in Figure 5.14 (a).

This pulse solution structure is qualitatively similar to that predicted by the

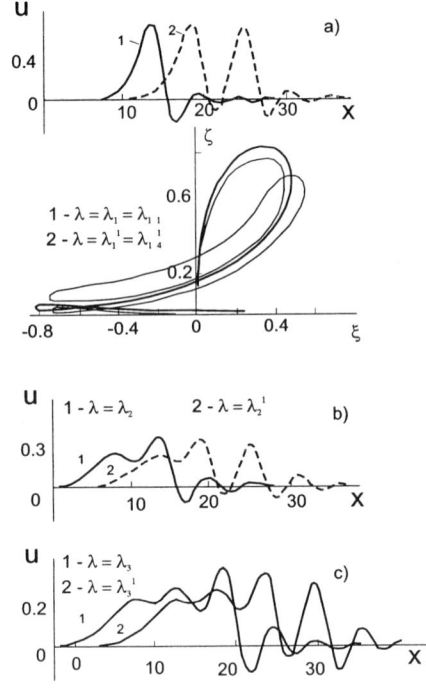

Figure 5.14: Pulses and their phase properties in the boundary points of a) the first segment C_1, $\lambda = \lambda_1$, solid line and $\lambda = \lambda_1'$, dashed line lower boundary; b) the second segment C_2, $\lambda = \lambda_2$, upper boundary of segment, solid line and $\lambda = \lambda_2'$ lower boundary, dashed line, see also Figure 5.13.

Shilnikov theorem, in particular, the segment structure of Figure 5.10. On the other hand, secondary humps do not repeat the primary hump.

Each of the segments C_n contains a complicated structure. Lengths of the segments are small compared to those of domains in which solitary waves do not exist. For example, $(\lambda_1 - \lambda_1')/(\lambda_1' - \lambda_2) = 7.5 * 10^{-2}$. While the primary solitary wave Γ_0 is a one-hump pulse, the generated $u(X)$ solutions are n-hump pulses, second and successive humps are in fact a repeated first one. solitary waves differ in hump number and the distance between them rather than their profiles, Shilnikov mechanism. Numerical analysis of the segments shows that despite their small intervals, the solution structure is even more complex than expected. This can be attributed to the existence of homoclinic orbits inside the segments. Our analytical investigation is based on the fact that a negligible variation of the parameter λ causes primarily variations in the neighborhood of the saddle - point O_1. If a structurally unstable heteroclinic contour appears which includes saddle - point O_2 or cycle γ, such contour can also generate a homoclinic orbit Γ that is doubly asymptotic to O_1, if the parameter λ is varied. The formation of such homoclinic trajectories will differ from Shilnikov's analysis in that the perturbation of the parameter will be present not only about O_1, but also about other saddles. There are more general theories which contain

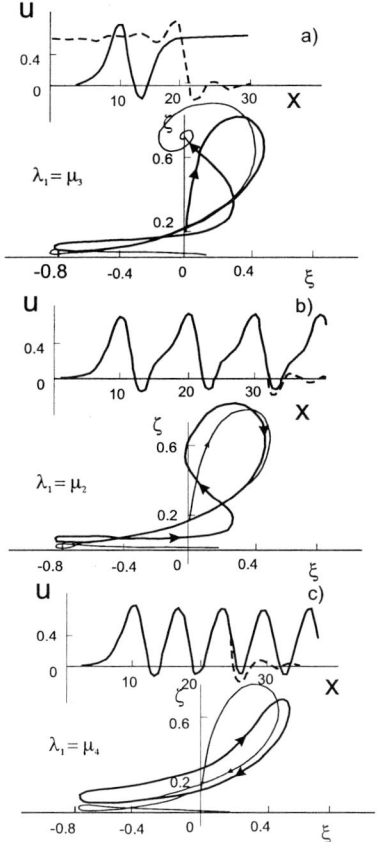

Figure 5.15: Four heteroclinic contours inside C_1: a) of the stationary points O_1 and O_2; b) of O_1 and the cycle γ_1, c) of O_1 and the cycle γ_2.

the main Shilnikov idea about the saddle value S. This value now is calculated not only near each saddle fixed point but also near saddle cycles. If S for a cycle is positive, we again have multi-hump pulses in which the limit-cycle profile are now repeated before decaying.

It was found, for example, four structures within C_1 at the values of $\lambda = \mu_1, \mu_2, \mu_3, \mu_4 \in C_1$. Each of these structures can generate its own multi-hump pulses:

(1) The original homoclinic curve having one loop in the phase space, $\mu_1 = \lambda_1 \simeq 1.2162$, refer to Figure 5.5.

(2) At $\mu_3 \simeq 1.20140$ there exists a heteroclinic contour of the following type: a structurally unstable heteroclinic curve coming from O_1 into O_2 is described by the relation (5.17); structurally unstable heteroclinic curve is determined

numerically. These curves are illustrated in Figure 5.15(a) by a continuous and a dashed line, respectively, see also Figure 5.6.

(3) At $\mu_2 \simeq 1.207956$ (and at $\mu_4 \approx 1.18203$) a heteroclinic contour exists consisting of two heteroclinic curves: a structurally unstable one tending to O_1 at $X \to -\infty$ and to periodic solution $\gamma_1(\gamma_2)$ at $X \to +\infty$, and a structurally unstable curve tending to O_1 at $X \to +\infty$ and to $\gamma_1(\gamma_2)$ at $X \to -\infty$. At $\lambda = \mu_3, m_4$, a structurally unstable heteroclinic trajectory, $\gamma_{12}(\gamma_{21})$, also exists which is asymptotic to $\gamma_1(\gamma_2)$ at $X \to -\infty$ and to $\gamma_2(\gamma_1)$ at $X \to +\infty$ The calculated projections of these contours and profiles $u(X)$ are shown in Figure 5.15(b) and (c). Heteroclinic curves are shown by dashed lines and periodic movements by continuous lines.

Homoclinic orbits generated by the varying parameter λ must lie close to the initial heteroclinic contour at $\lambda = \mu_1, \mu_2, \mu_3, \mu_4$, and be formed by the trajectory decomposition about the saddle - points $O_{1,2}$ and saddle cycles $\gamma_{1,2}$. Therefore, despite the complicate picture of homoclinic bifurcation, the homoclinic orbits consist of four primary structures.

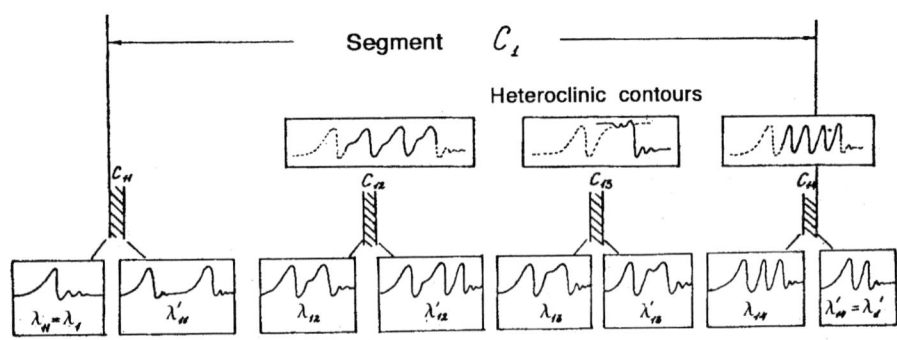

Figure 5.16: Subsegments C_{1j} inside C_1, caused by decomposition of the heteroclinic contours.

Numerical analysis also reveals new segments or subsegments within C_1: $\mu_1 \in C_{11} = [\lambda_{11}, \lambda'_{11}]$, $\mu_2 \in C_{12} = [\lambda_{12}, \lambda'_{12}]$, $\mu_3 \in C_{13} = [\lambda_{13}, \lambda'_{13}]$, $\mu_4 \in C_{14} = [\lambda_{14}, \lambda'_{14}]$, beyond which homoclinic orbits do not exist. The values of $\lambda_{ij}, \lambda'_{ij}$ are shown in Table 5.2 below. Figure 5.16 shows new segments inside C_1 caused by decomposition of the heteroclinic contour. Figures 5.17 illustrate phase projections and wave profiles obtained as a result of the calculations. Segment C_{11} begins with a one-hump pulse (as noted earlier) at $\lambda_{11} = \lambda_1$ (see Figure 5.14 (a)) and ends with a two-hump pulse whose second hump is an approximate repetition of the first one (see Figure 5.17(a)). The value λ_{12} corresponds to a two-hump pulse whose second turn in phase space is formed in the vicinity of γ_2 (see Figure 5.17(b)). Another value λ'_{14} also corresponds to a two - hump pulse which has its second turn about γ_1, as shown in Figure 5.17(a). At $\lambda = \lambda'_{12}$, a three - hump soliton exists having its second and third turns about

γ_2 and γ_1, respectively (see Figure 5.17(c)). At $\lambda = \lambda_{14}$, a three-hump solution exists having two turns about γ_1 (see Figure 5.17(f)). At $\lambda = \lambda_{13}, \lambda'_{13}$, (Figure 5.17(d),(e)), a solitary wave appears as a result of homoclinic decomposition about saddle - point O_2.

Table 5.2

k	1	2	3	4
λ_{1k}	1.2165	1.20815	1.20233	1.18207
λ'_{1k}	1.21532	1.20773	1.11988	1.18052

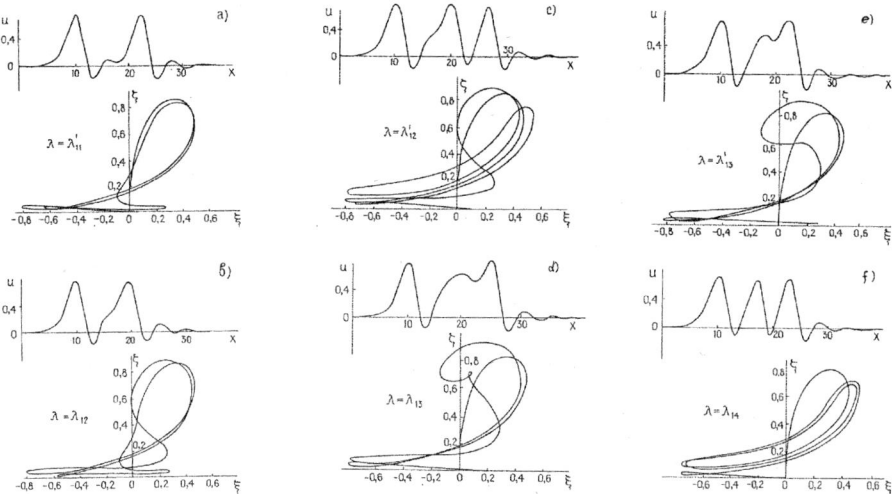

Figure 5.17: Pulses and their phase projections in the boundary points of subsegments of C_1: $\lambda_1 = \lambda_{11}$, the upper boundary, see Figure 5.14, a) lower boundary of C_{11} is presented by a two-hump pulse whose both loops are counterparts of the primary loop Γ_0, b) upper boundary of C_{12}, first loop of the pulse is near the primary loop Γ_0 and the second is near the cycle γ_1 before returning into O_1, c) lower boundary of C_{12}, solutions whose phase-trajectory makes first loop near the primary one; second - near the cycle γ_1 and third one - near the cycle γ_2; d) moves along the primary, then moves towards O_2 along stable manifold of O_2, repels by the unstable manifold of O_2 and returns to O_1.

Each of segments C_{1j} has its unique inner structures. While extending the segments, the division into new ones, C_{1jk}, is observed again. Out of these segments, no homoclinic curves exist. The boundary points of the segments are pulses consisting of the above - mentioned four-hump or fewer elements. In further extensions, segments C_{1jkm}, etc. are formed by removing open intervals. In the limit we obtain a Cantor set of wave speeds λ corresponding to bounded travelling waves. Pulses existing on the end of the segments form a countable subset. Indeed, the final construction of the segments is close the one predicted by the Shilnikov theorem.

Thus, the most general behavior of a homoclinic curve can be characterized as follows: 1) the trajectory starts along a stable manifold and makes its first turn about O_2 ; 2) the trajectory starts to be "drawn" into O_2 along some neighborhood of a stable manifold and bypass the second fixed point into some attracting neighborhood of O_1 . Further, the trajectory "skip over" the first fixed point to exit through the upper branch of W_1^u; 3) the trajectory is brought to the basin of attraction of one of saddle-type periodic movements and after making m turns on it ($m = 1, 2, \ldots$), exits from its neighborhood along the unstable manifold; 4) the trajectory either becomes asymptotic to O_1 or on skipping over the fixed point, repeats the events in this or another order.

5.4 Bifurcations of spatially periodic travelling waves and their stability

Travelling waves at velocity C are defined by equations (5.11), which is more convenient for periodic waves. They satisfy the conditions

$$H(0) = H(2\pi/s), \; H'(0) = H'(2\pi/s), \; H''(0) = H''(2\pi/s) \qquad (5.26)$$

and the normalization condition of (5.12). The parameter of choice will be the wave number s. Wave form $H(X)$, velocity C, and flow rate Q are determined as functions of this wave number s. We shall use λ sometimes in our presentation which is connected with our parameters by (5.14).

Instead of seeking cycles in the phase space, we construct the periodic travelling waves explicitly by a Galerkin expansion,

$$H = \sum_{k=1}^{\infty} a_k \cos skX + b_k \sin skX \qquad (5.27)$$

By substituting (5.27) into (5.11), and separate coefficients for different harmonics, we obtain a system of $2N$ algebraic equations for the to $2N + 1$ unknowns : C, Q, a_k and b_k, $k = 1, \ldots, N$. Without loss of generality, we can assume $a_1 = 0$ and close the system. Let C denote solutions with non zero speed $C \neq 0$. Stationary standing waves, $C = 0$, $a_k = 0$, $k = 1, \ldots N$ are odd functions of X with only sinusoid functions in Fourier expansion. Let us denote these solutions by S.

These equations are solved using Newton's method by continuing over the parameter s. Within a small neighborhood of each bifurcation points s_*, where the Jacobi matrix is singular, a travelling wave solution of a specific wavenumber bifurcates supercritically with respect to $|s - s_*|$.

The bifurcation tree can be suitably presented in projection on the parameter space of s and C, see Figure 5.18. Waveforms $H(X)$ are illustrated in Figure

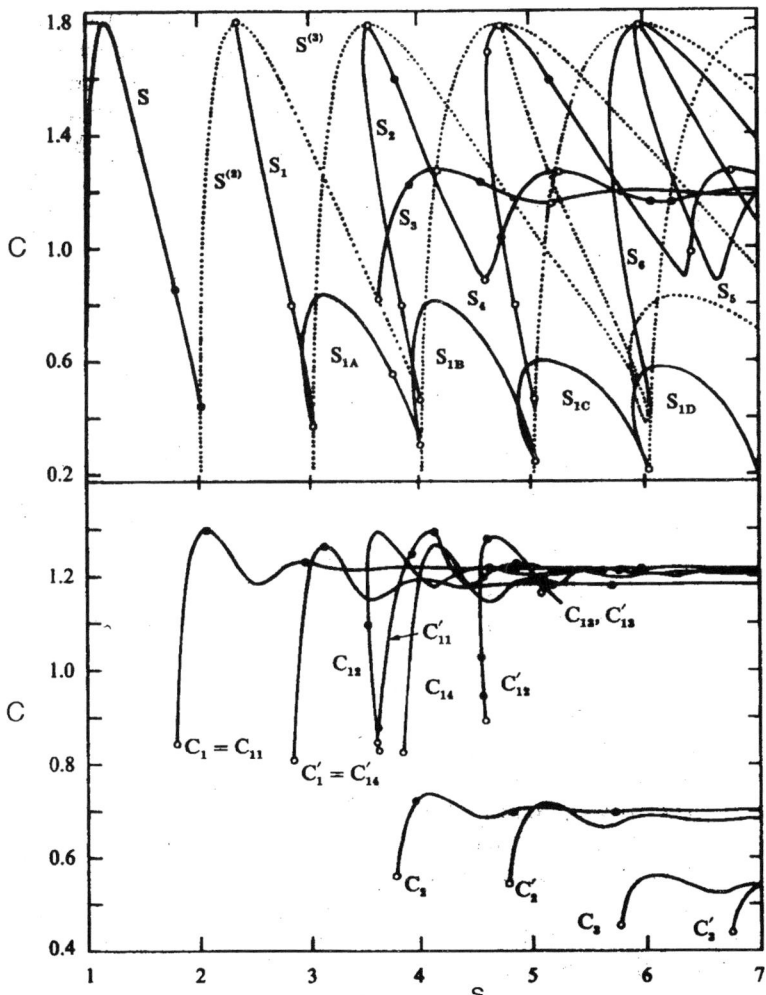

Figure 5.18: Stationary solution branches of the Kuramoto-Sivashinsky equation. The C_n families bifurcate from the circles at the top half of diagram. New solutions (tori) bifurcate from C_n at the market points.

5.19, 5.21, 5.22. Due to symmetry (5.11), a mirror image of Figure 5.18 exists with negative C.

For $s > 1$, the only solution is the trivial one: $H(X) = 0$ At $s = 1$ toward smaller s values, a stationary standing wave family $S^{(1)}$ branches from the trivial solution with wavelength 2π. The primary branch $S^{(1)}$ is a standing

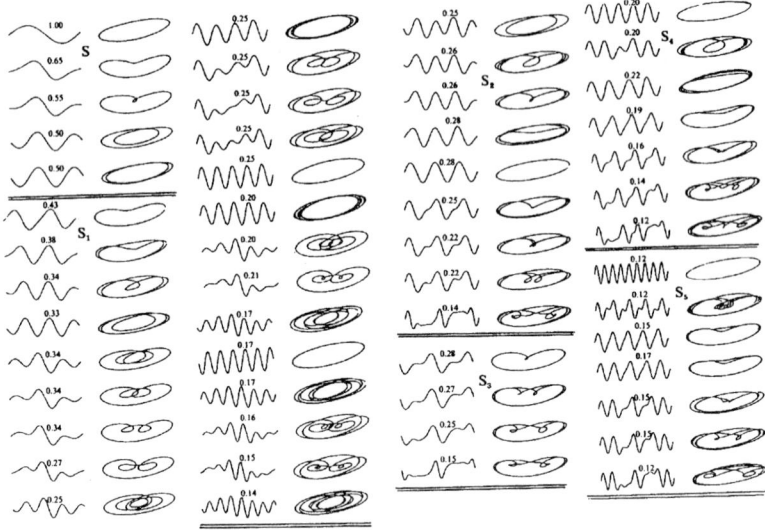

Figure 5.19: Wave profiles $H(X)$ of KS equation for the constant-thickness formulation and their corresponding closed trajectories in the phase space of (H, H', H''). The numbers are the normalized wavenumber s.

wave solution with $C = 0$. It hence represents a family of waves that travel at a constant speed equal to $c_0 = 3$ as the KS equation is derived in a moving frame with this phase speed. In the original laboratory x coordinate, this standing wave is a travelling wave with non-zero speed $c = 3$ that bifurcates from its bifurcation point $(C, (\alpha/\alpha_0)) = (0, 1)$ where α is the wave number in the original x-coordinate and not the normalized X-coordinate. A twin branch with speed $-C$ exists by virtue of (5.16). For the $S^{(1)}$-family, the wave number lies within the bound $s = (\alpha/\alpha_0) \in [1, 0.49775]$. At 0.49775, the $S^{(1)}$ branch of (5.9) coalesces with a second branch $S^{(2)}$ which bifurcated from $s = (\alpha/\alpha_0) = 1/2$, and the original family S disappears. Accordingly, $S^{(n)}$ also disappears after it merges with $S^{(2n)}$ at $s_n = 0.49775$. The $S^{(2)}$ branch is identical to $S^{(1)}$ except two waves are contained in one wavelength which is twice as long as the corresponding one on $S^{(1)}$. Likewise, $S^{(n)}$ branches bifurcate from $s = (\alpha/\alpha_0) = 1/n$ and they are all identical to $S^{(1)}$ except n units exist in one wavelength. Top index, $^{(1)}$, denoting that the solution has a primary period $2\pi/s$ will be omitted in further description, i.e. $S = S^{(1)}$. For $S^{(n)}$-family, only amplitudes of harmonics a_{nk}, b_{nk}, $k = 1, 2, \ldots$ differ from zero. In the vicinity of bifurcation point $s = 1/n(1 - \varepsilon^2)$, $\varepsilon \to 0$, $n = 1, 2, \ldots$, solutions $S^{(n)}$ are

$$H(X) = \pm \varepsilon \sqrt{6} \sin nsX - \varepsilon^2 \sin 2nsX + \ldots \tag{5.28}$$

We first focus on S. All along the S branch, the waves are nearly sinusoidal with longer waves having a larger second harmonic content and hence a steeper front edge (see Figure 5.19). These waves are the ones near the neutral curve and they are the ones amenable to the Stuart-Landau normal mode expansion formalism suggested by Benney (1966) and subsequently carried out by Gjevik (1970), Lin (1974), Nakaya (1975) and Chang (1989). However, as pointed out by Chang (1989), waves longer than twice $2\pi/\alpha_0$ contain many Fourier modes that cannot be easily resolved with a normal mode expansion. These waves are contained in two families of secondary branches. One family consists of the S_n branches that come off the $S^{(n)}$ branches. As is evident in Figure 5.19, at the bifurcation point on $S^{(n)}$, the waves on S_n are still quite sinusoidal but subsequent ones develop very rich structure with a broad Fourier content. Some of them (S_2, S_3, S_5, S_6 etc) extend to $\alpha = 0$ and terminate at solitary waves. Unlike the $S^{(n)}$ branches, these branches do not resemble each other. A subset of the S_n family are the branches S_{1A}, S_{1B}, etc. A second family of waves consist of C_n and C'_n branches as seen in Figure 5.18. In contrast to the $S^{(n)}$ and S_n families, these are travelling wave solutions of (5.9) with c in excess of 3. As shown in Figure 5.21, these travelling waves have unique solitary wave shapes that we have classified in a different report (Chang et al., 1993). The primary one is C_{11}, which will also be called C_1, bifurcates off S at $s = (\alpha/\alpha_0) = 0.5547$. This forwarding propagating branch is the only important branch in the C_n family that will survive at finite Reynolds numbers. In contrast, all inverted ones C_{-n} participate in the pertinent wave branches at finite Reynolds numbers. This C_1 branch has been studied by Armbruster et al. (1988) and, in their formulation, it corresponds to a travelling wave which is annihilated at a homoclinic bifurcation. See also the analysis of Chen and Chang (1992). The homoclinic bifurcation corresponds to an infinitely long solitary wave here as the C_1 branch extends to vanishing α in Figure 5.18. As evident in Figures 5.21 and 5.22, both the S_n and C_n families and their inverted twins with negative λ consist of waves with a large band of Fourier modes. Chang (1986, 1989) suggested that these waves can be described by elliptic functions provided by an analysis in the solitary ($\alpha \to 0$) limit, as will be detailed in section 5.5. For the constant-flux formulation, the homoclinic orbit corresponding to the solitary wave can be approximated by a heteroclinic cycle joining the fixed point corresponding to the Nussel flat film $u_1 = 0$ to the conjugate flat film $u_2 = \frac{1}{2}\lambda$. Hence, an estimate of the amplitude of the solitary wave is simply $u_{max} = \frac{1}{2}\lambda$. However, by multiplying (5.13) by u and integrating from $X = -\infty$ to $+\infty$, we obtain

$$\int_{-\infty}^{\infty} u^2(2u - \lambda)dX = 0, \tag{5.29}$$

which stipulates that $u_{max} > \frac{1}{2}\lambda$, and the above estimate must be considered as a tight lower bound of the amplitude of a solitary wave in the constant-flux formulation. In the solitary wave limit, the deviation flux $Q = \langle 2H^2 \rangle$ approaches zero for the constant-thickness formulation in the limit of vanishing amplitude. Consequently, (5.12) suggests that all solitary waves of the constant-thickness formulation of the KS equation should approach the amplitude-speed correlation $H_{max} = \lambda/2$ as λ and H_{max} approach zero. A more detailed analysis arriving at this same estimate will be presented in section 5.5. The estimate is confirmed to within our numerical accuracy in the solitary wave limits of the C_n branches in Figure 5.20. This, of course, also ensures agreement with the C_{-n} families by the symmetry argument. The only branches of importance at finite δ are S, C_1, C_{-n} and the S_{1A}, S_{1B}, S_{1C} series.

At $s = 0.42115$ and at a local maximum in λ, an S^2 solution bifurcates from $S^{(1)}$ and inherits its instability. However, soon afterwards, $S^{(2)}$ becomes a stable travelling wave. After making a loop in the coordinates $s - \lambda$ and touching the $S^{(3)}$ solution, solution S_1 disappears after merging with $S^{(4)}$.

Solution S_2 can also regarded as branching from $S^{(4)}$ at $s = 0.2490$ towards greater s. At $s = 0.2845$, the family turns back toward smaller s. (This point can be regarded as the birth point of a couple of branches of S_2 through the "birth-death" turning point bifurcation.) After turning backward, S_2 touches family $S^{(3)}$ at a maximum λ point.As a result of this bilateral bifurcation, $S^{(3)}$ acquires stability and unstable family S_2 extends to any arbitrary small values of s to transform into anti-kink at $s \to 0$.

At $s = 0.276$, solution S_3 branches from $S^{(2)}$. The solution exists at whatever small values of s and at $s \to 0$ it asymptotically transforms into a kink with an infinite period: as $X \to +\infty$, $H(X) \to const < 0$.

The S_4-family is similar to the S_2 in its behavior. It branches from $S^{(5)}$ at $s = 0.1990$, touches $S^{(4)}$-family at $s = 0.2096$, and tends to a kink family at $s \to 0$. At $s = 0.2095$, a multiple period family $S_1^{(2)}$ bifurcates from $S_1^{(4)}$ and acquires stability thereafter.

Families $S_1^{(2)}$ and S_3 lose their stability via a couple of conjugate complex eigenvalues of (2.2.4); $S_1^{(3)}$, $S_1^{(4)}$, and $S_1^{(5)}$ lose their stability via two couples of conjugate complex eigenvalues; here, $\mu_{1,2} = \mu_{3,4}$.

Families such as $S_1, S_2, S_3, S_4, S_5, S_6$, put an end to the solution set type of spatial - periodic stationary waves at $s \in [0.2; 1]$. We have given general properties of stationary travelling waves C. These properties are the generalization from extensive calculations.

(1) Each of periodic stationary travelling waves of C - type either branches from an S - type wave or "paired birth" of C - families occurs through the "birth-death" turning point bifurcation.

(2) At $s \to 0$, each C-type solution transforms into a one- or multi-hump pulse.

Major results of the calculations are present in Figures 5.18 and 5.21. The dash line in the second of family C_{13}, C'_{13} is used to denote that those solutions after the pulse solutions merge together as s - values increase. Otherwise, as

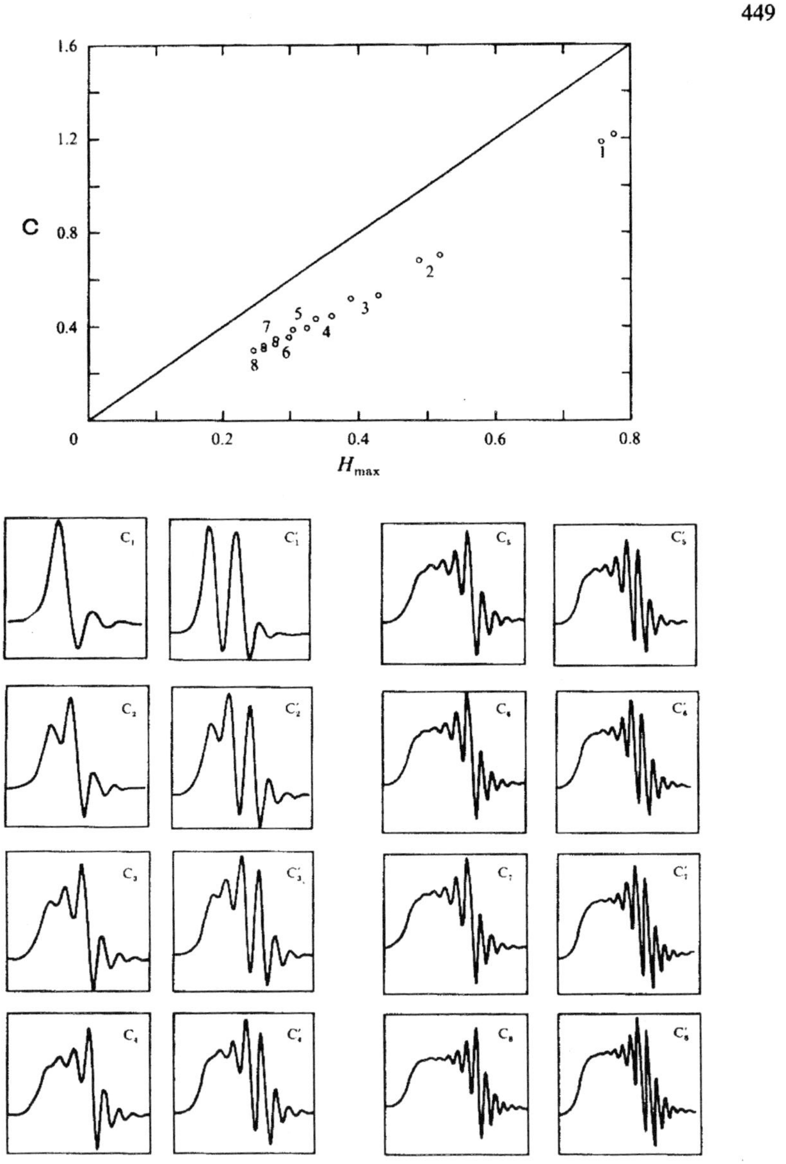

Figure 5.20: Comparison between the amplitude-speed correlation for solitary wave limits of the C_n branches and the analytical estimate of (5.61). The higher C_n and C_{-n} solitary waves are to the right of Figure 5.18.

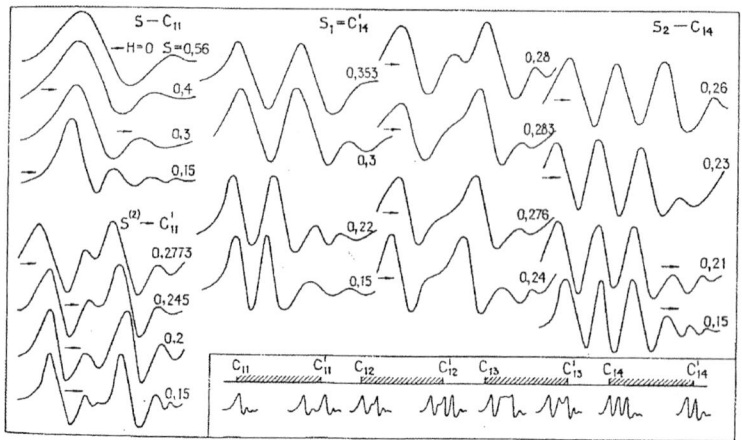

Figure 5.21: The stationary running periodical waves and their connection with solitary pulses.

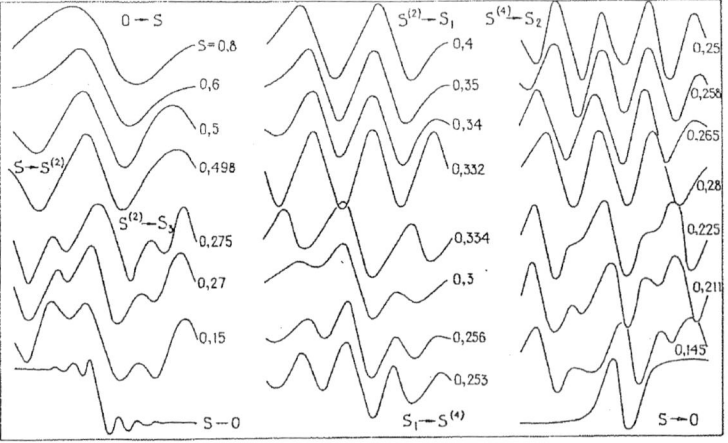

Figure 5.22: The stationary standing wave profiles and their connection with kinks and anti-kinks.

s - values decrease, a couple of C_{13}, C'_{13} isolated pulse branches appear via the "birth-death" bifurcation at $s = 0,1150$. Figure 5.22 shows profiles of travelling waves, $H(X)$, at various wave number values.

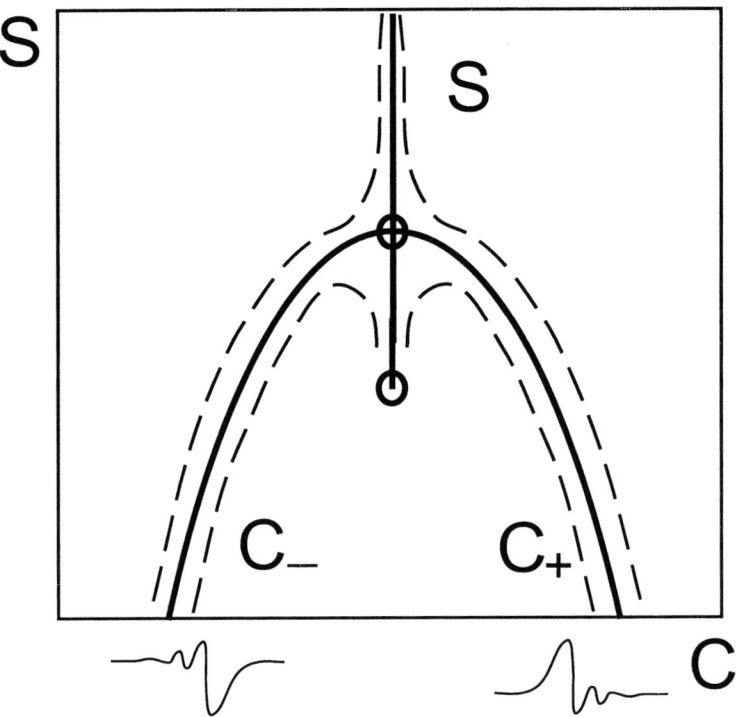

Figure 5.23: Main families of KS-equation which have imperfect pitchfork bifurcation.

Instead of such a complex bifurcation picture, however, only a few families are pertinent for finite Reynolds numbers. The most important families are standing waves S and two families of stationary running waves $C_{+1} = C_{+11}$ and $C_{-1} = C_{-11}$. In Figure 5.23, summarizing our results, we skip subscripts. At $s \to 0$ C_+-waves tend to one hump positive soliton running with a positive speed and C_--waves - to one - hollow negative soliton running with a negative speed. Such a bifurcation of two symmetric families with velocities $\pm C$ from a standing wave is structurally unstable and is possible only because of symmetry (5.14). Small perturbation of the equation (5.11) coming from higher-order terms missing in the KS equation will remove this symmetry and change the bifurcation diagram, as is shown in Figure 5.23. Two possible new solution structures are depicted there as dashed lines.

Our basic solution is S. Consider a branching condition of new periodic solutions branched from S in order to understand previous complex picture. Let us impose an infinitesimal disturbance, $H \to H + f$, to the solution of the first family, and having substituted it into equation (5.13) and linearized, we can obtain a linear equation with $2\pi/s$ - periodic coefficients :

$$f''' + f' - cf + 4Hf = 0 \tag{5.30}$$

The three fundamental solutions of this equation according to Floquet theorem are as follows:

$$f = F(X)\exp(is\nu_k X), \quad f_k(X + 2\pi/s) = \rho_k f_k(X) \tag{5.31}$$

Here, $F(X)$ - is a $2\pi/s$-periodic function; ρ_k are the Floquet multipliers. Since $f_3 = H'$ is a $2\pi/s$ periodic solution of equation (5.30), $\rho_3 = 1$. In equation (5.30) the term with a second-order derivative is not present, hence, $\rho_1\rho_2\rho_3 = 1$ or $\rho_1\rho_2 = 1$. This is a direct result of the Liouville theorem (5.21) for reversible system - the linearized Jacobian should have zero trace in the dynamical system (5.18) to (5.20) and the Floquet multiplier should multiply to unity.

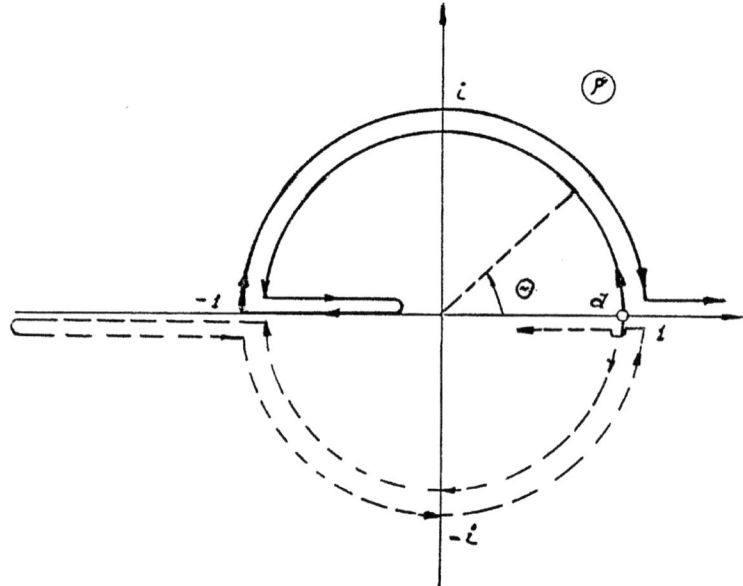

Figure 5.24: Multipliers $\rho_{1,2}$: start out from the point d $(1, \pm 0)$ as a complex conjugate pair and move along the complex unite circle till the point $(-1, \pm 0)$ where they become real numbers with $\rho_1\rho_2 = 1$ (hyperbolic case).They then move in different directions for a while and then approach each other and meet at $(-1, \pm 0)$ for the second time.

Since the coefficients in (5.30) are real, there are two possible cases: either the multipliers are complex conjugate numbers lying on a unit circle (elliptical case), $\rho_{1,2} = e^{\pm i\theta}$ being the case required for branching, or ρ_1, ρ_2 are purely real numbers (hyperbolic case), $\rho_2 = 1/\rho_1$ and then (5.30) has no solution at $X \to \pm\infty$ and no branching takes place. In the elliptic case, cycle γ appears to be surrounded invariant tori (corresponding to quasi-periodic travelling waves) in the phase space, as is observed in Hamiltonian systems. For our system, this is possible because of the reversibility condition (5.16) . In the hyperbolic case, γ-cycle is of saddle type, i.e. two invariant surfaces, W_γ^u and W_γ^s , exists in the phase space corresponding to stable and unstable manifolds of the cycle, see Figure 5.7.

Numerical calculations of the Floquet multipliers reveal that, as the value of ε increases starting from zero ($s = 1 - \varepsilon^2$), a pair of multipliers $\rho_{1,2} = e^{\pm i\theta}$ branch from point (1,0) of the complex plane and pass through the unit circle at the top and bottom, see Figure 5.24 At $s = s_1$, $\varepsilon^2 = \varepsilon_1^2 = 4.2 * 10^{-3}$ these multipliers merge at point $(-1,0)$. When ε increases further , a double root is separated into two and here, $\rho_{1,2}$ are negative numbers lying on a real axis both in and out of the limits of the unit circle (hyperbolic case). At $s = s_2$, $\varepsilon^2 = \varepsilon_2^2 = 7.6 * 10^{-2}$ multipliers return to point $(-1,0)$, merge into a single root and with further increase of ε they move along the unit circle, whereas at $s = s_3$, $\varepsilon^2 = \varepsilon_3^2 = 8.8 * 10^{-2}$ the roots merge together at point $(1,0)$. At $\varepsilon > \varepsilon_3$, multipliers appear to be positioned on the positive real semi-axis.

In some neighborhood of $\varepsilon = 0$ we could develop an analytical solution of (5.30). We will seek it in the form of an asymptotic expansion:

$$F \sim F_0 + \varepsilon F_1 + \varepsilon^2 F_2 + \varepsilon^3 F_3, \ \nu \sim \nu_1 + \varepsilon^3 \nu_3 \tag{5.32}$$

It is easy to obtain the expansion for Floquet coefficient:

$$\nu = \pm 2\sqrt{6}\varepsilon(1 + 871/24\varepsilon^2), \tag{5.33}$$

The analytical solution is in good agreement with the numerical one, except for a small neighborhood of $\varepsilon = \varepsilon_1$.

If we recalculate ε to s and find corresponding periodic solution u and its λ, we find out that for the branch S for $\lambda \in [0.449, 1.791]$ these periodic solutions are saddle type while in the remaining region - elliptic, see subsection 5.1.1.

Then, at $\varepsilon \in (0, \varepsilon_1] \cup [\varepsilon_2, \varepsilon_3]$ f_k in general are doubly periodic functions. However, at $\nu = 1/2, 1/3, \ldots, 1/m, \ldots$ f is a periodic function with periods, $4\pi/s, 6\pi/s, \ldots, 2\pi m/s, \ldots$, respectively, i.e. in respective points, $s = s^{(2)}, s^{(3)}, \ldots, s^{(m)}, \ldots$, the necessary condition for branching is satisfied for the periodic solutions of the generating equation. It is essential that the intervals where branching is possible be two. It reflects the fact that all bifurcation points of the families $S^{(n)}$ are located near maximum λ or at small values λ, see Figure 5.18.

In the limits $s \to 0$, doubly periodic solutions branch from $S^{(n)}$. Doubly periodic solutions, being invariant tori in the phase space, can be built analytically at $Q \to 0$ of (5.11).

We shall seek now infinitely long antisymmetric quasi-periodic solutions of (5.11) with $C = 0$ and are bounded within $X \in (-\infty, +\infty)$. Moreover, these travelling waves have zero mean thickness over the entire domain :

$$\lim_{l \to \infty} \frac{1}{2l} \int_{-l}^{l} H \, dX = 0 \qquad (5.34)$$

Let us compress the variable $X \to sX$, introduce a small value ε and in the neighborhood of $s = 1$, seek a solution by using the multi-scale method described in Van Dyke (1975) and Nayfeh (1973):

$$s = 1 \pm \varepsilon^2, \qquad (5.35)$$

$$H \sim \varepsilon H_1 + \varepsilon^2 H_2 + \varepsilon^3 H_3, \qquad (5.36)$$

$$Q \sim \varepsilon^2 Q_2 + \varepsilon^3 Q_3 \qquad (5.37)$$

$$H_k = H_k(X_0, X_1, X_2), \quad X_k = \varepsilon^k X \qquad (5.38)$$

By substituting (5.35) - (5.38) into (5.11) and collecting terms having identical powers of ε, we obtain:

$$H_1''' + H_1' = 0 \qquad (5.39)$$

$$H_2''' + H_2' = -\frac{\partial}{\partial X_1}(3H_1'' + H_1) - H_1^2 + Q_2 \qquad (5.40)$$

$$H_3''' + H_3' = -\frac{\partial}{\partial X_1}(3H_2'' + H_2) - 2H_1 H_2 - \frac{\partial}{\partial X_2}(3H_1'' + H_1) - 3\frac{\partial^2 H_1}{\partial X_1^2} \pm 2H_1''' + Q_3 \qquad (5.41)$$

Here Q_1 and Q_2 are unknown constants. A derivative with respect to X_0 is denoted by prime. The expansion is valid until $|x| = O(\varepsilon^{-3})$. Taking into account the symmetry of equation (5.11), we choose the solution in the form of:

$$H_1 = A_1 \sin X_0 + B_1, \quad A_1 = A_1(X_1, X_2), \quad B_1 = B_1(X_1, X_2) \qquad (5.42)$$

By substituting (5.42) into (5.40) we derive from the solvability condition :

$$\frac{\partial A_1}{\partial X_1} - A_1 B_1 = 0, \quad \frac{\partial B_1}{\partial X_1} + \frac{1}{2}A_1^2 + B_1^2 - Q_2 = 0 \qquad (5.43)$$

From the last equations:

$$\frac{\partial^2 B}{\partial X_1^2} + 2B(Q_1 - B^2) = 0 \qquad (5.44)$$

Bounded solution of (5.43) with respect to (5.44) are presented in the analytic form using elliptic functions:

$$B_1 = b \cdot sn(\xi, k), \quad A_1^2 = (a \cdot dn(\xi, k) - b \cdot ch(\xi, k))^2, \quad Q_2 = 1/2(a^2 + b^2)$$

Here, $b < a$, $k^2 = b^2/a^2$, $\xi = aX_1$. Among parameters a, b, k, one of them can be taken arbitrarily. Modulus k characterizing the second wave number of a biperiodic solution is chosen as such parameter. In this case, $a = a(k, X_2)$, $b =$

$b(k, X_2) = ka$; to determine k-dependence the third approximation is necessary. The expression for H_2 from (5.40) and that for

$$H_2 = A_2 \sin X_0 + D_2 \cos X_0 + B_2 - 1/12 A_1^2 \sin 2X_0$$

are substituted in (5.41). From the solvability condition, we obtain:

$$\frac{\partial A_1}{\partial X_1} - A_1 B_2 - B_1 A_2 = -\frac{\partial A_1}{\partial X_2}, \qquad (5.45)$$

$$\frac{\partial B_2}{\partial X_1} - A_1 A_2 - 2 B_1 B_2 - Q_3 = -\frac{\partial B_1}{\partial X_2}, \qquad (5.46)$$

$$\frac{\partial D_2}{\partial X_1} - B_1 D_2 = \frac{1}{2} A_1 \left(-\frac{19}{12} A_1^2 + 3 Q_2 \pm 2 \right) \qquad (5.47)$$

For the existence of multi-periodic solutions of (5.45)-(5.47), it is necessary and sufficient that the solution of an adjoint homogeneous equation, $\partial V/\partial X_1 + B_1 V = 0$, be orthogonal with respect to the right of (5.47), the Fredholm alternative. It is easy to find that $V = 1/A$. Then, on the basis of conditions of orthogonality and after appropriate transformations, we obtain:

$$a^2 = \pm 24/(38 \mathcal{E}(k)/\mathcal{K}(k) + k^2 - 37) \qquad (5.48)$$

Here, $\mathcal{K}(k)$ and $\mathcal{E}(k)$ are complete elliptic integrals of the first and second type. (Plus sign corresponds to branching down from $s = 1 - \varepsilon^2$; minus sign corresponds to branching up from $s = 1 + \varepsilon^2$). Figure 5.25(a) shows the behavior of $H(X)$ at various k - values. The dependence of $|a|$ on k is shown in Figure 5.25 (b). At $k = k_1 = 0.236$ denominator in (5.48) is zero and the expansion becomes invalid. For the first sign, $k \in [0, k_1]$. At $k = 0$, the solution transforms into a periodic solution

$$H \sim \pm \sqrt{6} \varepsilon \sin X - \varepsilon^2 \sin 2X, \quad Q \sim 12 \varepsilon^2.$$

which coincides with (5.28). For the second sign, $k \in (k_1, 1]$. At $k = 1$ there exist two limiting expansions which correspond to a phase space heteroclinic trajectory:

(1) $H = \sqrt{\frac{2}{3}} \left\{ \operatorname{sech} \left(2 \sqrt{\frac{2}{3}} \varepsilon X \right) \sin X - \tanh \left(\sqrt{\frac{2}{3}} \varepsilon X \right) \right\}, \quad Q = \frac{2}{3} \varepsilon^2, \quad (5.49)$

(2) $H = \sqrt{\frac{2}{3}} \tanh \left(2 \sqrt{\frac{2}{3}} \varepsilon X \right), \quad Q = \frac{2}{3} \varepsilon^2 \qquad (5.50)$

Analyzing higher expansions one can demonstrate that case (2) is not a converging asymptotic solution of (5.11).

These analytical results are complemented by direct numerical integration of (5.11). It is supposed that $H(0) = H''(0) = 0$; H' and Q are taken at the elliptic values. The plane $H' = 0$ is taken to define the domain and range of the Poincare

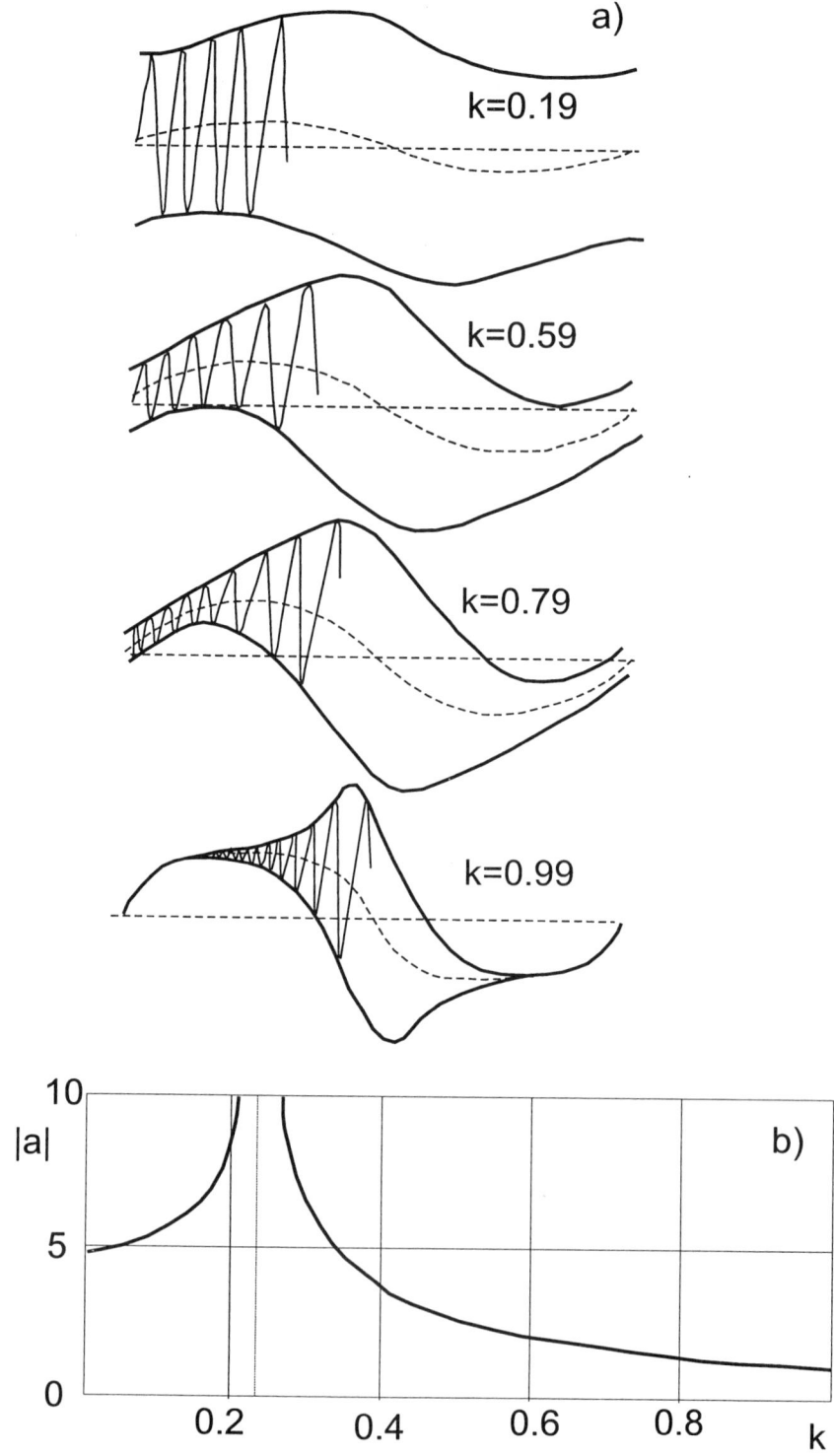

Figure 5.25: Wave profiles a) and amplitude (b) of two-periodical solutions.

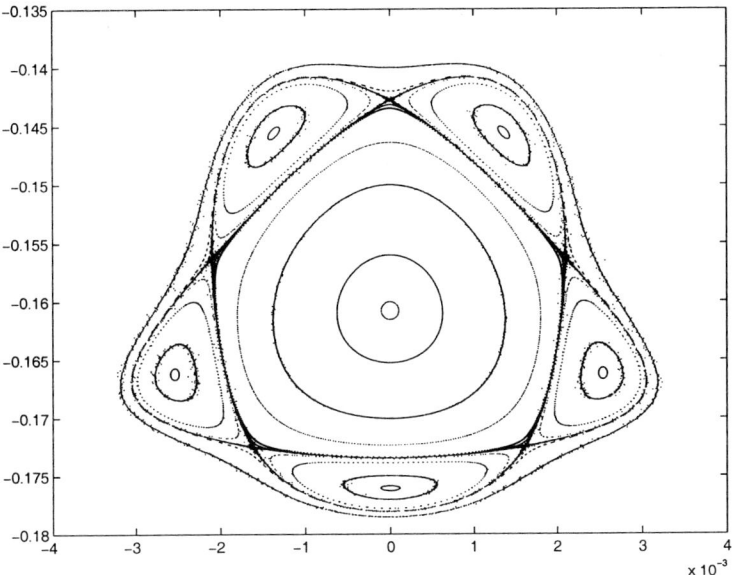

Figure 5.26: Poincare map in the plane H and H' for the elliptic case. One can see KAM - tori and, in the vicinity of the hyperbolic points, we can find more tori (compare this picture with the famous Henon map).

map T, see Figure 5.26. Periodic solutions appear as fixed points and tori as a circular collection of iterates in the figure. Most interestingly, a continuous family of tori exist and a countable infinite number of them are unstable to procedure hyperbolic and elliptic fixed points of T, as seen in the figure. Ergotic iterates persist in the homoclinic tangle created by this bifurcation. Each torus correspond to a quasi-periodic wave of Figure 5.25. However, the solvability condition of the previous construction assumes periodicity which is not obeyed by the chaotic trajectories if T which bifurcate from certain tori within the family. Unlike the KAM bifurcation, we are unable to decipher which tori undergo such bifurcations to yield hyperbolic and elliptic intersections of tori in Figure 5.26.

5.5 Normal Form analysis for the Kawahara equation.

Now let us consider gKS or Kawahara equation for $\delta \neq 0$. We shall examine the solitary wave solutions of the gKS equation, see Cheng and Chang (1986), Chang (1987) and Chang et al. (1993)

$$u''' + \delta u'' + u' - \lambda u + 2u^2 = 0 \tag{5.51}$$

which is invariant to

$$\delta \to -\delta,\ x \to -x,\ u \to -u,\ \lambda \to -\lambda. \tag{5.52}$$

We shall hence investigate either the case of $c > 0(\lambda > 0)$ or $\delta > 0$. It is clear from the reflection symmetry across the origin of the $\lambda - \delta$ parameter space that for every periodic or solitary wave that propagates in one direction for $\delta > 0$, it has a counter-propagating dual with an inverted profile for $\delta < 0$. For the Kuramoto-Sivashinsky equation with $\delta = 0$, there is an additional symmetry that implies that for every forward propagating solution, there is a backward propagating dual with an inverted profile. In fact, at $\delta = 0$ the KS equation is a "reversible" system (see (5.14) and (5.15)) which has been shown to behave like Hamiltonian systems. This curve represents pulse solutions. This point

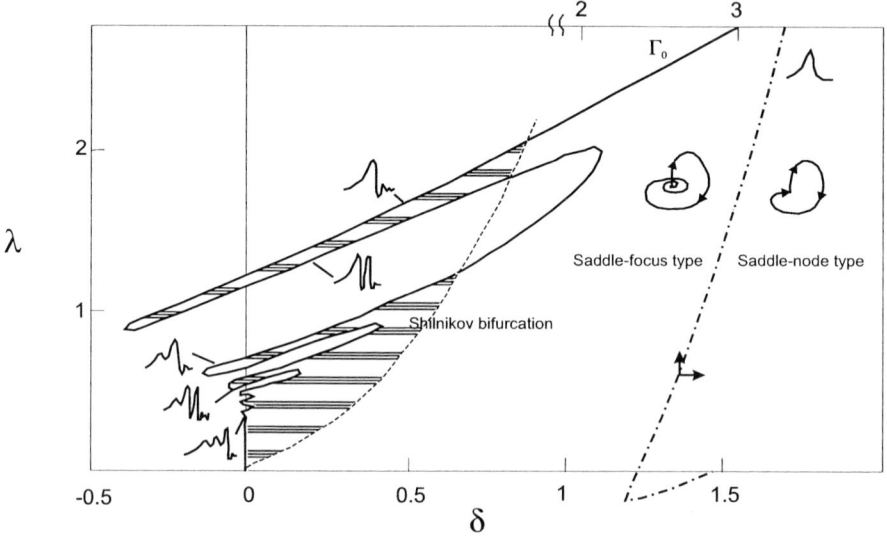

Figure 5.27: Bifurcation diagram for solitons of gKS (Kawahara) equation.

corresponds to our one- hump generic pulse of Figure 5.5 at $\delta = 0$. If one goes along δ to the left, the curve then makes a U-turn and the one-hump pulse will transform into the two-hump, then - three-hump, and so on, into a multi-hump pulse. Zero is an accumulation point. In the vicinity of this point it is possible to find the solution analytically.

Our strategy here is not to track all the solution branches as we did for the KS equation. Instead, we shall use the normal form theory to remove the fine structures and capture a carricature of the pulse solution only.

We can write (5.51) as a dynamical system by defining $u = (u, u', u'')$,

$$u' = \begin{pmatrix} 0 & 1 & 0 \\ 0 & 0 & 1 \\ 0 & -1 & 0 \end{pmatrix} u + \begin{pmatrix} 0 & 0 & 0 \\ 0 & 0 & 0 \\ \lambda & 0 & -\delta \end{pmatrix} u - \begin{pmatrix} 0 \\ 0 \\ 1 \end{pmatrix} 2u_1^2, \tag{5.53}$$

and study the stationary waves in the phase space. Two fixed points, $O_1(0,0,0)$ and $O_2(\lambda/2,0,0)$, exist for (5.53). Closed trajectories (limit-cycles) of (5.53) represent spatially periodic stationary waves and some of them can be shown to bifurcate from O_1 and O_2 via Hopf bifurcations. It is hence pertinent to analyze the phase-space stability of O_1 and O_2. Linearizing (5.53) about O_1 and O_2, one obtains the characteristic equations

$$\sigma^3 + \delta\sigma^2 + \sigma \pm \lambda = 0, \qquad (5.54)$$

where - and = correspond to O_1 and O_2 respectively. It is clear that O_1 and O_2 undergo a Hopf bifurcation at

$$\lambda = \pm\delta \qquad (5.55)$$

respectively with a frequency of $\sigma_i = 1$. Spatially periodic stationary waves with unit wavelength are hence expected near these points. The two Hopf lines are depicted in Figure 5.27.

Stationary solitary waves correspond to homoclinic trajectory Γ of O_1. For such a trajectory to exist, O_1 cannot be stable, $Re\{\sigma_i\} < 0$, or "absolutely" unstable, $Re\{\sigma_i\} > 0$ for $i = 1, 2$ and 3. A simple analysis of (5.54) yields that solitary waves can exist only in sectors

$$\psi \in (0, \frac{3}{4}\pi) \cup (\pi, \frac{7}{4}\pi) \qquad (5.56)$$

in the $\lambda - \delta$ parameter space as shown in Figure 5.27. Within the first sector, the fixed point O_1 has a stable manifold W_1^s corresponding to the stable eigenvalue σ_3 and σ_2 and an unstable manifold W_1^u corresponding to σ_1. One is a two-dimensional surface and the other a one-dimensional curve. Using the reflection symmetry across the origin of (5.52), we can restrict ourselves only to $\lambda > 0$. (In this region, dim $W_1^u = 1$ and dim $W_1^s = 2$.) The unstable eigenvalue of O_1 is necessarily a real one but the stable pair can be a complex pair or two real numbers, each case corresponding to a saddle-focus and a saddle-node, respectively. A forward travelling solitary wave in the top half of fig. 1, which corresponds to a homoclinic orbit connected to a saddle-focus, has small decaying "bow" waves in front of the main humps. One that is connected to a saddle-node exhibits smooth entry in the positive x-direction. We hence distinguish these two types of solitary waves by the criterion

$$S_0 = \lambda^2 - \delta\left(\frac{4}{27}\delta^2 - \frac{2}{3}\right)\lambda + \frac{1}{27}(4 - \delta^2), \qquad (5.57)$$

with a positive S_0 corresponding to a saddle-focus and a negative S_0 corresponding to a saddle-node. It is clear that S_0 is negative at large δ and we hence expect the small bow waves to be eliminated by dispersion. This is consistent with Kawahara's asymptotic estimate of the solitary wave at large δ

$$u \sim \frac{21\delta}{20\cosh^2\sqrt{\frac{7}{20}}x}, \qquad (5.58)$$

$$\lambda = \frac{7}{5}\delta, \tag{5.59}$$

which has the shape of a symmetric KdV solitary wave that exhibits no bow waves. We note that this limit yields a δ_0 value of 3.7 as the upper bound for solitary waves front-running bow waves from (5.57).

Another piece of information that can be extracted readily comes from Shilnikov's theory for homoclinic orbits in section 5.3. Since solitary waves in the top half of Figure 5.27 correspond to homoclinic orbits Γ which arise due to intersection of a one-dimensional W_1^u and two-dimensional W_1^s, they are structurally unstable. A small movement of one parameter can break this intersection. Consequently, for a given δ, there is only a discrete set of $\{\lambda_i\}$ that correspond to solitary waves. However, given a homoclinic orbit Γ, Shilnikov's theory stipulates that a countable infinite number of other homoclinic orbits can be generated by perturbing the speed parameter λ if the eigenvalues σ_i of O_1 satisfy the condition that

$$S_1 = \sigma_1 + Re\{\sigma_2\} = \lambda - \delta - 2\delta^3 \tag{5.60}$$

is positive. This curve is shown in Figure 5.27. Note that as $\delta \to 0$, the curve $S_1 = 0$ reduces to the Hopf line $\lambda = \delta$ of O_2 in Figure 5.27. These new homoclinic orbits consist of repeated traverse of the original Γ before being captured by O_1. Hence, for every solitary wave we uncover in the first quadrant with $\lambda < \delta$, a countable infinitely many others can be generated whose shapes can also be deduced from the above argument.

Finally, as a preliminary analysis before the numerical study, we offer a normal form analysis of the dynamical systems (5.53) in the neighborhood of $(\delta, \lambda) = (0, 0)$. In an earlier study (Chang, 1986), we have carried out a normal form analysis of the KS equation and obtained a correlation between amplitude and speed for solitary waves

$$u_{max} = \lambda/2 \tag{5.61}$$

and a normalized correlation between amplitude u_{max} and wave number α for periodic waves given by the following implicit formulae:

$$\frac{u_{max}}{\lambda} = z + 2r, \tag{5.62}$$

$$1280r^8 - 64r^6 + 32Cr^4 + C^2 = 0, \tag{5.63}$$

$$\left(\frac{1}{2}z - z^2 - r^2\right)16r^2 = C, \tag{5.64}$$

$$\frac{2\pi\lambda}{\alpha} = \tau(C), \tag{5.65}$$

$$\tau(C) = (2\sqrt{2}/R_1)K(\kappa), \tag{5.66}$$

$$\kappa = [1 - (R_0/R_1)^2]^{1/2}, \tag{5.67}$$

$$R_{0,1}(C) = (1 \pm \sqrt{1 - 64C})^{1/2}/2, \tag{5.68}$$

where K is the Legendre normal elliptic integral of the first kind and C is a convenient parameter which parameterizes the periodic solutions. we shall show that these results are still valid with finite but small dispersion. We begin by diagonalizing the Jacobian of (5.53) at $(\lambda, \delta) = (0,0)$ with the following similarity transform:
$$u = Tv, \tag{5.69}$$
where
$$T = \begin{pmatrix} 1 & 1 & 1 \\ 0 & i & -i \\ 0 & -1 & -1 \end{pmatrix}. \tag{5.70}$$

The transformed equation then becomes

$$v' = \begin{pmatrix} 0 & 0 & 0 \\ 0 & i & 0 \\ 0 & 0 & -i \end{pmatrix} v + \begin{pmatrix} \lambda & \lambda + \delta & \lambda + \delta \\ -\frac{1}{2}\lambda & -\frac{1}{2}(\lambda + \delta) & -\frac{1}{2}(\lambda + \delta) \\ -\frac{1}{2}\lambda & -\frac{1}{2}(\lambda + \delta) & -\frac{1}{2}(\lambda + \delta) \end{pmatrix} v + $$
$$+ \begin{pmatrix} -2 \\ 1 \\ 1 \end{pmatrix} (v_1 + v_2 + v_3)^2 = Av + \hat{A}v + g(v). \tag{5.71}$$

It is clear from (5.71) that at $(\lambda, \delta) = (0, 0)$, the eigenvalues of O_1 are $(0, \pm i)$. This is a codimension 2 singularity that can be analyzed with a normal form analysis. We shall restrict ourselves to an $O(\mu)$ neighborhood of $(\lambda, \delta) = (0, 0)$, viz. $\lambda, \delta \sim O(\mu)$. We seek a near-identity linear transformation
$$v = (T + B)z, \tag{5.72}$$
where $B \sim O(\mu)$ to simply (5.70). If $|z| \sim |v| \sim O(\mu^{1/2})$, this linear transformation will not affect the quadratic nonlinear terms. It can be readily shown that if
$$B = \begin{pmatrix} 0 & -(\lambda + \delta)i \\ -\frac{1}{2}\lambda i & 0 \end{pmatrix}, \tag{5.73}$$
(5.71) is transformed to
$$z' = \begin{pmatrix} 0 & 0 & 0 \\ 0 & i & 0 \\ 0 & 0 & -i \end{pmatrix} z + \begin{pmatrix} \lambda & 0 & 0 \\ 0 & -\frac{1}{2}(\lambda + \delta) & 0 \\ 0 & 0 & -\frac{1}{2}(\lambda + \delta) \end{pmatrix} z + \begin{pmatrix} -2 \\ 1 \\ 1 \end{pmatrix} (z_1 + z_2 + z_3)^2. \tag{5.74}$$

We note that the perturbation matrix $\hat{A}$ has now been diagonalized and (5.74) indicates that there is a simple bifurcation at $\lambda = 0$ and a Hopf at $\lambda = -\delta$ at O_1. This is, of course, consistent with Figure 5.27. A near-identity nonlinear coordinate transformation to remove nonresonant nonlinear terms is next carried out. It is the same one we used for the KS equation and we simply present the transformed equation here
$$z' = \lambda z - 2z^2 - 4r^2, \; r' = -\frac{1}{2}(\lambda + \delta)r + 2rz, \; \theta' = 1, \tag{5.75}$$

where $z \sim z_1$, $r^2 \sim z_2^2 + z_3^2$ and $\theta \sim \cos^{-1}[(z_2 + z_3)/2r]$. The KS equation is also transformed into (5.75) but with $\delta = 0$. We note that like (2.2), (5.75) is also invariant to transformation (5.52) which now corresponds to

$$\delta \to -\delta, \ \lambda \to -\lambda, \ z \to -z, \ x \to -x, \ \theta \to -\theta, \qquad (5.76)$$

viz. a reflection symmetry about the origin of the (λ, δ) plane yields the inverted, backward travelling dual. The fixed points O_1 and O_2 now correspond to $(z, r) = (0, 0)$ and $(\mu_1/2, 0)$. However, there is an additional fixed point p to (5.75) that is given by

$$\begin{pmatrix} z \\ r \end{pmatrix} = \begin{pmatrix} \frac{1}{4}(\lambda + \delta) \\ (\lambda^2 - \delta^2)^{1/2}/4\sqrt{2} \end{pmatrix}. \qquad (5.77)$$

Since $r \neq 0$ for this fixed point, the rotation of the invariant $z - r$ plane in the phase space of (5.74) by a unit angular velocity ($\theta' = 1$) indicates that this new fixed point p corresponds to a closed limit-cycle trajectory in the phase space and a periodic stationary wave in the real space. It exists for

$$\lambda^2 > \delta^2 \qquad (5.78)$$

viz. supercritical bifurcations occur at the Hopf lines of (5.55). The phase portraits of (5.75) are shown in fig. 2 which is equivalent to these of (5.53) in a small neighborhood near $(\lambda, \delta) = (0, 0)$. Because of the reflection symmetry across the origin, only one half of the $\lambda - \delta$ parameter space is shown.

The normal form analysis hence yields the following predictions for small δ. Two families of periodic solutions exist at every δ, one family with speed λ larger than δ and the other with speed λ less than $-\delta$. Hence, unlike the KS equation, standing waves ($\lambda = 0$) do not exist with dispersion. We also note that homoclinic orbits, if they exist, must lie within a small neighborhood of the λ axis that cannot be resolved by the resolution of (5.75). To "fan" out the degeneracy of the phase portrait, one would have to go on to the cubic or higher order terms. This will not be carried out here. We simply note that all solitary waves must be confined to a small cone near the λ axis. Physically, this implies that the solitary waves are mostly nested in a region near $\delta = 0$ and dispersion reduces the number of solitary waves. Nevertheless, such solitary waves and the multi-peak periodic waves corresponding to the closed orbits on the $r - z$ plane along the λ axis should still be described by correlations (5.61) and (5.66). Dispersion has negligible effect within this small cone. In particular, the primary homoclinic orbit Γ_0 in Figure 5.5 should be a good template for all solitary wave solutions near $\delta = 0$. Each solitary wave corresponds to different numbers of traverses around Γ_0 before returning to O_1 and to different numbers of loops around the fixed points O_1 and O_2 in the three-dimensional phase space. In fact, even for δ large, we shall always find a primary loop Γ_0 from which we can build the other solitary wave at the same δ. Of course, for δ away from zero, (5.61) and (5.66) do not apply while all solitary waves near $\delta = 0$ obey these two estimate based on Γ_0. Estimate (5.61) is favorable compared to the KS solitary pulses in Figure 5.20. For these pulses, $Q = 0$ and the two formulations of (5.11) and (5.13) are identical.

5.6 Nonlinear waves far from criticality – the Shkadov model.

Nonlinear waves on a falling fluid film at moderate values of Reynolds number R and large capillary forces, $\gamma \gg 1$, are described by Shkadov's model (3.88)-(3.89) for thickness $h(x,t)$ and flow rate $q(x,t)$. Stationary waves travelling with velocity c are described by

$$h^3 h''' + \delta[6(q_0 - c)^2 - c^2 h^2]h' + [h^3 - q_0 - c(h-1)] = 0 \qquad (5.79)$$

Here, q_0 is an integration constant. For periodic waves, as in (5.11), it corresponds to the flowrate for the periodic wave and must be determined in the course of solution. The parameter δ is the modified Reynolds number,

$$\delta = 3^{-7/9} 5^{-1} \gamma^{-1/3} R^{11/9}$$

where γ is the Kapitza number of the (2.24). As described in Chapter 2, the modified Reynolds number δ replaces two parameters R and W in the original Navier-Stokes equation. This simplification was achieved by stretching the longitudinal coordinate x κ times, where κ is determined by the relationship (2.40):

$$\varkappa = 3^{-2/9} \gamma^{1/3} R^{-2/9} \qquad (5.80)$$

At all δ, (5.79) has a trivial solution $h(x) \equiv 1$ and $q_0 \equiv 1$ corresponding to the Nusselt waveless flat film. At all δ, this solution is unstable for infinitesimal sinusoidal disturbances with wave-number α within the range $0 < \alpha < \alpha_0 = \sqrt{15\delta}$. The neutral wave number $\alpha_0 = \sqrt{15\delta}$ can be converted into the scales of the original coordinate x in the form of α_1:

$$\alpha_0/\varkappa = \alpha_1 = 3^{1/3} \gamma^{1/2} R^{5/6} = \sqrt{\frac{3}{W}} \qquad (5.81)$$

which agrees with the relationship of Benjamin (2.103) at $\theta = \pi/2$.

When $\delta \to 0$, the instability interval $\alpha \in (0, \alpha_0)$ shrinks towards the origin as $\alpha_0 \to 0$. It is convenient to introduce, for small α_0, new stretched variables $H(X), Q, \lambda$ and X from the relationships:

$$h = 1 + \alpha_0^3 H(X), \quad q_0 = 1 + \alpha_0^6 Q$$
$$c = 3 + \alpha_0^3 \lambda, \quad x = X/\alpha_0, \quad \alpha_0 = \sqrt{15\delta} \qquad (5.82)$$

Then, as $\delta \to 0$, (5.79) becomes the model KS equation:

$$H''' + H' - \lambda H + 3H^2 = Q \qquad (5.83)$$

Periodic travelling wave solutions of (5.79) exist for $\alpha \in (0, \alpha_0), \alpha_0 = \sqrt{15\delta}$ In particular the first family of such solutions bifurcates from the trivial one at the upper branch of the neutral stability curve $\alpha = \alpha_0(1-\varepsilon)$, in the limit of $\varepsilon \to 0$:

$$h = 1 + 2(\varepsilon)^{1/2}\beta \cos \alpha x + \varepsilon\beta^2(-7/5 \cos 2\alpha x + 1/\alpha_0^3 \sin 2\alpha x),$$
$$c = 3 - 123/10\varepsilon\beta, \quad q_0 = 1 + 6\varepsilon\beta^2, \quad \beta^2 = 2\alpha_0^6/(3 + 621/50\alpha_0^6) \quad (5.84)$$

Far from α_0 we numerically construct this solution in the form:

$$h = 1 + \sum_{k=1}^{N} h_k \exp(ik\alpha x) + h_k^* \exp(-ik\alpha x) \quad (5.85)$$

Without loss of generality, we can suppose that the imaginary part of h_1 is equal to zero – this corresponds to specifying the spatial phase of the periodic wave. If we substitute (5.85) to (5.79), we get a nonlinear system of $2N+1$ equations with $2N+1$ unknowns ($q_0, c, h_1^{(R)}, h_2^{(R)}, h_2^{(I)}, \ldots, h_N^{(R)}, h_N^{(I)}$). If the solution is given at an initial value α, it can be continued to the other parameter values using the Newton procedure.

The first wave family travels at a speed slower than the phase speed of inception waves, $c < 3$, thus the family will be called the family of slow waves, or γ_1. A typical γ_1 family at $\delta = 0.04$ is shown in Figure 5.28. As their wavelength increases, they evolve into dimple-like structures with back-running capillary waves. These, in fact, approach the negative - hump solitary waves of the C_- family of the KS equation in Figure 5.23. At small δ, the amplitude and the nonlinear distortion of the flowrate have two maximum while the phase velocity has one minimum. As δ increases, the smaller maximum $q_0(\alpha)$ vanishes, as does the unique minimum of the phase velocity. Following Shkadov (1967), we trace the wave with the maximum possible flowrate q_0 as a function of δ in Table 5.3. These waves are typically shorter than those with the maximum linear growth rate.

Table 5.3

δ	0.02	0.04	0.06	0.08	0.1	0.2	0.4	1.0
α	0.455	0.627	0.721	0.789	0.845	1.00	1.25	1.97
c	2.983	2.901	2.781	2.699	2.569	2.287	2.094	2.068
q_0	1.007	1.040	1.082	1.118	1.145	1.205	1.226	1.232

There are other periodic wave families. The one corresponding to the C_+ family of the KS equation of Figure 5.23 is of most interest – as it yeild the positive hump solitary wave observed in experiments. However, since symmetry (5.16) is broken for the Shkadov model, this new family is no longer antisymmetric to the C_- family as in the pitchfork bifurcation of the Figure 5.23. Instead, it bifurcates off the γ_1 branch and will be called the γ_2 family.

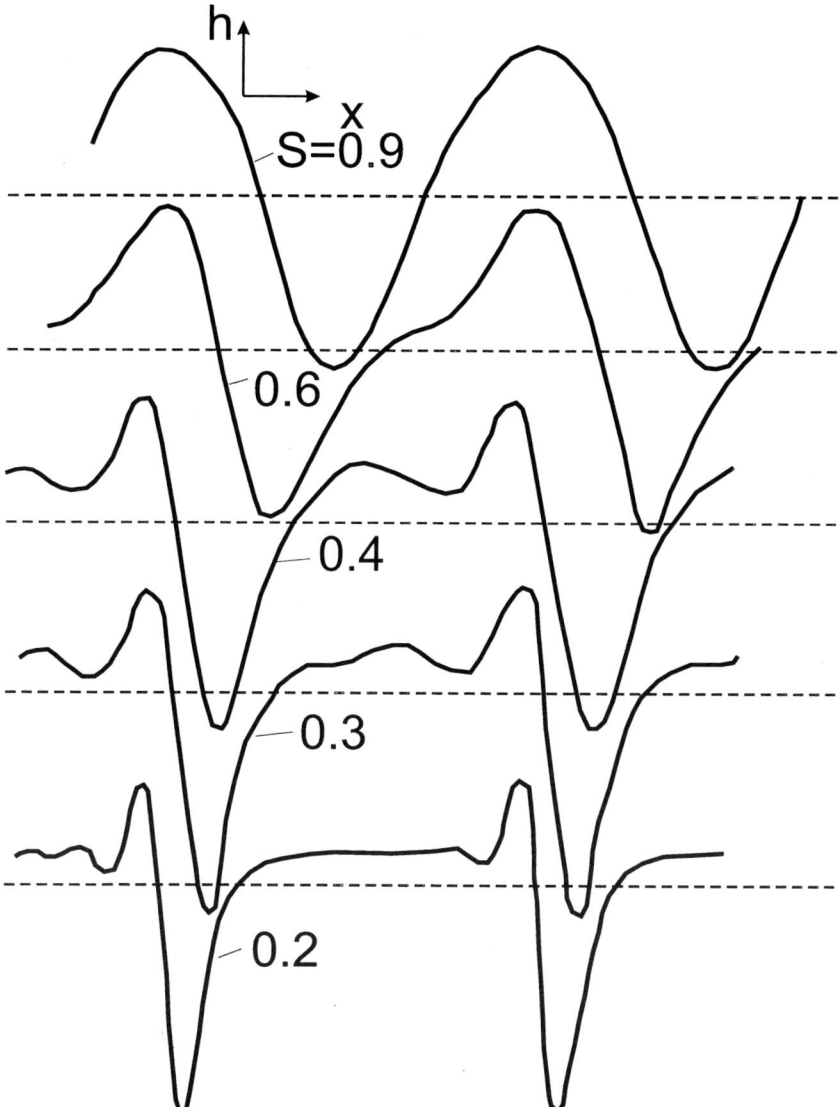

Figure 5.28: Wave profiles of the γ_1 family for different wavenumbers, $s = \alpha/\alpha_0$, at $\delta = 0.04$.

This γ_2 family exists for $s = \alpha/\alpha_0 < s_m (s_m \simeq 0.53 - 0.58)$. As we move away from the bifurcation point, the normalized wave number first increases to s_m and then decreases to zero as shown in Figure 5.29.

As for the KS equation, multi-hump solitary wave solutions exist for the Shkadov model. Calculations show that at any fixed δ, there is a countable set of the segments $[c_1, c'_1], [c_2, c'_2], ...$ out of which bounded solutions for $x \in (-\infty, +\infty)$, including the solitary waves, are absent. The ends of each segment $c = c_i$, $c = c'_i$ are the eigenvalues corresponding to the multi-hump solitary waves. The segments have the accumulation point $c = 3$ and form stripes in the $c - \delta$ plane of Figure 5.30. Here c_1 corresponds to one-hump pulses, c'_1 and c_2 - two-hump pulses, c'_2 and c_3 - three-hump ones, ... , c'_{k-1}, c_k - k-hump pulses. The values c_k, c'_k for $k = 1, 2, 3, 4$ are given at the Table 5.4.

Table 5.4

δ	0.05	0.0565	0.06	0.07	0.12	0.15	0.20
c_1	6.872	7.241	7.362	7.561	7.706	7.689	7.679
c'_1	6.339	6.826	6.997	7.303	7.664	7.680	7.678
c_2	3.883	4.287	4.551	5.277	6.623	6.860	7.065
c'_2	3.815	4.172	4.387	5.065	6.564	6.634	7.063
c_3	3.537	3.721	3.842	4.281	6.009	6.387	7.729
c'_3	3.516	3.687	3.800	4.188	5.923	6.358	7.722
c_4	3.413	3.542	3.619	3.789	5.663	6.057	6.491
c'_4	3.403	3.519	3.597	3.752	5.477	6.006	6.477

The stripes come out from the point $\delta = 0$ and $c = 3$. At small δ, the stripes approach the asymptote

$$c_k = 3 + 58.1\delta^{3/2}\lambda_k$$

Here λ_k are the eigenvalues of the model equation, see Table 5.2. The profiles of one-hump c_1 pulses and of three-hump c_3 pulses are shown in Figure 5.28 and 5.31. Interestingly, all multi-hump pulses approximately obey the following speed -amplitude correlation

$$c - 3 \approx 1.8a \tag{5.86}$$

where a is the wave amplitude. This is again consistent with (5.61) and Figure 5.20.

Let us now examine the negative solitary pulses, including the one at the end of the γ_1 family. In domain II (Figure 5.32a) there is a countable set of segments outside which there are no solitary waves : $[c_{-1}, c'_{-1}], [c_{-2}, c'_{-2}], ...$ The limit point of all segments is $c = 3$. The strips formed by the segments in the plane $c - \delta$ are depicted in Figure 5.32a). These strips are so narrow that look like a line in the figure. The values of $c_k, c'_k, k = -1, -2, -3, -4$ for some δ

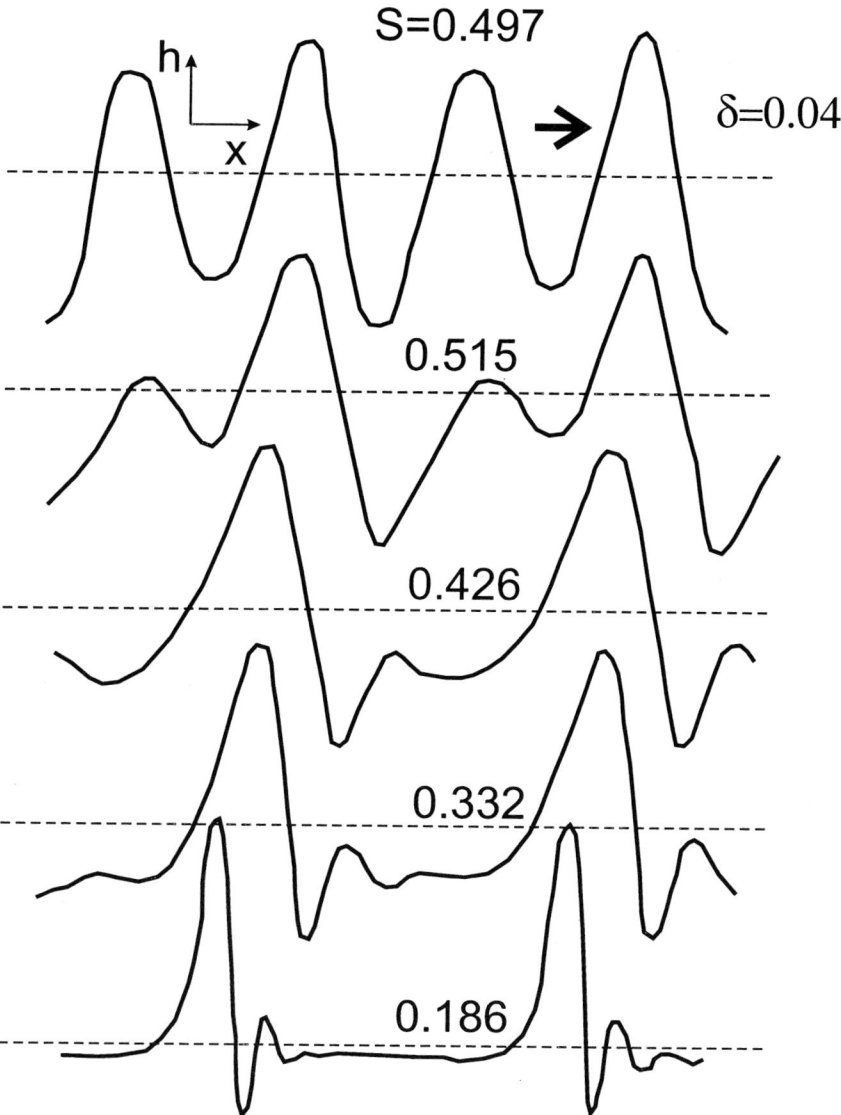

Figure 5.29: Wave profiles of the γ_2 family for different wavenumbers, $s = \alpha/\alpha_0$, at $\delta = 0.04$.

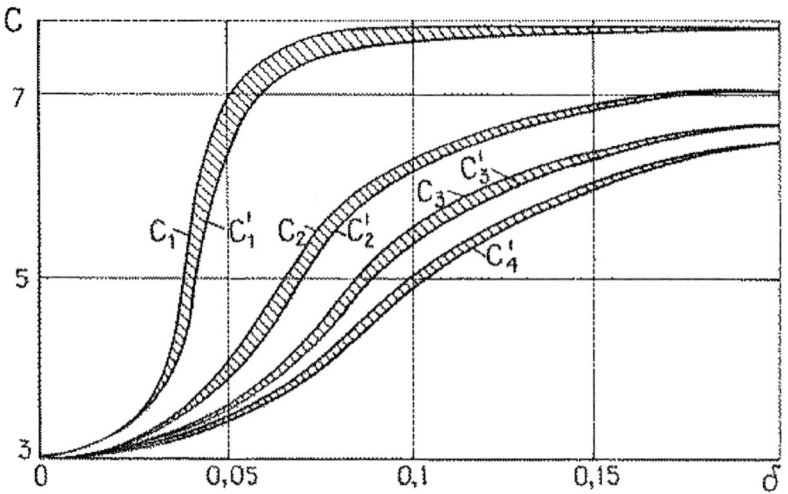

Figure 5.30: Positive multi-hump solutions existence areas.

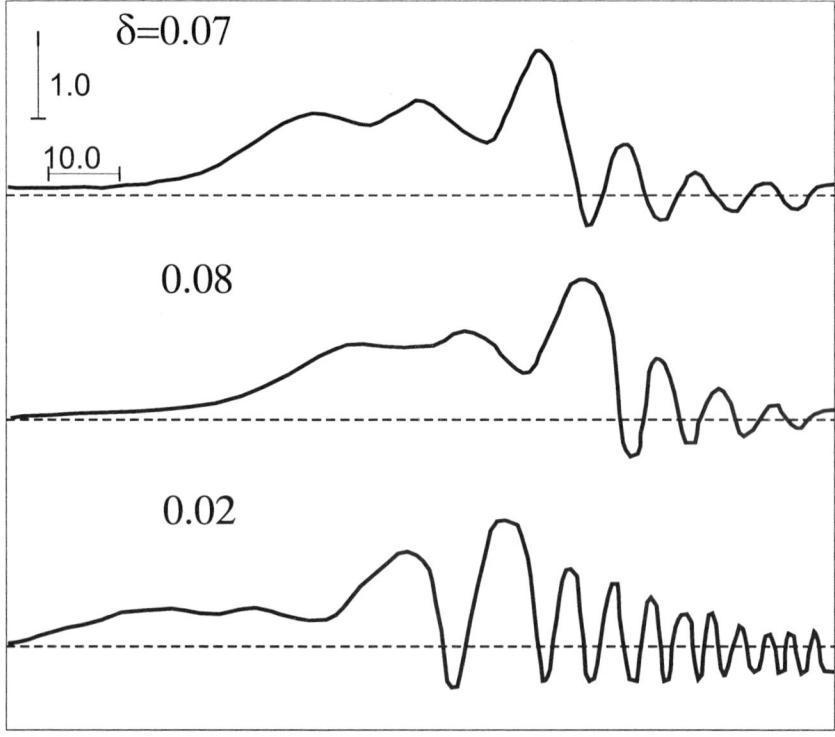

Figure 5.31: Three - hump solutions, C_3-family, for different δ.

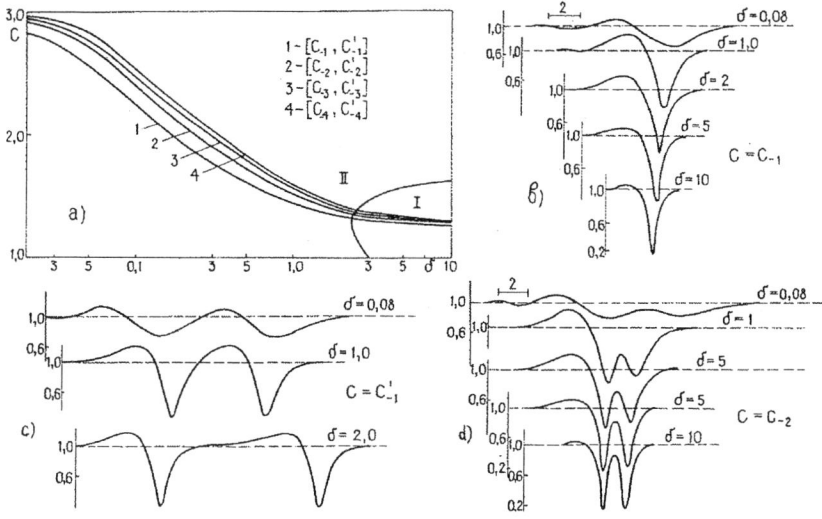

Figure 5.32: Multi-hump negative solutions: a) existence area; b), c), d) -types of solitons.

are presented in the Table 5.5. For small δ, the velocities of the negative humps can be correlated to the velocities of the model equation by the formula :

$$c = 3 - 58.1\delta^{3/2}\lambda_k$$

where λ_k are the eigenvalues of the model equation, see Table 5.2.

Table 5.5

δ	0.08	1.00	2.00	5.00	10.00
c_{-1}	2.3351	1.4322467	1.3370616	1.2777978	1.2610216
c'_{-1}	2.3406	1.4323212	1.3370617	is absent	is absent
c_{-2}	2.4916	1.50301	1.403010	1.380192	1.2702112
c'_{-2}	2.5001	1.50349	1.403108	is absent	is absent
c_{-3}	2.5672	1.55129	1.4235366	1.3117892	1.2789561
c'_{-3}	2.5731	1.55193	1.4236112	is absent	is absent
c_{-4}	2.6124	1.58814	1.44511811	1.346515	1.2821549
c'_{-4}	2.6117	1.58991	1.44520150	is absent	is absent

The profiles of the negative solitons c_{-1}, c'_{-1}, c_{-2} are presented in Figure 5.32, b)-d). The velocity of the negative humps can be approximated by a correlation to the amplitude

$$c - 3 \simeq 1.7a \tag{5.87}$$

Let us consider travelling jumps that approach two different asymptotes at the two infinities. We normalize the film thickness at $+\infty$ to unity and the one at $-\infty$ $1+a$. A positive b corresponds to a compression jump and a negative b corresponds to a decompression jump. These boundary conditions immediately yield.

$$c = 1 + h_1 + h_1^2 = 3 + 3a + b^2 \tag{5.88}$$

The case $b = \varepsilon \to 0$ can be investigated analytically. Let us compress for the study convenience independent variable $x \to \alpha_0 x$. We shall seek the solution by the multiscale method, see Nayfeh (1973) and Van Dyke (1975):

$$h = 1 + \varepsilon H_1 + \varepsilon^2 H_2 \varepsilon^3 H_3, \quad H_k = H_k(X_0, X_1, X_2)$$

$$X_k = \varepsilon^k x, \quad \frac{d}{dx} = \frac{\partial}{\partial X_0} + \varepsilon \frac{\partial}{\partial X_1} + \varepsilon^2 \frac{\partial}{\partial X_2} \tag{5.89}$$

By substituting (5.89) to (5.79) we obtain :

$$H_1''' + H_1' = 0 \tag{5.90}$$

$$H_2''' + H_2' = -3 \frac{\partial H_1''}{\partial X_1} - \frac{\partial H_1}{\partial X_1} - \frac{18}{5} H_1' + \frac{21}{5} H_1 H_1' + \alpha_0^{-3}(3H_1 - 3H_1^2) \tag{5.91}$$

A derivative with respect to X_0 is denoted by prime. The solution of the first equation

$$H_1 = A \exp(iX - 0) + A^* \exp(-iX_0) + B \tag{5.92}$$

where A and B are the functions of slow variables X_1, X_2. By substituting (5.92) to (5.91) we get from the solvability condition:

$$B = 1/2[1 - \tanh(3X_1/(2\alpha_0^3))], \quad A = |A| \exp(i\theta),$$

$$|A| = \pm 1/2 \, sech(3X_1/(2\alpha_0^3)), \tag{5.93}$$

$$\theta = 3/4 X_1 + 7/10 \alpha_0^3 \log \cosh(3X_1/(2\alpha_0^3)) + \theta_0(X_2), \tag{5.94}$$

$$\theta_0 = O(\varepsilon^2) \tag{5.95}$$

Besides this solution , there is another one of the form

$$B = 1/2[1 + \tanh(3X_1/(2\alpha_0^3))], \quad A = 0$$

The picture of the shock wave at $b \to 0$ is presented in Figure 5.33 (c) where $x = 3X_1/2\alpha_0^3$. The dotted curves are the analytical $B(x)$ and $B(x) + A(x)$. At $b = 0.2$, the wave profiles obtained numerically are indistinguishable from the analytical ones.

Decompression jumps with $b < 0$ and $c < 3$ are structurally unstable and exist only at discrete set of pairs $\{b_k, \delta_k\}$ which are now the eigenvalues. One pair of such values has been found for the model equation (5.22).

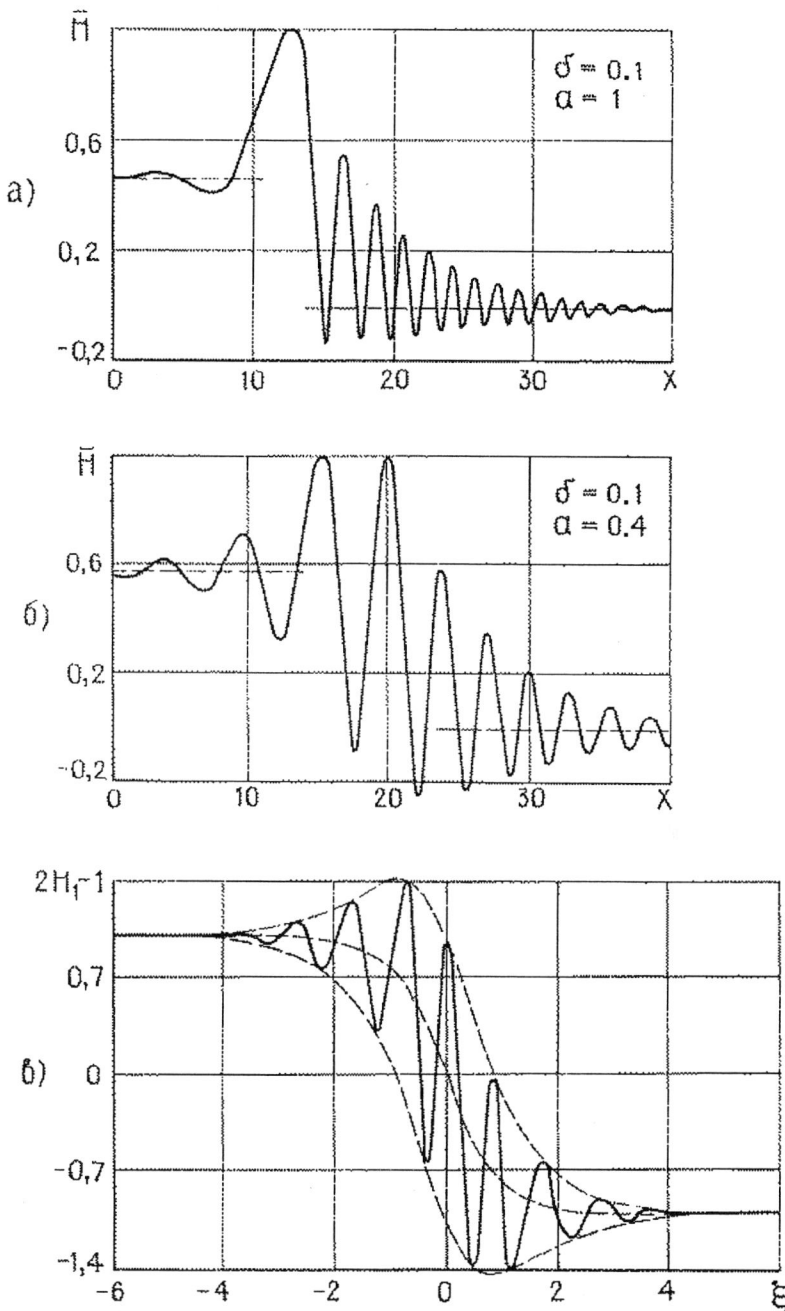

Figure 5.33: Jumps of thickness of different amplitude.

5.7 Stationary waves of the boundary layer equation and Shkadov model

Unlike the limiting case of vanishing Reynolds numbers for the Kuramoto-Sivashinsky equation, solution of the boundary layer equation is considerably more difficult because of the need to resolve the flow field under the film and to track the free surface $h(x)$. Bach and Villadsen (1984) have used a Lagrangian finite-element formulation to track the free surface and less than 3 elements in y to resolve the flow field. However, a Lagrangian formulation requires a rezoning procedure to untangle the convoluted meshes and is hence difficult to implement at high Reynolds number or δ when more y-elements are required. Kheshgi and Scriven (1987) and Hu and Patera (1990) use a different approach where the Lagrangian formulation is applied only in the y-direction to allow accurate front tracing without mesh entanglement. We shall use the same mixed Lagrangian-Euler method here. However, because the third derivative curvature term h_{xxx} that appears in our boundary-layer equation is numerically undesirable, we do not apply a direct finite-element decomposition. Instead, we divide the film into N layers in the y-direction and by manipulating the projected equations, eliminate the h_{xxx} term from all but one equation. The decomposition in the x-direction is a spectral Fourier expansion. The result of this mixed Euler-Lagrangian and spectral-element formulation is that the exponential convergence (with respect to mode number) of the spectral method is combined with the numerical advantages of the finite-elements method such as the elimination of the undesirable h_{xxx} term, higher order bases that mimic the averaged equation formulation and a narrow banded projected differential operator, all of which facilitate the iteration step. With this formulation, this difficult free-surface problem requires about three to seven layers to achieve convergence. This implies only a factor of 10 increase in the number of equations compared to the KS equation. If the full Navier-Stokes equation is used, the strong non-linearities in h at the interface and the flow field will render the solution even more difficult, especially during the Newton iteration step.

For the boundary layer equation (2.41) to (2.44), a travelling wave which propagates at a stationary speed c is described by

$$\frac{\partial}{\partial x}[u(u-c)] + \frac{\partial}{\partial y}(uv) = \frac{1}{5\delta}\left(h_{xxx} + \frac{1}{3}\frac{\partial^2 u}{\partial y^2} + 1\right), \qquad (5.96)$$

$$\frac{\partial u}{\partial x} + \frac{\partial v}{\partial y} = 0 \qquad (5.97)$$

$$\frac{\partial}{\partial x}\int_0^h (u-c)\,dy = 0, \qquad (5.98)$$

$$y = h(x): \quad \frac{\partial u}{\partial y} = 0, \qquad (5.99)$$

$$y = 0: \quad u = 0,\ v = 0. \qquad (5.100)$$

We shall be interested in stationary waves with wavelength $l = 2\pi/\alpha$, $(h, u, v)(x) = (h, u, v)(x + l)$. As the wavenumber α approaches zero, the waves become solitary waves. In constructing these waves, one could assume either that all waves correspond to the same flow rate

$$q = \frac{1}{l} \int_0^l \int_0^h u \, dx \, dy \qquad (5.101)$$

or that all of them have the same average thickness h_N

$$<h> = \frac{1}{l} \int_0^l h \, dx = 1. \qquad (5.102)$$

If the constant-flux condition (5.101) is imposed, the average flux should be equal to the Nusselt velocity and q is unity identically. If the constant thickness condition (5.102) is used, q is typically higher than unity. The two conditions yield similar but not identical results. The results are also mutually transformable with some effort. We choose to impose (5.102) for reasons that will become clear later.

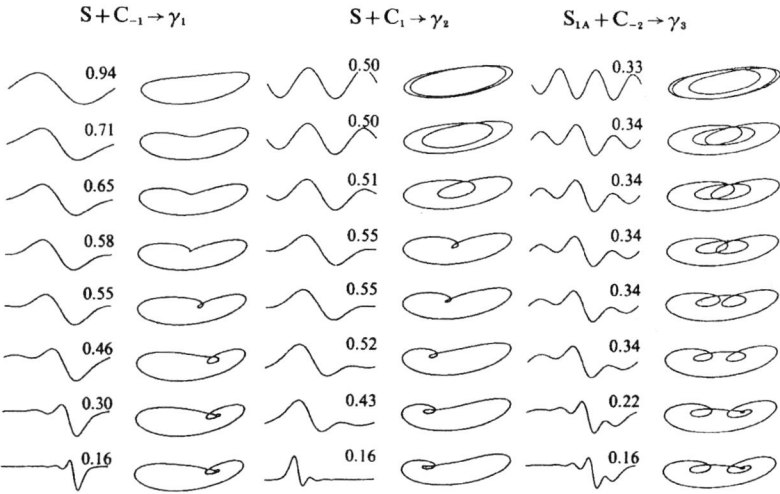

Figure 5.34: The imperfect bifurcation that occur at $\delta = 0^+$ which coalesce several branches of the KS equation to form the new γ_n branches. Only one branch approaches a one-hump positive positive solitary wave limit as (α/α_0) approaches zero. The γ_1 family ends as one-hollow negative solitary wave and the γ_n families ($n \geq 3$) culminate in an $n - 2$ hollow negative solitary wave.

Using the one-mode expansion of the velocity field for the boundary layer equation in Chapter 3, we can reduce the boundary layer equation to the

Shkadov model (3.85)-(3.87). Straight forward calculation shows that two-dimensional travelling wave solution to the Shkadov model obeys

$$h^3 h''' + \delta[6(q-c)^2 - c^2 h^2]h' + [h^3 - q - c(h-1)] = 0 \qquad (5.103)$$

This equation must be solved in conjunction with the constant thickness condition (5.102). Stationary waves of this averaged equation has been studied by Bunov et. all (1982, 1984, 1986), Demekhin and Shkadov (1985), Trifonov and Tsvelodub (1985, 1991)and Tsvelodub and Trifonov (1992). All our stationary waves branches obtained in the full equation are in quantitative and qualitative agreement with the stationary wave solutions of (5.103) for $\delta < 0.06$. Some wave profiles on specific branches are shown in Figure 5.34. Beyond this value, considerable difference is seen. The unique bifurcation sequence of γ_2 in Figure 5.35 is notably missing from the averaged equation. This then implies the averaging is strictly valid for extremely low Reynolds numbers, $R < 10$ for water.

The origin of their symmetry can be seen in (5.11) and (5.13). This constant flux condition actually allows two flat-film solution, $u_1 = 0$ and $u_2 = \lambda/2$. The former is simply the deviation form of the Nusselt flat film while u_2 is a unique "conjugate" solution that arises because, in a Lagrangian frame, there is another flat film solution that can sustain the same flow rate. Returning to (5.96), it is clear that the velocity profile for a flat film is $U = 3y(h - \frac{1}{2}y)$ and inserting this into the integral from of the kinematic condition (5.98) with the flow-rate definition (5.101), one obtains $h(c - h^2) = q = c - 1$. Hence, for c larger than $\frac{3}{4}$, which is always true, the conjugate flat film solution exists in addition to $h = 1$,

$$h = \frac{1}{2}[-1 + (4c-3)^{1/2}] \sim 1 + \frac{1}{3}(c-3) \qquad (5.104)$$

which is exactly $u_2 = \frac{1}{2}\lambda$ after the proper transformations. This multiplicity of flat film solution is also seen in the drag-out problem in coating flow (Tuck, 1983). The existence of a conjugate solution is essential since shock-like stationary solutions, which are observed in falling films (Alekseenko et al., 1985) and other coating flow, require two flat-film solutions. Linearizing (5.104) about the conjugate solution, one obtains that waves that appear on this conjugate flat film travel at a speed -2λ different from those on the Nusselt flat film. Hence, for every slow wave bifurcating from the Nusselt flat film with speed $-|\lambda|$, one expects a symmetric fast wave with speed $|\lambda|$ from the conjugate flat film. This is also seen in the reversibility condition (5.17) for the KS equation. We recall that standing waves are not permissible in the constant-flux formulation. If one transforms the two travelling wave families, with symmetric speeds about $\lambda = 0$ or $c = 3$, of the constant-flux formulation to the constant-thickness formulation by (5.101), these two families collapse to a single standing wave family for α close to α_0. There is, however, a pitchfork bifurcation at smaller a which again gives rise to two travelling wave families with positive and negative μ (see Figure 5.35) which can be traced back to the soliton structure of the KS equation in Figure 5.23. Hence, although the conjugate flat-film basic state does not exist for the constant-thickness for obvious reasons, formulation the two symmetric families of travelling waves that appear in this formulation can still be traced back to the existence of two flat-film basic states of the constant flux formulation. The origin of the fast γ_2 wave family via a bifurcation off the conjugate flat-film solution is the preferred physical explanation. Both families persist at larger δ in both formulations.

We first sketch the changes in the solution structure qualitatively. If plotted on the $c - \alpha$ parameter space, the stationary wave solutions of the KS equation (5.11) in Figures 5.18 and 5.19 collapse into a very simple structure. All the $S^{(n)}$ families can now be represented by the primary S branch which spans $s = (\alpha/\alpha_0) \in (1, 0.49775)$ along the $c = 3$ line in Figure 5.35(a). With finite Reynolds numbers, the reversibility symmetry is broken and waves are not exactly symmetric about $\lambda = 0$ or $c = 3$. However, the S family and the C_{-1} family coalesce to form the slow γ_1 family and C_1 and S interact to form a fast γ_2 family (see Figure 5.35). Since the S family of standing waves correspond to travelling waves in the constant-flux formulation, this division into γ_1 and γ_2 travelling wave families was already in place at $\delta = 10$ for that formulation. At larger δ, interactions with the other standing and travelling wave branches lead to hybrid branches of γ_2. Otherwise, the division of wave families into the slow and fast γ_1 and γ_2 families remains intact. These two families of stationary waves share a common feature. Their amplitude increases with their wavelength such that the solitary wave limit of each family is the largest wave. In fact, due to symmetry (5.11), while the solitary wave limit of γ_2 (depicted as C_1 in Figure 5.21 resembles the observed solitary wave, the solitary wave limit of γ_1 is an inverted version of C_1 (negative solitary wave) and is not observed. Because the wave amplitude is small only for short waves on γ_1 near the neutral wave number α_0, the weakly nonlinear Stuart-Landau formalism and the weakly nonlinear evolution equations like the KS equation are strictly applicable for these short waves on γ_1. They are typically invalid for the γ_2 family of fast waves except for near critical conditions when the wave amplitude on both families are small. Even then, the classical Stuart-Landau formalism must be replaced by high order bifurcation theories.

The S_n branches appear at lower wave numbers but they do not participate in the creation of important branches at finite d and hence will be omitted in Figure 5.35. The remaining standing waves S_{1A}, S_{1B}, S_{1C} etc are also marked in Figure 5.35. They are drawn disjoint to show that the subsequent bifurcations will separate them. We note that since these branches bifurcate subcritically with respect to α^{-1} before turning around, each point on the indicated segments may represent two waves. The primary travelling waves C_1 and C_{-1} that bifurcate from S and its inverted counterpart are also shown in Figure 5.35 as a pitchfork bifurcation with speeds in excess and less than 3, respectively. For a small but finite δ, the pitch fork undergoes an imperfect bifurcation such that two travelling waves branches, γ_1 and γ_2, are born as shown in Figure 5.35(a). Likewise the C_{-n} branches coalesce with the S_1 branches to form $\gamma_3, \gamma_4, \gamma_5$ etc. for $\delta \neq 0$. The C_n branches also give rise to new solutions but they will not participate in the future evolution of the solution branches. The only pertinent ranches will be shown to be γ_n for $\delta \neq 0$. From the wave profiles in Figure

5.34, it is clear that only γ_2 terminates at $\alpha = 0$ with a positive solitary wave. The remaining branches all culminate in an inverted solitary wave with a hollow instead of a hump. The branch γ_2 is clearly faster than the critical linear speed $c_0 = 3$ while the other branches are either slower or close to 3. This relative speed with respect to 3 will change as δ increases but γ_2 remains the fastest branch. It should also be pointed out that $\gamma_3, \gamma_4, \gamma_5$ etc. retain the topologies of C_{-2}, C_{-3}, C_{-4} etc. and exhibit in the solitary wave limit 2 hollows, 3 hollows etc. (Chang et al., 1990). The actual profiles of these branches at low δ are shown in Figure 5.34.

The above bifurcation features for small δ are consistent with those exhibited by the Shakov model (5.103), see Figure 5.36. However, for $\delta > 0.06$, the γ_2 branch begins to wrinkle as γ_3 and γ_4 move towards it. This and all subsequent evolution are distinct from the averaged equation. We also note that for $\delta > 0.06$, the velocity profile of the waves definitely deviates from the parabolic profile assumed in the averaged equation. At $\delta_1^* \approx 0.09$, the turning point of γ_3, originating from the imperfect bifurcation at $\delta = 0$, coalesce with a turning point of γ_2 to yield an isolated branch γ_2' in a unique "pinching" isolate point. As shown in Figure 5.35, this pinching occurs successively as δ increases as γ_2 coalesce with γ_4, γ_5, etc. to yield γ_2'', γ_2''' respectively. The γ_2'' branch is born at $\delta \approx 0.15$ as shown in the wave profiles of Figure 5.37. The $\gamma_2^{(n)}$ branches resemble their mother branches $C_{-(n+1)}$ of the KS equation with $(n+1)$ hollows in the solitary wave limit. However, prior to the solitary wave limit, they can take on shapes that resemble a positive solitary wave with one big hump as the small hollows resemble capillary bow waves. In the limit of infinite δ, the original γ_2 branch has given rise to an infinite family of these $\gamma_2^{(n)}$ branches. In Figure 5.36, the actual computed solution branches are shown. It is seen that the symmetry about $c = 3$ of the KS equation is broken. The γ_2 family increases in amplitude and speed as δ increases much more than the γ_1 family. This is also evident in Figure 5.37 which shows the positive near-solitary waves of γ_2 are much larger and much faster in absolute deviation speed from 3 than their negative solitary wave counterparts on γ_1.

We note that the periodic waves in Figure 5.34 correspond to closed trajectories (limit cycles) in the phase space of (h, h_x, h_{xx}). In the solitary wave limit ($\alpha \to 0$), these limit cycles approach a fixed point corresponding to the Nusselt flat film $(h, h_x, h_{xx}) = (1, 0, 0)$. They hence approach a homoclinic orbit connected to the fixed point. All stationary wave branches in our study at finite δ approach homoclinicity and hence have a solitary wave limit. An intriguing property of dynamical systems that possess homoclinic orbits is that all bifurcations near homoclinicity can be qualitatively discerned with only information on the eigen spectrum of the fixed point. We can then use the elegant theorem of Silnikov to explain the pronounced wrinkling of the γ_2 branch (and the weaker wrinkling of the γ_1 branch) in Figure 5.36. Each wrinkle corresponds to a saddle node bifurcation of the limit cycle solution branch as it approaches homoclinicity. In all cases studied here, the fixed points corresponding to the KS equation in (5.11) or (5.13) or the Shkadov model (5.103) for $c > 3$ all yield one real eigen-

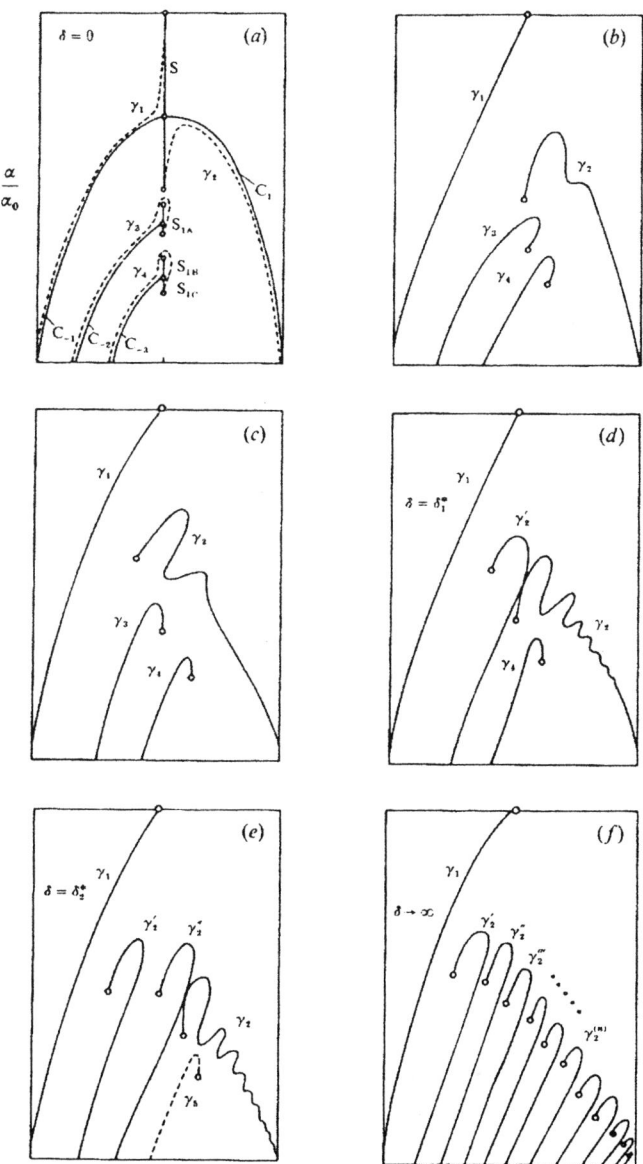

Figure 5.35: Schematic depicting the bifurcation sequence as δ increases. The γ_1 family survives intact. It consist of short, near-sinusoidal waves for $(\alpha/\alpha_0) \to 1$ and a one-hollow inverted solitary wave for $(\alpha/\alpha_0) \to 0$. the $\gamma_2^{(n)}$ families, born by successive coalescence of the γ_2 branch with γ_n in a pinching bifurcation, are detached from the neutral wavenumber $(\alpha/\alpha_0) = 1$ and do not possess near-sinusoidal waves. The γ_2 branch ends as a one-hump positive solitary wave while $\gamma_2^{(n)}$ ends as an (n+1) hollow negative solitary wave. However, the $\gamma_2^{(n)}$ waves away from the solitary wave limit resemble positive solitary waves (see Figure 5.37).

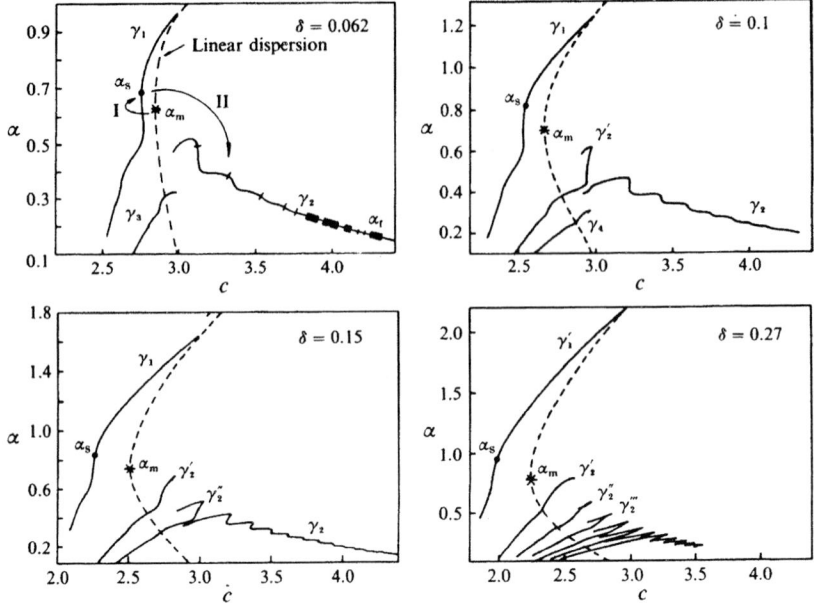

Figure 5.36: Actual computed wave branches for $\delta = 0.062, 0.1, 0.15$ and 0.27 and comparison to the linear dispersion curve. The fastest growing linearing waves α_m and the most stable waves on the wave branches are represented by stars, circles and blackened segments in some figures. The wave transition are indicated in the first figures.

value σ_1 and one complex conjugate pair with a real part σ_2. The signs on σ_1 and σ_2 are different and are dependent on whether the positive or solitary waves are involved. The eigenvalues are responsible for the capillary bow waves and the real eigenvalue the gentle slope on the other side of the solitary waves. One also reaches the same conclusion regarding the spectrum of the "fixed point" corresponding to the full boundary layer equation at $x \to \pm\infty$. According to Silnikov's theorem (see section 5.12), if $\sigma_1 + Re\{\sigma_2\}$ is positive, one expects a periodic orbit to undergo an infinite sequence of saddle node bifurcation with increasing period as it approaches the homoclinic orbit in the parameter space. The distance between two adjacent saddle nodes also decreases monotonically as the periodic solution branch winds towards homoclinicity. Since a periodic orbit corresponds to a periodic wave here, the saddle node bifurcations of the periodic solution branch are exactly the wrinkles seen on γ_1 and γ_2 as the solitary wave limit is approached. We have found $\sigma_1 + Re\{\sigma_2\}$ to be positive in all of our calculations.

Several comparisons to earlier weakly nonlinear analyses can be made here. We first note that, at large δ, the solitary wave of the slowest branch γ_1, approaches a speed of about $1.5u_0$ which is close to the $1.67u_0$ predicted by Prokopiou et al. (1991) and Alekseenko, Nakoryakov and Pokusaev (1994) using the Shkadov model. This shows the averaged equation may be reasonably accurate in the estimate of the wave speed of the slowest solitary wave on the γ_1 branch even though the local velocity profile beneath the hump and the bow waves is far from parabolic for δ in excess of 0.06. The averaged equation fails to describe γ_2 and its hybrid families $\gamma_2^{(n)}$ completely. The wave profiles of some of the branches are shown in Figure 5.37 to Figure 5.39. We note that, for δ in excess of say 0.1, the only near-sinusoidal waves appear on γ_1. Even the shortest waves on the $\gamma_2^{(n)}$ families are much longer than the near-sinusoidal ones on γ_1 and they possess wave profiles with large Fourier contents.

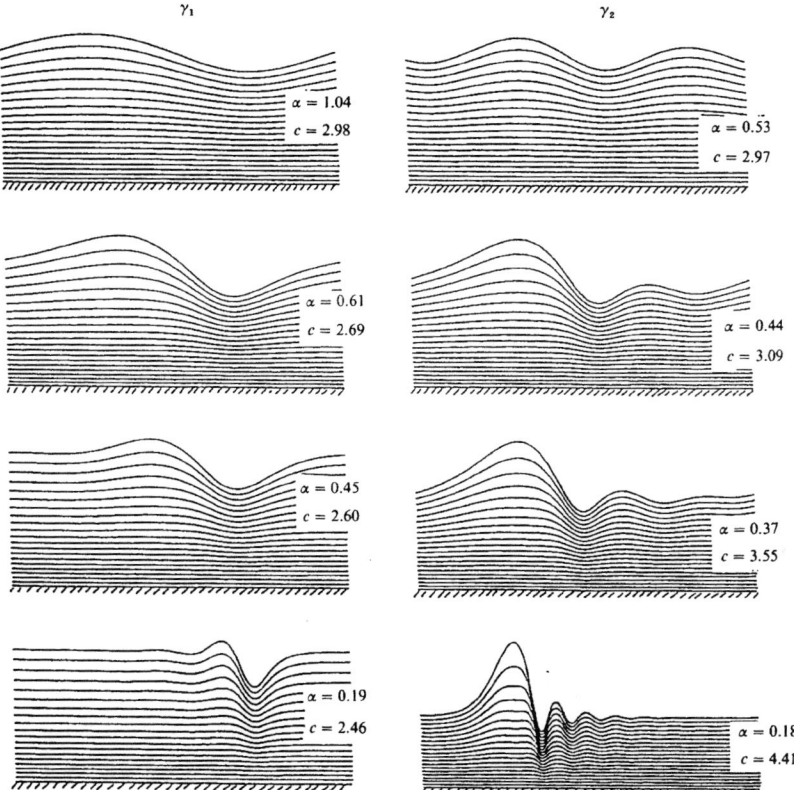

Figure 5.37: Small δ: family γ_1 branches off the trivial solution and disappears after saddle-node bifurcation (negative pulse); family γ_2 starts from γ_1 and ends after saddle-node bifurcation (positive pulse).

This implies that for moderate δ and beyond, the observed near sinusoidal waves just beyond the inception region must all lie on the slowest γ_1 branch while the solitary waves that evolve downstream lie on the faster γ_2 and $\gamma_2^{(n)}$ branches. This scenario will be confirmed in the next section and it explains the initial deceleration of waves out of the inception region and the subsequent acceleration as they evolve from the near-sinusoidal regime into the solitary wave region observed by Stainthorp and Allen (1965).

In comparing our results to experimental wave profiles, we shall sometimes select a wave of the same wavelength $l = 2\pi/\alpha$ as the measured one for a given δ and compare the wave speeds c and wave profiles $h(x)$. In some cases, only the wave period, $T = l/c$, is given on a strip chart and we shall then compare the wave period instead of c. If the wave speed c is measured accurately, we shall locate a wave of the same speed and compare l or T and $h(x)$.

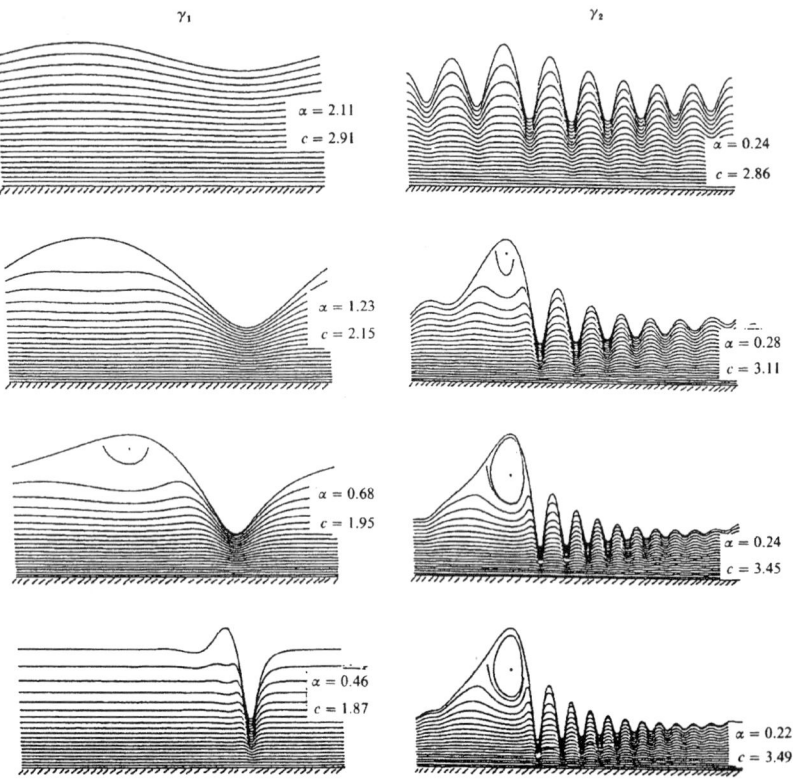

Figure 5.38: Large δ; see caption for Figure 5.37

Finally, the Reynolds number $\langle R \rangle$ of most reported data are based on the true flow rates for films with waves. It is not identical to our R based on the flat film h_N. An iteration is then necessary to compute q of (5.101) and relate the two Reynolds numbers. For a given wave profile with wavelength l measured at a specific $\langle R \rangle$, we begin by assuming $\langle R \rangle = R$ and obtain δ and α by the

following formulae

$$h_0 = (3\nu^2/g)^{1/3} R^{1/3}, \quad u_0 = (\frac{1}{3}g\nu)^{1/3} R^{1/3}, \tag{5.105}$$

$$\delta = 3^{-7/9} 5^{-1} \gamma^{-1/3} R^{11/9}, \quad \kappa = 3^{-2/9} \gamma^{1/3} R^{-2/9}, \quad \alpha = 2\pi \kappa h_N / l. \tag{5.106}$$

After completing the numerical calculation for the given δ and α, we obtain an

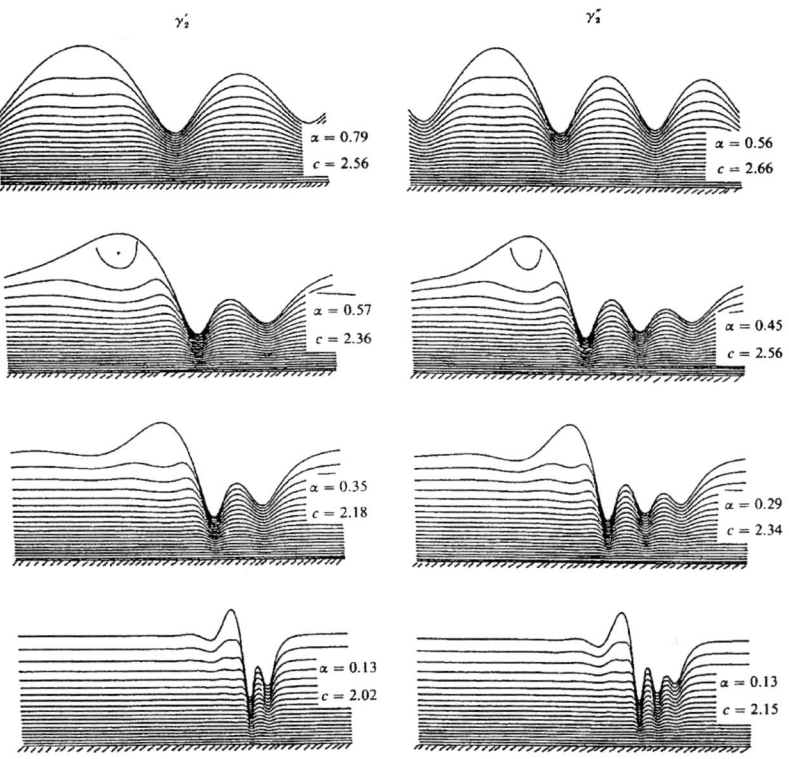

Figure 5.39: Other families which branch from periodical solutions. For (a) $\delta = 0.062$ and (b) $\delta = 0.27$.

improved estimate of R by setting

$$R = \langle R \rangle / q$$

Typically, three iterations suffice to yield an accurate estimate of R.

In Figure 5.40 and 5.41, we compare our computed wave tracings of the same period T to the experimental ones of Nakoryakov et al. (1985) at various $\langle R \rangle$. The computed values of δ, α and c are shown in Table 5.6 along with the wave family the wave profile belongs to. In some cases, two waves on different families, γ_2 and γ'_2, have the same wavelength as the measured one. It is clear that the wave on γ_2 is the selected one. All recorded waves are solitary waves and they appear on the positive one-hump branch γ_2 or on segments of the γ_2' branch which resemble a one or two hump solitary wave. (The γ_2' family eventually evolves towards a negative solitary wave with two hollows as is evident in Figure 5.40 and 5.41.) In Figure 5.42, we compare our results to the only wave tracing for water offered by Stainthrop and Allen (1965). It was taken at 5 cm below the distributor and corresponds to the spatial station at the beginning of the acceleration stage from near-sinusoidal waves to solitary waves. Since the speed of the profile was reported at $c = 0.22$ m/s at $\langle R \rangle = 15$, we shall locate waves of the same speed. Two waves, one on g1 and the other on γ_2', are found to have this wave speed. The one on γ_1 is slightly longer with $l = 1.10$ cm and the one on γ_2' is 1.03 cm long. Both are in satisfactory agreement with the value of 1.0 ± 0.2 cm estimated from the wave tracing (Figure 5.40 and 5.41 of their paper).

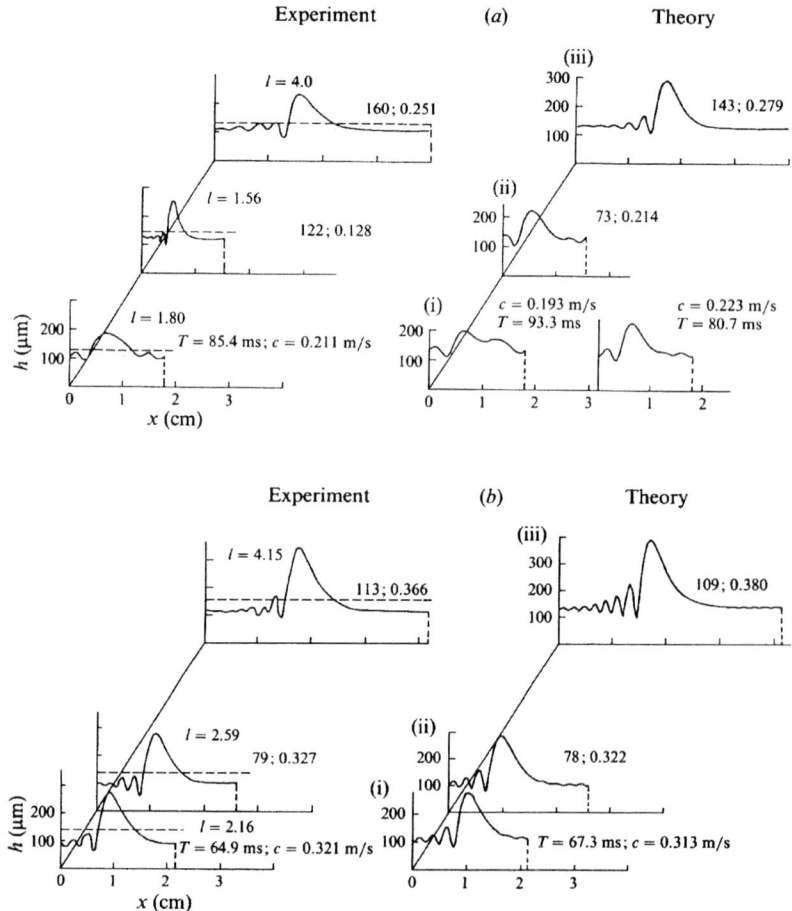

Figure 5.40: For the caption see Figure 5.41.

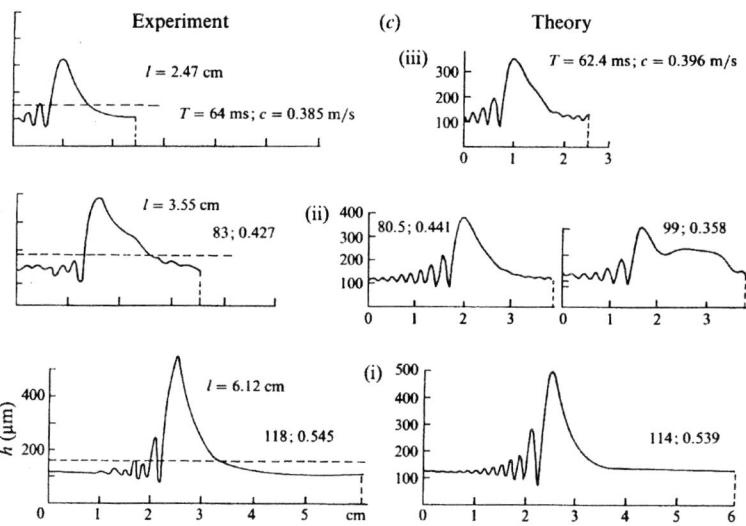

Figure 5.41: Comparison to the wave data of Nakoryakov et al. (1985) for the same wave period. The wave period and speeds in ms and m/s are indicated. The wave parameters are listed in table 3. All are near solitary waves on γ_1 and $\gamma_2'(a) <R>= 10$, (b) 18.8, (c) 31.

	(a) $<R>= 10$ (water)					(b) $<R>= 18.8$ (water)			
	δ	α	c	Wave family		δ	α	c	Wave family
(i)	0.09205	0.3673	2.8015	γ_2'	(i)	0.1484	0.3210	3.5026	γ_2
	0.08208	0.3661	3.4412	γ_2	(ii)	0.1448	0.2663	3.7606	γ_2
(ii)	0.08273	0.4299	3.2877	γ_2	(iii)	0.1380	0.1650	4.4211	γ_2
(iii)	0.07476	0.1628	4.5233	γ_2					

(c) $<R>= 31$ (water)

	δ	α	c	Wave family
(i)	0.2301	0.1218	4.7418	γ_2
(ii)	0.2465	0.2041	3.7389	γ_2
	0.3122	0.2085	2.6643	γ_2'
(iii)	0.2567	0.2942	3.2867	γ_2

Table 5.6. Wave parameters for Figure 5.41.

However, on closer examination of the detailed wave shapes, it is clear that the γ_2' wave is selected. This again confirms that the acceleration beginning at 5 cm from the distributor signals the departure from the nearly-sinusoidal γ_1 family and the approach towards waves on the $\gamma_2^{(n)}$ families. Finally, in Figure 5.43, we offer comparison to the classical photographs of Kapitza (1948, 1949). The parameters are listed in Table 5.7. Good agreement is again evident. Whenever there is a choice between a wave on the γ_2 family and one on γ_2', the γ_2 wave is always the chosen one. The somewhat lower amplitude of the photographed waves can be due to distortion by the tube curvature during the imaging.

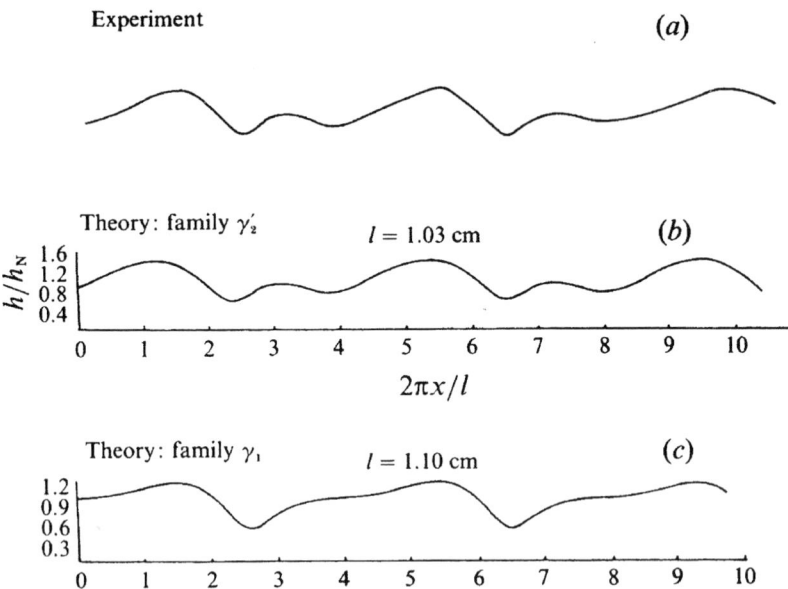

Figure 5.42: Comparison to the tracing Stainthorp and Allien (1965) of the same speed measured at the boundary between the near-sinusoidal and the solitary wave regimes. The height is normalized by h_0 and the length by $l/2\pi$.

The above favorable comparison to measure profiles verifies the boundary-layer approximation. The only remaining task is then to decipher why specific waves on certain branches are selected. We expect this selection mechanism to be related to the relative stability of the waves with respect to two-dimensional and three-dimensional disturbances of all wavelengths. The latter disturbance should also be more dominant on the $\gamma_2^{(n)}$ families since the two-dimensional solitary waves break up into three-dimensional nonstationary waves downstream. We shall confirm these speculations with a stability analysis in the next section. For now, however, we note that the solitary pulse on the γ_2 branch seems to be the final state of the evolution along the periodic families. The speed of the pulse is already depicted in Figure 3.5 and 4.38. Their profiles are now shown in Figure 5.44.

Plates	Q (cm^2/s)	$<R>$	l (cm)	R	q	δ	α	Exp c (cm/s)	Theory c (cm/s)	Family
6	0.084	7.368	1.	6.850	1.076	0.0627	0.639	17.2	15.4	γ_1
7	0.097	7.748	0.89	8.543	1.103	0.0725	0.728	16.3	16.3	γ_1
8	0.118	10.361	0.86	9.196	1.127	0.0898	0.767	17.7	17.7	γ_1
9	0.137	12.020	0.82	10.488	1.146	0.1055	0.817	18.2	18.8	γ_1
10	0.156	13.684	0.88	11.903	1.150	0.1232	0.772	18.6	19.6	γ_1
11	0.201	17.629	0.80	14.949	1.179	0.1628	0.871	21.7	21.9	γ_1
12	0.232	20.350	0.85	17.288	1.177	0.1944	0.833	23.2	23.0	γ_1
19	0.097	8.521	2.46	7.348	1.1597	0.0682	0.262	21.6	21.0	γ_2
20a	0.230	20.178	2.40	14.748	1.368	0.1600	0.290	31.5	33.2	γ_2
20b	0.230	20.173	2.40	17.159	1.176	0.1926	0.295	31.5	28.4	γ_2
21	0.156	13.684	1.31	11.642	1.175	0.1199	0.517	24.4	22.7	γ_2

Table 5.7. Kapitza's data for water in Figure 5.43

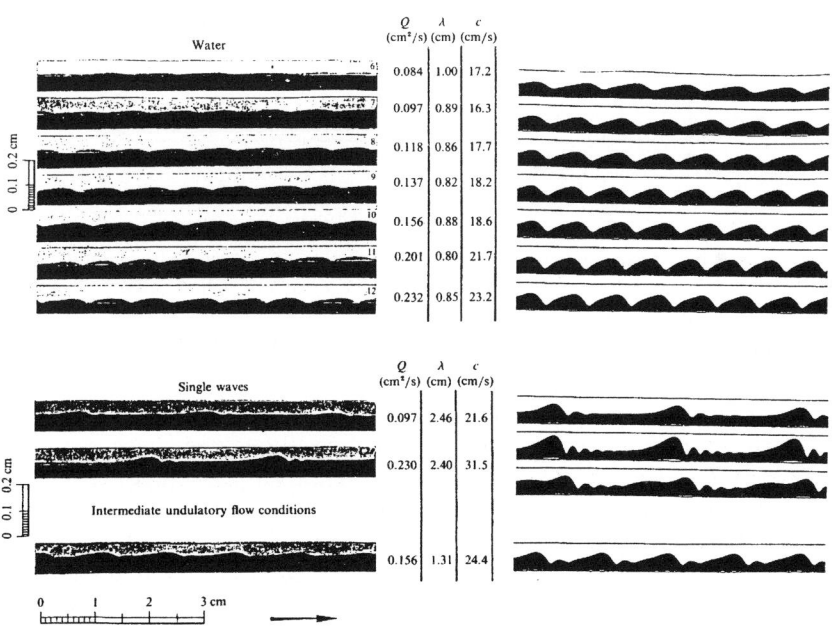

Figure 5.43: Comparison with the photographs of Kapitza. The wave parameters are listed in table 4.

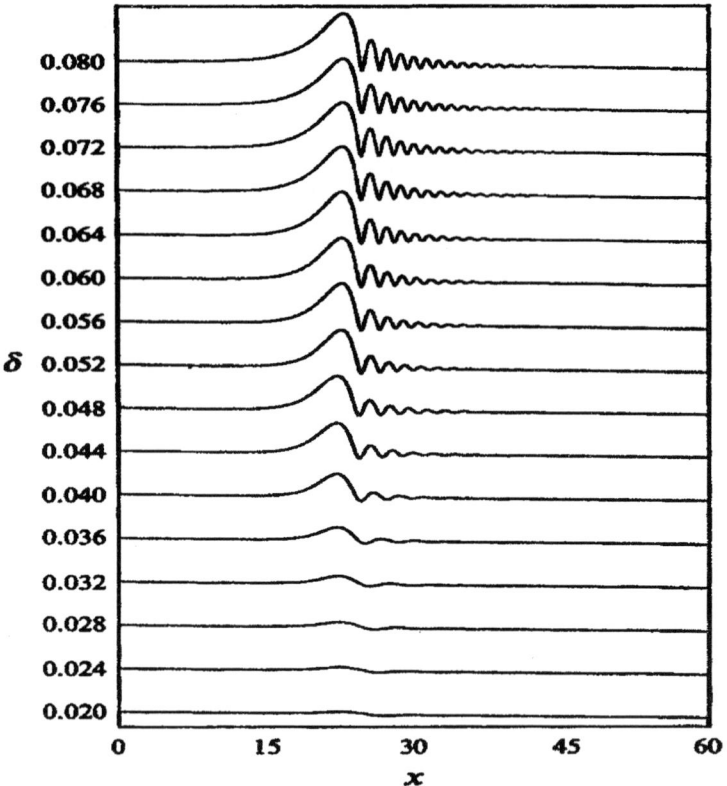

Figure 5.44: One-parameter family of pulses for Shkadov model.

5.8 Navier-Stokes equation of motion – the effects of surface tension

When the effect of surface tension decreases, the simplified Shkadov model is not applicable. This occurs for large Reynolds numbers; for small inclination angles; when the Kapitza number γ tends to zero for such liquids as oil or glycerine; and for waves in more complicated film-type flows with external active effects(see Jurman and McCready, 1989). We then need to consider the full Navier-Stokes equations (2.1) - (2.16).

Bach and Villadsen (1984) first used finite elements to solve wave dynamics with the Navier-Stokes equation. We shall only restrict ourselves to travelling wave solutions of the Navier-Stokes equation for vertical films.

The problem is described by three parameters : two external ones, γ and R, and one internal one, the wavenumber α. As is shown in chapter 2, at $\gamma \to \infty$ the problem is controlled by one external parameter δ. As initial guess, we take from section 5.6 the periodic wave solution found for large γ. The solution is then continued in the direction of decreasing γ from $\gamma = 2850$ to $\gamma = 0$ at fixed R and $s = \alpha/\alpha_0$. The solution is then continued with the parameter s at fixed R and γ.

Bifurcation diagram for the γ_1 and γ_2 periodic solutions in coordinates c and s is given in Figure 5.45. These calculations are carried out with initial guesses at $\delta = 0.06$. The γ value and the Reynolds number R are varied in such a way as to keep δ constant,

$$R = (5^9 3^7 \gamma^3 \delta^9)^{1/11}$$

For $\gamma > 250$, including water at $\gamma = 2850$, the solutions are practically independent of γ and are only functions of one parameter δ. The wave velocity is less than the phase speed 3, $c < 3$, and decreases monotonically with decreasing s on the γ_1 family. The γ_2 family bifurcates from the γ_1 family at $s \approx 0.5$ and at $s \to 0$ they resemble positive-hump solitary waves. Curves 1 and 2 on the bifurcation diagram correspond to members of the γ_1 and γ_2 families at $\gamma = 2850$, with $R = 6.577$.

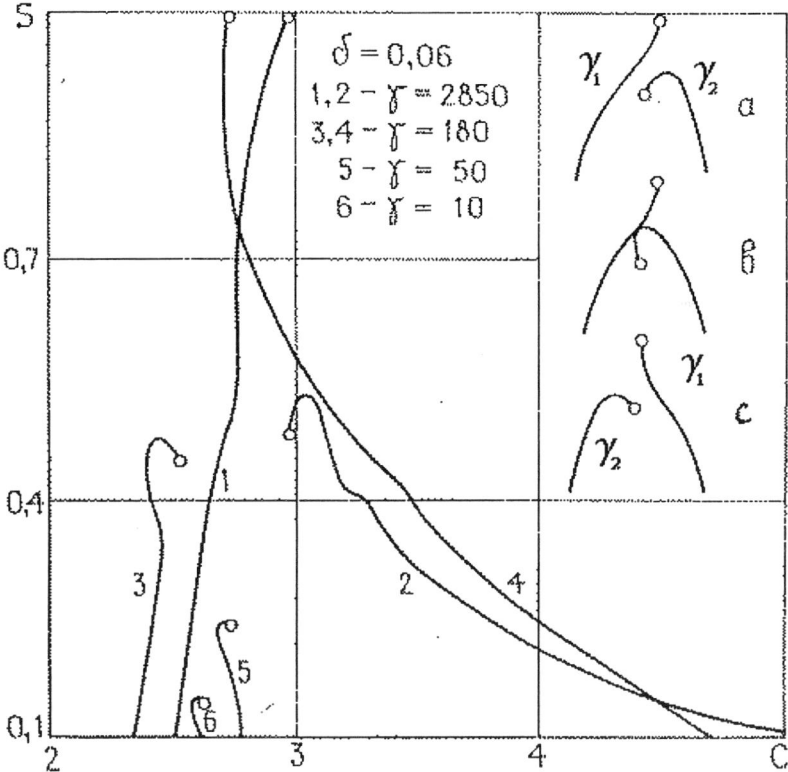

Figure 5.45: Bifurcation diagram which clarify the role of surface tension.

At fixed R and with decreasing γ, the branches of these two families are deformed and approach each other for some intervals of wave numbers. At $\gamma = \gamma_*(R)$, they touch tangentially at $s = s_*$. (At $R = 3.142, \gamma_* = 190$ and $\gamma_*(R)$ decreases with R). We shall still call γ_1 the family that branches from the trivial solution and γ_2 the family that bifurcates from γ_1. However, γ_2 may now be the slower family.

For $\gamma < \gamma_*$, the γ_1 family of periodic waves approaches negative-hump solitary waves in the limit of $s \to 0$. This asymptote remains for all calculated $\gamma < \gamma_*$. Curve 3 in Figure 5.45 for $R = 3.096$ and $\gamma = 180 < \gamma_*$ corresponds to this family. This behavior is consistent with earlier models.

The γ_2 family has a more complex solution branch different from the other models. After the branches merge at γ_* and for $\gamma_{**} < \gamma < \gamma_*$, the γ_2 family bifurcates from the γ_1 family with doubled period and at $s \to 0$ asymptotically evolves into the positive-hump solitary wave. Curve 4 in Figure 5.45 corresponds to this case at $R = 3.096$ and $\gamma_{**} < \gamma$.

With further decrease of γ, the γ_1 family does not undergo any qualitative modification but γ_2 is subjected to new bifurcation transformations. These transformations all take place at smaller s values. We find that, at $\gamma_{***} < \gamma < \gamma_{**}$, the γ_2 family bifurcates from the γ_1 family with quadruple period. This is curve 5 on the bifurcation diagram at $R = 2.184$ and $\gamma_{***} < \gamma = 50 < \gamma_{**}$. The wave profiles $h(x)$ at $R = 2.184$ and $\gamma = 50$ are shown in Figure 5.46. Profiles 1,2,3 correspond to the waves of the γ_1 family at $s = 0.864; 0.643; 0.437$; profiles

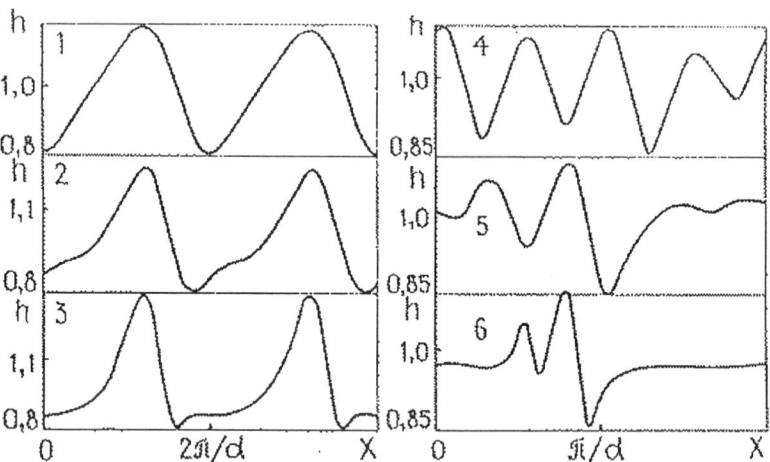

Figure 5.46: Wave profiles of the γ_1 and the γ_2 families at $\gamma = 50$ and $R = 2.184$; $1 - s = 0.864$; $2 - 0.643$; $3 - 0.437$; $4 - 0.243$; $5 - 0.231$; $6 - 0.107$.

4, 5, 6 correspond to the waves of the γ_2 family at $s = 0.243; 0.231; 0.107$. For the waves of the first family two periods were demonstrated and for the waves of the second family one period was shown.

By means of the above-described calculating way one more bifurcation of the family was found : at $\gamma_{****} < \gamma < \gamma_{***}$ branching of the second family solutions from the first family solutions takes place with period 8 - curve 6 in Figure 5.45 at $R = 1.408$ and $\gamma = 10$.

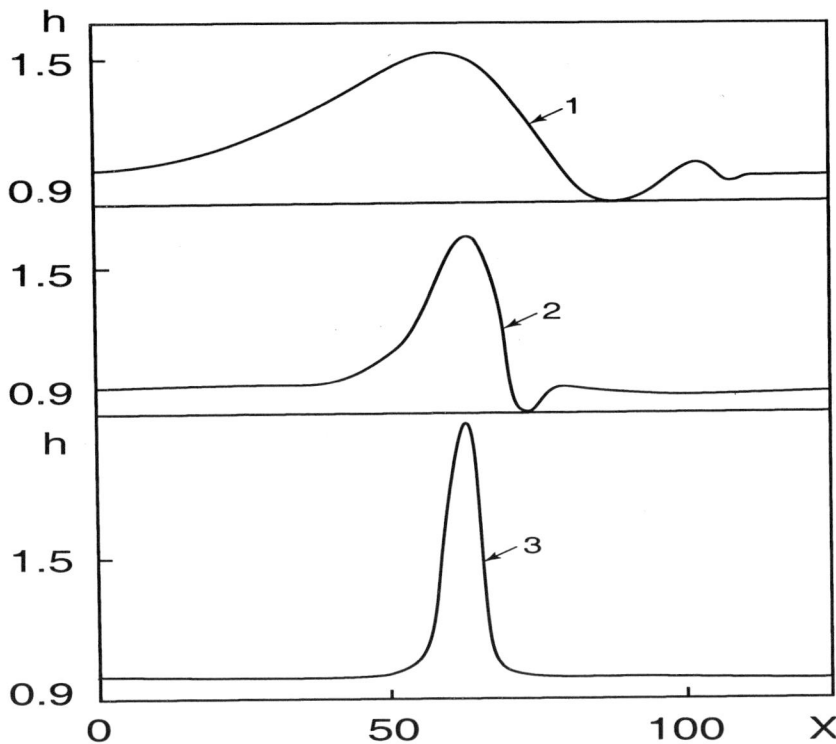

Figure 5.47: Pulse profiles at (1)$\gamma = 2850$; $R = 4.115$; $c = 4.278$, (2)100; 1.698; 4.548 and (3)1.00; 0.349; 4.670.

We are unable to trace the γ_2 family below $\gamma = 7$. The solitary wave limit of this family tends to lose its front-running capillary waves as seen in Figure 5.46.

Table 5.8

R	0.326	0.354	0.388	0.408	0.431	0.465
c	4.712	5.069	5.284	5.407	5.506	5.786
a	0.764	0.982	1.100	1.164	1.248	1.287

Table 5.8 presents values of velocity c and amplitude $a = h_{max} - h_{min}$ of the positive hump at $\gamma = 1$.

We note an interesting wave behavior at zero surface tension (Figure 5.48). In the vicinity of the bifurcation point of the periodic solution from the trivial one, the wave form is close to sinusoidal. As the wave number decreases, the crest of the wave sharpens and its amplitude increases. Our continuation scheme fail below $s = s_0(R)$. In the vicinity of $s = s_0$, the wave crest sharpens abruptly and cannot be relaxed with insertion of more Fourier harmonics. Pukhnachev

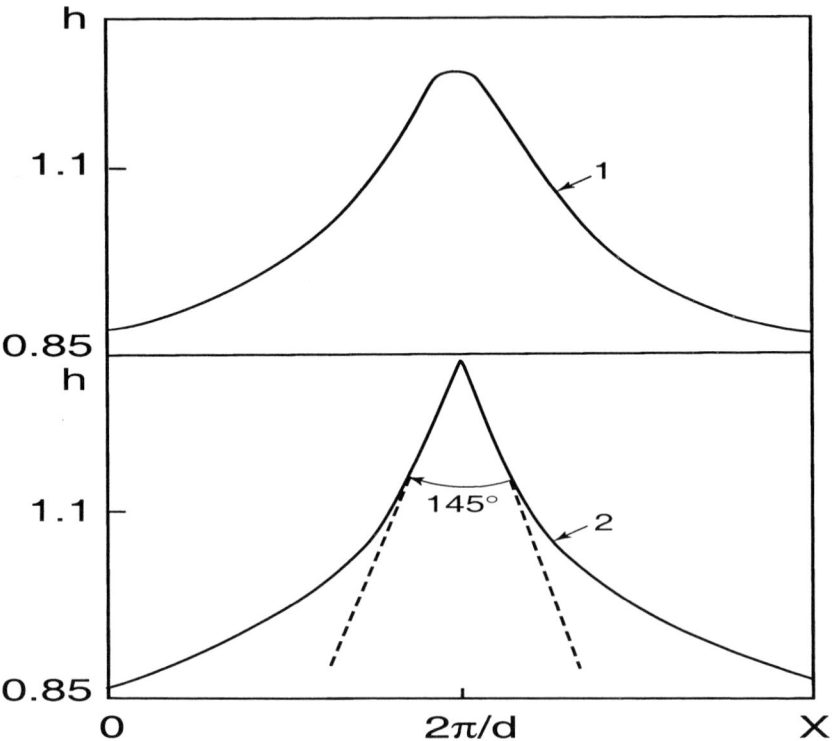

Figure 5.48: Wave profiles at zero surface tension: $R = 5$; $(1)s = 0.926$ and $(2)0.849$.

(1972) predicted a singularity in the wave solution at zero surface tension, but the type of singularity was not specified. Our investigation shows, so far as our numerical resolution allows, that at $\gamma = 0$ and at a rather small wave number, an angular wedge appears at the wave crest. The wedge angle is different from that of Stokes waves at $120°$ (Stokes, 1957) but is instead within $140°$ to $150°$.

CHAPTER 6

Floquet Theory and Selection of Periodic Waves

Many of the periodic waves constructed in Chapter 5 can only be observed in periodic forcing experiments with a specific forcing frequency. Such experiments have been carried out by the Kapitzas, Alekseenko's group and Liu and Gollub. Even in these periodically forced systems, many periodic waves often have frequencies close to the forcing frequency and it is the most stable (or least unstable) wave that is observed. Such selection becomes even more competetive in naturally excited waves where the noise spectrum is broadbanded. Eventhough periodic waves still appear transiently, as seen in the experiments and simulations of Chapter 4, they are selected from all frequencies in the spectrum. To further muddle the picture, the selection process is quite sensitive to the transients present during the selection dynamics. The selected wave is not unique and depends on the noise present. Since the transition is often from one periodic wave to another, the selected wave at one transition often depends on the previous wave or even the one before that, as the noise present during the transition is triggered by these earlier waves. In this chapter, we examine the stability of the periodic waves constructed in Chapter 5 to determine not only which waves will be selected at a particular transition but also the next selected wave to decipher the entire sequence of transitions. All waves will be shown to be unstable to a general three-dimensional noise field. However, for a given noise field, some are less unstable than the others and can even be stable. The wave dynamics hence approach these waves for a long transient but will eventually move away due to their instability. However, their instability generates another set of noise to select another set of least unstable waves. For relatively low R with weak primary and secondary instabilities and with relative weak and narrow-banded inlet noise, each wave transition is only determined by the previously selected periodic waves. A sequence of wave transitions can then be constructed from our theory. It is reminiscent of chaotic dynamics that evolve from one hyperbolic fixed point to another via heteroclinic orbits. Such a transition sequence will be shown to be most appropriate for periodically forced waves. For most naturally excited waves, a stochastic theory will be offered in Chapter 10.

6.1 Stability and selection of stationary waves.

We expect the waves exiting the inception region to be monochromatic waves with the fastest growing wave number α_m. However, slowly evolving finite-amplitude effects then suggest that a subset of the stationary waves we constructed be selected successively downstream. Since the first stationary waves are short, near-sinusoidal ones and the only wave family with short wave members is the slow family γ_1 (see Figure 5.35 and 5.45 for water with $\gamma = 2850$), we expect a short member of γ_1 to be selected first. This would then imply a deceleration of the wave speed immediately beyond inception which is consistent with the observation of Stainthorp and Allen for naturally excited waves shown in Figure 4.8. (Note the deceleration in the first 10 cm). Further downstream, near solitary waves resembling the solitary waves on γ_2 and not those on γ_1 are observed.

We hence expect the waves to evolve towards the solitary end of the fast family γ_2. This is again consistent with Stainthorp and Allen's observation of acceleration towards the solitary wave regime. We shall confirm and quantify these observations here by subjecting the stationary waves to three-dimensional disturbances of all wavelengths. Two secondary instability mechanisms dominate their stability–modulation (sideband) and subharmonic instabilites. The first involve disturbances wavelengths much longer than the periodic waves and the instability manifests itself as long modulating envelopes over the periodic waves. Periodic waves that suffer the subharmonic instability, on the other hand, would lose every other peak as a result of the instability. In some subharmonic instability theories, emphasis is placed on the phase difference between the disturbance and the periodic wave. Such a phase difference is related to resonant interaction in deep-water waves and the strength of the subharmonic instability. In the present context, this phase information may decide whether the waves resulting from the instability will be shifted from the original and whether their wave peaks/troughs would be sharp and symmetric (Cheng and Chang, 1995; Sangalli et al, 1997). However, we find such details to be relatively unimportant in falling-film waves and will simply present the stability result without detailed analysis. That the most stable or least unstable waves are selected in a particular periodic wave family has been speculated since the first paper of Kapitza (1948). He suggested that waves with the minimum average energy dissipation,

$$\dot{E} = -\left\langle \int_0^h \left[\left(\frac{\partial u}{\partial y}\right)^2 + \left(\frac{\partial v}{\partial x}\right)^2\right] dy \right\rangle, \qquad (6.1)$$

where $\langle \rangle$ denotes averaging over a wavelength, is the observed one. This argument originates from an inertialless energy stability analysis and we expect it to be valid only at low δ (or R) since inertia is surely also important at higher δ. However, this dissipation criterion will be shown to be quite accurate at low δ and it offers a very physical selection criterion in this limit. At low δ, α_0 is small and if one imposes the same small-amplitude, long-wave expansion carried out for the KS equation, it can be shown from (3.61) that

$$\frac{\partial u}{\partial y} \sim -3(y-h)$$

such that

$$\dot{E} \sim -3\langle h^3 \rangle$$

Since $\langle h^2 \rangle$ corresponds to the flow rate for a constant-thickness formulation, viz. the first term within the square bracket of the Benney lubrication equation, the above argument implies that at small amplitude and δ, the stationary wave selected among waves of the same average thickness is one with the highest flow rate. Shkadov (1968) also arrived at the above conclusion by subjecting the γ_1 family of the averaged equation to an elaborate two-dimensional sideband stability analysis. However, due to the limitation of the Shkadov model, he concluded that these waves with the maximum flow rates are stable to two-dimensional disturbances.

We shall show that, for $\delta < 0.037$, Kapitzas' dissipation criterion (6.1) offers an excellent estimate of the least unstable wave on γ_1 which is nevertheless, unstable to two-dimensional disturbances. The conclusion that the wave with the highest dissipation rate also carries the highest flux in criterion (6.2) with a small amplitude expansion and is expected to be valid only for small-amplitude waves like the short members of the γ_1 family. Since the γ_2 family begins with large amplitude waves, the longwave expansion is not expected to hold for this fast family but the full criterion (6.1) will be shown to remain as a good estimate.

Cheng and Chang (1992) have subjected the γ_1 family of the KS equation (with the constant flux formulation) to two-dimensional and three-dimensional sideband disturbances and concluded that a band of short waves with wave next to the neutral curve α_0 are unstable to two-dimensional disturbances while the ones close to the maximum-growing mode α_m are unstable to three dimensional disturbances.

There is a window of waves stable to all disturbances in between. The first result corrects the conclusion of the first attempt on finite-amplitude two-dimensional sideband instability by Lin (1974). The results by Nepomnyaschy for the S family of the KS equation suggests that even the long waves on γ_1 are unstable to two-dimensional sideband instability such that a narrow window of waves stable to sideband disturbances exists. However, in the solitary wave limit, one expects the stability of the finite-amplitude waves to be identical to that of a flat-film, viz. unstable to all two-dimensional disturbances with wave numbers within $(0, \alpha_0)$. Prokopiou et al. (1991) subjected the γ_1 waves of the averaged equation to subharmonic instability and found that a band of waves close to the neutral curve are also unstable to two-dimensional subharmonic disturbances. These analytical results then suggest that γ_1 waves close to the neutral curve are unstable to either two-dimensional subharmonic or sideband disturbances. Near solitary waves are unstable to the same two-dimensional disturbances of the flat film, as much of these waves consist of flat films. The intermediate waves are more susceptible to three-dimensional disturbances.

If there is any window of stable waves, it should exist only if these instability regions do not overlap in the intermediate region near α_m. From the result of Cheng and Chang (1992), for the KS equation, this stable window should exist only for vanishingly small δ. Demekhin and Kaplan (1989) confirmed this numerically and showed that at the KS limit of $\delta = 0$, the stable band is within $(\alpha/\alpha_0) \in (0.77, 0.84)$ which was first suggested by Nepomnyaschy (1974). Beyond $\delta = 0.037$, our numerical result below indicates that this stable window no longer exists.

The entire γ_1 family is hence unstable. Nevertheless, one expects the least unstable member of the γ_1 family in the intermediate α range and it should be most susceptible to three-dimensional disturbances from the earlier results. There have been no prior stability study of the γ_2 family of fast waves.

Linearizing the boundary layer (BL) equation of (3.77) to (3.80) in a frame moving with speed c about the periodic travelling wave solution (u, v, h) of (5.1) to (5.5) with speed c and introducing the disturbance vector $(\hat{u}(x,y), \hat{v}(x,y), \hat{w}(x,y), \hat{h}(x,y)) \exp(i\alpha\nu x + i\beta z + \alpha\mu t)$ where the hat variables have the same wavelength as the stationary waves and are hence $l = 2\pi/\alpha$ periodic in x, ν is a real number between $(-\frac{1}{2}, \frac{1}{2})$ to span the subharmonics of α, β is the transverse wave number and $\alpha\mu$ is the growth rate, one obtains

$$\alpha\mu\hat{u} + \frac{\partial}{\partial x}(2u-c)\hat{u} + i\alpha\nu\hat{u}(2u-c) + \frac{\partial}{\partial y}(u\hat{v} + v\hat{u}) + i\beta u\hat{w}$$

$$= \frac{1}{5\delta}\left[\left(\frac{d}{dx} + i\alpha\nu\right)^3 \hat{h} - \beta^2\left(\hat{h}_x + i\alpha\nu\hat{h}\right) + \frac{1}{3}\frac{\partial^2 \hat{u}}{\partial y^2}\right], \quad (6.2)$$

$$\alpha\mu\hat{w} + \frac{\partial}{\partial x}(u-c)\hat{w} + i\alpha\nu\hat{w}(u-c) + \frac{\partial}{\partial y}(v\hat{w})$$

$$= \frac{1}{5\delta}\left[i\beta(\hat{h}_{xx} + 2i\alpha\nu\hat{h}_x - \alpha^2\nu^2\hat{h}) - i\beta^3\hat{h} + \frac{1}{3}\frac{\partial^2 \hat{w}}{\partial y^2}\right], \quad (6.3)$$

$$y = h: \quad \frac{\partial \hat{u}}{\partial y} = -\frac{\partial^2 u}{\partial y^2}\hat{h}, \quad \frac{\partial \hat{w}}{\partial y} = 0, \quad (6.4)$$

$$y = 0: \quad \hat{u} = \hat{w} = 0, \quad (6.5)$$

$$\alpha\mu\hat{h} + \frac{\partial}{\partial x}\left[\int_0^h \hat{u}\,dy + u\hat{h}\right] + i\alpha\nu\left[\int_0^h \hat{u}\,dy + u\hat{h}\right] + i\beta\int_0^h \hat{w}\,dy = 0, \quad (6.6)$$

$$\hat{v} = -\int_0^y \left(\frac{\partial \hat{u}}{\partial x} + \frac{\partial \hat{w}}{\partial y} + i\alpha\nu\hat{u}\right)dy. \quad (6.7)$$

The parameter ν determines the nature of the secondary instability. For ν small, a modulation (sideband) instability exists with a disturbance wavelength much longer than that of the periodic wave. For ν equal to 1/2, the secondary instability is a subharmonic one and every other peak of the periodic wave vanishes.

We note that the introduced disturbances correspond to constant-thickness perturbations. It is not clear whether the stability of the waves to such disturbances is the same as its stability to constant-flux disturbances. In Poiseuille flow, for example, there are subtle differences in the stability with respect to these two disturbances. The stability of the stationary solution (u, v, h) is then determined by the growth rate $\alpha \mu_R$ where μ_R is the real part of μ. Since the transformation $\nu \to -\nu$ simply yields the conjugate equation, we can restrict the bound of ν to $(0, \frac{1}{2})$. We use the Petrov-Galerkin numerical method of section 2.6 for Orr-Sommerfeld equations here to span the y direction. Fourier expansion is carried out in the x direction. The end result is a generalized eigenvalue problem for complex matrices at a given α

$$|\mathbf{A} - \alpha\mu\mathbf{E}| = 0 \qquad (6.8)$$

which is solved by means of a QR algorithm.

Typical computed growth rates of the dominant disturbances of the γ_1 family are shown in Figure 6.1 and 6.2, and Table 6.1 for $\delta > 0.037$. Only the most unstable disturbance for every stationary wave is depicted. Many of the earlier predictions are confirmed. The short waves near the neutral curve are either unstable to two-dimensional sideband disturbances (ν small) or two-dimensional subharmonic disturbances ($\nu = 0.5$). The least unstable wave at α_s is slightly above α_m and it is dominated by three dimensional disturbances with small β and $\nu = 0.5$. The stationary waves near α_s remain unstable to sideband ($\nu \ll 1$) and subharmonic ($\nu = 0.5$) two-dimensional disturbances with growth rates close to but smaller than the three-dimensional ones. In fact, the wave least unstable to two-dimensional disturbances has a wave number α_2 only slightly higher than α_s.

The growth rate of the dominant disturbance at α_s is two to four times the dominant growth rate of the flat film (which is equivalent to that of the solitary wave at $\alpha = 0$ in Figure 6.1 and 6.2). This implies that the life time or length of the selected stationary wave is two to four times that of the flat inception region which is laready quite visible in a channel. The dominant growth rate of the stationary waves at α_s increases with δ. For δ in excess of unity, it is of the same order as the dominant flat-film growth rate. Beyond that ($\delta \geq 1$), one does not expect the wave evolution to be locally stationary. This is consistent with our numerical simulation results of section 4.4 where chaotic formation and destruction of pulses from the primary wave field is observed beyond $\delta = 0.6$.

Another requirement for the stationary assumption to hold is time-scale separation that stipulates the unstable disturbances to have a far smaller absolute value than the stable ones. This is related to the argument that the evolution involves two time-scale which is used in the Stuart-Landau formalism and the

Center Manifold theories (Cheng and Chang, 1990). For all our computed results, only one or two eigenvalues are unstable near α_s and their growth rates are at least 5 times larger than the growth rate of the most stable mode.

The results for the γ_1 family is also tabulated in Table 6.1 and compared to Kapitzas' prediction (6.1). It is clear that the latter is an extremely accurate estimate of the stability of the stationary waves to two-dimensional disturbances at small δ. The wave in γ_1 with the highest flow rate is also predicted to the third decimal place by Kapitza's criterion in Table 6.1, cofirming the ability of (6.1) to predict α_2 to within 10% for $\delta < 0.20$. The dissipation rate $\dot{E}$ is shown in Figure 6.3 for $\delta = 0.062$. A single minimum close to α_s is seen. The location of α_s is also marked in Figure 6.7.

It is clear that this selected wave is slower and shorter than the fastest growing wave at inception with wave number α_m. In Figur 6.4, we compare the inception speed data of Stainthorp and Allen (1965) and the γ_1 stationary wave speed data from naturally excited experiments of Stainthorp and Allen and Kapitza & Kapitza (1949) to the predicted speed from the linear dispersion relationship at α_m and the speed of the stationary wave at α_s. It is clear that, while both sets of data from Stainthorp and Allen are above the predicted values and a statistically significant separation exists between them, the γ_1 stationary waves of Kapitza & Kapitza is well predicted. This implies that the periodic forcing experiments of the Kapitzas do select a least unstable wave from the γ_1 family but the naturally excited waves of Stainthorp and Allen do not. Naturally excited waves must suffer from broadband residaul noise from the inlet through out its evolution sequence. Each transition in it sequence is hence not determined by the previous periodic wave in the sequence. Another theory will be developed for naturally excited waves in Chapter 10.

Typical growth rates of the dominant two-dimensional disturbances for the γ_2 family are shown in Figure 6.5. The growth rate again approaches the those of the flat-film Nusselt basic state at the solitary limit $\alpha \to 0$. The shorter waves remain unstable to sideband and subharmonic instabilities. However, in the intermediate region, multiple discrete regions of stability are seen.The distance between these intervals decreases with decreasing α. There exists a limiting stable interval with a lowest interval of α, centered at α_f. This lower bound decreases with increasing δ. This is in contrast with the growth rates of the γ_1 family where all waves are unstable to two-dimensional disturbances for $\delta > 0.037$ and there exists a unique minimum of the two-dimension growth rate.

In all these discrete intervals of stability, the waves are unstable to three-dimensional disturbances with a very small growth rate which is not shown in Figure 6.5. Outside the stable intervals, two-dimensional disturbances dominate and, within the intervals, the small but dominant three-dimensional disturbances have long transverse variation ($\beta \ll 1$) and subharmonic ($\nu = 0.5$) or sideband ($\nu \ll 1$) streamwise variations. As δ increases, the growth rate of the γ_2 family tends to increase as a whole with the shorter waves destabilizing more than the longer waves.

As a result, some of the stable intervals at higher α disappear with increasing δ. The stable intervals at lower α also shrink in size such that they can be

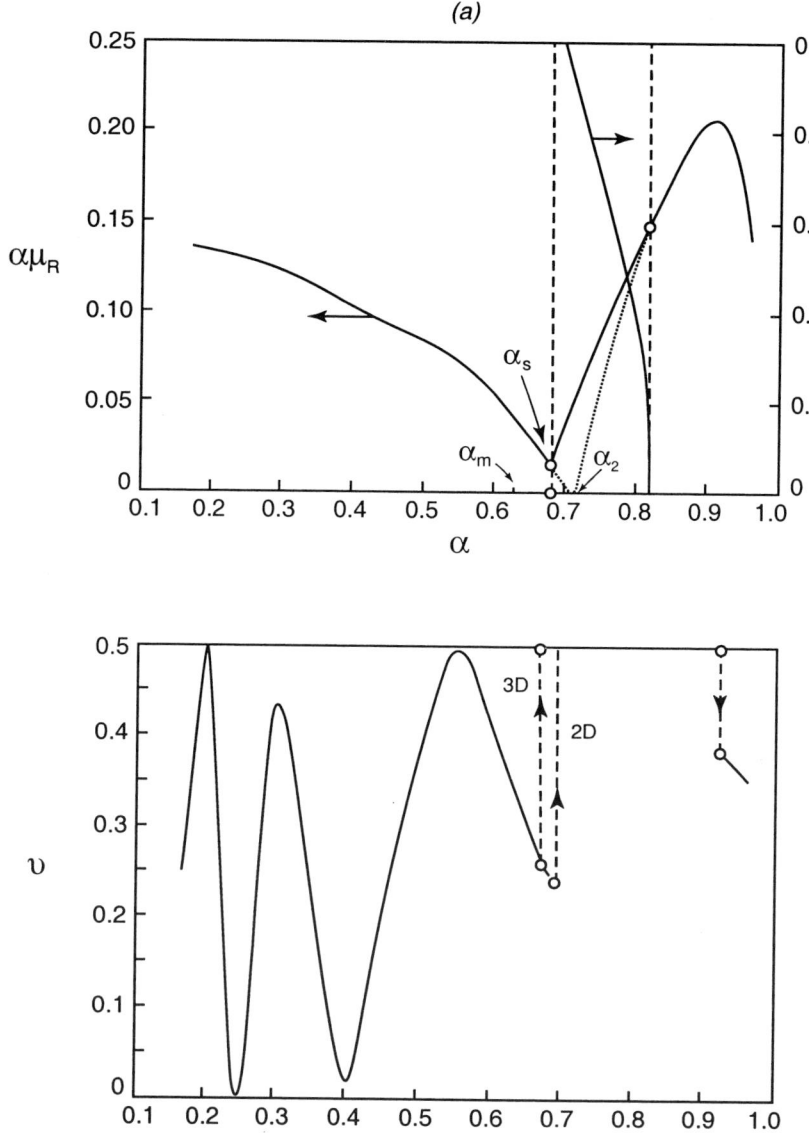

Figure 6.1: See the caption for Figure 6.2

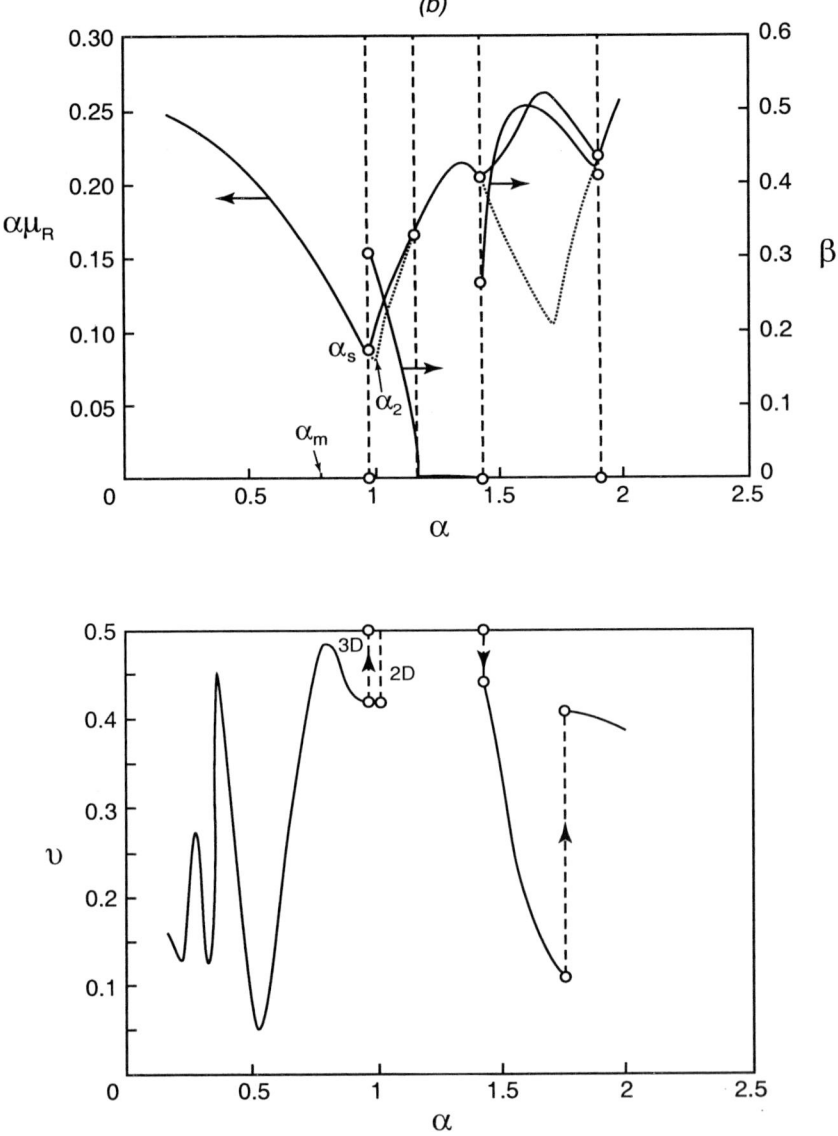

Figure 6.2: Typical computed rates of the dominant disturbances of the γ_1 family for $\delta = 0.062$ ($\alpha_2 = 0.711, \alpha_s = 0.675, \alpha_m = 0.623$) and $\delta = 0.27$ ($\alpha_2 = 1.006, \alpha_s = 0.975, \alpha_m = 0.78$). Three-dimensional distubances are only dominant in the indicated windows within some of which α_2 and α_s lie. The depicted ν corresponds to these three-dimensional disturbances. The growth rates of the two-dimensional distubances in these windows are only slightly lower and they are depicted by broken curves.(a) $\delta = 0.062$, (b) 0.27.

accurately represented by a single value of α. For example, the interval with the longest stable waves on γ_2 corresponds to a small gap around α_f. We will hence represent these intervals by the discrete wave numbers at their centers.

Kapitzas' dissipation selection theory is again extremely accurate for the γ_2 family at low δ as seen in Figure 6.3 and Table 6.2. The stable intervals appear as local minima in the dissipation and several of them are accurately captured.

Comparing to the dissipation rate of the γ_1 family, it appears that a global minimum of the dissipation rate exists and it corresponds to the lower accumulation point α_f of the stable intervals before the growth rate approach the flat-film limit of the solitary waves. At this small value of α_f, the waves resemble solitary waves with the large tear-drop hump occupying only about one-fourth or one-fifth of the entire wavelength. The remaining portion is covered by small to imperceptible capillary waves. Nevertheless, these capillary waves must provide sufficient interaction to suppress the flat-film instability of a true solitary wave. At higher δ, the capillary waves are larger and their stronger interaction can then sustain a larger separation between the humps.

The conclusion that the near-solitary wave at α_f is the global minimum of all stationary waves is, of course, only true at low δ. At higher δ values, inertia and surface tension effects must be considered. The stable intervals of the γ_2 family of water are shown in Figure 6.6. As δ (or $\langle R \rangle$) increases, some of the shorter waves begin to destabilize and the longer waves begin to disappear as minima in the dissipation rate.

The implied conclusion that the long waves near α_f are unstable at large δ is, however, erroneous since our linear stability result indicates that they remain stable. Hence, interaction between the capillary waves, which can involve the capillary or inertia forces ignored in the dissipation theory, remains stabilizing.

We have also investigated the stability of the $\gamma_2^{(n)}$ families and found them to all be unstable to two-dimensional disturbances. They hence will probably not be selected under sufficiently rich excitation.

6.2 Stable intervals from a Coherent Structure Theory

An intriguing qualitative explanation for the multiple intervals of long γ_2 waves that are stable to two-dimensional disturbances is offered by the coherent structure theory for solitary wave interaction (Elphick et al., 1988).

One notices in examining Figure 5.38 that the long periodic waves on γ_2 can be accurately approximated by separating the same solitary wave structure at vanishing α by different intervals corresponding to the wavelengths. (The wave tracings in Figure 5.38 are normalized by their wavelengths and this observation hence requires some care.) Consequently, even though the true solitary wave is unstable because of the large span of flat films away from its coherent structure (a tear drop hump preceded by front-running capillary "bow" waves), which occupies a relatively small interval of space, the same coherent structure can be

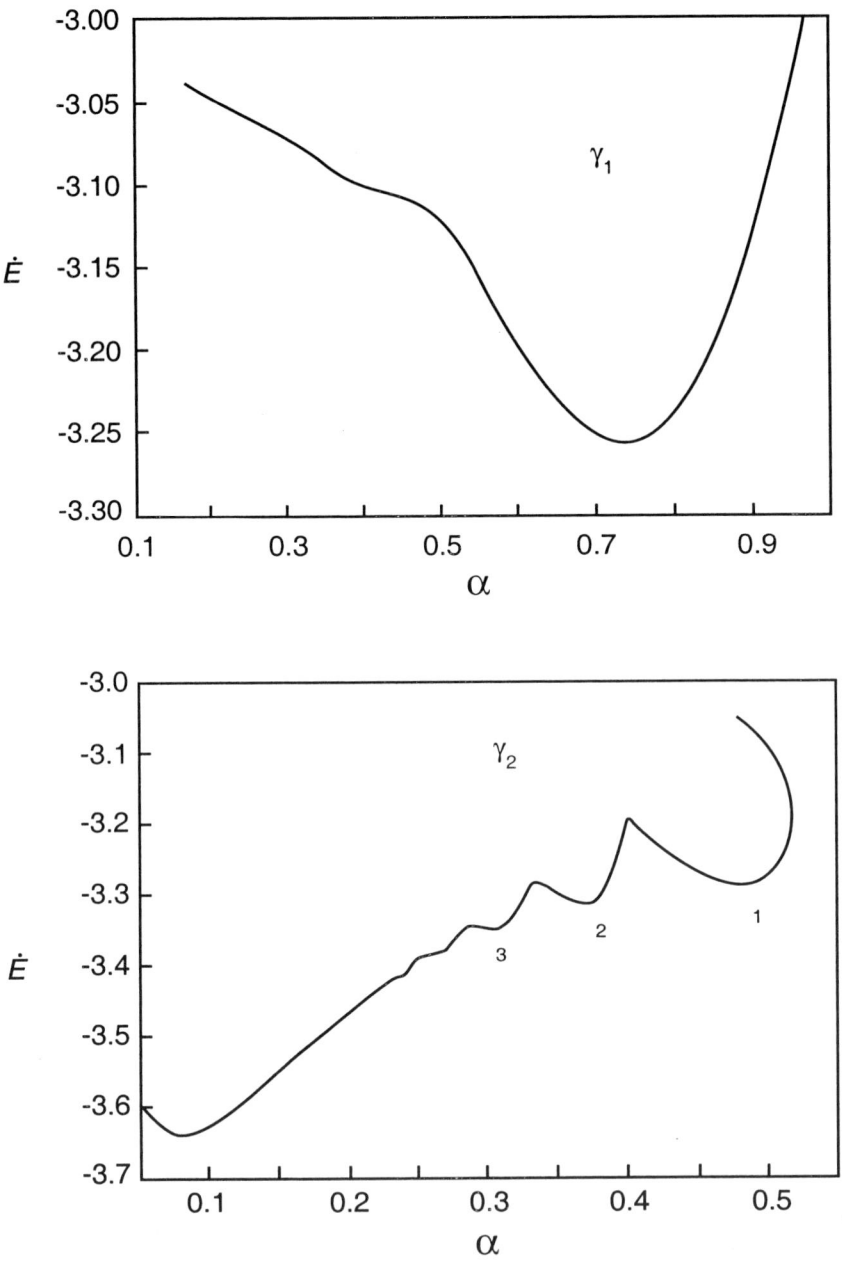

Figure 6.3: The viscous dissipation rate of the stationary waves on the γ_1 and γ_2 families at $\delta = 0.062$. The local minima for the γ_2 family are also listed in table 6.2

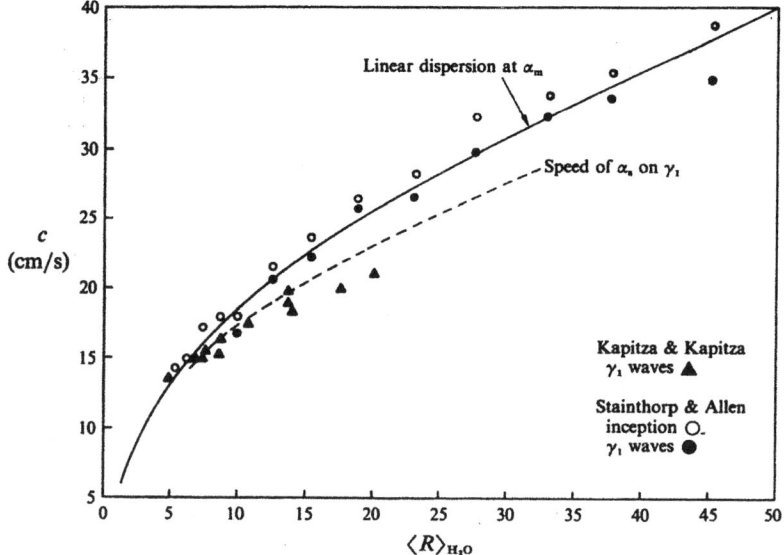

Figure 6.4: Difference in wave speeds in the inception region and the finite-amplitude capillary wave region corresponding to the γ_1 wave family.

placed in a periodic train to yield a stable periodic wave. We again demonstrate this concept with the simple KS equation (5.1) even though the stable waves appear only at finite δ. Let us approximate $H(\xi, \tau)$ by

$$H(\xi, \tau) = U_i(\xi_i) + \sum_{k \neq i} U_k(\xi_k), \qquad (6.9)$$

where U_i is a solitary wave solution satisfying either (5.9) and (5.10) or (5.11), depending on the formulation. The function U_k represents the other solitary wave coherent structures which have identical shapes but different locations and possibly different speeds. The local coordinates $\xi_i = \xi - \int c_i(\tau) d\tau$ has origin at the maxima of the ith solitary wave which moves at speed c_i.

(Since (5.1) has been properly reduced, its solitary wave solution approaches zero at both infinities and (6.9) does not require base-line corrections.) We shall only be interested in the interaction of the ith solitary wave with its two nearest neighbors, namely $k = i-1$ and $i+1$. The solitary waves are also assumed to be sufficiently apart such that the interaction occurs only at the small-amplitude ends of the solitary waves which can be described by the tails of section 5.5

$$U_f \sim A e^{-\sigma_1 \xi} \cos(\omega \xi + \phi_0), \qquad (6.10)$$

$$U_b \sim B e^{\sigma_2 \xi} \qquad (6.11)$$

where A, B and ϕ_0 are positive constants that must be obtained from the full solitary wave solution but σ_i and ω can be easily determined from the analysis

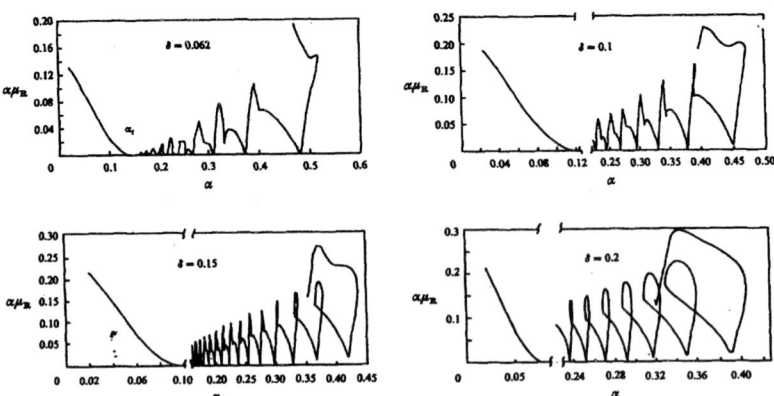

Figure 6.5: Typical computed growth rates of the dominant two-dimensional distubances of the γ_2 family at $\delta = 0.062, 0.1, 0.15$ and 0.2.

in section 5.5. Equation (6.10) for the front end describes the front-running capillary waves. Knowing that μ approaches 1.2 from Figure 6.5 and inserting $H \sim e^{\sigma \xi}$ into the linearized version of (5.11), one obtains the eigenvalue problem

$$\sigma^3 + \sigma - 1.2 = 0, \qquad (6.12)$$

which yields $\sigma_1 = 0.38$, $\omega = 1.20$ and $\sigma_2 = 0.76$.

Multiplying (5.1) by H and integrating from $\xi = -\infty$ to $\xi = +\infty$ yields an evolution equation for the deviation flux Q:

$$\frac{1}{2}(d/dt)<H^2> = <H_\xi^2> - <H_{\xi\xi}^2>. \qquad (6.13)$$

Inserting (6.9) into (6.13) and retaining only the leading order mutual interaction terms (all self interaction terms vanish by definition), one obtains

$$\frac{1}{2}\frac{d}{dt}<H^2> = 2\sum_{k \neq i}\left\langle \frac{\partial U_i}{\partial \xi_i}\frac{\partial U_k}{\partial \xi_k} - \frac{\partial^2 U_i}{\partial \xi_i^2}\frac{\partial^2 U_k}{\partial \xi_k^2}\right\rangle = 2\sum_{k \neq i}\left\langle [\mu U_i - 2U_i^2]\frac{\partial U_k}{\partial \xi_k}\right\rangle, \qquad (6.14)$$

where (6.9) has been invoked after integration by parts.

The excess flux of a stationary periodic train of solitary waves does not vary in time, $(d/dt)<H^2> = 0$, and the separation between the solitary waves is constant at s,

$$\xi_{i-1} = \xi_i + s, \quad \xi_{i+1} = \xi_i - s. \qquad (6.15)$$

Equation (6.14) can then be rewritten for a periodic wave as

$$\langle [\mu U_i - 2U_i^2]A(-\sigma_1 + i\omega)\exp(-\sigma_1 + i\omega)\xi\rangle \exp\{(-\sigma_1 + i\omega)s + i\phi_0\} + c.c.$$
$$+\langle [\mu U_i - 2U_i^2]B\sigma_2 \exp(\sigma_2 \xi)\rangle \exp(-\sigma_2 s) \equiv F(s) = 0, \qquad (6.16)$$

where c.c. denotes complex conjugate.

The zeros of $F(s)$ are hence the wavelengths of stationary periodic waves and the stable periodic waves form a subset of these zeros. We shall refrain from an exact evaluation of the necessary integrals in (6.16) here but observe that in the long wave limit, $s \to \infty$, $F(s)$ oscillates about zero with a frequency ω corresponding to the wave frequency ω of the front-running capillary waves. Consequently, an infinite number of equally spaced zeros exist at large s,

$$\lim_{n \to \infty} s_n = 2\pi n/\omega. \tag{6.17}$$

(There is another subset corresponding to $n\pi/\omega$ but these correspond to unstable periodic waves.) Equivalently, (6.17) predicts that the wave number α_n of the long stable waves on γ_2 varies as n^{-1},

$$\alpha_n \sim \omega/n \tag{6.18}$$

and the interval between the stable waves vary as n^{-2},

$$\Delta \alpha_n = \alpha_n - \alpha_{n-1} \sim \omega n^{-2}. \tag{6.19}$$

Both predictions are consistent with our linear stability results of Figure 6.5. For example, for $\delta = 0.062$, the stable waves are located at approximately $\alpha = (0.48, 0.38, 0.31, 0.26, ...)$.

This yields an s_n of approximately $13 + 3.8n$, after invoking the length scale of the KS equation $\xi = \alpha_0 x = (18\delta)^{1/2} x$, which is close to the (6.17). The variations of α_n and $\Delta \alpha_n$ for all δ in Figure 6.5 are also close to the predictions of (6.19).

The above theory also measures the interaction among the nearest coherent structures in a train of solitary waves. It ignores the primary instability of a flat film when the solitary waves are too far apart. Consequently, the series α_n of (6.19) does not approach zero in Figure 6.5. Instead, below α_f, the neglected primary instability of the flat film begins to dominate and destabilizes all waves with wavelengths longer than $2\pi/\alpha_f$.

Strong coherent-structure interaction reduces this flat-film primary instability and hence lowers α_f. This explains the decrease in α_f with increasing δ since the solitary wave amplitude increases with δ. It may also explain an apparent inconsistency of the coherent-structure theory. Due to symmetry (5.14) and (5.15), the theory can also be applied to the γ_2 (actually C_{-1}) family of the KS equation to yield the conclusion that the negative solitary wave of Figure 5.23 can also interact favorably to form a stable periodic train. This stability is not found in long wave members of the γ_1 family.

This is probably due to the far smaller (and slower in absolute speed) solitary wave limit of the γ_1 family relative to the γ_2 family evident in Figure 6.7. These weak solitary waves of γ_1 simply cannot overwhelm the flat-film instability of long periodic stationary waves. This is evident in the growth rates of the γ_1 family in Figure 6.1.

The flat film instability begins to dominate at relatively high $\alpha(< 0.6)$ compared to $\alpha_f(< 0.2)$ of the γ_2 family. This argument alos explains why the C_1 family of the KS equation is found to be unstable. The positive solitary waves at vanishing δ are also too weak to sustain a periodic train.

The stable intervals hence exist at small but finite δ on the γ_2 family because of the dramatic increase in amplitude and speed of the γ_2 solitary wave with respect to δ.

Finally, in spite of the surprising accuracy of (6.19) in the above theory, it must be reminded that the theory remains speculative since the KS equation invoked in the analysis does not yield stable periodic waves near the solitary-wave limit. The only stable waves of the KS equation lies within the near-neutral wavenumber band $(\alpha/\alpha_0) \in (0.77, 0.84)$ on the S family. The dispersive effect of δ somehow validates the above coherent structure theory. A more detailed theory which includes δ will appear in the next chapter.

6.3 Evolution towards solitary waves

A selection mechanism has now appeared from our analysis. As is sketched in Figure 6.6, a system forced with a weak narrow banded inlet noise which includes α_m will select two-dimensional monochromatic waves with wave number α_m after the inception region. It should then approach finite-amplitude capillary waves on γ_1 with wave numbers close to α_s or α_2 (part I) and then near solitary waves close to α_f on γ_2 (part II).

The second transition from γ_1 to γ_2 can be a subharmonic instability if the suprpressed three-dimensional disturbances have grown to a significant amplitude. In this case, a stable wave on γ_2 with a wave number approximately half that of as will be approached.

If the disturbances remain two-dimensional, the dominant two-dimensional disturbance on γ_1 is either a sideband instability ($\nu \ll 1$) or a subharmonic instability ($\nu = 0.5$) and a long stable wave on γ_2 near α_f may also be selected. Further downstream, these near-solitary waves on γ_2 will eventually succumb to long transverse variations.

If periodic forcing with two-dimensional disturbances are introduced, any of the stable waves on γ_2 can be observed since the initial transitions (parts I and II) have been suppressed. If the forcing frequency is higher than the shortest stable wave of γ_2, then the waves will most likely evolve towards as on γ_1 before undergoing transitions to the long waves on γ_2 as in a naturally excited system. This would then imply that the observed waves for a forced experiment are bounded between α_s and α_f. This was indeed reported by Alekseenko *et al.* (1985) and we favorably compare their experimental bounds to α_s and α_f from our theory in Figure 6.7. We also depict the typical Floquet growth rates of the γ_2 waves of water in Figure 6.9. Kapitza's prediction (6.1) which works well for the γ_1 waves of Table 6.1, also works well for the γ_1 waves of Table 6.1 also works well for γ_2 waves until R exceeds 8. A more detailed comparison is offered in Table 6.2.

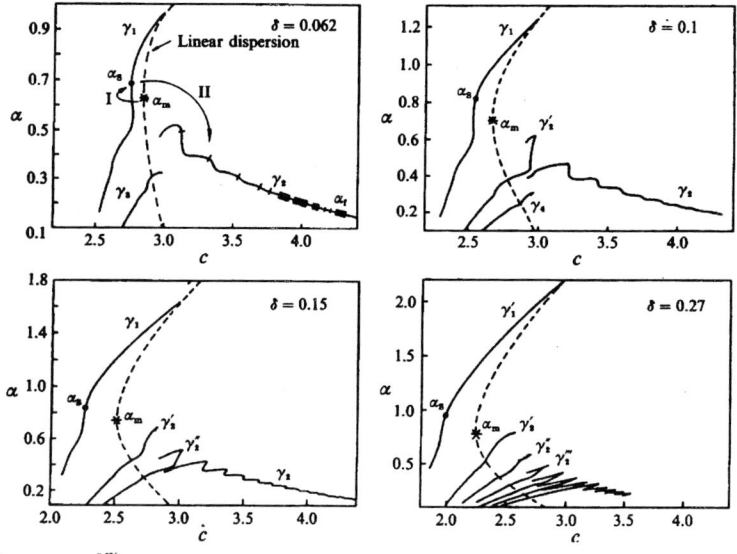

Figure 6.6: Actual computed wave branches for $\delta = 0.062, 0.1, 0.15$ and 0.27 and comparison to the linear dispersion curve. The faster growing linear wave α_m and the most stable waves on the wave branches are represented by stars, circles and blackened segment in some figures. The wave transition are indicated in the first figure.

Finally, we report a two-dimensional simulation that supports this scenario of the wave evolution process. Simulation of wave evolution on a falling film is hampered by the exceedingly long evolution length of all the transitions. One solution is to carry out the simulation in a Lagrangian frame moving at a speed of 3 and to invoke periodic boundary conditions. To include several near solitary waves in the domain, however, the spatial period of the computational box must be large. Moreover, periodicity usually introduces non-stationary interaction among the waves that does not appear in the real system. We overcome these obstacles by ignoring the linear inception region and begin our simulation with a wave packet with the local wave number α_{loc} close to α_m in the middle of a computation box in the Lagrangian frame.

The wave packet initial condition removes the boundary effects of the periodic boundary condition and it also minimizes the interaction between the γ_1 and γ_2 waves since they will propagate towards the flat films at two different ends of the wave packet. A total of 256 complex Fourier modes are used with a fourth order Runga-Lutta time integration scheme with a step size of $\alpha \Delta t = 10^{-4}$.

As seen in Figure 6.9, one sees an initial approach towards a shorter and

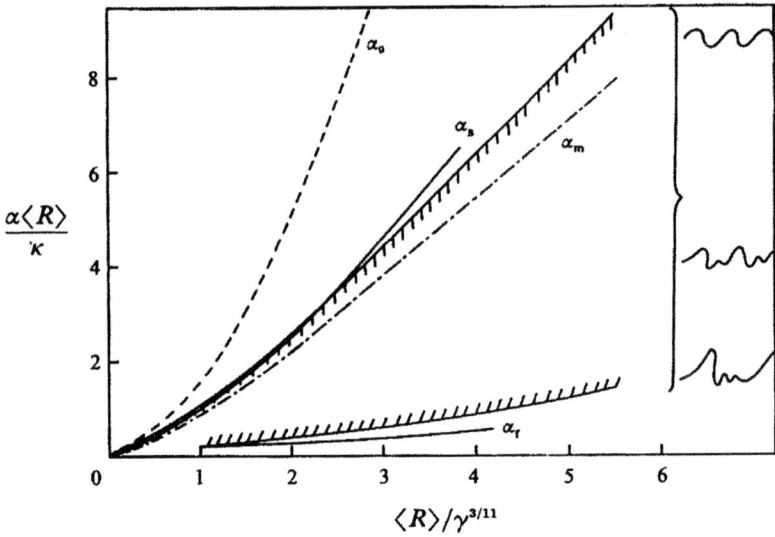

Figure 6.7: The shaded curves are bounds on the wavenumbers of the observed waves in the forced experiments of Alekseenko *et. al.* (1985) for water. They are favourably compared to our predictions of α_s and α_1.

slower γ_1 capillary wave as these waves move toward the back of the packet. A secondary transition towards the near-solitary waves of γ_2 near af is then seen as the unique solitary wave shapes begin to evolve and separate from the front of the wave packet. Our simulations out $\delta < 0.037$ show that the second transition to γ_2 waves does not occur. This is, of course, due to the existence of stable waves on γ_1 near as. Changing the width or amplitude of the initial wave packet does not alter the evolution process qualitatively. Reducing aloc of the initial wave packet, however, can initiate a direct transition to the γ_2 family without approaching the stationary waves on γ_1. This then corresponds to slowly forced experiments that bypass the intial deceleration state (part I in Figure 6.6) towards the γ_1 waves.

Evolution triggered by white natural inlet noise is, however, different from both periodic forcing and wave packet evolution described above. As seen in section 4.4, there seems to be a dominant modulation frequency in this case, where as Figure 6.1 indicates a Floquet growth rate that grows with wavenumber of frequency near α_s. Because of this characteristic modulation frequency, some wave members of the initial periodic wave patch, with wavenumber close to α_m or α_s on the γ_1 branch, are transformed first to the solitary pulses on γ_2. The others are individually transformed subsequently to the same pulses on γ_2 — but with a much smaller amplitude. The initial ones are called "excited" pulses and the later ones "equilibrium" pulses. This non-uniform transformation into pulses in a noise-driven system will be examined with a different theory in Chapter 10.

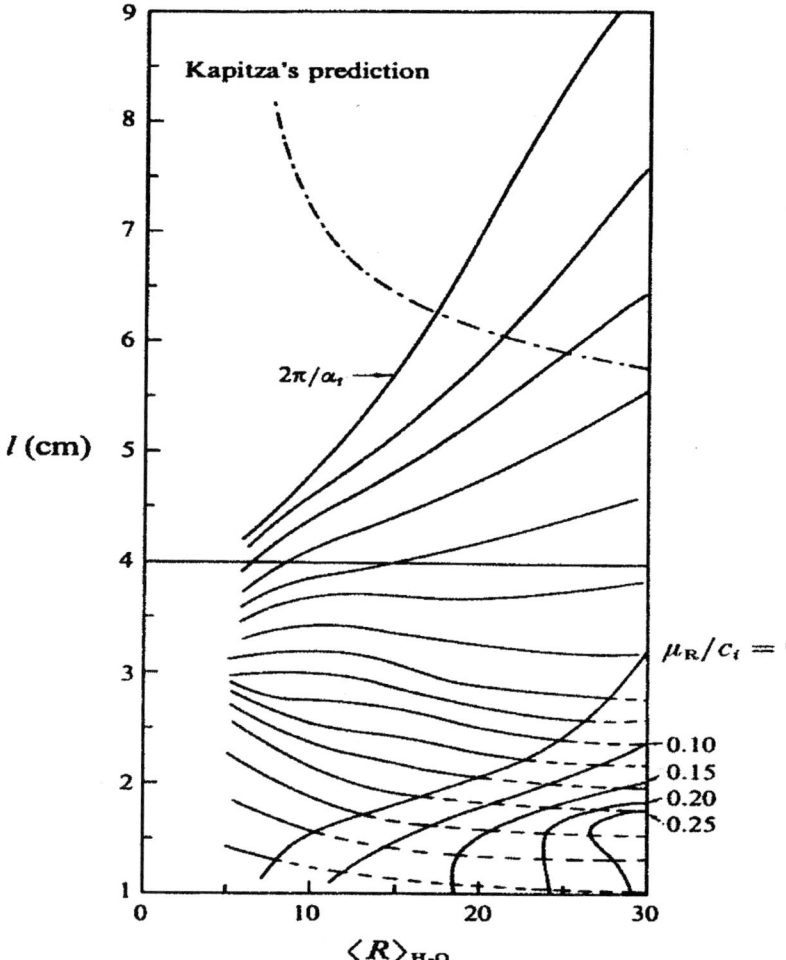

Figure 6.8: The predicted stable waves on the γ_2 family for water. The shorter ones destabilize at higher Reynolds numbers. The longer ones are very robust in spite of Kapitza's prediction of eventual destabilization at high Reynolds numbers.

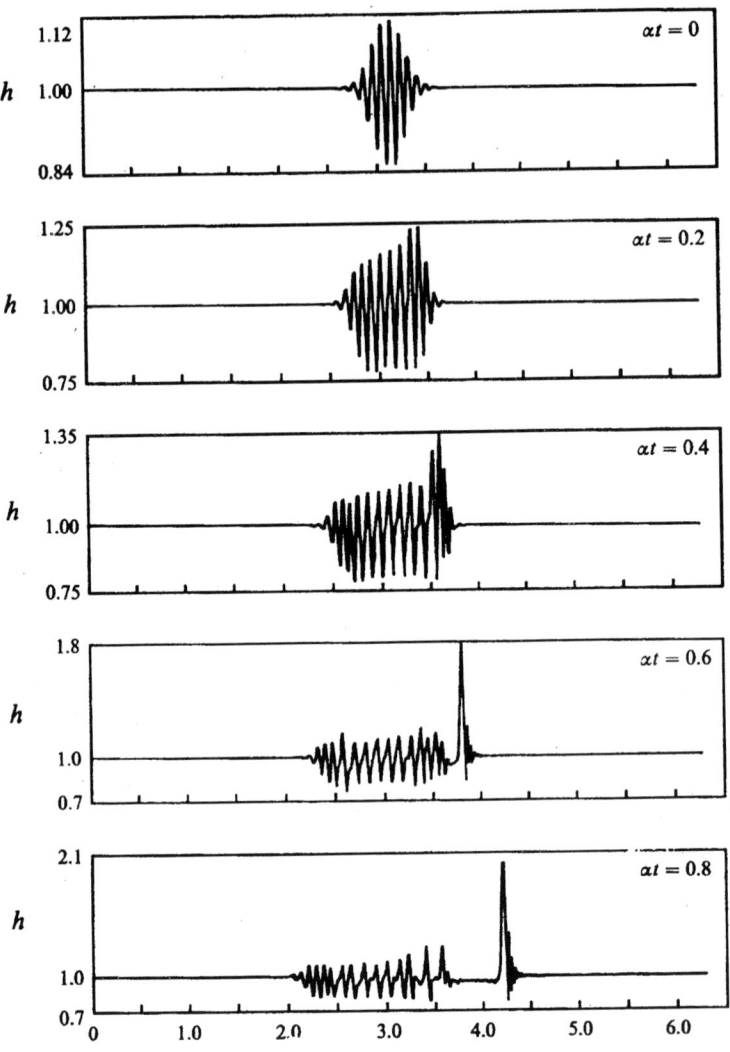

Figure 6.9: Evolution simulation in the Lagrangian frame with speed 3 at $\delta = 0.05$. An initial packet with a local wave number equal to $\alpha = 0.612$ is introduced. Slower with shorter wavelengths corresponding to α_2 on γ_1 are seen to fall back from the wave packet during path I of Figure 6.6. this is followed by the creation of long solitary waves near α_f on the γ_2 family which separates from the packet during path II of the transition. The spatial coordinate is αx where $\alpha = 0.00897$.

2-dimensional disturbances

$\langle R \rangle_{H_2O}$	l (cm)	c (cm/s)	$2\pi/\nu\alpha$ (cm)	$\alpha\mu_R/\alpha c$	δ	α_s
7.36	0.897	15.38	7.89	0.0106	0.062	0.711
11.45	0.822	18.48	3.24	0.0993	0.10	0.811
16.47	0.778	21.38	2.29	0.198	0.15	0.889
21.13	0.751	23.69	1.94	0.267	0.20	0.945
27.22	0.726	26.41	1.71	0.316	0.27	1.006

3-dimensional disturbances

$\langle R \rangle_{H_2O}$	l (cm)	c (cm/s)	$2\pi/\nu\alpha$ (cm)	$2\pi/\beta$ (cm)	$\alpha\mu_R/\alpha c$	δ	α_s
7.32	0.961	15.34	1.92	1.138	0.126	0.062	0.675
11.42	0.848	18.42	1.70	1.38	0.186	0.10	0.783
16.40	0.819	21.30	1.64	1.71	0.238	0.15	0.868
21.10	0.764	23.60	1.53	2.49	0.247	0.20	0.928
27.16	0.742	26.37	1.48	2.97	0.334	0.27	0.984

Kapitza's prediction

$\langle R \rangle_{H_2O}$	l (cm)	c (cm/s)	δ	α_s
7.36	0.882	15.39	0.062	0.724
11.46	0.763	18.71	0.10	0.874
16.51	0.710	21.81	0.15	0.971
21.21	0.693	24.2	0.20	1.025
27.4	0.621	27.7	0.27	1.173

TABLE 6.1 Most stable member of the γ_1 family

	Two-dimensional linear stability		Kapitza's criteria	
$\langle R \rangle_{H_2O}$	l (cm)	c (cm/s)	l (cm)	c (cm/s)
7.42	1.33	17.4	(1) 1.33	17.4
7.49	1.71	18.6	(2) 1.72	18.6
7.57	2.06	19.6	(3) 2.11	19.7
7.65	2.37	20.5	Absent	—
8.07	4.51	25.2	(4) 7.83	29.5

TABLE 6.2 Selected wave members on γ_2 at $\delta = 0.062$

CHAPTER 7

Spectral Theory for gKS Solitary Pulses

It is clear from the simulations of Chapters 4 and 6 that the final evolution is towards the solitary travelling waves on the γ_2 family, regardless of how the waves are triggered at the inlet. This is an important feature that allows a comprehensive analysis of the subsequent wave dynamics. For a vertical water film with natural inlet noise, this transition to solitary waves occurs as soon as 20 cm from the inlet. Yet the wave texture and wave speed continue to evolve downstream for at least another meter, as seen in Chapter 4. Hence, almost all the spatio-temporal wave dynamics on the falling film involve solitary waves. These waves are strictly unstable in the Floquet sense, as we saw from the last chapter. Yet they survive for long durations on the flat film substrates. In fact, they are only destroyed by mutual coalescence and by a much later three-dimensional front instability. A new pulse stability clearly needs to be formulated to replace the Floquet theory that assumes periodic disturbances. In fact, since the solitary wave or pulse is so localized in space, it is only sensitive to localized wave packet disturbances within a very narrow region. The Floquet instability in the previous chapter actually reflects the instability of the flat substrate and not the localized pulse. This observation implies that the correct stability theory must involve the relative speed of the pulse and the wave packet, viz. pulse speed vs group velocity. The localized shape of the pulse also means that one must consider a convective stability theory–the growth of the disturbance at a specific location (the pulse) and not everywhere as in the Floquet theory. These features are captured in the new spectral theory we introduce in this chapter. For simplicity, we demonstrate the ideas with the simple gKS equation. Application to more sophisticated equations, like the Shkadov model, will be presented in the next chapter. This new theory will more explicitly capture how the solitary waves interact with the wave packets on the substrate. Not only does it determine how the pulse is affected by the wave packet disturbance, it also describes how the pulse affects the wave packet. In the case when the wave packet is not stabilized by the pulse, a "turbulent spot" can actually be created by the unstable film. Furthermore, this new theory shows that the dynamics of the pulses are governed by just a few discrete modes due to certain physical symmetries. Such dynamics include the important drainage dynamics–how the pulse drains excess liquid. Such excess liquid is introduced during a coalescence event and the drainage dynamics determine whether the coalescence cascade of Chapter 4 will continue. Most impressively, the discrete mode that describes drainage actually arises from a continuum of modes. As such, nonlinear theories like Center Manifold theory and Invariant Manifold theory, which seek to reduce the model equations of an infinite-dimensional system in an unbounded domain, can be very fruitfully applied to falling-film dynamics. Its spatio-temporal dynamics can indeed be described by low-dimensional dynamical systems.

7.1 Pulse spectra

To examine the stability of a localize pulse solution travelling with a constant speed c, we first move onto a coordinate translating with the same speed c and then linearize the equation in the new coordinate about the pulse solution. In particular, we are interested in the primary pulse solution branch of the gKS equation in Figure 5.27 and reproduced in Figure 7.1. In general, this yields a scalar equation of the form

$$\frac{\partial u}{\partial t} + \mathcal{L}u = 0 \qquad (7.1)$$

We shall restrict ourselves to disturbances u that do not blow up away from the pulse. More specifically, the asymptotes of u away from the pulse are either finite — amplitude oscillations or zero asymptotes.

For the gKS equation, the linear operator L has the form

$$\mathcal{L} = \frac{\partial^4}{\partial x^4} + \delta\frac{\partial^3}{\partial x^3} + \frac{\partial^2}{\partial x^2} + \frac{\partial}{\partial x}(4h(x) - C) \qquad (7.2)$$

where $h(x)$ is the pulse solution. At the infinities the operator L has constant coefficients, because the pulse tends to the trivial solution at the infinities.

The solution of (2.1) can be carried out by an expansion with the eigenfunctions of the operator $\mathcal{L}$

$$(\lambda + \mathcal{L})\psi = 0 \qquad (7.3)$$

We also have an adjoint problem,

$$(\bar{\lambda} + \mathcal{L}^*)\varphi = 0 \qquad (7.4)$$

with respect for this scalar product,

$$(u, v) = \int_{-\infty}^{+\infty} u\bar{v}dx \qquad (7.5)$$

and with orthogonality conditions

$$(\psi_k, \varphi_m) = \delta_{k,m}, \qquad (7.6)$$

We shall avoid the more mathematical issues of completeness and use the eigenfunctions of $\mathcal{L}$ to span all disturbances. Nevertheless, it is clear that the growth and decay of the disturbances are determined by the eigenvalues of the operator L and we hence seek its spectrum.

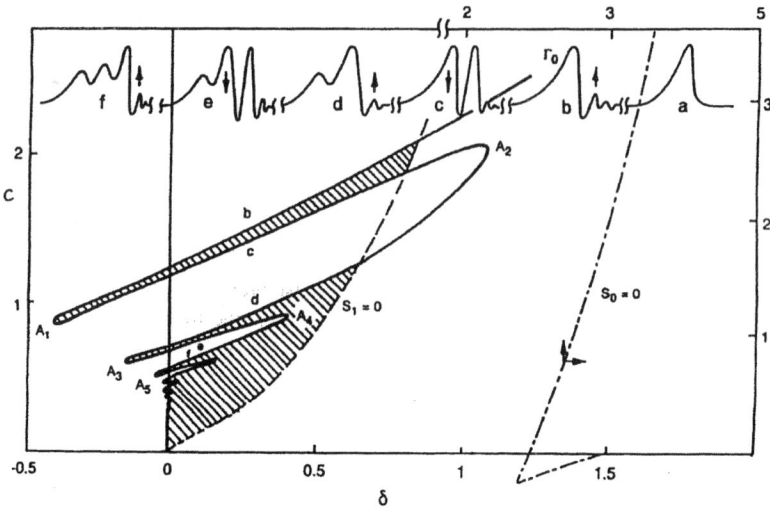

Figure 7.1: The pulse solution branch of the gKS equation. Branch b and its near-KdV extension a correspond to the one-hump pulses. Branches c and d correspond to two-hump pulses while the multi-hump pulses reside on subsequent branches. There are additional pulse branches in the shaded region.

The spectrum of the operator $\mathcal{L}$ in the infinite domain has two parts, discrete and continuous. The discrete spectrum λ_k has a countable or finite number of point eigenvalues.

The discrete eigenfunctions are local; they decay exponentially to zero far from the hump. If any discrete eigenvalue lies on the unstable side of the complex plane, the pulse is unstable. The system always has a zero eigenvalue because of the translational invariance of the governing equation– if $h(x)$ is a solution of our system, $h(x+l)$ is also a solution . Since $h(x+l) \sim h(x) + h'(x)l$ for small translates and inserting this leading-order expansion into the linearized equation (7.1), we see that this degeneracy implies that $\mathcal{L}h' = 0$ Hence, $h'(x)$ is a null eigenfunction or a zero mode of $\mathcal{L}$. Other zero modes occur due to other symmetries of the problem, such as mass conservation. If all other modes are stable, these zero modes then dominate the dynamics of the pulse. The gKS-equation also has one stable real eigenvalue, λ_2. For the one hump pulse of the gKS-equation, zero and stable eigenvalues make up the whole discrete spectrum. If the pulse-solution decays slowly at the infinities, not exponentially fast, as $1/x$, or $1/x^2$, then the number of discrete eigenvalues can form a countable set. In particular, this occurs for 2-dimensional and 3-dimensional localized structures.

Because of the infinite domain, there is also a unique essential or continuous part to the spectrum. The essential eigenvalues are defined by the linear dispersion relationship for the trivial solution in the moving frame. The essential spectrum is represented by some continuous curve Γ in the complex λ-plane (see Figure 7.2). The essential spectrum for a dispersive medium coincides with the imaginary axis (for the KdV-equation, for example). A typical spectrum for an active medium has an unstable part, because of instability of the trivial solution and a stable part, for the large wavenumbers. The dispersion relationship for the essential spectrum of the gKS-equation, parameterized by the wave number, is

$$\lambda = i\alpha c + i\delta\alpha^3 + (\alpha^2 - \alpha^4)$$

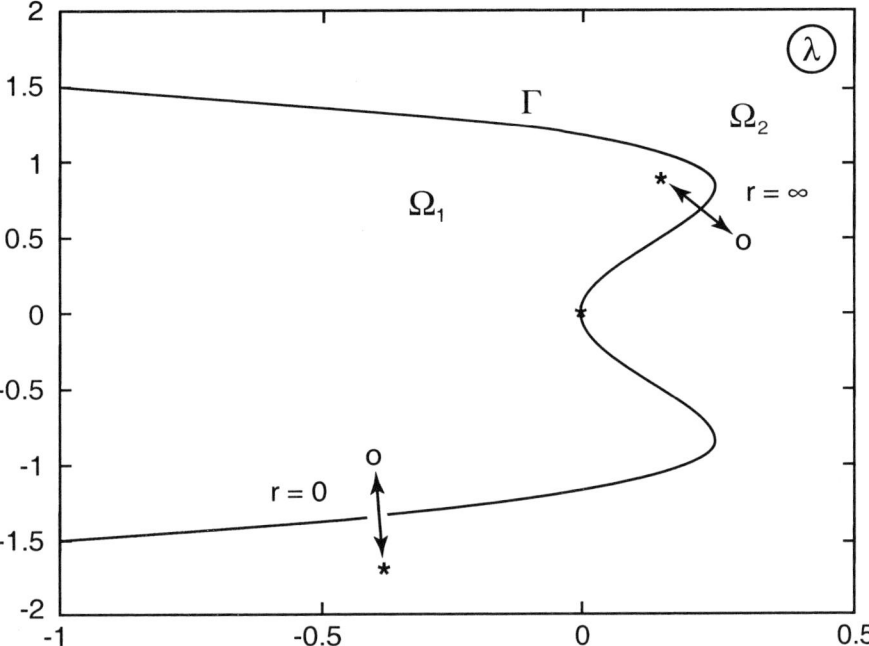

Figure 7.2: A typical essential spectrum Γ and the discrete spectrum of a pulse. The two regions Ω_1 and Ω_2 are shown. The possible resonance pole/eigenvalue transition across Γ are sketched.

The eigenfunctions of the essential spectrum are merely bounded at the infinities, viz. they exhibit constant-amplitude oscillations far from the hump, with the same wave number, but with different amplitude to the right and left hand sides . Their amplitude decay or grow exponentially in time, depending on which part of the spectrum they belong. At the same time, because of the nonzero imaginary part of λ, they propagate with some speed, different from the pulse speed. They move into the pulse and come out of the pulse, diminished or amplified in the process. The essential spectrum contains an infinite number of modes in its continuum, including a large section in the unstable region. This continuum of unstable modes would ordinarily defeat any attempt to construct low-dimensional models. However, with a special weighted spectral theory, we shall show that the dominant (most unstable) portion of this continuum can actually collapse into a single discrete mode !

7.2 Some numerical recipes to construct eigenfunctions and obtain spectra.

The numerical recipe for spectra calculation has the same flavor as that for the construction of a stationary solitary pulse in section 5.3. It is convenient to rewrite the eigenvalue problem, like (7.3) of the gKS equation, as a first order system

$$\frac{d\mathbf{y}}{dx} = A(x,\lambda)\mathbf{y} \tag{7.7}$$

where

$$\mathbf{y} = (\psi, \psi', \psi'', \psi''') \tag{7.8}$$

The behavior of $\mathbf{y}$ at the infinities s determined by the constant matrix A_∞ obtained by setting h constant, in our case zero,

$$A_\infty = \lim_{x \to \pm\infty} A(x,\lambda) \tag{7.9}$$

Thus, the behavior of the eigenfunctions on the infinities is exponential and is determined by the roots of the matrix A at the infinities. For the gKS equation, these roots are determined by the characteristic polynomial, for gKS-equation

$$\sigma^4 + \delta\sigma^3 + \sigma^2 - C\sigma + \lambda = 0 \tag{7.10}$$

There are four complex roots for a given λ in the complex plane. We are interested especially in the signs of their real parts. The curve of the essential spectrum Γ is a locus where the real part of one of the roots , let us say σ_2, changes its sign. The eigenfunctions of the essential spectrum do not decay at the infinities but oscillate in sinusoidal way. For the gKS-equation in Ω_1 there is one root with a positive real part and there are three other, with negative real parts. In region Ω_2, there are two roots with positive and two roots with negative real parts. Our eigenfunctions have to vanish at the infinities, and hence an algorithm must be developed to link the appropriate modes.

In Ω_1 for large negative x only one root is pertinent $\mathbf{y} \to \mathbf{v}_1 e^{\sigma_1 x}$, at $x \to -\infty$ where $\mathbf{v}_1$ is an eigenvector, associated with σ_1. If we integrate our system numerically from $-\infty$, the behavior at $+\infty$ will be a superposition of the four fundamental solutions

$$\mathbf{y} \to \sum_{k=1}^{4} b_{k1} \mathbf{v}_k e^{\sigma_k x}, \quad at \quad x \to +\infty \tag{7.11}$$

where v_k are the eigenvectors of A at the infinities, so we ignite all the fundamental solutions. One of these roots corresponds to the growing mode at plus infinity solution; namely, the one we started with. In order to suppress this root, one has to make the coefficient b_{11} zero, in the same way as for the algorithm for a stationary pulse solution. Now the complex coefficient b_{11} is a function of the complex eigenvalue λ. And here let us define a function of a complex variable, so called Evans' function, see Evans (1972) and Swinton (1992).

$$D_1(\lambda) = b_{11} \tag{7.12}$$

Evans' function is nothing more than a "transition coefficient" across the nonlinear region.

Zeros of this function are our discrete eigenvalues. Then in Ω_2, σ_1 and σ_2 are the roots with positive real parts, and thus at minus infinity there are two pertinent fundamental solutions

$$\mathbf{y}_1 \to \mathbf{v}_1 e^{\sigma_1 x}$$

$$\mathbf{y}_2 \to \mathbf{v}_2 e^{\sigma_2 x}$$

and after the numerical integration through the nonzero pulse region these two initial conditions will give two trajectories. Again, at $x \to +\infty$ we excite all the modes including unbounded ones:

$$\mathbf{y}_1 \to \sum_{k=1}^{4} b_{k1} \mathbf{v}_k e^{\sigma_k x} \tag{7.13}$$

$$\mathbf{y}_2 \to \sum_{k=1}^{4} b_{k2} \mathbf{v}_k e^{\sigma_k x} \tag{7.14}$$

and our solution must be a linear combination of them

$$\mathbf{y} = C_1 \mathbf{y}_1 + C_2 \mathbf{y}_2 \tag{7.15}$$

Now we have to suppress two growing roots and hence

$$\begin{pmatrix} b_{11} & b_{12} \\ b_{21} & b_{22} \end{pmatrix} \begin{pmatrix} C_1 \\ C_2 \end{pmatrix} = 0 \tag{7.16}$$

and this is possible only if the determinant of the system is zero. We can then define the Evans' function for the region Ω_2

$$D_2(\lambda) = \begin{vmatrix} b_{11} & b_{12} \\ b_{21} & b_{22} \end{vmatrix} = 0 \tag{7.17}$$

Hence we have in our complex plane λ two Evans' functions, D_1 and D_2. It is then obvious how to define the Evans' functions for more complex problems. These functions have the following properties:

(a) $D_i \to 1$ as $|\lambda| \to \infty$.
(b) D_i are real on the curve of the essential spectrum Γ.
(c) D_i do not have poles in Ω_i.

The number of discrete eigenvalues within any region in Ω_1 or Ω_2 can be estimated by using the argument principle of an analytical function

$$n - p = \frac{1}{2\pi i} \oint_C \frac{D_i'(\lambda)}{D_i(\lambda)} d\lambda = \frac{1}{2\pi} \Delta_C \arg D_i(\lambda)$$

In our case the number of poles p is zero, and one has to calculate only the number of zeros. The right-hand side of the expression is the winding number of C', which is the image of C through the map D_i, around the origin. The symbol $\Delta_C arg$ denotes change in the argument over the contour C. The number of the zeros is determined solely by the winding number of C'. This procedure is possible to do also for numerically defined functions and it works as well as if an

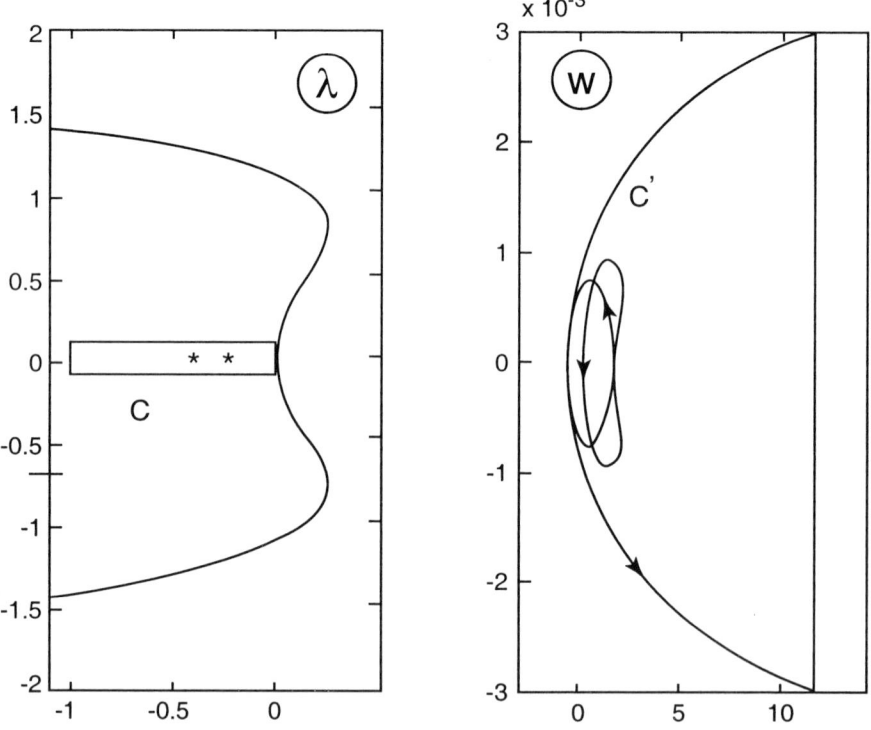

Figure 7.3: Winding number of C' gives us number of eigenvalue inside C.

analytical expression is available. A typical calculation is shown in the Figure 7.3 where two stable eigenvalues in Ω_2 found by the shooting method is consistent with the winding number of C'. One can also use this method to estimate the location of the eigenvalues by shrinking the closed contour successively. Properly using the properties of the Evans' function in combination with the argument principle and Newton procedure, one can easily find the discrete spectrum.

It is easy to see that the essential spectrum functions can also be obtained by a modified numerical scheme. On Γ, since $Re\sigma_1 > 0$ and $Re\sigma_2 = i\alpha$, the asymptotic behavior of y at minus infinity is a linear combination of y_1 and y_2

$$\mathbf{y} = C_1\mathbf{y}_1 + C_2\mathbf{y}_2 \tag{7.18}$$

where

$$y_1 \to v_1 e^{\sigma_1 x} \tag{7.19}$$

$$y_2 \to v_2 e^{i\alpha x} \tag{7.20}$$

and one can find the transmission coefficient as the ratio of these amplitudes and so

$$r(\lambda) = \frac{\begin{vmatrix} b_{11} & b_{12} \\ b_{21} & b_{22} \end{vmatrix}}{b_{11}} = \frac{D_2(\lambda)}{D_1(\lambda)} \tag{7.21}$$

We also define the adjoint problem

$$\frac{d\mathbf{z}}{dx} = -A^T \mathbf{z} \tag{7.22}$$

It is easy to see that

$$\frac{d}{dx}(\mathbf{z}^T\mathbf{y}) = (-\mathbf{z}A^T)\mathbf{y} + \mathbf{z}(A\mathbf{y}) = 0 \tag{7.23}$$

and hence the scalar product is independent on x.

7.3 Stability of gKS pulses.

Linearizing the gKS equation about any member of the pulse branch in the frame moving with the pulse, whose various solution branches are shown in Figure 7.1 one obtains the following equation for the localized disturbance H

$$\frac{\partial H}{\partial t} + \mathcal{L}H = 0$$

$$t = 0 \quad H = H_0(x) \tag{7.24}$$

$$\lim_{x \to \pm\infty} H_0(x) = 0$$

where

$$\mathcal{L} = \frac{\partial^4}{\partial x^4} + \delta \frac{\partial^3}{\partial x^3} + \frac{\partial^2}{\partial x^2} + \frac{\partial}{\partial x}(4h - c).$$

After a Laplace transform, $\hat{H}(x,p) = \int_0^\infty H(x,t)e^{-pt}dt$, one gets

$$(p + \mathcal{L})\hat{H} = H_0 \tag{7.25}$$

$$\hat{H}(x \to \pm\infty) = 0$$

which can be inverted with the introduction of a Green's function $G(x;\xi,p)$

$$\hat{H} = \int_{-\infty}^{\infty} G(x;\xi,p)H_0(\xi)d\xi \qquad (7.26)$$

where

$$(p+\mathcal{L})G(x;\xi,p) = \delta(x-\xi)$$

inverting (7.26), one gets the evolution of the disturbance

$$H = \frac{1}{2\pi i}\int_{a-i\infty}^{a+i\infty}\hat{H}(x,p)e^{pt}dp$$

$$= \frac{1}{2\pi i}\int_{a-i\infty}^{a+i\infty}\int_{-\infty}^{\infty} G(x;\xi,p)H_0(\xi)d\xi dp \qquad (7.27)$$

where a is chosen such that the path integral lies to the right of all singularities for causality reasons. There are two types of singularities associated with the spectrum of $\mathcal{L}$ (discrete eigenvalues and the continuous essential spectrum),

$$\mathcal{L}\psi = -\lambda\psi$$

$$\psi \text{ bounded as } x \to \pm\infty \qquad (7.28)$$

Deforming the path integral in (7.27) appropriately and applying the residue theorem, one gets

$$H(x,t) = \sum_{k=1}^{N} A_k \psi_k(x)e^{\lambda_k t} + \int_{\Gamma} A(\lambda)\psi(x,\lambda)e^{\lambda t}d\lambda \qquad (7.29)$$

where $\{\psi_k, \lambda_k\}$ correspond to the N discrete eigenvalues with $\psi_k(\pm\infty) = 0$. The essential spectrum lies on the locus parameterized by $\alpha\epsilon(-\infty,\infty)$

$$\Gamma = \{\lambda : \lambda = -\alpha^4 + i\delta\alpha^3 + \alpha^2 + i\alpha c\} \qquad (7.30)$$

in the complex λ plane as shown in Figure 7.2. The corresponding eigenfunctions $\psi(x,\lambda)$ approach bounded sine waves at $x = \pm\infty$ with wavenumber α. If one decomposes $\psi(x,\lambda)$ as $\psi(x,\lambda) = K(x;\lambda)e^{i\alpha x}$ then the transmission coefficient is defined as

$$r(\lambda) = r(\alpha) = K(\infty,\lambda)/K(-\infty,\lambda) \qquad (7.31)$$

This transmission coefficient is a complex number in general due to the phase shift in the two sine waves at the infinities. Its modulus $|r|$, however, measures the ratio of the two limiting amplitudes. Physically, if $|r|$ is larger than unity, sinusoidal waves are dampened as they pass through the pulse from the front to the back. This transmission coefficient is connected with the well-studied phenomenon of radiation transmission through a localized potential well in quantum mechanics as described by the Schrödinger equation. However, for the conservative dynamics of the Schrödinger equation, a radiation mode produces both a reflected wave and a transmitted wave, such that the total energy is conserved. This conservation is lost here and there is no counterpart to the reflected wave. The amplitude of the wave is attenuated due to dissipation during its passage through the pulse, our analog of the localized potential well.

It is clear from (7.29) that the evolution of the disturbances is completely determined by the discrete and essential spectra of $\mathcal{L}$. The discrete spectrum with decaying eigenfunctions, $\psi_k(\pm\infty) = 0$, has a finite number N of discrete eigenvalues. Two members of this set can be identified readily. Due to the translational invariance of the gKS equation, it is clear that if $h(x)$ is a pulse solution so must its translate $h(x + l)$. There is hence a family of possible solutions for a given pulse as generated by the translation. Since $h(x + l) \sim h(x) + h'(x)l$ for small translates, this degeneracy implies that $\mathcal{L}h' = 0$ and hence the translational mode

$$\psi_1 = h'(x) \tag{7.32}$$

where $\psi_1(\pm\infty) = 0$ is always a null eigenfunction corresponding to a simple zero eigenvalue λ_1 for any δ value. This simple zero eigenvalue is permanently embedded in the essential spectrum and it exists for all δ values. It can only be shifted by perturbations to $\mathcal{L}$ that break the translational symmetry. The excitation of this translational mode is the key component of coherent structure theory.

One expects $\mathcal{L}$ to have another zero eigenvalue at the specific δ values in Figure 7.1 where the saddle-node bifurcations take place. Differentiating (2.1) with respect to x and δ, one obtains

$$\mathcal{L}h_\delta = \frac{dc}{d\delta}h' - h''' \tag{7.33}$$

where $h_\delta = \frac{\partial}{\partial \delta}h(x;\delta)$. Defining $\psi_2 = \frac{h_\delta}{\frac{dc}{d\delta}}$, one obtains

$$\mathcal{L}\psi_2 = h' - h'''/(\frac{dc}{d\delta}) = \psi_1 - h'''/(\frac{dc}{d\delta})$$

It is then clear that, at the saddle-node points, ψ_2 becomes a generalized null eigenfunction $\mathcal{L}\psi_2 = \psi_1$ associated with an additional generalized zero eigenvalue λ_2. It can also be readily shown that the transversality condition $\frac{d\lambda_2}{dc} \neq 0$ and in fact $sgn\{\frac{d\lambda_2}{dc}\} = sgn\{\frac{d^2\delta}{dc^2}\}$ at the saddle-node points. Consequently, at every saddle-node point, an eigenvalue crosses the origin on the real axis and the direction of each crossing alternates for each subsequent saddle-node bifurcation. An additional piece of information regarding a real eigenvalue of the discrete spectrum can be obtained from the KdV limit at large δ by the scalings $t \to t/\delta$ and $h \to \delta h$. Other than the translational null mode ψ_1, the symmetry of the KdV equation associated with its integrability stipulates the existence of another simple zero eigenvalue. A standard spectral perturbation analysis on the transformed equation shows the third eigenvalue varies as $-0.379\delta^{-1}$ such that the corresponding eigenvalue for (7.28) approaches a constant at large δ,

$$\lambda_3 \sim -0.379 \qquad (7.34)$$

The discrete eigenvalues of the near KdV pulses are hence either zero or stable.

Since $\mathcal{L}$ is not self-adjoint with respect to the L_2 inner product, the discrete eigenvalues can be complex. Hence, although we have exhausted all possible simple bifurcations of the pulse solution branch at $\delta = \infty$ and at the saddle points by the above study of zero eigenvalues, we have not covered the possibility of a complex pair of discrete eigenvalues crossing the imaginary axis for a Hopf bifurcation. To complicate matters, the number of discrete eigenvalues can change as they vanish or emerge from the essential spectrum Γ. This represents an exchange of identity between the discrete eigenvalues and "resonance poles" Pego and Weinstain (1992, 1994) which will be scrutinized in the next section.

The destabilization of the discrete spectrum via a saddle-node or a Hopf bifurcation corresponds to the excitation of a localized mode, since $\psi_k(\pm\infty) = 0$, which travels at the same speed c as the pulse. They hence represent one instability mode for the pulse. However, even if the discrete spectrum is stable, Figure 7.2 indicates that the essential spectrum Γ is always unstable for all δ since the gKS system is an active medium. Since the essential eigenfunctions $\psi(x;\lambda)$ do not decay to zero at the infinities, these are actually spatially global instabilities reflecting the instability of the homogeneous states away from the pulse. However, if these growing instabilities are in the form of radiation wavepackets which convect away from the pulse, the pulse will remain intact. This concept is similar to the classical convective stability theory for the homogeneous state of an active medium with respect to localized disturbances Bers (1975). We extend this theory to the pulse by assuming the discrete spectrum in (7.29) are stable and examining only the portion corresponding to the essential spectrum, Chang, Demekhin and Kopelevich (1995, 1996). Writing it as

$$\lim_{t\to\infty} H(x,t) = \lim_{t\to\infty} \int_{-\infty}^{\infty} A(\alpha)K(x;\alpha)e^{i[\alpha\frac{x}{t}+\alpha c-\omega]t}d\alpha \qquad (7.35)$$

where we have chosen to use the real wavenumber α to parameterize Γ and

$$\omega = i\alpha^2 - i\alpha^4 - \delta\alpha^3 \qquad (7.36)$$

is the complex wavefrequency. The conversion of $\psi(x;\lambda)$ to $K(x;\alpha)e^{i\alpha x}$ allows a connection to the classical convective stability theory for a localized disturbance on the homogeneous state. Since the pulse is localized in the medium, one expects the large-time asymptotic behavior of a localized disturbance to be the same with or without the pulse. As such, $A(\alpha)K(x;\alpha)$ is expected to vary less rapidly with α, as is the case in the classical theory, than $A(\lambda)\psi(x;\lambda)$ does with λ. Hence, the large-time behavior of the Fourier integral (7.35) can also be determined by the stationary phase (steepest descent) technique. For a given (x/t) ray, the behavior of $H(x,t)$ is determined by the saddle pinch point

$$\frac{\partial}{\partial\alpha}(\alpha c - \omega) = c - \frac{\partial\omega}{\partial\alpha} = -(x/t) \qquad (7.37)$$

Hence, the pertinent behavior near the pulse at $(x/t) = 0$ is dominated by the saddle point α_* where

$$\frac{\partial \omega}{\partial \alpha}(\alpha_*) = c \tag{7.38}$$

viz. the group velocity which is equal to the pulse speed c. Hence, the pulse at $x = 0$ is not destroyed at large time only if the growth rate of α_* is negative

$$Im\{\alpha_* c - \omega(\alpha_*)\} < 0 \tag{7.39}$$

Otherwise, the growing disturbances simply convect away, behind the pulse for the gKS equation.

Conditions (7.38) and (7.39) can also be derived from the perspective of the wave packet in the fixed frame x. According to the classical convective stability theory Bers (1975), a localized disturbance $\delta(x)\delta(t)$ on the trivial basic state will grow into a wave packet in the x frame and the wave number selected along the (x/t) ray is defined by

$$\frac{\partial \omega}{\partial \alpha}(\alpha_*) = (x/t) \tag{7.40}$$

and whether the disturbance will grow or decay along this ray is determined by the sign of $Im\{\alpha_* x/t - \omega(\alpha_*)\}$. This is derived from the appropriate Fourier integral with a similar stationary phase argument. Hence, the two boundaries of the wave packet are defined by the rays $(x/t)_\pm$ where

$$\frac{\partial \omega}{\partial \alpha}(\alpha_\pm) = (x/t)_\pm \tag{7.41}$$

$$Im\{(\alpha x/t)_\pm - \omega(\alpha_\pm)\} = 0 \tag{7.42}$$

Comparing (7.38) and (7.39) to (7.41), it is clear that if the solitary wave speed c is such that

$$(x/t)_- < c < (x/t)_+ \tag{7.43}$$

the solitary pulse is unstable. Conversely, if c is larger than $(x/t)_+$ or smaller than $(x/t)_-$, the pulse is stable. This then links the stability of a pulse to a wave packet disturbance to the classical convective instability theory for a localized disturbance on the trivial state. While the former involves expansion with the essential eigenfunctions and the latter Fourier expansion, the pertinent saddle points governing the large-time asymptotics are identical for both cases since the pulse is localized and the essential eigenfunctions of the trivial state are exactly the Fourier modes.

It is far easier to use (7.41) to (7.43). For the gKS, (7.41) becomes

$$4\alpha^3 - 3i\delta\alpha^2 - i(x/t) = 0 \tag{7.44}$$

for any given x/t. This complex polynomial has three roots for $\delta \neq 0$ and two roots for the KS limit at $\delta = 0$. Only one of the roots corresponds to the true saddle point and we utilize the classical complex pinch-point analysis method

Bers (1975) to determine which is the true saddle point. The growth rate along this particular (x/t) ray

$$\gamma = Im\{\alpha x/t - \omega(\alpha)\}$$

is then evaluated at the proper root of (7.44). The constructed $\gamma(x/t)$ curve for the KS and gKS equation (for $\delta = 0.5$) are shown in Figure 7.4. It is clear that $(x/t)_\pm$ for the symmetric KS case are ± 1.622 and they bound $c = 1.216$ of the KS pulse. Due to the reflection symmetry of the KS equation, a localized disturbance will evolve into a symmetric radiation packet whose envelopes expand at the speeds ± 1.622, larger than the solitary pulse speed $c = 1.216$. With the introduction of dispersion, the phase speed at every unstable wavenumber is reduced and we expect the two envelope speeds $(x/t)_\pm$ to be shifted in the negative direction. This is clearly seen in Figure 7.4 where $(x/t)_+$ is reduced to 0.704 while $(x/t)_-$ is shifted to -2.728 at $\delta = 0.5$. The single-hump solitary pulse with a speed of $c = 1.709$ at this δ value can now escape the expanding radiation packet and is hence convectively stable.

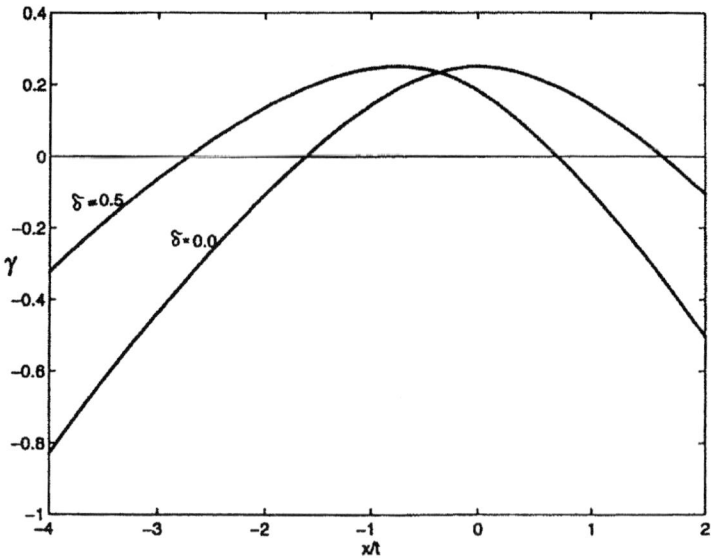

Figure 7.4: The growth rate of the wave packet along different (x/t) rays for the one-hump pulse on branch b at $\delta = 0$ and 0.5. The waves grow within the envelop $(x/t)_\pm$ where γ is positive.

The neutral convectively stability criterion, an equality version of (7.39), is also shown against the pulse solution branch in Figure 7.5. The critical value $\delta_c = 0.17$ below which the one-hump pulse solution is unstable is in agreement with earlier simulation results which do not exhibit coherent pulses for the KS limits Frish, Zhen and Thual (1986), Conrado and Bohr (1994), Kawahara and Toh (1987). The large δ segment ($\delta > \delta_c$) of the one-hump branch is hence convectively stable to radiation modes. This is also consistent with earlier simulations. However, in Figure 7.5 indicates that parts of the multi-hump branches c, d, e and f are also convectively stable but they have never been observed in simulations. This suggests their discrete spectrum is unstable and we shall construct the discrete spectrum in the next section to finalize the stability analysis and to determine why the one-hump pulse is selected. The discrete spectrum will also be shown to interact with the essential spectrum Γ in an intricate manner.

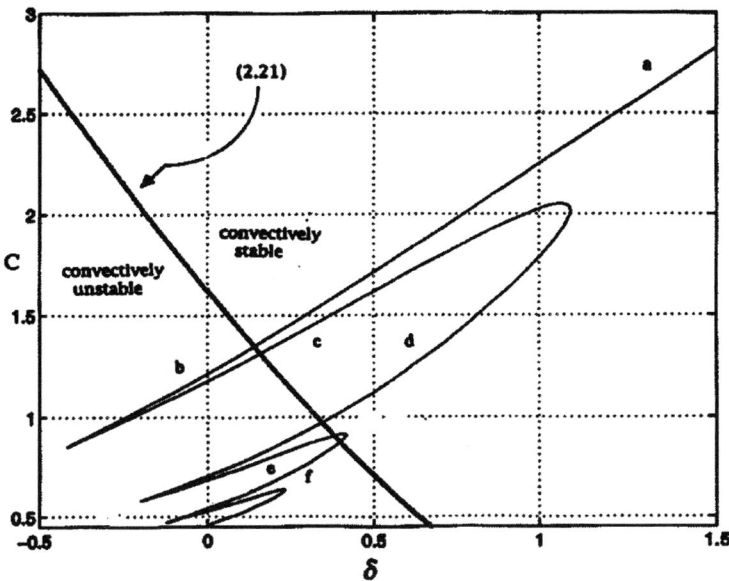

Figure 7.5: The pulse speed c for pulses vs. the upper envelope speeds $(x/t)_+$ of the radiation mode. The lower envelope speed (x/t) is negative and is not shown.

We have carried out extensive numerical experiments to verify the above stability theory for pulses. In Figure 7.6, two isolated KS pulses at $\delta = 0$ are placed in the laboratory frame with some localized initial noise between them. The noise quickly develops into an expanding wave packet whose envelopes with speeds ± 1.622 are shown in dotted lines. The one-hump KS pulses travel at the speed of $c = 1.216$ as shown by the dashed lines. They cannot escape the ex-

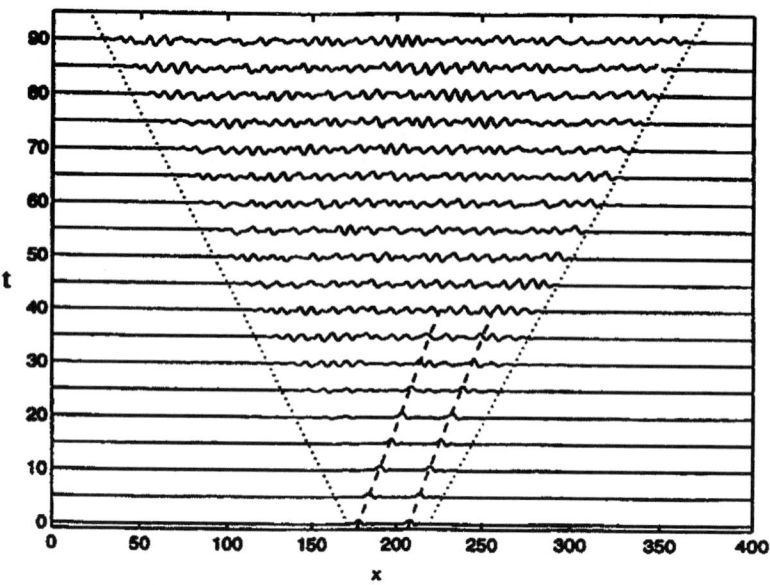

Figure 7.6: The convective instability of KS one-hump pulse at $\delta = 0$.

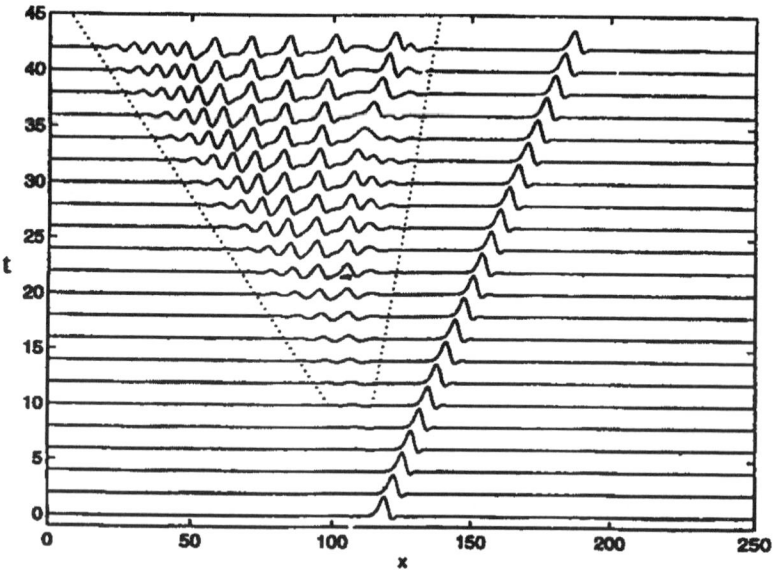

Figure 7.7: The convective stability of a stable one-hump pulse at $\delta = 0.5$.

panding wave packet and are destroyed within 25 units of time. (The maximum growth rate γ of the wave packet disturbance is 0.25 per unit time for the KS limit as shown in Figure 7.4). The destruction of both pulses generates more noise and the region between the envelopes is soon filled with irregular structures that do not resemble any of the pulses in Figure 7.1. This demonstrates that the KS chaos is caused by the expanding radiation packets near the unstable pulses. The situation changes dramatically at $\delta = 0.5$ when the pulse is convectively

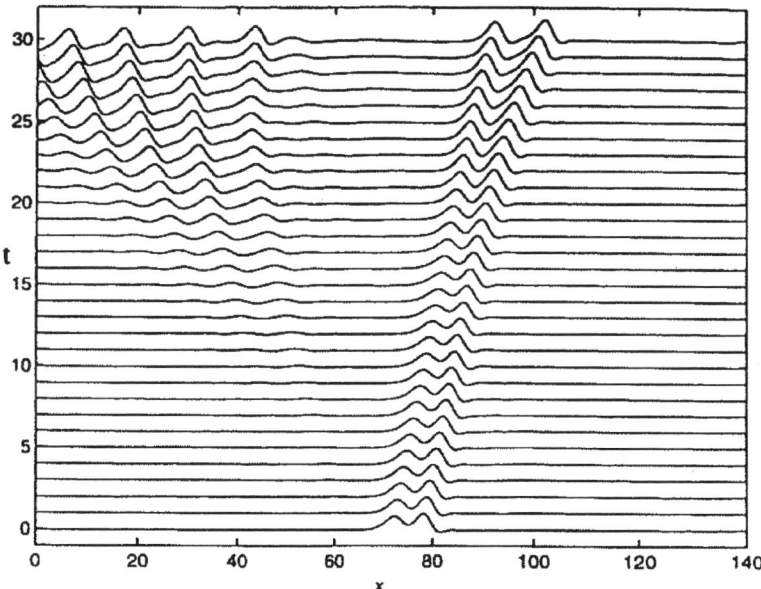

Figure 7.8: The oscillatory local instability of two-hump pulse on branch d leading to a one-hump pulse at $\delta = 0.5$.

stable to the radiation instability. Figure 7.7, a single one-hump pulse is placed in the laboratory frame with some localized noise. The noise quickly develops into an expanded packet whose envelopes travel at speeds 0.704 and -2.728 respectively, as shown in Figure 7.4. Dispersion has shifted both speeds to smaller values. As a result, the one-hump pulse with speed $c = 1.709$ escapes the radiation packet unscathed. The wave packet left behind by the pulse continues to grow and eventually triggers the birth of more one-hump pulses which also escape the turbulent packet where they are created. If a trailing pulse is not far behind the original one, it will suppress some of the growing radiation and fewer or no new pulses will be created from the wave packet. This ability of the stable one-hump pulses to swallow and suppress radiation packets in front is related to the fact that its transmission coefficient has modulus in excess of unity, as will shown in the next section. Eventually, the new pulses and the original ones will combine to reach a critical density (separation between pulses) such that the radiation wave packets will always be absorbed by a trailing pulse without creating any additional pulses. In contrast, if the pulses are unstable to radiation modes, they are rapidly destroyed by even the smallest of wave packets when the two collide. Hence, stable pulses with large transmission modulus also serve to sweep and suppress the radiation modes such that the asymptotic dynamics is driven almost entirely by pulse-pulse interaction. This implies that for $\delta > \delta_c$, the usual coherent structure theory should be quite accurate a model for the large-time dynamics, except when the pulse separation is excessive such that the radiation mode has time to nucleate new pulses. Of course, radiation mode can corrupt the initial transient and the duration of this transient before pulse-pulse interaction becomes the dominant dynamics is a function of the pulse separation, the radiation attenuation factor determined by the transmission modulus and the radiation growth rate. Some of the effects of the radiation mode on transient and large-time dynamics are revealed in the numerically constructed regime diagram of Balmforth et al. (preprint).

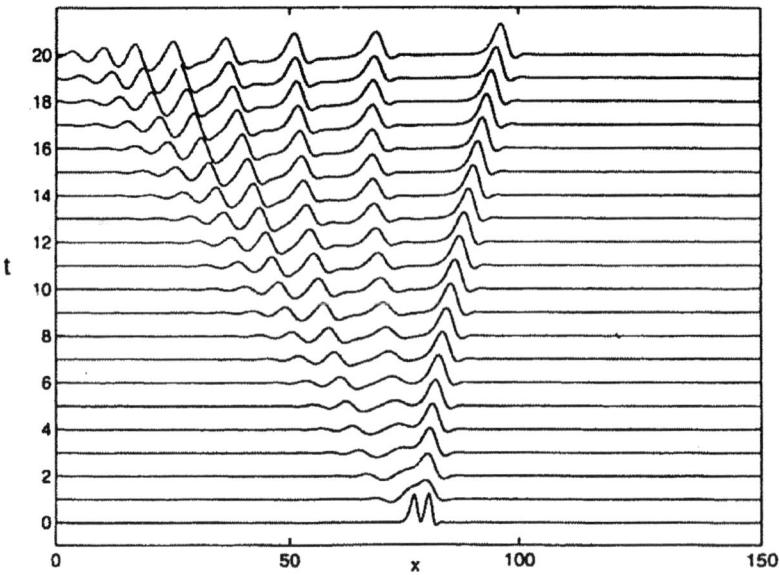

Figure 7.9: The evolution of a two-hump pulse on branch c to a one-hump due to a localized static instability induced by a real, unstable discrete eigenvalue at $\delta = 0.5$.

We also examine the selection of the one-hump pulse by the localized instability of the discrete spectrum by placing a two-hump pulse at $\delta = 0.5$ on branch c in Figure 7.8 in a frame moving with the same speed as the two-hump pulse. This pulse is convectively stable to the radiation mode and the wave packet from localized noise is seen to be convected behind and trigger the formation of a row of one-hump pulses in the back. However, as seen in Figures 7.12 and 7.13, this two-hump pulse has an unstable real discrete eigenvalue. As a result, the valley between the two humps quickly fills up due to the instability and a one-hump pulse is reached. This one-hump pulse travels faster than the two-hump one and it is seen to move forward in the moving frame of Figure 7.8. This localized instability creates additional noise during its inception which is again convected to the back. In Figure 7.9, the two-hump pulse on branch d, which has an unstable complex pair of eigenvalues, is subjected to the same test. This time the localized oscillatory instability separate the two humps in a growing oscillation (not very apparent in the figure) until two isolated one-hump pulses are formed.

Although we have not exhaustively analyzed the stability of all pulses of the gKS equation, especially those on the isolated branches within the shaded region in In Figure 7.1, it is safe to say that, for $\delta > \delta_c$, the one-hump pulse is the only pulse stable to both the radiation mode and the localized instability. As such, it is the only pertinent coherent structure and the coherent structure theory can be appropriately applied. The gKS equation also admits kink solutions but we have not observed any sustained presence of these structures in our numerical experiments and suspect they are unstable to both instabilities. The local stability theory of kinks, however, is more involved due to the presence of two different homogeneous states.

7.4 Attenuation of radiation wave packet by stable pulses.

The attenuation effect of a stable pulse on a wavepacket passing through it as shown in In Figure 7.10 can be deciphered without considerable numerical effort. Consider a wave packet with an initial Fourier content of

$$H(x,0) = \int_{-\infty}^{\infty} A(\alpha) e^{i\alpha x} d\alpha \qquad (7.45)$$

we stipulate that this wave packet passes through a stable pulse of width w since the group velocity $(x/t)_-$ of the trailing edge of the wave packet is slower than the pulse speed according to (7.43). The speed and time for each Fourier mode of the packet to pass through the pulse is different, with a speed $c + Re\{\omega\}/\alpha$ for each of its Fourier component α. The transmitted packet which exits the pulse at about $t \sim w/c$ then has the Fourier content

$$H(x+w, w/c) \sim \int_{-\infty}^{\infty} \frac{A(\alpha)}{r(\alpha)} e^{i\theta} e^{(\alpha^2 - \alpha^4)w/c} d\alpha \qquad (7.46)$$

where $\theta = \alpha x + \alpha c t - Re\{\omega\}t$ is the phase with unit modulus. Hence, the attenuation of Fourier mode α is $|e^{(\alpha^2-\alpha^4)w/c}/r(\alpha)|$. For large α, condition (a) implies that $r(\alpha)$ approaches unity but since the growth rate $\alpha^2 - \alpha^4 \sim -\alpha^4$ is very negative, all high wavenumber Fourier modes are essentially attenuated by the primary damping of the homogeneous state as the packet travels on it towards the pulse. As α approaches zero, the eigenfunctions of the essential spectrum approach the constant-amplitude Fourier normal modes of the homogeneous spectrum. These long waves are much longer than the pulse width w and hence are oblivious to the presence of the pulse. As a result, $|r(\alpha)|$ approaches unity in this limit-these long-waves modes are hence barely attenuated. A simple expansion yields $|r(\alpha)|-1$ is of order $O(\alpha^3)$. Consequently, there exist an intermediate range of $\alpha \sim O(1)$ where $|r(\alpha)|$ exhibits a maximum, which is of the order of 10 for pulses on branch b. For these intermediate modes, radiation attenuation is again very effective but it now proceeds via a "nonlinear" mechanism as the wave packet passes through the pulse, since these modes would otherwise grow in a homogeneous medium without the pulse. Consequently, the only unattenuated modes are the ones with $\alpha \ll 1$ and if the initial wave packet contains very little of these low wavenumber modes, it will be effectively damped as it passes through the pulse. Since the pulse width w introduces a low-wavenumber cutoff, all wave packet modes with wavenumber less than $2\pi/w$ need to be discarded in any case since they will only serve to adjust the base line of the pulse.

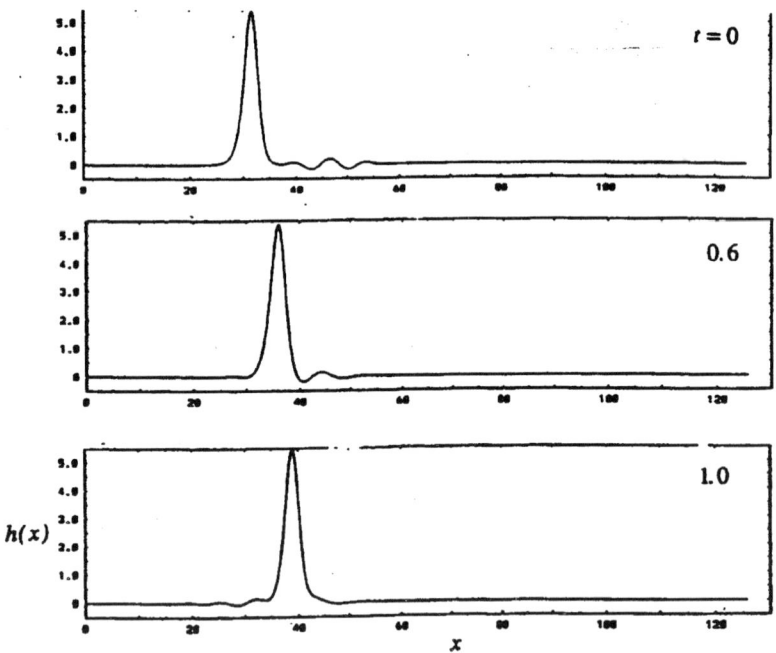

Figure 7.10: Suppression of localized packet when it goes through an equilibrium pulse, which is relatively unaffected by the the radiation transmission.

Equation (7.59) immediately suggests three possible scenarios concerning the generation and destruction of discrete eigenvalues to form resonance poles (and vice-versa) at the essential spectrum Γ. If an eigenvalue in Ω_1 crosses Γ at λ to become a resonance pole in Ω_2 or the reverse, $D_1(\lambda) = 0$ and $D_2(\lambda) \neq 0$. Consequently, $r(\lambda) = \infty$ and the essential eigenfunction at the crossing point has an infinite transmission coefficient. If an eigenvalue in Ω_2 becomes a resonance pole in Ω_1 or the reverse, $r(\lambda) = 0$. If, however, a degenerate case occurs when an eigenvalue crosses Γ and remains an eigenvalue, $D_1(\lambda) = D_2(\lambda) = 0$ but $r(\lambda)$ is bounded. However, the amplitude of the eigenfunction $\psi(x, \lambda)$ is unbounded near $x = 0$ in this case. For the gKS pulse solution, we find the last degenerate case to occur at every saddle-node bifurcation of the $c(\delta)$ pulse solution branch in Figure 7.1. Across each saddle-node point, a real eigenvalue crosses Γ. In Figure the modulus of the complex transmission coefficient along Γ for various values of δ along the one-hump branch are shown. The transmission coefficient along Γ for the stable segment of the one-pulse branch b has a modulus exceeding unity, indicating waves that pass through this pulse from the front to the back are attenuated in amplitude. Approaching the saddle-node point with the two-hump branch c, transmission coefficient begins to dip below unity on portions of Γ as shown in Figure 7.11. It is clear that at $\delta_1 = -0.4081(c(\delta_1) = c_1 = 0.8572)$, a pair of resonance poles from Ω_1 migrate across Γ to become an unstable pair of complex eigenvalues in Ω_2 as $|r|$ becomes zero. The corresponding essential eigenfunction with a zero transmission coefficient is shown in Figure 7.13. At $\delta_3 = -0.4159(c_3 = 0.8526)$, the saddle-node bifurcation from the one-hump branch b to the 2-hump branch c occurs as an eigenvalue goes from Ω_2 to Ω_1 along the real axis. This standard static bifurcation can occur across the branch cut Γ without causing the eigenvalue to go on another Riemann sheet and become a resonance pole because there is always a discrete eigenvalue at the origin due to the translational invariance described by (2.14). There is hence a "hole" in the spectral branch cut at the origin through which a real eigenvalue can stabilize or destabilize to induce a static bifurcation of the pulse. At $\delta_6 = +0.173(c_6 = 0.8298)$ on branch c, another zero is observed in $r(\lambda)$ and this time a complex pair of eigenvalues migrates from Ω_2 and turns into resonance poles in Ω_1. At $\delta'_6 = 0.159(c'_6 = 0.8191)$ on branch c, the transmission coefficient approaches infinity at a point on Γ where a pair of resonance poles in Ω_2 become eigenvalues in Ω_1.

Using the above numerical schemes, we are able to obtain the discrete eigenvalues of all the pulse solutions in Figure 7.1. Some of the pertinent configurations for the one-hump branch and the two-hump c and d branches and higher multi-hump branches are shown in Table 7.1 and Figures 7.12 and 7.13. All told, no more than 6 eigenvalues and resonance poles are responsible for the discrete spectrum of the first four pulse branches. The various bifurcations and resonance pole/eigenvalue transitions are detailed in Table 7.1. It is clear that near the saddle-node point between branches c and d, where the two-hump pulses are convectively stable, branch c has an unstable real eigenvalue while branch d has an unstable complex pair. As is consistent with expansion (7.34), the one-hump branch a has a stable discrete spectrum for positive δ. Consequently, for $\delta > \delta_c = 0.17$, the one-hump pulse is the only pulse stable both to the convective radiation instability of the continuum and the localized instability

Table 6.1
Discrete spectra and bifurcations

Branch	δ	c	Fig. 10	Fig. 11	Type of transition
				a	
One-hump b	−0.4081	0.8572	1		$\binom{0}{0}(\Omega_1) \to \binom{x}{x}(\Omega_2), r = 0$
				b,c	
One-hump b	−0.4156	0.8532	2		$\binom{x}{x}(\Omega_2) \to (\times\times)(\Omega_2)$
				d	
One-hump b	−0.4159	0.8526	3		$\times(\Omega_2) \to \times(\Omega_1)$
				e	
Two-hump d	1.0899	2.0173	4		$\times(\Omega_1) \to \times(\Omega_2)$
				f	
Two-hump d	1.0727	1.9399	5		$(\times\times)(\Omega_2) \to \binom{x}{x}(\Omega_2)$
				g	
Two-hump d	0.1732	0.8298	6		$\binom{x}{x}(\Omega_2) \to \binom{0}{0}(\Omega_1), r = 0$
				h	
Two-hump d	0.1591	0.8191	6'		$\binom{0}{0}(\Omega_2) \to \binom{x}{x}(\Omega_1), r = \infty$
				i	
Two-hump d	−0.1930	0.5849	7		$\binom{0}{0}(\Omega_1) \to \binom{x}{x}(\Omega_2), r = 0$
				j,k	
Two-hump d	−0.1978	0.5816	8		$\binom{x}{x}(\Omega_2) \to (\times\times)(\Omega_2)$
				l	
Two-hump d	−0.1983	0.5813	9		$\times(\Omega_2) \to \times(\Omega_1)$
Three-hump e	0.1501	0.7740	10		$\binom{x}{x}(\Omega_1) \to \binom{0}{0}(\Omega_2), r = \infty$
Three-hump e	0.1698	0.7859	10'		$\binom{0}{0}(\Omega_1) \to \binom{x}{x}(\Omega_2), r = 0$
Three-hump f	0.4166	0.9090	11		$\times(\Omega_1) \to \times(\Omega_2)$
Three-hump f	0.4069	0.8837	12		$(\times\times)(\Omega_2) \to \binom{x}{x}(\Omega_2)$

Notations: $\binom{0}{0}(\Omega_i) \to (\Omega_k)$ $(i \neq k, i, k = 1, 2)$ a complex conjugate pair of resonance poles converts into a complex pair of eigenvalues across Γ. $\binom{x}{x}(\Omega_i) \to (\times\times)(\Omega_i)$ a complex conjugate pair of eigenvalues converges and become two real roots. $\times(\Omega_i) \to \times(\Omega_k)$ $i \neq k$ – one real eigenvalue crosses Γ at $\lambda = 0$, $D_1 = D_2 = 0$.

of the discrete spectrum. It is hence the selected coherent structure and the spatio-temporal dynamics for $\delta > \delta_c$ is driven by interaction among these one-hump pulses. For $\delta < \delta_c$, there are no stable pulses and the coherent structure description fails.

7.5 resonance pole–a discrete culmination of the continuous spectrum.

A simple integration of (7.3) shows that all discrete eigenfunctions integrate to zero,

$$\int_{-\infty}^{\infty} \psi_k(x) dx = 0 \qquad (7.47)$$

This indicates that the total area (mass) of the pulse cannot be increased or decreased by the excitation of the discrete modes.

Since all the discrete modes do not carry mass, the mass draining from an excited pulse, previously associated with the speed mode of the Galilean symmetry, must be carried by the radiation continuum of the essential spectrum. However, the fact that the essential spectrum contains a band of unstable modes seems to suggest that it cannot describe a decaying pulse in drainage. Actually,

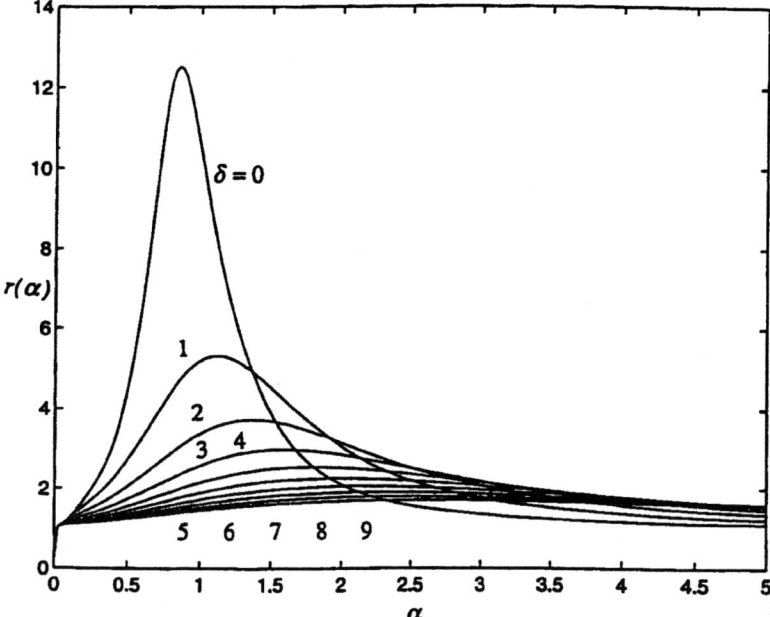

Figure 7.11: The transmition coefficient $r(\alpha)$ for singe-hump pulses at various values of δ.

the growing modes are convected away from the pulse while their cumulative drainage effect in the vicinity of the pulse can be conveniently represented by a single stable mode - the resonance pole. The description of them we can find in functional analysis, see for example , Reed and Simon (1978).

The radiation that drains out of the pulse, after appropriate attenuation, will accelerate and grow on the substrate. However, Pego and Weinstein (1992) pointed out that if one places a weight that decays exponentially in the direction of negative x on the disturbance u, this distant growth can be filtered away and one could focus on the decay effect near the pulse. We hence use the e^{ax} weight of Pego and Weinstein (1992) on the disturbance u in (3.1), $v = e^{ax}u$, and define a corresponding eigenvalue problem

$$\mathcal{L}_a \phi = e^{ax} \mathcal{L}(e^{-ax}\phi) = \lambda \phi \qquad (7.48)$$

where $\phi = e^{ax}\psi$ is the weighted eigenfunction that spans the weighted disturbance $w(x,t)$. It is clear that the essential spectrum Γ_a of $\mathcal{L}_a$ is defined by shifting that of $\mathcal{L}$ in (3.3) by the transformation $\alpha \to \alpha + ia$.

$$\lambda_a = i(\alpha + ia)c + i\delta(\alpha + ia)^3 + [(\alpha + ia)^2 - (\alpha + ia)^4] \qquad (7.49)$$

The net result is that the essential spectrum is shifted to the left in the complex

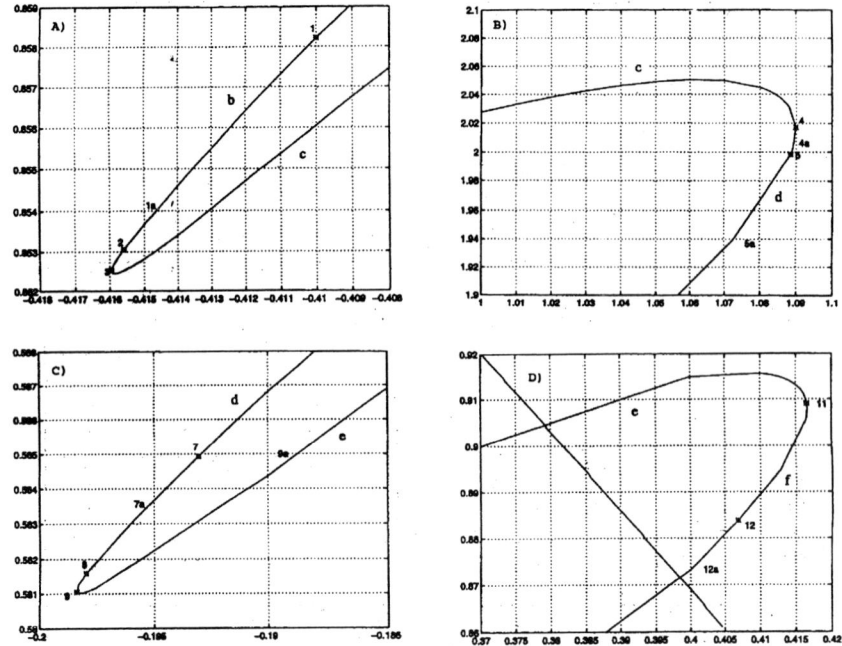

Figure 7.12: The pertinent bifurcation point and resonance pole/eigenvalue crossing on branches b, c, d, e and f.

plane as the real parts of λ decrease. This is demonstrated in Figure 7.14. The fact that a particular value of a exists that can shift the entire essential spectrum to the left half plane implies that the pulse is stable. This is the essence of Pego and Weinstein's stability theory.

There is however, a limit to how deep into the stable left-half plane Γ_a can be shifted. There are certain locations in the complex λ plane that Γ_a can not pass through. As a result, such points become part of Γ_a as a increases and represent the most unstable portion of the asymptotic Γ_a as a approaches infinity. These singularities are the saddle points defined by

$$\frac{d\lambda}{d\alpha}(\alpha_*) = 0 \qquad (7.50)$$

where λ is given at the end of section 7.1. The complex saddle point α_* yields a complex growth rate

$$\lambda_* = \lambda(\alpha_*) \qquad (7.51)$$

The shifted spectrum Γ_a cannot pass through λ_* such that a λ_* lying to the left of the unshifted spectrum Γ cannot lie on the right of the shifted spectrum Γ_a in the complex λ plane. Hence, if there exists a λ_* that lies to the left of Γ but is in the right half plane, there is no weight that can yield a stable Γ_a. In other

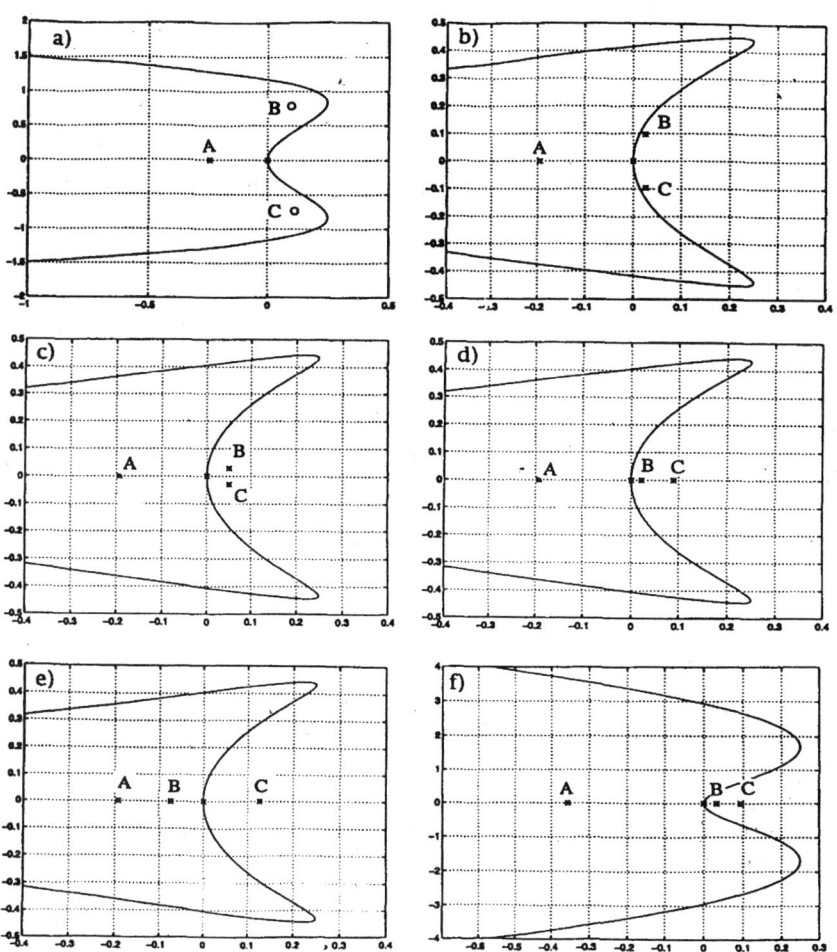

Figure 7.13: See below.

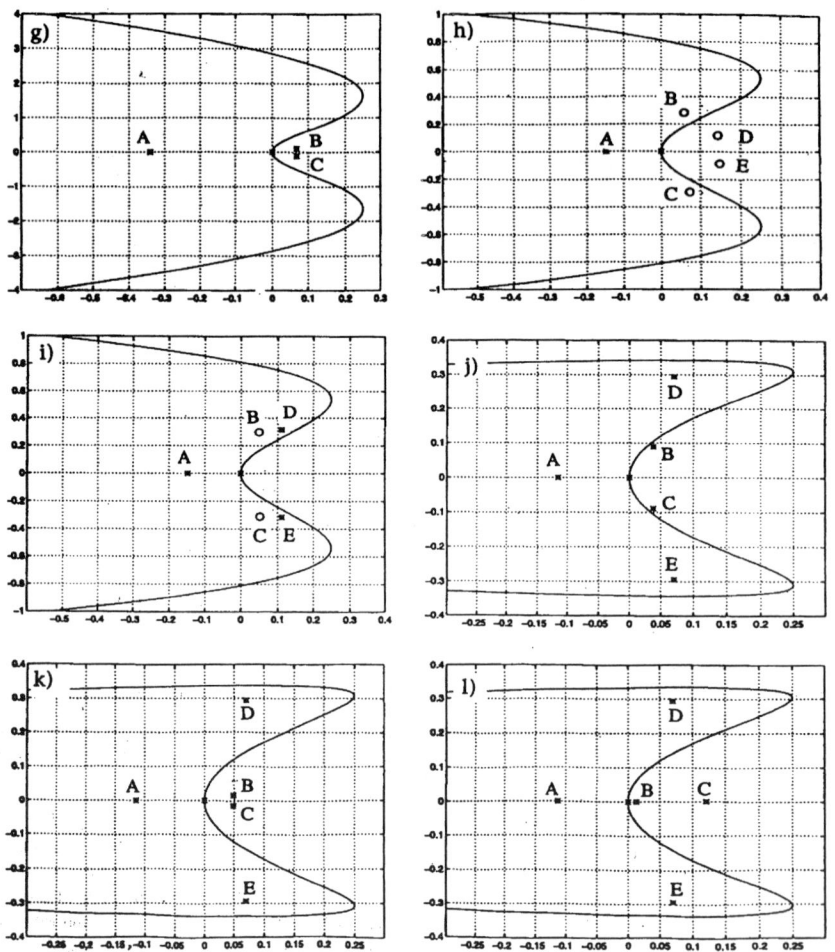

Figure 7.13: The essential spectrum, discrete spectrum (x) and the resonance poles (0) at points indicated in Table 7.1. The same eigenvalue/resonance pole is tracked and labeled.

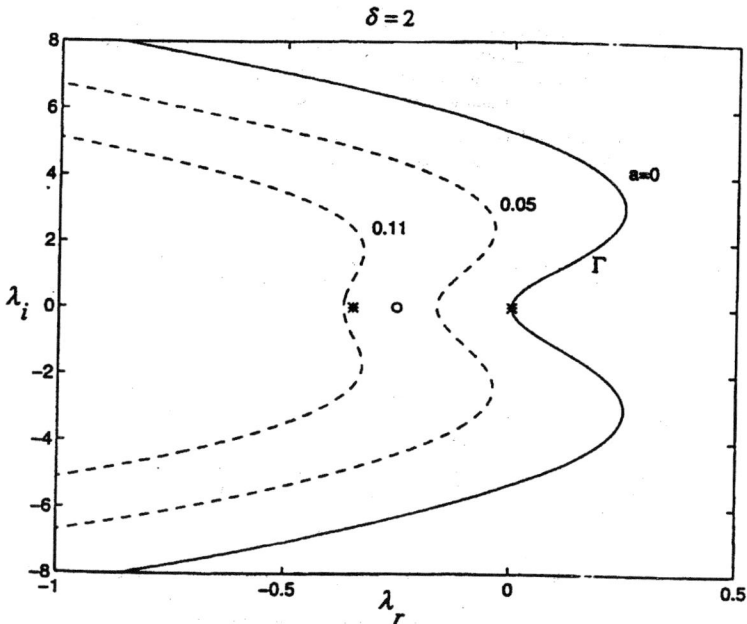

Figure 7.14: The typical spectra of a one-hump gKS pulse at $\delta = 2$ with a simple zero λ_1, corresponding to the translational mode, a stable discrete resonance pole $\lambda_\#$. The eigenvalues are represented by stars and the resonance pole by a circle. The essential spectrum Γ is shifted by the weight and passes both the resonance pole and the stable mode λ_2 to exchange their roles.

words, one cannot find a weight to ensure convective stability for the pulse with the theory of Pego and Weinstein. However, we have shown in section 7.3 by a steepest descent analysis of a direct Laplace Transform solution of the linearized initial value problem that the saddle-point wavenumber α_* corresponds to the dominant wavenumber along the ray traveled by the equilibrium pulse in the laboratory frame. The condition $Re\{\lambda_*\} < 0$ is, in fact, the criterion we have derived for the pulse to escape unstable radiation. This then clarifies how Pego and Weinstein's stability theory for pulses in a purely dispersive medium can be connected to ours for an unstable medium. Our theory yields the result that only single-hump gKS pulses are stable and only for δ in excess of 0.17.

It can be easily shown that, unlike the essential spectrum which is shifted by the weight, the discrete spectrum of $\mathcal{L}$ is also a discrete spectrum of $\mathcal{L}_a$ - with an important exception. The discrete eigenfunctions ψ_k of $\mathcal{L}$ decay to zero as $|x| \to \infty$ with an exponential rate determined by the roots α of the spatial characteristic polynomial

$$P(\alpha) = \lambda_k - \lambda(\alpha) \qquad (7.52)$$

where λ_k is the discrete eigenvalue corresponding to ψ and $\lambda(\alpha)$ is given by the dispersion relationship. The "unstable" spatial roots with positive real parts determine the decay rate of $x \to -\infty$ and "stable" ones with negative real parts the rate at $x \to \infty$. There is only one unstable spatial root α_1 to the left of Γ, while the other three have negative real parts. One then requires the discrete eigenfunction ψ_k to decay as $\exp(\alpha_1 x)$ as $x \to -\infty$ ans as a linear combination of $\exp(\alpha_i x)$ ($i = 2, 3, 4$) as $x \to +\infty$. The vanishing of three modes at $x \to -\infty$ and one mode at $x \to +\infty$ then provides four equations for the fourth order problem such that λ can be specified. However, one can, in principle, also construct functions which satisfy $\mathcal{L}\psi_\# = \lambda_\# \psi_\#$ but do not decay to zero as $x \to -\infty$ on the left side of Γ. For example, one can require ψ_k to grow as $\exp(\alpha_2 x)$ as $x \to -\infty$ instead of decaying as $\exp(\alpha_1 x)$. Three conditions still result at this infinity but they are now a different set of conditions. Since these eigenfunctions do not decay to zero, they are not part of the discrete spectrum and are called resonance poles Reed and Simon (1978). We have found only one real and negative resonance pole ($\lambda_\# < 0$) for all one-hump gKS pulses with positive δ, as shown in Figure 7.13.

These resonance poles also satisfy the weighted operator $\mathcal{L}_a \phi_\# = \lambda_\# \phi_\#$ where $\phi_\# = \psi_\# e^{ax}$ but their decay properties can now change. Since the essential spectrum Γ_a for $\mathcal{L}_a$ is defined by $\lambda_a(\alpha)$ for α real, when there is a discrete eigenvalue on the essential spectrum Γ_a, one of the roots to the spatial characteristic polynomial (7.10) is purely imaginary. Consequently, as Γ_a is shifted across the resonance pole $\lambda_\#$, the negative root α_2 crosses the imaginary axis and its real part becomes positive. Hence, as Γ_a is shifted across the resonance pole, it becomes a bona fide discrete eigenvalue of $\mathcal{L}_a$. Conversely, if Γ_a is shifted across the discrete eigenvalue λ_2, is becomes a resonance pole to $\mathcal{L}_a$. The translational zero mode λ_1 remains an eigenvalue as Γ_a passes by due to a hole in the Riemann surface. Other rules on the exchange between eigenvalues and resonance poles can be found in Pego and Weinstein. For every stable one-hump pulse, we are able to shift Γ_a such that $\lambda_\#$ becomes a discrete eigenvalue of $\mathcal{L}_a$ and λ_2 a resonance pole as seen in Figure 7.14.

The presence of these resonance poles introduces a rare exception to the stability criterion $Re\{\lambda_*\} < 0$. If there are unstable resonance poles to the left of Γ, even if Γ_a can be shifted entirely to the left half plane, the uncovered resonance poles are unstable and can destabilize the pulse. While we found some rare examples of this instability, the resonance pole of the single-hump pulse is stable for positive δ and hence only represents decay dynamics. There are hence three possible mechanisms for a pulse to destabilize – unstable saddle points λ_*, unstable resonance poles $\lambda_\#$ (typically complex conjugates) and unstable discrete eigenvalues λ_k.

Due to the exponential growth, $<\psi_\#> = \infty$ while its weighted counterpart $\phi_\#$ contains finite mass $<\phi_\#> = <e^{ax}\psi_\#> \neq 0$. As a result, the discrete spectrum of $\mathcal{L}_a$ is now mass-carrying. The shifted essential spectrum Γ_a is also mass-carrying but due to its far-left position in the complex plane, it represents a very fast transient. Hence, the asymptotic decay rate of the weighted disturbance $v(x,t)$ is governed solely by $\lambda_\#$. Since the original disturbance $u = e^{-ax}v$ is only slightly affected by the weight near the origin $x = 0$, the dynamics of the disturbance u around the pulse is also governed by $\lambda_\#$ - the local drainage rate is conveniently described by the resonance pole at large time. That the local cumulative effect of the original unstable essential spectrum Γ on drainage can be described at large time by a stable discrete mode is remarkable. It requires significant dispersion to convect the radiation backwards and sufficient growth to counter this drainage to produce a finite decay rate. The KdV equation, for example, has no resonance pole.

Actually, the resonance pole concept can be introduced in a more physical manner without the weights of Pego and Weinstein. We can use the discrete and essential spectra $\mathcal{L}$ to expand any localized disturbance to the pulse. By localized we mean the disturbance must decay sufficiently rapidly from zero at a rate to be specified later. That the two spectra form a complete basis for localized disturbances is also known from standard spectral theory. We hence expand the disturbance $u(x,t)$ as

$$u(x,t) = \sum_{k=1}^{2} A_k \psi_k(x) e^{\lambda_k t} + \int_{-\infty+ia}^{+\infty+ia} A(\alpha) K(\alpha, x) e^{i\alpha x + \lambda_a(\alpha) t} d\alpha \qquad (7.53)$$

where A_k and $A(\alpha)$ are the coefficients of expansion for the discrete and essential spectral respectively. While the original expansion with the continuous eigenfunction would be carried out over the real wavenumber α, we can deform the contour of integration in the complex α plane by shifting it above the real line by a distance a. The expansion coefficient $A(\alpha)$ is also dependent on the shift but we shall not include this dependence in our notation for convenience. This shift in the contour of integration corresponds to the weight of Pego and Weinstein (1992), (1994) and shifts the essential spectrum $\lambda(\alpha)$ to that of λ_a in (7.49). In the complex λ space, this shift in the contour of integration then corresponds to a shift of Γ to Γ_a — shifting the essential spectrum to the left.

For $\delta > 0.137$, λ_* is in the left half of the complex λ plane and an a can always be found that shifts Γ_a to the left half plane as seen in Figure 7.14. There is hence no pole in the term $\exp\{\lambda_a(\alpha)t\}$ that prevents the deformation of the contour of integration in (7.53). With a properly localized disturbance, there is also no singularity in the coefficient $A(\alpha)$. Since $K(\alpha, x)$ blows up at $-\infty$ for the resonance pole, the only pole in the integral of (7.53) is the resonance pole. Hence, by the residue theorem, if one shifts Γ_a beyond $\lambda_\#$, the expansion (3.9) becomes

$$u(x,t) = \sum_{k=1}^{2} A_k \psi_k(x) e^{\lambda_k t} + B\psi_\#(x) e^{\lambda_\# t}$$

$$+ \int_{-\infty+ib}^{\infty+ib} A(\alpha) K(\alpha, x) e^{i\alpha x + \lambda_b(\alpha) t} d\alpha$$

(7.54)

where b is such that Γ_b is to the left of $\lambda_\#$ and is completely in the left half plane.

If Γ_b is shifted across λ_2 such that the latter is transformed into a resonance pole, the eigenfunction ψ_2 can be omitted in the above expansion.

The use of $\psi_\#$ in (7.54) seems problematic at first glance since it diverges as $x \to -\infty$. However, the shifted essential eigenfunctions $K(\alpha, x)$, corresponding to the shifted essential spectrum $\lambda_b(\alpha)$, also diverge as $x \to -\infty$ and act collectively as a cut-off for the resonance pole eigenfunction $\psi_\#$ such that a bounded $u(x,t)$ still results. While the shifted essential eigenfunctions diverge as $x \to -\infty$, they are also stable – the real parts of $\lambda_b(\alpha)$ are negative. As a result, they decay locally near the pulse. Hence, the local large-time dynamics near the pulse are essentially independent of the essential spectrum. The large-time restriction arises from the fact that $\lambda_\#$ is closer to the imaginary axis the $\lambda(\alpha)$ in the complex λ space. Hence, unlike Landau damping, the resonance pole here only dominates the local dynamics. Perhaps because of this, the present resonance pole, unlike Landau damping, is separable in x and t. However, due to the localized nature of the pulses, the local dynamics are exactly the ones of interest. We have addressed the linearized initial value problem with a direct Laplace Transform and have connected the saddle points (7.50) with large-time evolution of a localized disturbance. The low-dimensional description of the present drainage dynamics is hence necessarily local in the current theory.

Transformation of (7.53) to (7.54) by distorting the contour of integration simplifies the expansion drastically. Since Γ_b is more stable than $\lambda_\#$, the continuum in (7.54) decays much more rapidly than $\lambda_\#$ and can in fact be neglected at large time. If we likewise omit the very stable λ_2 mode, the large time dynamics can be approximated by

$$u(x,t) \sim A_1 \psi_1(x) e^{\lambda_1 t} + B \psi_\#(x) e^{\lambda_\# t} \qquad (7.55)$$

That the dynamics of a continuum of modes, including unstable ones, can be described at large time by a single stable mode is extremely convenient! Note, however $\psi_\#(x)$ blows up as $x \to -\infty$ and the decay dynamics is meaningless far from the origin. The local validity of (7.55) will be scrutinized in more detail in the next section but it is clear that the dominant dynamics of a stable gKS pulse is driven by two modes, a neutrally stable translation mode and a stable resonance pole mode. We also note that the stable continuum in the integral of (7.54) decay as e^{-bx} in the positive x direction and the resonance pole and discrete modes decay in a manner dictated by (7.52). Such decay rates of the eigenfunctions in the shifted expansion then stipulate that the disturbance must decay sufficiently rapidly in the $+x$ direction. We shall avoid a detailed estimate of the precise condition here. Nevertheless, we observe that (7.54) implies that a localized disturbance is expanded by a basis $\psi_\#$ which blows up at $-\infty$ but decays in time. Actually, as mentioned earlier, the continuum in (7.54) also blows up negatively at $-\infty$ and essentially serves as a cutoff to truncate the growing tail of $\lambda_\#$. We shall demonstrate this and use the essential spectrum and the resonance pole to construct the shelf behind a draining pulse in the next section. Since this requires a continuum of modes, we will not be able to obtain an estimate of its full profile but only its length.

There is also a subtle connection between the pulse spectra of the gKS and the KdV equations. The transformation

$$h = w\delta/4$$
$$t = \tau/\delta \tag{7.56}$$

converts the gKS equation (1.1) to a perturbed form of the KdV equation,

$$w_\tau + ww_x + w_{xxx} + \delta^{-1}(w_{xx} + w_{xxxx}) = 0 \tag{7.57}$$

The solitary pulses of (7.57) are given by (2.3) in the limit of large δ and after proper transform. The same transform shows that the eigenvalues μ of pulses w^* in (7.57) to the operator

$$L_\delta = w^* \frac{\partial}{\partial x} + \frac{\partial w^*}{\partial x} \cdot + \frac{\partial^3}{\partial x^3} + \frac{1}{\delta}(\frac{\partial^2}{\partial x^2} + \frac{\partial^4}{\partial x^4}) \tag{7.58}$$

are related to λ by

$$\mu = \lambda/\delta \tag{7.59}$$

Hence, in the limit of large δ, the discrete eigenvalue λ_2 and the resonance pole $\lambda_\#$ correspond to eigenvalues of (7.58) that both collapse to the origin. This is true since, as seen in Figure 7.15, both λ_2 and $\lambda_\#$ approach constant asymptotes at large δ. As is well known, the KdV operator $L_\infty = \underset{\delta \to \infty}{Lim} L_\delta$ has three zero modes,

$$L_\infty \frac{\partial w^*}{\partial x} = 0$$
$$L_\infty \frac{\partial w^*}{\partial c} = -\frac{\partial w^*}{\partial x} \tag{7.60}$$
$$L_\infty 1 = \frac{\partial w^*}{\partial x}$$

where the first zero is the translational mode, the second generalized zero is a speed mode due to a symmetry of the KdV operator and the third one is again the improper Galilean mode. These three modes are the singular limits of λ_1, λ_2 and $\lambda_\#$ and the KdV soliton does not have a resonance pole. Nevertheless, since the last two modes of the KdV operator L_∞ are generalized zero modes, the eigenfunctions ψ_2 and $\psi_\#$ near the origin must approach ψ_1 to create the degeneracy. Although we cannot offer a rigorous proof, it seems that in the "singular" KdV limit, the resonance pole moves up to the axis, appears in the branch cut and emerges as a real eigenvalue.

In essence, δ and the shift b or the weight a break the degeneracy of (7.60) and "unfold" the generalized eigenvalues. As a result, for δ large, we expect both ψ_2 and $\psi_\#$ to be very close to $\psi_1 = h_x$ and that λ_2 and $\lambda_\#$ are small eigenvalues. This is clearly evident in Figure 7.15 and 7.16 where $\psi_\#$ is compared to ψ_1. Although $\psi_\#$ blows up exponentially as $x \to -\infty$, its shape is very close to ψ_1

for x near zero and $x \to +\infty$. This is true even for moderately large δ of order unity.

This proximity to degeneracy allows us to approximate $\psi_\#$ near the origin by

$$\psi_\# \sim h_x + \lambda_\# f(x) \qquad (7.61)$$

where $\lambda_\#$ is close to zero as seen in Figure 7.15. Although (7.61) is strictly valid for large δ, we find it to be a good approximation, viz. $\psi_\#$ is close to $\psi_1 = h_x$, even for small δ in our numerically constructed spectrum. The difference $f(x)$ also carries mass but blows up as $\psi_\#$ at $-\infty$. For positive x, (7.61) is actually valid even far from the origin as seen in Figure 7.16. It is only at $-\infty$ where $\psi_\#$ blows up while h_x decays that the difference grows.

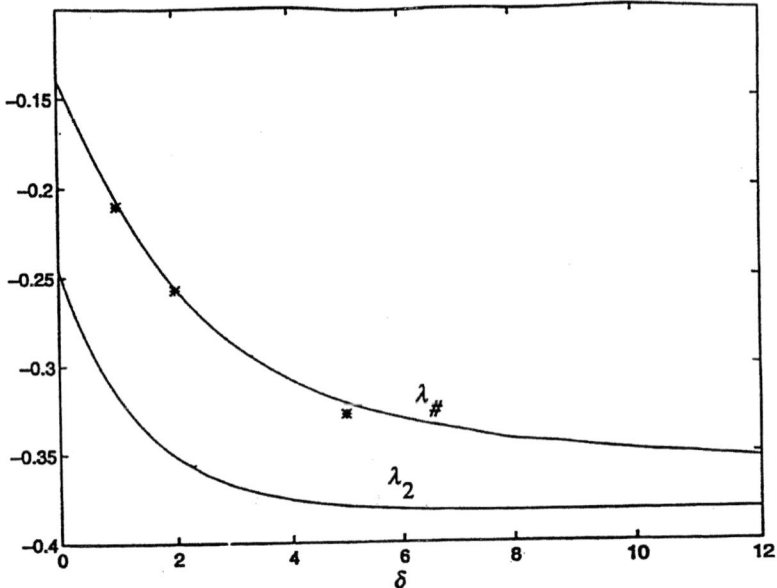

Figure 7.15: The resonance pole and stable eigenvalue for the single-hump pulse as a function of δ. The stars are the decay rates measured from simulation as in Figs. 7.17 and 7.18.

7.6 resonance pole description of mass drainage

We shall first describe the drainage dynamics of Figure 7.16 and 7.18 with the weighted projection of Pego and Weinstein and then augment it with the more detailed expansion of (7.54) and (7.55). While the asymptotic drainage rate is immediately given by the resonance pole $\lambda_\#$, the property of $\lambda_\#$ in (7.61) also provides important information about the drainage process. At large time,

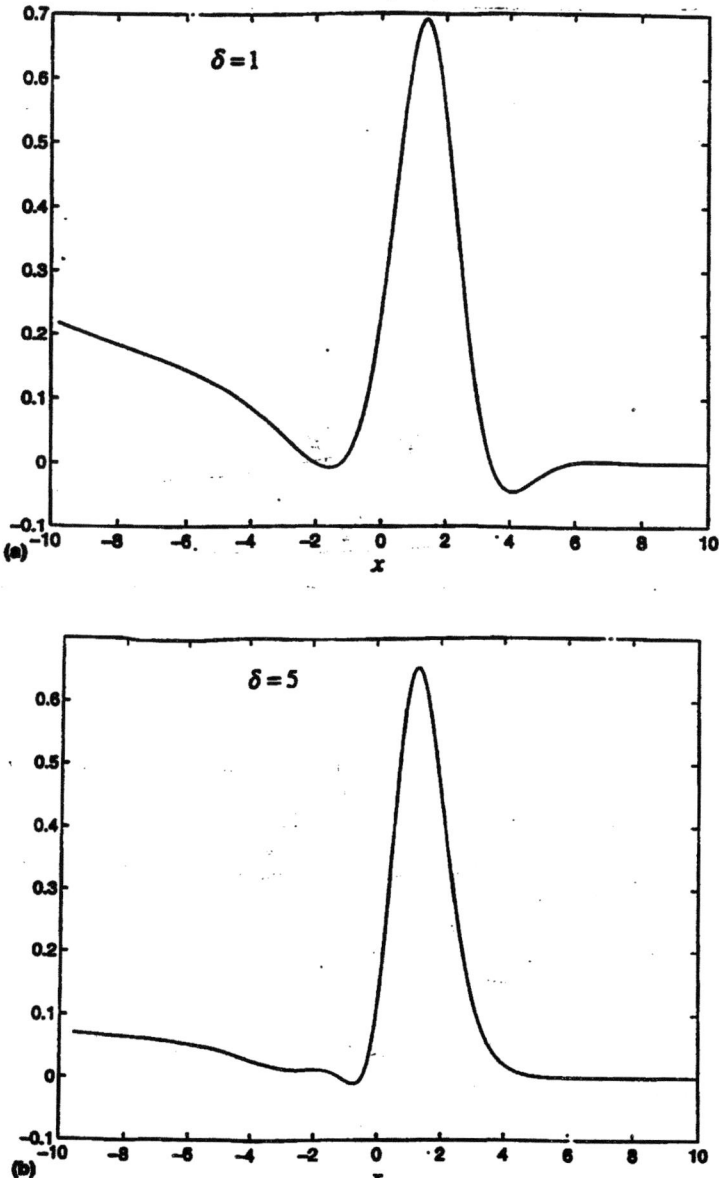

Figure 7.16: The local structure of draining excited pulses at $\delta = 0.1$ and 0.5 are shown in (a) and (b). A flat neck region is most obvious for $\delta = 5.0$ behind the back dimple. The draining mass grows exponentially into the shelf behind the neck. This tail region is driven by the resonance pole mode $\psi_\#$.

the behavior of $u(x,t)$ near $x = 0$, as measured by $v(x,t) = e^{ax}u(x,t)$, can be estimated by expanding $v(x,t)$ in terms of the discrete spectrum of L_a, ϕ_1 and $\phi_\#$, and the essential spectrum Γ_a. The essential spectrum decays very rapidly and since ϕ_1 is a simple zero mode, its coefficient remains constant in time. Hence, at large time and near the origin

$$v(x,t) \sim \xi_0 \phi_1 + a_0 \phi_\# \, e^{\lambda_\# t}$$

$$= \xi_0 \psi_1 e^{ax} + a_0 e^{ax} \psi_\# \, e^{\lambda_\# t} \tag{7.62}$$

$$\sim \xi_0 h_x e^{ax} + a_0 (h_x + \lambda_\# f(x)) e^{\lambda_\# t} e^{ax}$$

where a_0 is the initial value for the residue of the resonance pole mode. Transforming back to the original unweighted disturbance, we get

$$u(x,t) \sim \xi_0 h_x + a_0 (h_x + \lambda_\# f(x)) e^{\lambda_\# t} \tag{7.63}$$

which is simply (7.55) if one substitutes $\lambda_1 = 0, \psi_1 = h_x$ and (7.61). Since h_x corresponds to a translation as seen in (7.63), the excited resonance pole mode causes a shift in the position ξ as well as drainage. This shift from the final equilibrium position ξ_0 is given by

$$\hat{\xi} = a_0 e^{\lambda_\# t}$$

Hence, the deviation speed decays as

$$\hat{c} = \frac{d}{dt}\hat{\xi} = a_0 \lambda_\# e^{\lambda_\# t} \tag{7.64}$$

Although both the weighted projection and the complete expansion yields the same speed decay dynamics, they differ in the description of the amplitude decay of Figure 7.17. The weighted disturbance $v(x,t)$ in (7.62) can be integrated immediately to yield

$$\frac{d<v>}{dt} = \frac{d}{dt} <ue^{ax}>$$
$$\sim a_0 \lambda_\#^2 <f(x)e^{ax}> e^{\lambda_\# t} \tag{7.65}$$

where $<fe^{ax}> \; > 0$. Although $<v> = <ue^{ax}>$ is close to $<u>$, it nevertheless represents only a weighted projection of the disturbance $u(x,t)$ and not the true mass $<u(x,t)>$. The value $<v>$ is also dependent on the exponent a of the weighting function. One sees, however, from (7.65) that

$$<\hat{v}> = <v>(t) - <v>(\infty) \sim a_0 \lambda_\# <fe^{ax}> e^{\lambda_\# t} \tag{7.66}$$

and the excess weighted mass $<\hat{v}>$ also decays exponentially in time as the speed (7.64). As a result,

$$\hat{v}/\hat{c} \sim <fe^{ax}> \tag{7.67}$$

The integral on the right is bounded since f decays more rapidly than e^{ax} in the positive x direction and although $f(x)$ blows up to infinity in $-\infty$ as seen in (7.61), the weight e^{ax} suppresses this growth in the negative x direction as $\phi_\# = e^{ax}\psi_\#$ is bounded in both directions. One can approximate the deviation mass (area) $\hat{s} = <u>$ by $<v>$ but the tightness of the approximation is often dependent on the weight a.

It is quite inaccurate, on the other hand, to approximate the measured deviation height $\hat{h}$ by $u(0,t)$ in (7.63). The mass-carrying resonance pole mode $\psi_\#$ resembles h_x near the origin (the maximum of the equilibrium pulse) as seen in (7.61). However, the maximum of h_x (and hence $\psi_\#$ near the origin) is away from the origin as seen in Figure 7.16. As a result, when the disturbance is large, the maximum of the excited pulse lies not at the origin but shifts some place between the origin and the maximum of h_x as seen in Figure 7.16. For this reason, it is easier to measure the excess mass (area) $\hat{s}$ instead of the excess amplitude $\hat{h}$ of the excited pulse although the two are clearly proportional to each other because of the localized nature of $\psi_\#$ near the origin stipulates a constant width.

To obtain a better estimate of the true area, however, we must return to the full expansion of (7.54) and examine how the shifted essential spectrum cuts off the growing tail of the resonance pole function $\psi_\#$ to form a shelf. We shall ignore the translation mode ψ_1 here and write (7.54) as

$$u(x,t) = a_0 \psi_\# e^{\lambda_\# t} + \int_{-\infty+ib}^{\infty+ib} A(\alpha, x) e^{i\alpha x + \lambda_b t} d\alpha \tag{7.68}$$

We shall focus on the region behind the pulse ($x < 0$) and set $K(\alpha, x)$ to its unit asymptotic value. We shall now seek an estimate of the Fourier coefficient $A(\alpha)$ for a given a_0 to investigate how the essential spectrum cuts off the exponential growth of $\psi_\# \sim e^{\eta x}$ in the negative x direction with $\eta < 0$ being the growth exponent of $\psi_\#(x)$ in $-\infty$. It is defined by (7.10),

$$\lambda_\# = c\eta - \eta^2 - \delta\eta^3 - \eta^4 \tag{7.69}$$

As is evident in Figure 7.16 and explained in (7.61), $\psi_\#(x)$ is close to $h_x(x)$ near the pulse. Far behind the pulse, $\psi_\#(x)$ grows exponentially by $e^{\eta x}$ and departs form $h_x(x)$. We determine the amplitude ϵ of this exponential tail by extrapolating it backwards to the point where $\psi_\#$ exhibits a minimum. This minimum (neck) of $\psi_\#$ can be chosen as the origin for x and its tail then behaves as $\epsilon e^{\eta x}$. This amplitude ϵ is a function of the pulse and hence a function of δ and must be estimated numerically. There is hence a neck to $\psi_\#$ which separates the pulse region where it can be approximated by $h_x(x)$ and the exponentially growing tail. This neck is evident in Figure 7.16. The neck is several pulse widths to the left of the origin and the approximation $K(\alpha, x) = 1$ used in (7.68) can still be safely imposed for the neck region and behind it. In fact, $\psi_\#(x)$ in this exponentially - growing tail can be approximated by

$$\psi_\#(x) \sim \epsilon e^{\eta x} H(-x) \tag{7.70}$$

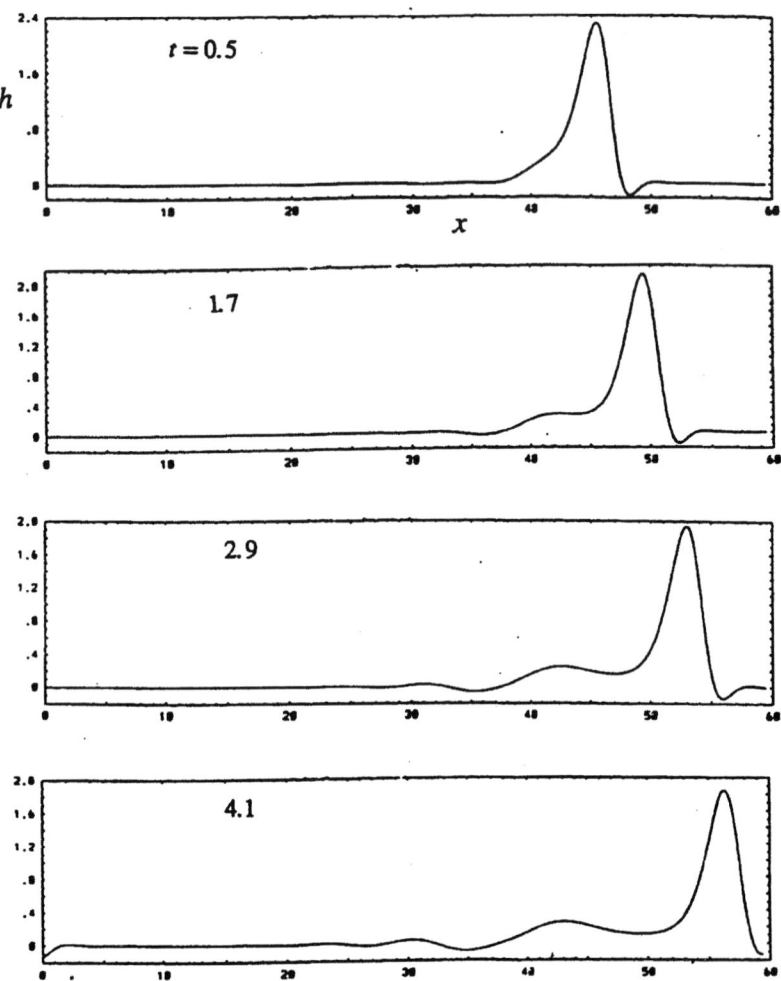

Figure 7.17: Drainage from an excited pulse when excess mass is added to equilibrium pulse for $\delta = 1$. The draining mass in the back forms a shelf, which breaks up into an oscillatory tail. The snapshots begin at $t = 0.5$ with $\Delta t = 1.2$ increments.

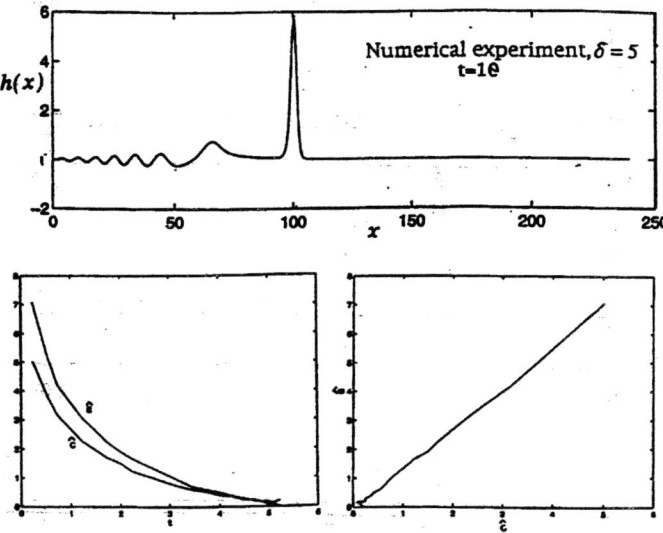

Figure 7.18: The excess speed $\hat{c}$ (deviation from the equilibrium pulse speed) and excess mass(area) $\hat{s}$ of the draining pulse both decrease exponentially in time and correlate linearly to each other.

where we have shifted the origin $x = 0$ to the minimum of $\psi_\#$ for convenience. The results do not depend on the location of the origin.

Since the disturbance $u(x,t)$ vanishes in the tail region where (7.70) is valid, (7.68) can be written at time $t = 0$ as

$$a_0 \epsilon e^{\eta x} H(-x) + \int_{-\infty+ib}^{\infty+ib} A(\alpha, x) e^{i\alpha x} d\alpha = 0 \qquad (7.71)$$

It is then clear that the localized initial disturbance can be decomposed as an exponential tail and a cut-off represented by the integral in (7.71) to ensure $u(x \to -\infty, 0)$ vanishes. This cut-off actually specifies $A(\alpha, x)$. A simple change of variable $\beta = \alpha - bi$ converts (7.71) to

$$a_0 \epsilon e^{\eta x} H(-x) + \int_{-\infty}^{\infty} A(\beta) e^{i\beta x - bx} d\beta = 0$$

or

$$-a_0 \epsilon e^{(\eta+b)x} H(-x) = \int_{-\infty}^{\infty} A(\beta) e^{i\beta x} d\beta \qquad (7.72)$$

for $x < 0$. It is then clear that $A(\beta)$ is the Fourier coefficient of the function on the left hand side of (7.72). A simple Fourier integral then yields

$$A(\beta) = \frac{a_0 \epsilon i}{2\pi[\beta(\eta + b)]}$$

or

$$A(\alpha) = \frac{a_0 \epsilon i}{2\pi(\alpha + i\eta)} \qquad (7.73)$$

Inserting this into (7.68), one can then resolve the tail evolution and, in particular, the mass spreading in the form of a thinning shelf with an oscillatory edge,

$$u(x,t) \sim a_0[\psi_\#(x)e^{\lambda_\# t} - \epsilon F(x,t)] \qquad (7.74)$$

for $x \leq 0$. The cut-off function

$$F(x,t) = \frac{i}{2\pi} \int_{-\infty+ib}^{\infty+ib} \frac{e^{i\alpha x + \lambda_b(\alpha)t}}{\alpha + i\eta} d\alpha \qquad (7.75)$$

is a positive cut-off function that blows up as $e^{\eta x}$ in $-\infty$ to cut-off the exponential tail of $\psi_\#$. Initially, the localized disturbance representing the drained mass from the pulse is in the form of a discontinuous shock (shelf) as seen in (7.70). However, the step function $H(-x)$ relaxes in time as seen in (7.75) as the shelf decreases in amplitude and its jump relaxes into the decaying oscillations seen in Figure 7.17 and 7.18. The draining mass spreads over the substrate in the form of a thinning shelf with an edge relaxed by decaying oscillations.

The length $l_s(t)$ and the thickness (width) $w_s(t)$ of this shelf can also be estimated. Returning the origin to the position where the maximum of equilibrium pulse is located and denote x_* as the position of the neck of $\psi_\#(x)$ in Figure 7.16. We also invoke approximation (7.61) for $x > x_*$. Initially, the shelf length $l_s(t)$ is zero and the shelf-edge is located at x_* or $x = 0$ in the previous coordinate of (7.70). The total excess mass (area) $\hat{s}$ at this instant is simply

$$\hat{s}_0 = <u> = a_0\lambda_\# \int_{x_*}^{\infty} f(x)dx = a_0 \|f\| \lambda_\# \qquad (7.76)$$

where $f(x)$ is defined in (7.61). For larger time, the edge moves behind x_* to $x_* - l_s(t)$. Although the edge now relaxes into decaying oscillations, we shall approximate it still as a jump. From (7.61) and (7.74), it is clear that the excess mass contained in the pulse in front of x^* now decays as

$$\hat{s} = \hat{s}_0 e^{\lambda_\# t} = a_0 \lambda_\# \|f\| e^{\lambda_\# t} \qquad (7.77)$$

By mass conservation, this mass must drain into the shelf between x_* and $x_* - l_s(t)$. Approximating this region by (7.70), the area of the shelf is then

$$\hat{s} = \frac{a_0 \epsilon}{\eta} e^{\lambda_\# t} \left(1 - e^{-\eta l_s(t)}\right) \qquad (7.78)$$

Equating $\hat{s}$ to $\tilde{s}$, one obtains

$$l_s(t) = \frac{1}{\eta} ln \left\{ 1 + \frac{\eta \parallel f \parallel \lambda_\#}{\epsilon} \left(1 - e^{-\lambda_\# t}\right) \right\} \quad (7.79)$$

such that for $t << |\lambda_\#|^{-1}$

$$l_s(t) \sim |\lambda_\#^2 \parallel f \parallel /\epsilon| t \quad (7.80)$$

and for $t >> |\lambda_\#|^{-1}$

$$l_s(t) \sim |\lambda_\#/\eta| t \quad (7.81)$$

Hence, the length of the shelf increases linearly but with different rates at small and large times.

Similarly, by using (7.77) to determine how much mass has drained into the shelf, one concludes that the thickness $w_s(t)$ of the shelf remains relatively constant at $a_0\epsilon$ for $t << |\lambda_\#|^{-1}$ since the mass comes into the shelf at a linear rate while the shelf length also increases linearly. For $t >> |\lambda_\#|^{-1}$, however, the thickness decreases as t^{-1}

$$w_s(t) \sim a_0 \parallel f \parallel |\eta/\lambda_\#| t^{-1} \quad (7.82)$$

Comparing (7.77) to the weighted excess mass of (7.66), it is clear that we have replaced the weighted norm $< fe^{ax} >$ with the cut-off norm $\parallel f \parallel = \int_{x_*}^{\infty} f(x) dx$ which is independent of the weighting function e^{ax}. In real and numerical experiments, it is simple to measure the decrease in the area of the pulse captured by (7.77). One can also estimate the excess amplitude by defining a "virtual height" $\hat{h}$

$$\hat{h} = \hat{s}/\sigma \quad (7.83)$$

where σ is the pulse width. Both the excess area and the excess amplitude clearly decay exponentially in time with the same exponent $\lambda_\#$ as the speed in (7.64). Combining (7.64) with (7.64), one obtains

$$\hat{s}/\hat{c} = \parallel f \parallel \quad (7.84)$$

which verifies the linear correlation between excess speed and excess mass (or amplitude) observed in many experiments, Liu and Gollub (1994), Brock (1969) and simulations, Kawahara and Toh (1988) of decaying pulses.

We have carried out an extensive numerical study of how the gKS pulse drains fluid as seen in Figure 7.17 and 7.18. The excess area and speed are all observed to decay exponentially in time with exponent $\lambda_\#$ as seen in Figure 7.15. The linear correlation between $\hat{c}$ and $\hat{s}$ predicted by (7.84) also fits the numerical data reasonably well, as seen in Figure 7.19, despite some arbitrariness in the neck position x_*.

To show how the essential spectrum cuts off the growing tail of the resonance pole eigenfunction $\psi_\#$, we have subtracted the solution $h(x,t)$, in our simulation of the gKS equation in Figure 7.17, from the pulse solution to obtain the

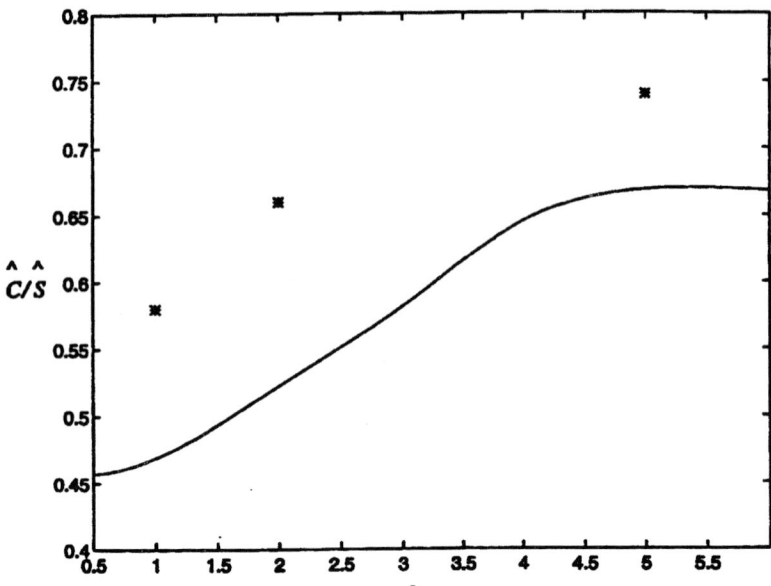

Figure 7.19: Comparison of the measured ($\hat{c}/\hat{s}$) ratio of a draining pulse to the predicted one of (7.84) for different values of δ.

disturbance $u(x,t)$. At a particular time, we estimate a_0 and then use (7.70) to estimate the coefficients $A(\alpha)$ of the essential spectrum. In Figure 7.20 and 7.21, we compare the reproduced signal according to (7.74) to the simulated disturbance evolution. It is clear that the full drainage evolution is faithfully reproduced. This decaying resonance pole mode in the form of a shelf and the cut-off exercised by the essential spectrum as the shelf shifts to the left are also evident. A lump of mass is simply spread over the substrate behind the pulse. For the small-time drainage depicted, the shelf thickness remains relative constant while its length $l_s(t)$ clearly increases linearly in time as predicted by (7.79). This finite-volume drainage mechanism, with one side of the draining shelf bounded by the growing resonance pole mode and the other radiation represented by the oscillating edge, is characteristic of draining pulses whose decay dynamics is described by a resonance pole.

The unique drainage dynamics with a linear scaling between the instantaneous speed and amplitude, when both are reduced by the equilibrium value, results mathematically from the near-degeneracy between the resonance pole $\psi_\#$ and the translational mode $\psi_1 = h_x$ in (7.61). To underscore this degeneracy, we depict both eigenfunctions in Figure 7.22. Except for $x \to -\infty$, these two eigenfunctions are indistinguishable! This near-degeneracy is observed for all falling-film pulses and hence their excess amplitudes are always proportional

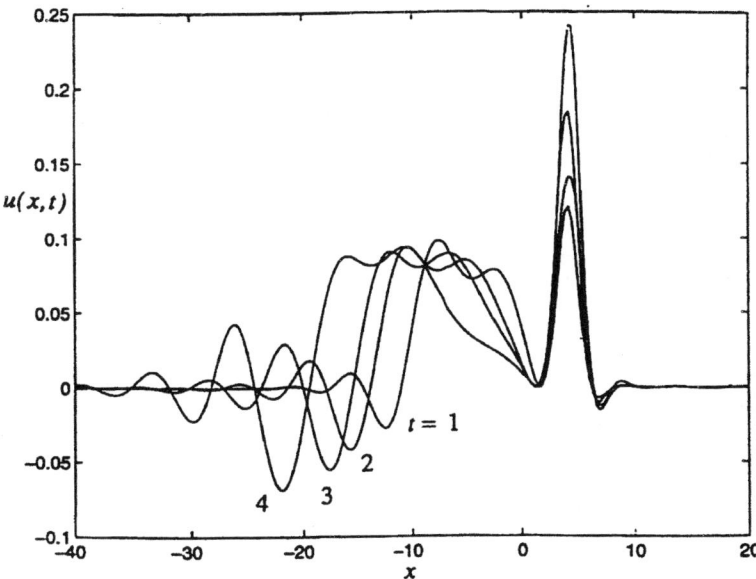

Figure 7.20: Simulated evolution of the disturbance $u(x,t)$ of a draining pulse at $\delta = 0.1$ for $t = 1.0$, 2.0, 3.0 and 4.0. The shelf retains its width and recedes roughly linearly in time. The deviation from perfectly linear spreading is due to the stable shifted essential spectrum, the stable ψ_2 mode, and nonlinear effects.

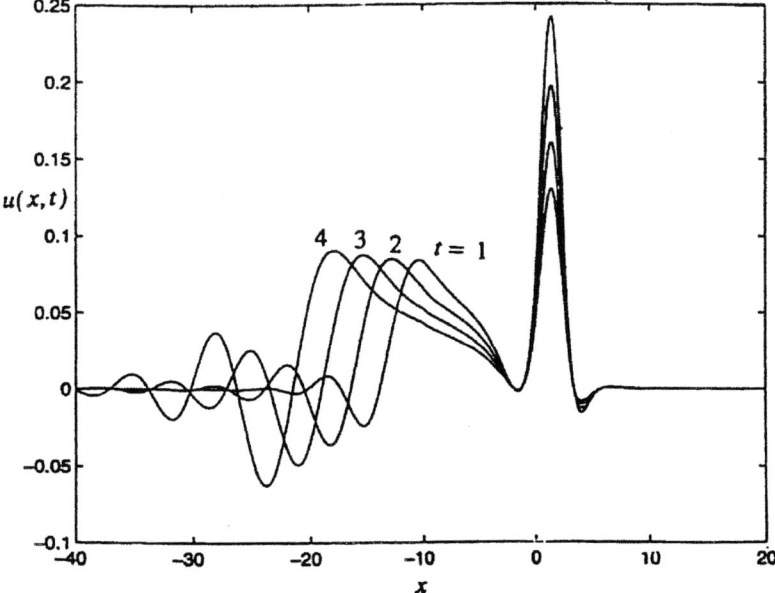

Figure 7.21: Analytical estimate using the coefficients of the resonance pole and the essential spectrum at time zero. Linear spreading is observed here.

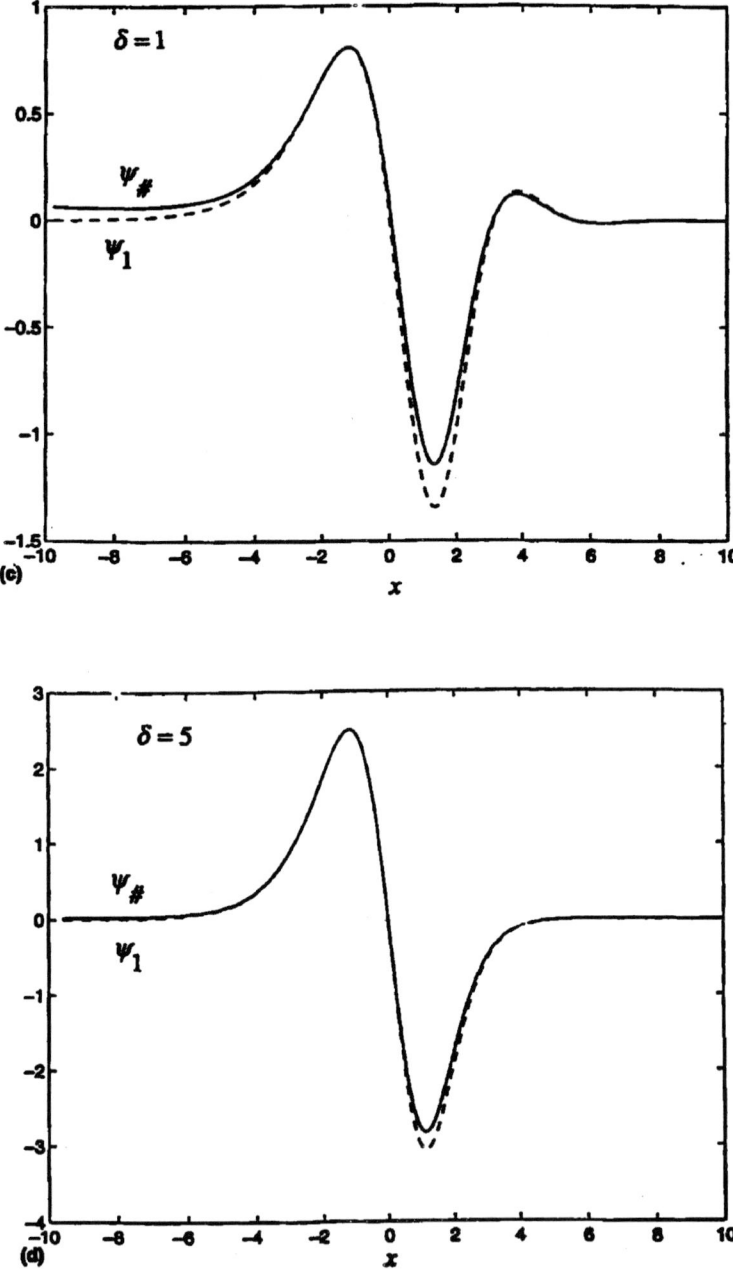

Figure 7.22: Near degeneracy between the resonance pole eigenfunction $\psi_\#$ and the translational discrete mode ψ_1

to their excess speed during drainage. We shall exploit this unique property in the next chapter to formulate a coalescence model.

7.7 Suppression of wave packets by a periodic train of pulses.

The radiation emitted by the drainage shelf in Figure 7.20 and 7.21 will grow into pulses if a trailing pulse does not absorb it. Since this absorbing pulse will again emit the disturbance behind it, the radiation can only be suppressed if its amplitude is attenuated after each absorption in spite of the growth on the active substrate. The amplitude gained on the substrate is measured by the spatial growth rate Frish, Zhen and Thual (1986) and it is very sensitive to the phase speed since the latter determines the time a particular Fourier mode spends on the substrate. As such, for a given δ, there is a critical separation l_c between two identical pulses beyond which radiation will not be suppressed in amplitude as it passes from the leading pulse to the trailing one. In this connection, we observe that, although the high δ pulses are less absorbing and suppresses the radiation less, as seen from the transmission coefficient in Figure 7.11, the high dispersion on the substrate at large δ also implies that the radiation propagates faster and has less time to grow on the substrate. We shall show that the latter effect dominates and l_c actually increases with δ such that in the KdV limit l_c approaches infinity and the separation can become infinitely large.

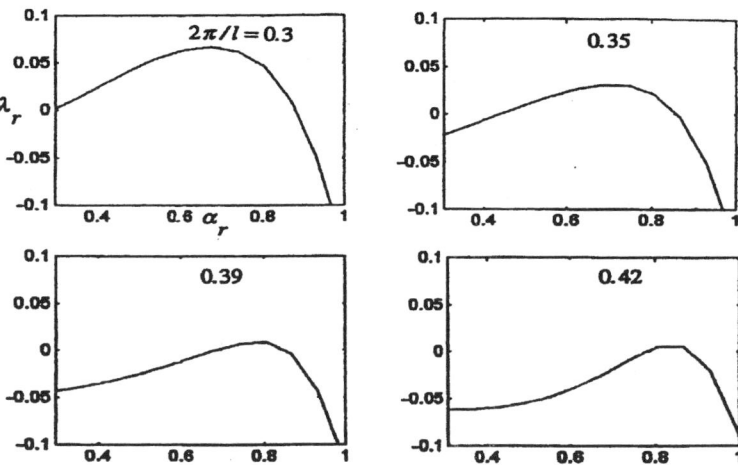

Figure 7.23: The temporal growth rate $\lambda_r(\alpha_r; l)$ of a spatially neutral Fourier mode wavenumber α_r with the radiation packet various values of $2\pi/l$. All modes are temporally stable for $2\pi/l > 2\pi/l_c = 0.40$ for $\delta = 0.13$ here.

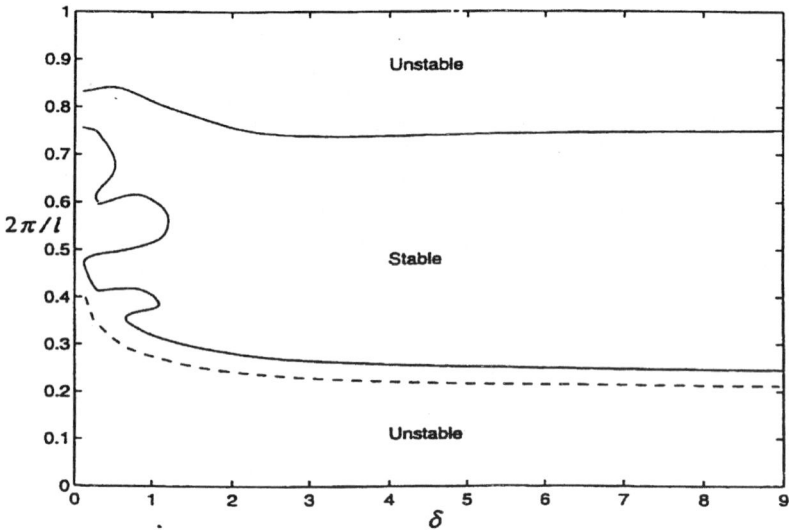

Figure 7.24: Comparison of $l_c(\delta)$ in dotted line to the results of Floquet theory for periodic pulse trains. The 'holes' at low δ are due to subharmonic instabilities not consider here. The upper bound α_s for sideband stability from Chapter 7 also shown.

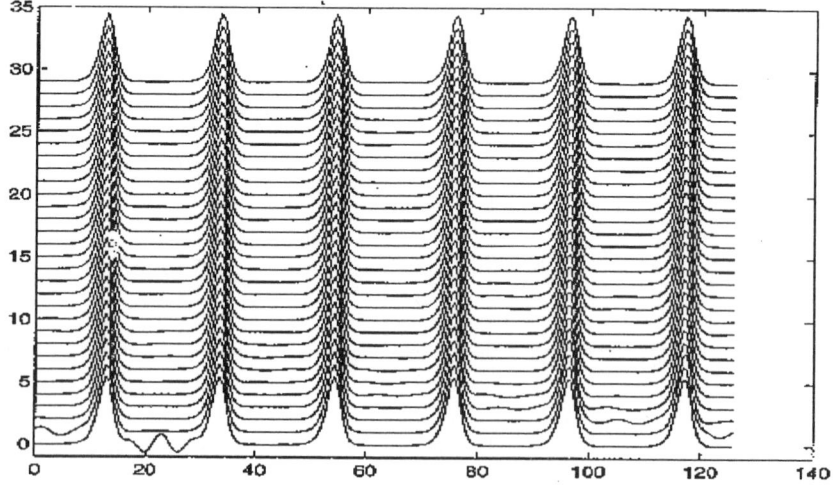

Figure 7.25: See below.

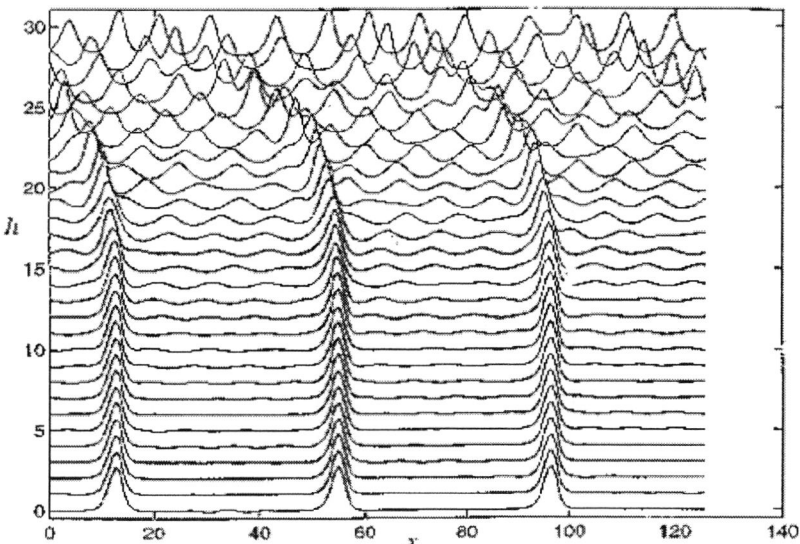

Figure 7.25: Suppression of localized radiation by unevenly spaced pulse train at $\delta = 5$. Four spacing are with wavenumber $\alpha_1 = 0.3 > \alpha_c = 0.2$, while one spacing is with $\alpha_2 = 0.15 < \alpha_c$. The predominance of short spacing can still suppress the radiation growth.

To render these arguments more explicit, we first note that the Fourier mode with frequency $\omega = \lambda_i(\alpha_r)$, where λ_i is the imaginary part of the eigenvalue in (3.3) for a real wavenumber α_r, emitted from the leading pulse will grow in amplitude by the amount $\exp[\alpha_i(l - \sigma)]$ over the length of substrate $l - \sigma$ where α_i is the imaginary part of the complex wavenumber, l is the separation between the two pulses and σ the pulse width. The same mode, though, was attenuated in amplitude by a factor of $1/r(\alpha_r)$ when it passed through the front pulse. Hence, for the critical condition when the amplitude of the Fourier mode entering the trailing pulse is equal to its amplitude entering the front pulse,

$$\exp[\alpha_i(l - \sigma)] = r(\alpha_r) \qquad (7.85)$$

or

$$\alpha_i = \frac{\ln r(\alpha_r)}{l - \sigma} \qquad (7.86)$$

which relate the real and imaginary parts of the complex wavenumber through the transmission coefficient of Figure 7.11. However, the same Fourier mode with frequency λ_i and wavenumber α_r when it first leaves the leading pulse must also not grow temporally,

$$\lambda_r(\alpha_r, \alpha_i) < 0 \qquad (7.87)$$

Since α_i is related to α_r by (7.86) for a mode that is spatially neutral, its temporal growth rate λ_r is a function only of α_r and l. In Figure 7.23, we depict the computed temporal growth rate λ_r as a function of α_r for several functions of l. As is evident, all Fourier modes are spatially neutral and temporally stable for l in excess of approximately $l_c = 2\pi/0.40$. (The numerical estimate of l_c is about $2\pi/0.43$.) For l less than l_c, all modes are spatially and temporally stable and any radiation mode will be suppressed by pulse absorption. Since unstable pulses cannot attenuate any growing radiation, we expect l_c to approach zero as δ approaches 0.137, the limit of stable pulses. For the purely dispersive KdV limit at infinite δ, we expect l_c to approach infinity as mentioned earlier since the substrate is not active.

In Chapters 5 and 6, we have considered the Floquet stability of periodic waves with wavenumber α of the gKS equation. In the limit of small α, these periodic waves resemble a periodic lattice of equally separated single-hump pulses. It was found then that if the wavelength exceeds a critical value (α less than critical value α_c), the periodic waves are unstable and the Floquet growth rate resembles the substrate growth rate of (3.3). It is clear that this α_c is simply $2\pi/l_c$ of the above absorption theory and in Figure 7.24, we favorably compare the $2\pi/l_c$ determined from the transmission coefficient of the essential spectrum Γ to the Floquet result. The critical α_c goes to infinity at $\delta = 0.17$ and zero at infinite δ as expected. In Figure 7.25, we show two sets of simulations where a small finite-mass disturbance is placed between the first two pulses of two periodic pulse trains whose separations are below and beyond l_c, respectively. It is clear that the radiation is swept clean by one and the other produces additional dynamics as new pulses are generated from the radiation. In Figure 7.25, the former scenario applies. However, since the initial periodic train is not quite the asymptotic periodic state with a locked separation and amplitude, the suppression of the initial radiation packet also triggers an evolution towards the final periodic train with a lower speed as seen in the final six frames of the figure. There is also some residual nonlinear interaction with the radiation that seems to last for a long time but the final periodic pulse train is largely impervious to this interaction.

CHAPTER 8

Spectral Theory and Drainage Dynamics of Realistic Pulses

We are now ready to apply the pulse spectral theory do more realistic model equations than the gKS model, like the Shkadov model. We are especially interested in the drainage dynamics of an "excited pulse" that results when two pulses coalesce. As seen before, this is the key that determines whether the coalescence cascade is perpetuated. To this end, we reproduce Figure 5.44 in Figure 8.1 for the solitary pulses of the Shkadov model. We have seen in Figure 4.38 that the simulated and observed pulses are, on the average, members of this one-parametric family. We shall attempt to describe this drainage dynamics with the single resonance pole discrete mode. However, the resonance pole theory is strictly valid for small excited pulse with an infinitesimally small amount of excess liquid to drain. We hence also present an approximate theory for large excited pulses for comparison. Such a theory for dispersive media is developed in the work by Kath and Smith (1995). See also Kadomtsev and Karpman (1971) and Karpman (1975).

8.1 Role of drainage in pulse coalescence.

Liu and Gollub (1994) and later Bontozoglou (private pubicaition) performed the most definitive experiments on the basic mechanism driving wave coarsening on a falling film. This mechanism is the irreversibly coalescence between two pulses — a larger back pulses and a smaller front one. Since the back pulse is larger and travels faster, it chases down the front one to precipitate the irreversible coalescence. The irreversible coalescence event again produces a large pulse, which chases the next smaller pulse. Our model of this coalescence event is a uniformly flat substrate with a front pulse which is the true equilibrium

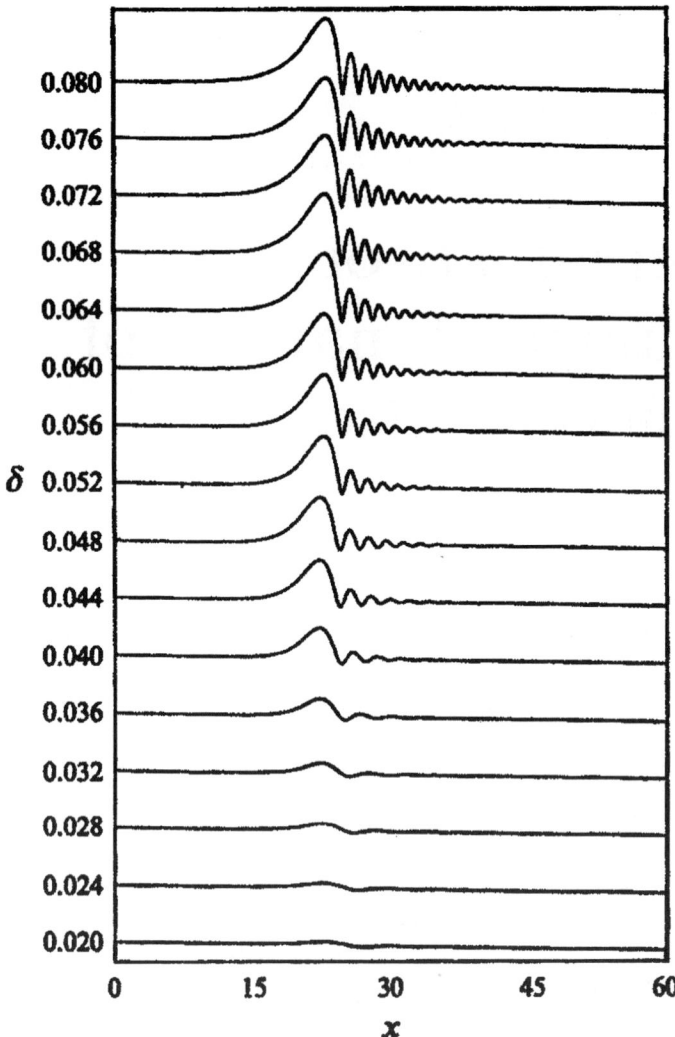

Figure 8.1: One-parametrer family of pulses for Shkadov's model.

pulse for that substrate. The back pulse, however, is an "excited" pulse with an amplitude and a speed larger than the equilibrium one. Most importantly, the excited pulse drains parts of its liquid during its pursuit of the front one. Its a result, its amplitude and speed decay as it propagates. Whether and when it captures the front pulse is then mostly determined by this drainage dynamics of the excited pulse. This drainage dynamics can be captured by the resonance pole theory of section 7.9 but we shall offer a large-amplitude extension of the linear theory of Chapter 7.

We begin, however, with a detailed numerical scrutiny of the coalescence dynamics as described by the Shkadov model. We place an equilibrium pulse on a substrate in a frame moving with its speed. As a result, this pulse remains stationary in this frame.

As done in the experiment of Liu and Gollub (1994), a larger solitary pulse is placed behind the origin stationery one. As shown in Figure 8.2(a), when the separation between the pulses is smaller than a critical value, the two pulses will attract each other and coalesce irreversibly to form a single large pulse. If the initial separation is large, there will be some transient interaction but the two pulses will eventually repel each other and yield two independent pulses, as shown in Figure 8.2(b).

These two pulses will be separated by a constant separation and are known as a "bounded pair" as shown in Figure 8.3(a). After some careful numerical experiment, it is possible to obtain the critical initial separation delineating coalescence and repulsion for a given pair of excited and stationary pulses, and one sees that the two pulses from a far closer "bounded pair" than the repulsion case, which resembles the two-hump solitary pulse shown in Figure 8.3. This particular bounded pair hence represents the transition state for coalescence. Identical stationary pulses coalesce if their separation is smaller than the critical distance l^* represented by the separation between the two humps of the transition-state bounded pair.

It is important to note that the critical distance l^* for the coalescence of two identical pulses on the same substrate is different from the critical initial separation observed in Figure 8.2 between a given pair of excited pulses and its stationary front neighbour. The latter critical separation is a function of the initial difference in the amplitudes of the two pulses and it approaches l^* as the difference approaches zero. The objective of our theory is to estimate how the critical distance depends on the amplitude difference. Using the front stationary pulse as the reference pulse, this difference can then be termed the "degree of excitation" of the back pulse which can be measured by $\hat{h}, \hat{c}$ or $\hat{\chi}$, the difference in amplitude, speed or substrate thickness, respectively. Each measure is equivalent since the excited back pulse is a member of the one-parameter solitary pulse family at every instant in its decay. The dependence of the critical separation on the degree of excitation of the back pulse explains why some near encounters far from the inlet in the world lines of Figures 4.20 and 4.34 in Chapter 4 do not lead to coalescence even though the two pulses come as close as those do coalesce further upstream. Each prior coalescence gives rise to a pulse of roughly equal degree of excitation but a coalesced pulse far downstream

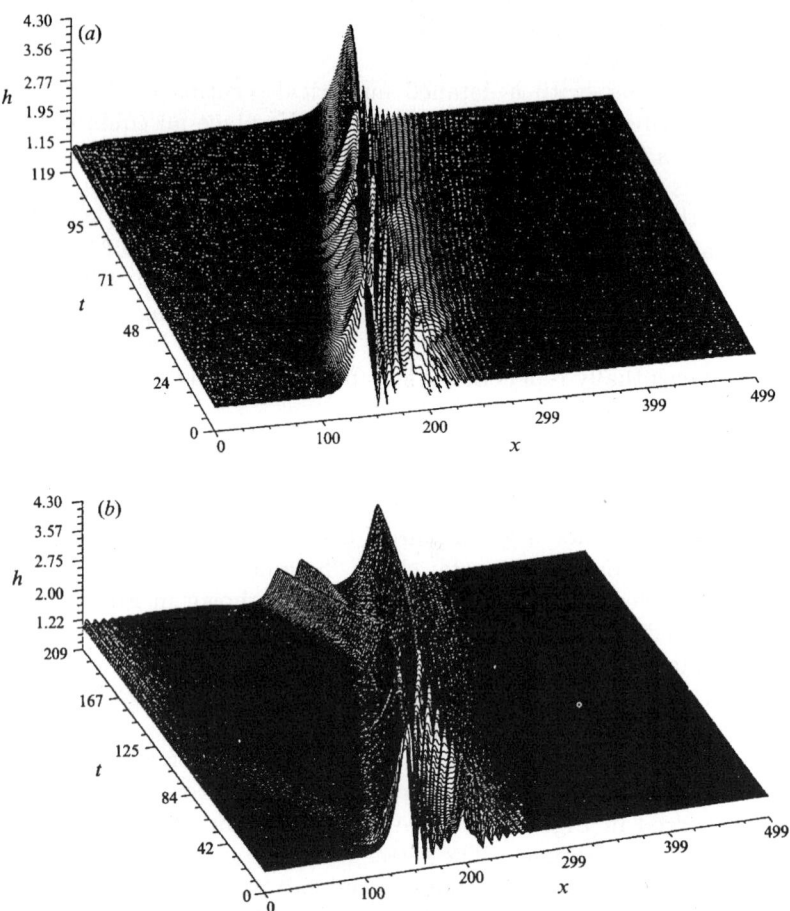

Figure 8.2: Two interaction experiment showing (a) coalescence and (b) repulsion for $\delta = 0.05$ in a frame moving with the speed of the front stationary pulse $c^*(\delta)$. In both cases, the initial amplitudes are the same with the back pulse excited but the initial separations are different.

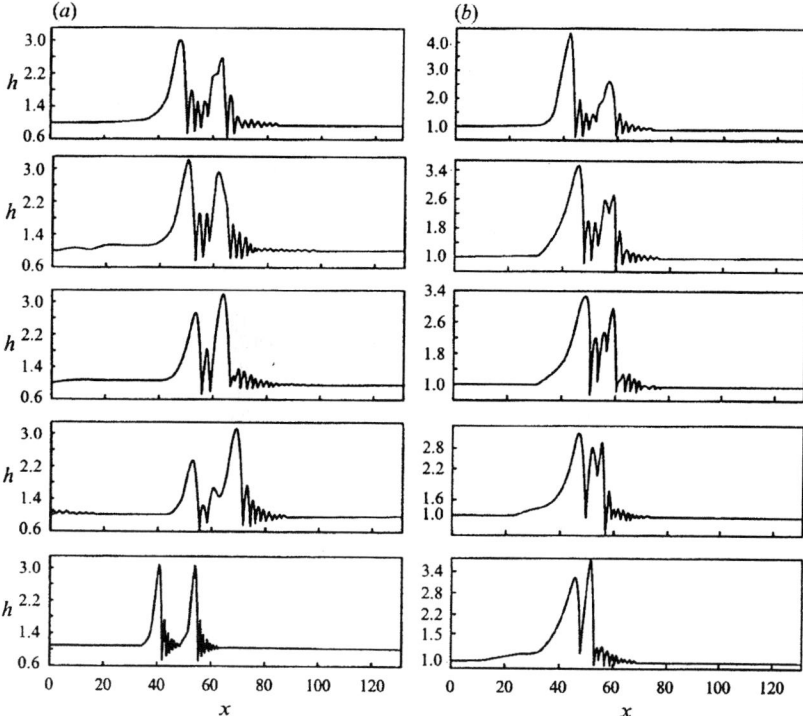

Figure 8.3: (a) The repulsion case Figure 8.2(b) resultion finally in a bounded-pair with large separation and (b) the critical innitial separaration that result in a two-hump pulse.

is simply further from its front neighbour. The next coalescence may not occur even if the two pulses approach each other since the back pulse has drained most of its fluid, viz. the initial separation exceeds the critical value. A more detailed depiction of the coalescence even from our numerical experiments is shown in Figure 8.4. It is seen that, at the initial large separation, the excited pulse 1 at the back is oblivious to the stationary pulse 2 in front and its amplitude decays in a unique fashion, which will be scrutinized in a latter section, as it catches up with the stationary pulse. The stationary pulse 2 is also unaffected by the decay dynamics of the excited pulse during the transient period. When two pulses are sufficiently close, however, the short-range attractive interaction takes over and causes the two pulses to coalesce to form pulse 3.

The decay dynamics of the excited pulse at the back in Figure 8.4 is extremely important. At any given instant, it resembles a solitary pulse of a larger δ in Figure 8.1. The approach towards the stationary pulse of the actual δ in the experiment roughly corresponds to a quasi-steady evolution along the family of solitary pulses in Figure 8.1 towards the one corresponding to the actual δ value. This is further corroborated in Figure 8.5 where the decay dy-

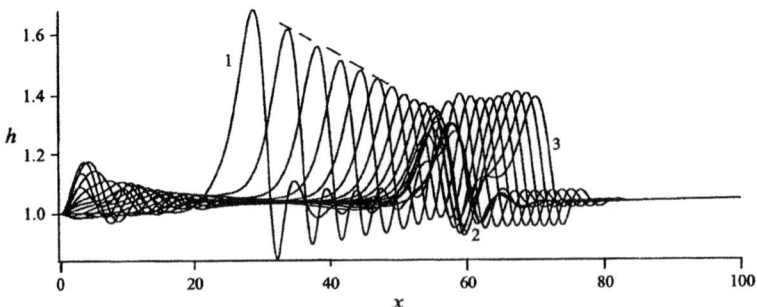

Figure 8.4: A different depiction of the interaction in Figure 8.2(a) that leads to coalescence. Note the linear decay of the excited back pulse. Pulse 1 is the initial excited pulse and pulse 2 the intial stationary pulse. Pulse 3 is the coalesced pulse.

namics is compared to the solitary pulse branch represented by $c^*(\delta)$ and $h_m(\delta)$. (Note the similarly in the shapes of the two curves $c^*)\delta)$ and $h_m(\delta)$. This is another manifestation of the fact that $(c^* - 3)/(h_m - 1)$ is approximately constant.) Ideally, the instantaneous δ should be determined from the measured instantaneous film thickness behind the pulse. However, we have found it easier to determine δ by measuring the area A under the pulse, which is a larger number than the thickness, and using table1 fund the corresponding δ. It suffices to view δ as a measure of the decay time and c^* and h_m seem to track the c^* and h_m of the solitary pulse family in Figure 8.1 very closely. The excited solitary pulse is observed to perch on a back substrate layer of thickness χ that is thicker than the unit value corresponding to the actual δ. (See the coalesced pulse 3 of Figure 8.4 where the back substrate is at its thickness value.) An exaggerated caricature of this excited pulse is depicted in Figure 8.6.

As the excited pulse decay, the raised substrate thins towards unity while liquid drains out of the pulse as it approaches the true stationary one. A small shock is created behind due to this drainage and it propagates backward a shock actually corresponds to an interaction of the solitary pulse with radiation modes spanned by its continuous spectrum. However, in this preliminary study, we shall approximate the drainage mechanism with a different formulation involving the point spectra.

A qualitative understanding of the evolution dynamics shown in Figure 8.3 and 8.4 has now emerged. After each coalescence, an excited pulse which resembles a stationary pulse at a higher δ is created and travels faster than its smaller front neighbour. It will hence shorten the separation from its stationary front neighbour. However, given sufficient separation from its front neighbour, the excited pulse will decay quasi-steady in amplitude and speed towards a true stationary pulse and the separation will eventually reach an asymptotic value when the two pulses travel at the same speed. If this asymptotic value is smaller

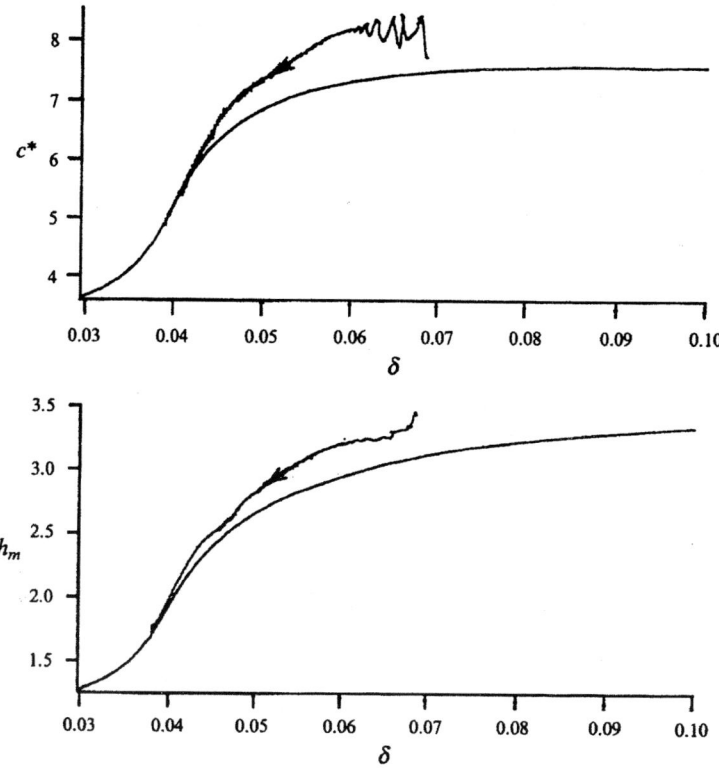

Figure 8.5: Decay of the excited pulse in Figure refch8:2 along the solitary pulse family of Figure 8.1. The instantaneous δ values are computed from transformation (8.6) by measuring the area A under the pulse and using Table 8.1.

than a critical separation, short-range attractive forces will cause the two pulses to coalesce. Otherwise, two independent stationary pulses will results. Since the initial separation between an excited pulse and its front neighbour will increase downstream on average as more and more pulses are annihilated by coalescence while the "degree of excitation" due to the coalescence of two nearly stationary pulses remains the same, an equilibrium separation will eventually be reached such that additional coalescence is not possible. It is then clear that the most important key to the understanding of the pulse dynamics is the binary interaction between an excited pulse and a stationary pulse in front. We shall develop an analytical theory that describes the long-range decay of the excited pulse and the relatively short-range interaction of the two pulses. In particular, an estimate of the critical separation l^* will be constructed. This binary theory promises to elucidate the more complex dynamics in figures 8.2, 8.3 and 8.4.

δ	w	A	c*	h_m	m	β	γ	R	S	θ (rad)	Γ	l_2	c^*_2	$h_c(0)$ (theory)	$h_c(0)$ (measured)
0.03	15.6	1.38	3.65	1.27	0.277	1.044	0.021	2.37	2.12	1.11	9.4	8.61	3.57	0.14	0.26
0.04	15.9	4.39	5.19	1.93	0.311	1.853	0.018	44.09	17.6	2.29	8.0	6.54	4.68	0.13	0.52
0.05	21.4	9.35	6.87	2.68	0.236	2.855	0.047	80.86	41.3	2.66	7.2	5.59	6.34	0.35	0.89
0.06	26.7	12.7	7.36	2.98	0.190	3.380	0.056	58.17	49.0	−1.07	5.3	5.37	7.00	0.46	0.94
0.07	31.2	15.4	7.56	3.17	0.161	3.762	0.058	37.54	50.9	2.18	4.9	5.25	7.30	0.49	0.94
0.08	34.5	17.8	7.65	3.28	0.140	4.077	0.057	24.38	48.8	0.57	4.9	5.15	7.51	0.48	
0.09	36.9	20.1	7.69	3.36	0.124	4.348	0.056	16.30	44.0	−1.46	5.1	5.09	7.65	0.46	
∞			7.71	3.42			0.056				5.0			0.46	

TABLE 8.1 Coefficients for the single-hump and double-hump solitary pulse families with $\chi = 1$

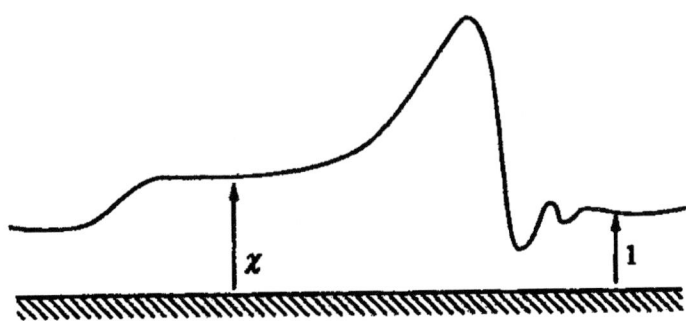

Figure 8.6: Schematic of an excited pulse with a back substrate thickness χ that is larger than the unit thickness in front.

8.2 Spectrum of the solitary pulse.

The eigenvalue problem that determines the stability of a solitary pulse to localized disturbances is now the eigenvalue problem

$$\mathbf{L}\psi = \lambda\psi \tag{8.1}$$

$$\psi \text{ bounded as } x \to \pm\infty \tag{8.2}$$

where the components of the vector function ψ denote the deviation flow rate $\hat{q}$ and the deviation height $\hat{h}$.

$$\begin{aligned}
\psi &= (\hat{q}, \hat{h}), \\
\mathbf{L} &= \begin{pmatrix} L_1 & L_2 \\ -\frac{d}{dx} & c^* \frac{d}{dx} \end{pmatrix}, \\
L_1 &= c^* \frac{d}{dx} - \frac{1}{5\delta h^{*2}} - \frac{6}{5}\frac{d}{dx}(2q^*/h^*), \\
L_2 &= \frac{1}{5\delta}\left[h^* \frac{d^3}{dx^3} + h^*_{xxx} + 1 + 2q^*/h^{*3}\right] + \frac{6}{5}\frac{d}{dx}[(q^*/h^*)^2].
\end{aligned} \tag{8.3}$$

The spectrum defined by (8.1) contains two zero eigenvalues due to the symmetries of the solitary pulse. Perturbations to the solitary pulse that break these symmetries will shift these two eigenvalues but their real parts remain small. These perturbations are system perturbations and not the transient disturbances in the classical linear stability theory. They correspond to the effect of a perturbations operator L' on a simple zero eigenvalue of L, viz. we would like to resolve the near-neutral spectrum of $L + L'$ given that L has some zero eigenvalues. In our case, the perturbation operator is in the form of another solitary pulse or a change in the substrate thickness in front of the pulse (a hydraulic jump). Provided that the other point eigenvalues are stable and bounded away from the imaganary axis in the complex plane, these two modes will dominate the dynamics of a pulse even after they are shifted by the imposed perturbation.

The first pertinent symmetry is the translation symmetry — if $(q^*, h^*)(x)$ is a solution, so should $(q^*, h^*)(x - x_0)$ be for any arbitrary shift x_0. Hence, for a known pulse solution, a one-parameter family of pulses parameterized by x_0 can be generited simply by shifting. Put another way, if one perturbs a stationary solitary pulse by translating it slightly, the degeneracy that arises from the continuous family of solutions implies that this motion is in the direction of an eigenfunction with a zero eigenvalue.

More precisely, for x_0 small

$$\begin{pmatrix} q^* \\ h^* \end{pmatrix}(x - x_0) \sim \begin{pmatrix} q^* \\ h^* \end{pmatrix}(x) - x_0 \begin{pmatrix} q^*_x \\ h^*_x \end{pmatrix}(x) \tag{8.4}$$

and

$$\psi_1 = \begin{pmatrix} q^*_x \\ h^*_x \end{pmatrix}(x) \tag{8.5}$$

is a null eigenfunction of L. Since q^* and h^* approach unity at the infinities, the vanishing-derivative boundary conditions are obviously satisfied. One component h^*_x of ψ_1 is shown in Figure 8.7 for a particular pulse.

The second symmetry that generated "zero dynamics" is more subtle. It corresponds to the degeneracy related to the family of solitary pulses in Figure 8.1. For the KS equation, this degeneracy results from the Galilean symmetry (5.4). As mentioned earlier, each member of the family is normalized such that $h^*(\pm\infty)$ and $q^*(\pm\infty)$ are unity. The substrate thickness and its corresponding flow rate are then represented by δ. In turn, one can fix δ and allow $h^*(\pm\infty)$ and $q^*(\pm\infty)$ to vary. The asymptotic film height and flow rate cannot vary independently due to mass conservation in (8.4) and (8.5) and it is convenient to denote $h^*(\pm\infty)$ as χ. The value δ can be arbitrarily fixed but is usually given the flat-film value at the inlet, viz. based on the inlet flow rate. Physically, however, the δ-parametrized family and χ-parametrized family are identical and there must be a transformation that maps one family to the other. For a given flow condition, one simply cannot vary χ and δ independently. This also implies that in our numerical experiment in Figure 8.3 which integrates (8.1) at fixed δ value, there is actually a one-parameter family of solitary pulses parametrized by the substrate thickness χ and the spectrum of each solitary pulse member of the family must contain a zero eigenvalue due to this degeneracy that arises because

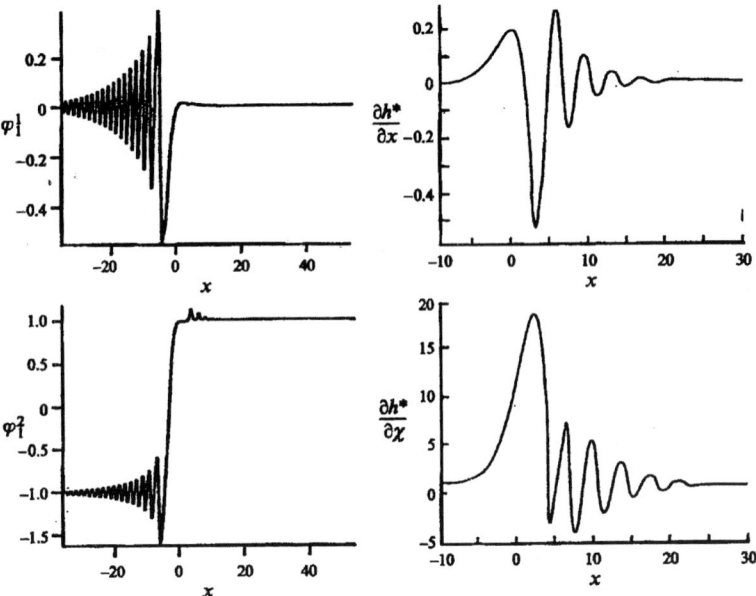

Figure 8.7: The two components of the adjoint eigenfunction ϕ_1 and the $\hat{h}$ components of ψ_1 and ψ_2 at $\delta = 0.07$.

the local substrate thickness and flow rate are not specified in the defining equations (8.4) for the solitary pulse. The transformation that transforms the δ-parametrized family of Figure 8.1 to a χ-parametrized family at fixed δ can be determined from a simple application of the π-theorem,

$$\left.\begin{aligned} h^*(x;\delta,\chi) &= \chi h^*(x\chi^{-1/3};\delta\chi^{11/3},1), \\ q^*(x;\delta,\chi) &= \chi^3 q^*(x\chi^{-1/3};\delta\chi^{11/3},1), \\ c^*(\delta,\chi) &= \chi^2 c^*(\delta\chi^{11/3},1), \end{aligned}\right\} \quad (8.6)$$

and the defining equations (8.4) for the solitary pulse with $h^*(\pm\infty) = \chi$ are invariant under this transformation. Note that the solitary speed c^* must also be transformed and the spatial coordinate x stretched. This transformation allows us to transform the δ-parametrized family of Figure 8.1 where $\chi = 1$ to a family parameterized by χ for any given δ. Since the simulation in Figure 8.1 is carried out for a fixed δ, each stationary solitary pulse observed there is then a member of the latter family. There is, of course, an infinite number of such solitary pulses possible with different substrate thickness. The variation in the substrate thickness from one pulse to another is quite apparent in Figure 8.1. We note that in the binary interaction experiment of Figure 8.2, 8.3 and 8.4, the final substrate thickness, regardless of whether coalescence takes place, is unity since there are only one or two solitary pulses on the film. This is, of course, not true in Figure 8.1 where numerous pulses exist. For a given δ, the translation invariance that generates a one-parameter family yields a zero eigenvalue. The invariance to transformation (8.5) also yields a one-parameter family and hence a zero eigenvalue of L related to this symmetry is expected. The corresponding eigenfunctions are

$$\psi_2 = \begin{pmatrix} q^*_\chi \\ h^*_\chi \end{pmatrix}(x,\chi) \tag{8.7}$$

one component of which is shown in Figure 8.7. It measures the changes in q and h as one moves through the pulse family of Figure 8.1. Since the decay of an excited pulse corresponds to such a quasi-steady motion through the family, this mode determines the decay rate. In fact, h^*_χ specifies how much fluid drains out of the excited pulse of Figure 8.4 as it decays. In the present theory, we shall set χ to unity in the deriavtive in (8.7). By taking the derivative of (2.3) with respect to c^* and by using (2.4) one immediately sees that

$$\mathbf{L}\begin{pmatrix} q^*_c \\ h^*_c \end{pmatrix} = \psi_1$$

where subscipt c denotes derivative with respect to c^*. It is more convenient to use ψ_2 and since

$$\psi_2 = \frac{\partial c^*}{\partial \chi}(\chi,\delta)\begin{pmatrix} q^*_c \\ h^*_c \end{pmatrix},$$

it immediately follows that ψ_2 is generalized null eigenfunction such that for a given χ (or δ),

$$\mathbf{L}\psi_2 = \alpha\psi_1, \tag{8.8}$$

where the parameter α is

$$\alpha = \frac{\partial c^*}{\partial \chi}(\chi,\delta)$$

and it corresponds to a derivative along the solitary pulse family. Note also that ψ_2 has vanishing derivatives at the infinities but decays to the unit constant itself. This is because there is a change in substrate thickness as one goes from one member of the family to another within the solitary pulse family parametrized by χ.

The fact that ψ_2 approaches unity at the infinities presents a special problems. This implies that the disturbance that we would like to expand with ψ_1 and ψ_2 can also approach constant values at the infinities. Such non-integrable disturbances are not physical and do not correspond to any definable functional space. Equivalently, we cannot find a generalized adjoint eigenfunction to allow projection onto the generalized zero mode. In section 7.6, we have replaced this ψ_2 mode of the gKS pulse by the resonance pole $\psi_\#$, which is also degenerate with the translational mode ψ_1. As seen in Figure 7.22, $\psi_\#$ and ψ_1 are only different far behind the pulse. We shall not construct $\psi_\#$ here but simply use these earlier results and omit the tail ends of ψ_2 away from the pulse. This corresponds to localized disturbances and the implicit assumption that the decay dynamics of Figure 8.4 involves mass drainage from the pulse only and not the substrate thickness. The change in the pulse area is still well approximated by

ψ_2 within the pulse width. It becomes a good approximation of the resonance pole mode.

There are no other zero modes in the spectrum of the solitary pulse and we expect ψ_1 and ψ_2 to dominate the dynamic reaction of the pulse to perturbations. This is consistent with our observations from the numerical experiments. The slow decay dynamics of an excited wave corresponds to an excitation of the ψ_2 speed mode due to a change in the front substrate thickness as shown in Figure 8.4. Although the translation invariance is also broken, the relative short range of the ψ_1 interaction implies that the position mode is not excited directly, as we shall demonstrate. The excited ψ_2 mode causes a transformation from a solitary pulse on substrate thickness χ to one on a unit substrate thickness in Figure 8.4. In Figure 8.3, this decay dynamics would connect solitary pulses on a thicker substrate χ_2 at the back to a thinner one χ_1 in the front. It is always a decay dynamics since the back substrate is always thicker than the front one after a coalescence event. The excited speed ψ_2 mode also shifts the position mode through the dependence between ψ_1 and ψ_2. This moves the back pulse closer to the front one until interaction between the two pulses through the position ψ_1 mode dominates over the ψ_2 decay dynamics — pulse-pulse interaction occurs on the same substrate thickness, unity in Figure 8.2 to 8.4. This final interaction determines whether the two pulses will coalesce.

Although there are no other zero modes, the spectrum of L still contains non-zero point eigenvalues that are unrelated to any symmetry. These eigenvalues must all be stable and, in fact, bounded from the imaginary axis such that the perturbed zero modes are always dominant. Otherwise, the solitary pulses will lose their coherence and break up. Unlike the two zero modes, which cause a pulse to shift or accelerate without significantly changing its shape, the others are strongly shape altering and if they are excitable, it is not meaningful to apply our approach which presumes dynamics with small departures from the solitary pulse dominated by the two zero modes. We have carried out a detailed numerical analysis of the spectrum of L using a shooting method. The leading non-zero point eigenvalue of the solitary pulse are shown in Figure 8.8. It is clear that for $\delta < 0.03$, this point eigenvalue is stable but of the same magnitude as the perturbed zero eigenvalues. It would hence dominate the pulse dynamics as much as the other two. While it is possible to include this mode so that there are three dominant modes, we shall develop a simple theory here with just the two dominant modes generated by the system symmetries. We hence require δ to be larger than 0.03 which encompasses most realistic conditions but excludes the use of low-δ evolution equations.

To project the pulse dynamics on to the zero modes, we require the null adjoint eigenfunctions. Using the L_2 inner product,

$$(\mathbf{u}, \mathbf{v}) = \int_{-\infty}^{\infty} (u_1 v_1 + u_2 v_2) dx.$$

the adjoint operator to **L** of (8.3) is defined as

$$(\mathbf{u}, \mathbf{L}\mathbf{v}) = (\mathbf{L}^+\mathbf{u}, \mathbf{v})$$

After standart integration by parts and index change, the adjoint operator is found to be

$$\mathbf{L}^+ = \begin{pmatrix} L_1^+ & \frac{d}{dx} \\ L_2^+ & -c\frac{d}{dx} \end{pmatrix}$$

where L_1^+ and L_2^+ are given in (8.3) with the same bounded boundary conditions as the eigenvalue problem (8.1). It is quite obvious that a null eigenfunction of the adjoint operator L^+ is

$$\varphi_2 = \begin{pmatrix} 0 \\ d \end{pmatrix}$$

where d is a constant. Another null eigenfunction L_1 was found numerically whose first component φ_1^1 goes to zero at the infinities but its second component φ_1^2 approaches -1 at $-\infty$ and $+1$ at $+\infty$. This null eigenfunction is shown in Figure 8.7.

We hence find two null adjoint eigenfunctions

$$\mathbf{L}^+\varphi_1 = 0, \quad \mathbf{L}^+\varphi_2 = 0 \tag{8.9}$$

The two components of φ_1 are shown in Figure 8.7 while $\varphi_2 = (0, d)^t$ where d is a constant. It is clear that, using the inner product over the infinitedomain, $(\psi_2, \varphi_2) = \int_{-\infty}^{\infty} \psi_2 \cdot \varphi_2 dx$ is unbounded and hence unacceptable. This again relates to the fact that ψ_2, as shown in Figure 8.7, is not square-integrable. However, if we define the inner product $(\cdot, \cdot)$ involving ψ_2 to be only over the pulse width, we can always select a d such that

$$\left. \begin{array}{c} (\psi_1, \varphi_1) = (\psi_2, \varphi_2) = 1, \\ (\psi_1, \varphi_2) = 0, \end{array} \right\} \tag{8.10}$$

where inner product involving ψ_1 is still over the infinite domain. We note that the biorthogonality conditions (8.10) does not include $(\psi_2, \varphi_1) = 0$. Even by restricting the domain of integration over the pulse width, this integral involving ψ_2 does not vanish. It is, however, a small number which we shall neglect. This anomaly arises because the operator L of (8.1) has a generalized zero mode while its "adjoint" (8.9) does not. This peculiarity, in turn, originates from the non-square-integrable nature of ψ_2, the substrate/speed mode which controls fluid drainage from the pulse. A negligible (ψ_2, φ_1) corresponds to a decoupling between the drainage dynamics at large separation and the pulse-pulse interaction dynamics, which is controlled by ψ_1, at short separation. This is also consistent with the shrinking of the domain of integration in inner products involving ψ_2 — the drainage is unaffected by other pulses as is evident in the numerical simulation of Figure 8.4.

We note that the second height component of the null adjoint eigenfunction φ_1 approaches constant value of opposite sign in opposite directions as seen in

Figure 8.7. This is consistent with the null adjoint eigenfunction constructed by Elphick *et al.* (1991) for the KS equation. The constant d for φ_2 such that $\psi_2, \varphi_2) = 1$ can be easly estimated from symmetry (8.6). Let $x \in (a, b)$ be the pulse domain where the inner product is evaluated. This normalization then becomes

$$d \int_a^b \frac{\partial h}{\partial \chi} dx = d \frac{\partial}{\partial \chi} \int_a^b h dx = 1.$$

The integral $\int_a^b h dx$ is simply the area under the pulse. However, from (8.6),

$$\int_a^b h(x; \delta, \chi) dx = \chi \int_a^b h(x\chi^{-1/3}; \delta\chi^{11/3}, 1) dx =$$

$$\chi^{4/3} \int_{a'}^{b'} h(\xi; \delta\chi^{11/3}, 1) d\xi = \chi^{4/3} A(\delta\chi^{11/3})$$

where $\xi = x\chi^{-1/3}$ and $(b', a') = (b, a)\chi^{-1/3}$. We have neglect the area within the film beneath the pulse which is negligible compared to the pulse area. The quantity $A(\delta\chi^{11/3})$ is simply the area A in Table 8.1 for the pulse family with $\chi = 1$. Hence, at $\chi = 1$

$$\frac{\partial}{\partial \chi} \int_a^b h dx = \frac{4}{3} A(\delta) + \frac{11}{3} \delta \frac{dA}{d\delta}$$

where $A(\delta)$ and $dA/d\delta$ can be obtained from Table 8.1. The constant d is then

$$d = \left[\frac{4}{3} A(\delta) + \frac{11}{3} \delta \frac{dA}{d\delta}(\delta) \right]^{-1}$$

We can now expand any solution to the averaged equations about the solitary pulse of an arbitrary χ (or δ) in the frame moving at the same speed as the reference pulse.

$$\begin{pmatrix} \hat{q} \\ \hat{h} \end{pmatrix} (x, t) = \begin{pmatrix} q \\ h \end{pmatrix} (x, t) - \begin{pmatrix} q^* \\ h^* \end{pmatrix} (x) \sim -x_0(t)\psi_1(x) + \hat{\chi}(t)\psi_2(x) \quad (8.11)$$

where $\hat{\chi}$ denotes deviation from the substrate thickness of the reference pulse and the amplitude of the zero modes have been represented by physically meaninfgul variables. The amplitude of the speed mode $\hat{\chi}(t)$ can also tramsformed to $\hat{c}(t)$ from (8.6) where both $\hat{\chi}$ and $\hat{c}$ denote deviations from χ and c^* of the reference pulse. If we transform the averaged equation to a coordinate moving with the speed c^* of the reference solitary pulse in (8.11), substitute (8.11) into the resulting equation, linearize and take inner product with the null adjoint eigenfunctions, we obtain

$$\frac{d}{dt}\begin{pmatrix} x_0 \\ \hat{\chi} \end{pmatrix} = \begin{pmatrix} 0 & \alpha \\ 0 & 0 \end{pmatrix}\begin{pmatrix} x_0 \\ \hat{\chi} \end{pmatrix} \quad (8.12)$$

in the absence of any perturbations. The dependence between the position and speed modes has yielded the double-zero Jacobian of the projected linear dynamics. Without any perturbation, the solitary pulse remains stationary and x_0 and $\hat{\chi}$ are zero exactly. The pertinent dynamics observed in the previous section is then trigged when perturbations are imposed.

8.3 Quasi-jump decay dynamics

We focus first on the decay dynamics corresponding to ψ_2. For a given δ, the solitary pulse on a unit substrate is taken to be the reference solitary pulse. The change in substrate thickness $(\chi - 1)$ behind the excited pulse, shown in Figure 8.6 and seen as a backward propagating shock in Figure 8.2 and 8.4, then represents a small and negligible step change in the substrate film but a significant mass sink. As the shock propagates backwards relative to the excited pulse, it coast the thinner substrate. The liquid mass required for this coating comes from the excited pulse. The coating rate is not constant since the capillary, inertial and vicius forces conspire in a quasi-steady manner such that the substrate thickness immediately behind the excited pulse decreases steadily. This explains why the back shock seen in Figure 8.4 increases in amplitude away from the pulse: it is a reflection of the ever-decreasing substrate thickness as liquid mass is drained from the pulse. The rate of this mass loss can be easily determined from the averaged equations (3.88)-(3.89), it is clear that the flow rate in the laboratory frame for a flat film of thikness χ is $q = \chi^3$. In a frame moving with the reference solitary speed, the flow rate becomes $q = \chi^3 - c^*\chi$. Consequently, the change in flow rate when a substrate exhibits a hydraulic jump from thickness χ to thickness unity is $\Delta q = \chi^3 - c^*\chi - (1-c^*) \sim (3-c^*)\hat{\chi}$ where $\hat{\chi} = \chi - 1 \ll 1$ is the "degree of excitation" of the back pulse measured by the deviation substrate thickness. Since c^* lager than 3, a net drainage from the pulse occurs. This sink can be written as $(3-c)\hat{\chi}\delta(x^*)$ where x^* is the location of the jump in Figure 8.6 and $\delta(x)$ is a Dirac delta function. The exact location of x^* is not important as long as it is not within the pulse. This perturbation to a single solitary pulse can then be descibed by the following perturbed and linearized version of the averaged equation in the moving frame.

$$\frac{d}{dt}\begin{pmatrix} \hat{q} \\ \hat{h} \end{pmatrix} = L\begin{pmatrix} \hat{q} \\ \hat{h} \end{pmatrix} + \begin{pmatrix} 0 \\ (3-c)\hat{\chi}\delta(x^*) \end{pmatrix}. \quad (8.13)$$

All other terms that arise due to the hydraulic jump are negligible compared to this points sink. Expanding $(\hat{q}, \hat{h})$ by ψ_1 and ψ_2 as in (8.11), substituting into (8.13) and taking inner product $(\cdot, \cdot)$ with φ_1 and φ_2, one obtains the perturbed version of (8.14),

$$\frac{d}{dt}\begin{pmatrix} x_0 \\ \hat{\chi} \end{pmatrix} = \begin{pmatrix} 0 & \alpha \\ 0 & 0 \end{pmatrix}\begin{pmatrix} x_0 \\ \hat{\chi} \end{pmatrix} - \begin{pmatrix} 0 & 0 \\ 0 & \gamma \end{pmatrix}\begin{pmatrix} x_0 \\ \hat{\chi} \end{pmatrix} \qquad (8.14)$$

where $\gamma = \varphi_2^2(x^*)(c^* - 3) = d(c^* - 3)$ where d is the second $\hat{h}$ component of φ_2 and

$$\gamma = \frac{3(c^* - 3)}{4A + 11\delta(dA/d\delta)}. \qquad (8.15)$$

The values of the decay factor γ are tabulated in Table 8.1. It also approaches a constant value for δ larger than 0.05.

In deriving (8.14), we have neglect the nonlinear terms in favour of the linear substrate perturbations term due to the hydraulic jump. In essemce, we are ignoring the nonlinear interaction between the solitary pulse and the hydraulic jump. This is obviously valid when the jump small such that $\gamma\hat{\chi}$ is lager than $O(|\hat{\chi}|^2)$ and $O(x_0|\hat{\chi}|)$ terms. This is generally true for the degrees of excitation involved but the omitted nonlinear terms are responsible for the larger errors and fluctuations seen in Figure 8.3 when the degree of excitation is relatively large. We have also neglect the small jump in front of the excited pulse. It is more precise to model the decay of the excited pulse as the interaction of two shocks, with the front one corresponding to a shock travelling wave (a heteroclinic orbit) like the one constructed by Pumir et al. (1983). Such an endeavour requires far more effort, however, and we have approximated the front shock by a solitary wave (a homoclinic orbit). This is consistent with the small-$\hat{\chi}$ approximation.

One of the zero eigenvalues of the solitary pulse has hence been shifted by the presence of the jump in the substrate thickness which breaks the corresponding symmetry. This jump then drains fluid from the pulse which causes the pulse to evolve quasi-steadily into another solitary pulse with a thinner substrate. This is the observed slow decay dynamics of the excited pulse. The translation symmetry is also broken but the shift in the position zero mode is negligible for a small jump. Nevertheless, the slow decay induced by the shifted speed zero mode will trigger a shift in the position due to their linear dependence. The magnitude of γ is small as seen from Table 8.1. The drainage ceases when the jump vanished at $\hat{\chi} = 0$ but the drainage rate is related to how the area under the pulse changes as it moves through the family in Figure 8.1. This change in area is captured by φ_2^2. The above slow decay requires that δ be in excess of 0.04 such that the omitted mode in Figure 8.8 is larger in magnitude than γ. To justify the omitted nonlinear terms, we should also require the excitation $\hat{\chi}$ to be smaller than γ, which is easily satisfied for δ in excess of 0.04.

If these stipulation hold, (8.14) indicates that the slow decaying dynamics is linear such that the thinning of the raised substrate χ in Figure 8.6 is governed by $\hat{\chi} \sim \exp{-\gamma t}$ where $\hat{\chi} = \chi - 1$. Since c^* and h_m can be expressed as a function of χ only during the quasi-steady decay, we have

$$\hat{c} = c^*(\chi) - c^*(1) \sim \frac{dc^*}{d\chi}(1)\hat{\chi} = \alpha\hat{\chi},$$

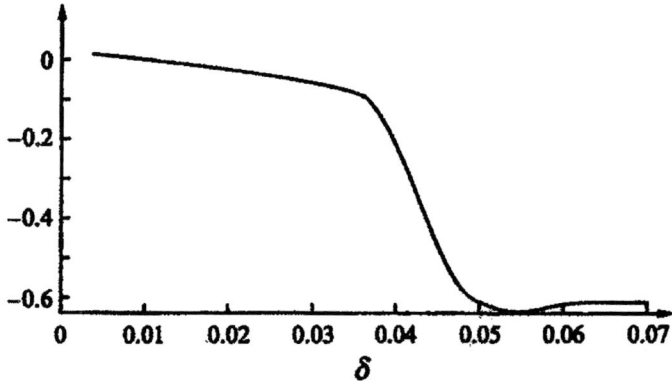

Figure 8.8: The leading non-zero discrete λ_2 eigenvalue of the solitary pulse family.

$$\hat{h} = h_m(\chi) - h_m(1) \sim \frac{dh_m}{d\chi}(1)\hat{\chi}$$

and

$$\hat{c} = \frac{dc^*}{dh_m}(1)\hat{h}.$$

As mentioned in the previous section, we find from the constructed solitary pulse family that $c^* - 3 \sim 2.4(h_m - 1)$ and hence the derivative dc^*/dh_m is approximately 2.4. To demonstrate this relationship more explicitly and to further verify that the decay dynamics evolves through the family of solitary pulses parametrized by δ, we show in Figure 8.9 the evolution in amplitude and speed of the excited back pulse and the new pulse created by coalescence. The speeds and amplitudes of both decay exponentially in time but when cross-plotted, they lie on the same straight line close to the estimated correlation. We have hence recovered the linear correlation between the excess speed and amplitude of a draining pulse, first predicted in (7.84) for the gKS pulse with a resonance pole theory. At the end of this section we shall also compare the linear speed - amplitude correlation with a resonance theory for Shkadov model.

We note here that, since the derivatives of c^* and h_m with respect to δ or χ are non-zero, the second equation in (8.14) also implies that the speed, substrate thickness and amplitude all decay exponentially in time by the same rate:

$$\hat{c}, \hat{\chi}, \hat{h} \sim \exp(-\gamma t).$$

In Figure 8.10, this unique decay dynamics is again verified by extracting the dynamics of the excited pulse in Figure 8.4 for $\delta = 0.05$ The decay exponent of $\gamma = 0.047$ provides a very accurate description of the decay of both $\hat{c}$ and $\hat{h}$.

Another prediction of this theory for the decay of an excited pulse is that $\hat{h}$ decays linearly in space. Integrating the position x_0 equation in (8.14) that is

260

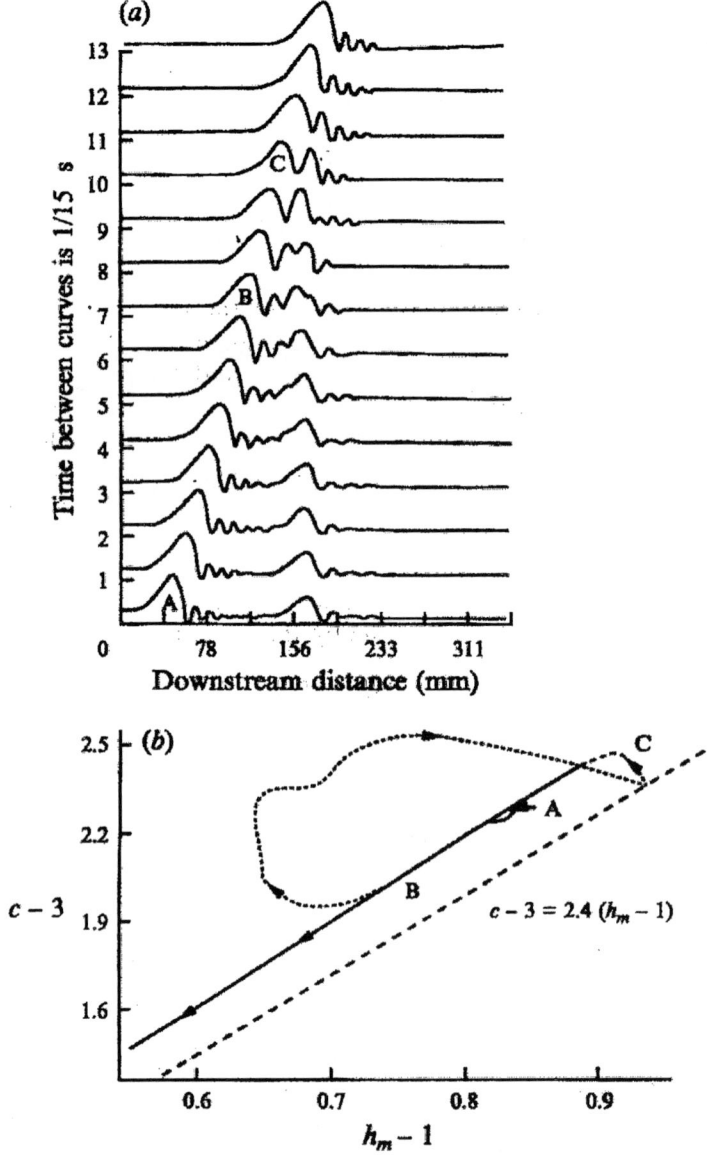

Figure 8.9: Decay dynamic of the excited back pulse and the new pulse generated by coalescence in real and dimensionless unit at $\delta = 0.09$. The labels A, B and C correspond to the indicated pulse of the 'snapshots' shown. Coalescence occurs between B and C. The amplitude h_m and speed c are cross-plotted in (b) with the dotted line representing the short interval during coalescence. Both excited pulses decay by the same decay rate in time and when cross-plotted, lie on the same $c - h_m$ straight line which is very close to the $c - 3 = 2.4(h_m - 1)$ dashed line. This verifies the quasi-steady decay through the pulse family. Note also the close encounter with two-hump transition state at $9/15s$ which is represented by * in (b).

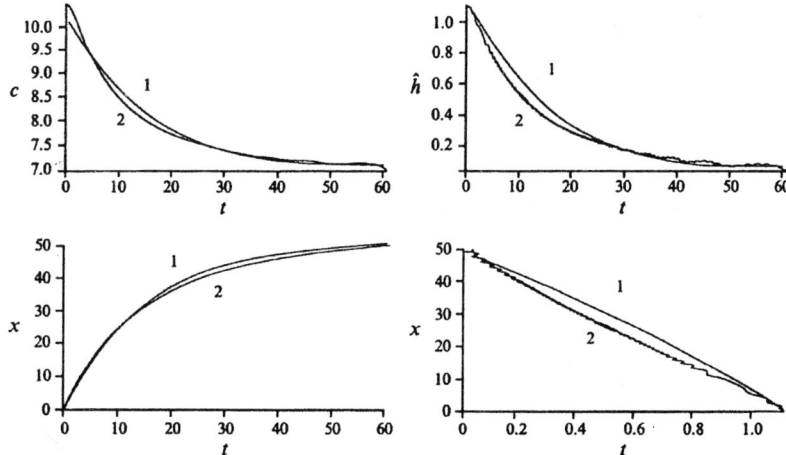

Figure 8.10: Decay dynamics of the excited pulse in Figure 8.2 and 8.4 are shown as curves labelled 2. The theoretical curves 1 from the linear decay theory correspond to a decay rate of $\gamma = 0.047$.

coupled to the $\hat{\chi}$ speed mode, one obtains

$$x_0 \sim \frac{\hat{c}(0)}{\gamma}(1 - e^{-\gamma t}) = x_\infty(1 - e^{-\gamma t}) \tag{8.16}$$

which is also in agreement with the experimentally measured position of the excited pulse as it decays, as seen in Figure 8.10. Since γ is small, the excited pulse is shifted significantly by an amount x_∞ from its original position in the moving frame. Combining the above equations, one obtains the linear amplitude decay in space:

$$\hat{h} \sim \hat{h}(0) - \gamma \frac{x_0 \hat{h}_0}{\hat{c}(0)} = \hat{h}(0)\left(1 - \frac{x_0}{x_\infty}\right). \tag{8.17}$$

In Figure 8.4, the initial excitation, as measured by $\hat{c}(0)$, is approximately 3.0 which yields a predicted shift of $x_\infty = 60$. This is in agreement with the measured x_∞ in Figure 8.10. This predicted x_∞ also suggests that the excited amplitude decays linearly in space at $\delta = 0.05$ by a slope of 0.017 which is in good agreement with the simulations shown in Figure 8.4 and 8.10, even when the excited pulse has come very close to the stationary pulse in front of it. The measured linear decay in Figure 8.4 has a slope of approximately 0.02. This implies that any excited pulse which is roughly twice the amplitude of the initial separation is more than three times the pulse width. This is true for all δ values and hence, as long as the initial separation is more than three times the pulse width, the final interaction is between two identical pulses on the same substrate. Conversely, if the initial separation is less than two or three pulse widths and the back pulse is excited by the same amount, coalescence will occur but it will not be descibed by the present theory.

In particular, the repulsion between two identical solitary pulses will most likely be masked by the dominant decaying dynamics of the back pulse during coalescence if the separation is not sufficiently far.

8.4 Essential and resonance pole spectra of the pulses

We shall examine the pulse decay with the rigorous resonance pole theory of Chapter 7. Although it is exact, it is strictly valid for small perturbations and its results will be compared to the empirical but presumably larger-amplitude theory of the last section. We shall also obtain the essential spectrum of these realistic pulses to examine their effectiveness in suppressing small amplitude wave packets (radiation modes) on the substrate. In particular, we shall determine the maximum separation between pulses, but now for realistic pulses of the Shkadov model. The formulation is identical to that for the gKS equation in the previous chapter and we shall describe only the final results.

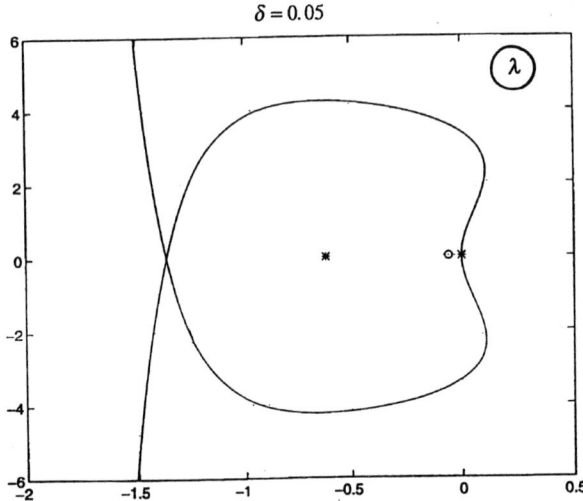

Figure 8.11: Discrete (stars) and essential (solid line) spectra of a positive pulse at $\delta = 0.05$. The open circle corresponds to a resonance pole.

Any small perturbation can be presented in the form of a superposition of discrete and essential modes

$$H(x,t) = \sum_{k=1}^{N} A_k \psi_k(x) e^{\lambda_k t} + \int_{\Gamma} A(\lambda) \psi(x,\lambda) e^{\lambda t} d\lambda \qquad (8.18)$$

where $\psi_k(x)$ and λ_k are eigenfunctions and eigenvalues of the discrete spectrum of the operator in (8.1), $k = 1, 2, \ldots, N$. The contour integral corresoponds to the essential part of spectrum and so the path of integration Γ is determined by the dispersion relation in the frame moving with pulse velocity speed,

$$\alpha^4 + \delta\left[(5\lambda - i\alpha c)^2 + 12i(\lambda - i\alpha c)\alpha - 6\alpha^2\right] + 3i\alpha + (\lambda - i\alpha c) = 0 \quad (8.19)$$

Another discrete eigenvalue λ_2 is found to be real and stable for all δ, $\lambda_2 < 0$. The dependence $\lambda_2(\delta)$ is given in Figure 8.8. In fact, $\lambda_1 = 0$ and $\lambda_2 < 0$ are the only discrete eigenvalues at all δ. Profiles of the null and stable eigenfunctions are shown in Figure 8.7. Typical spectrum for a positive pulse is shown in Figure 8.11 where the continuous curve corresponds to the essential part of the spectrum, stars to two discrete eigenvalues and the empty circle we shall consider later.

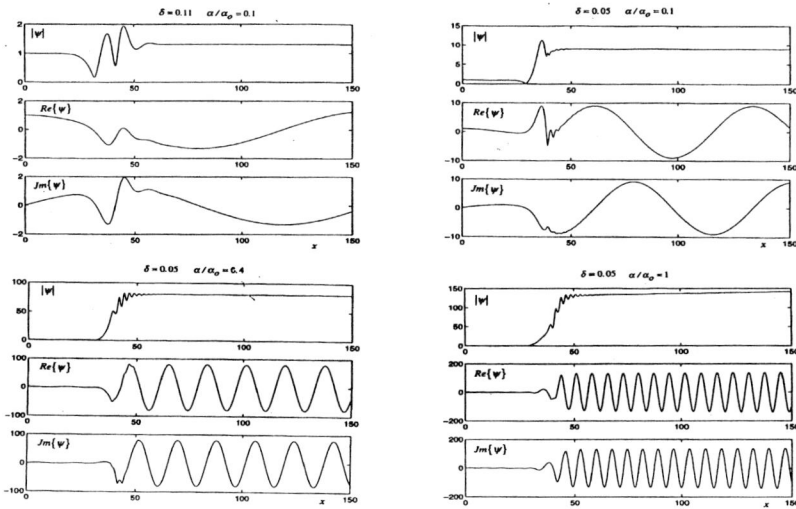

Figure 8.12: Real and imaginary parts of essential eigenfunctions and their absolute value at different wavenumbers and δ..

The essential eigenfunctions are expressed in the usual form

$$\psi(x, \alpha) = K(x, \alpha) e^{i\alpha x} \quad (8.20)$$

for the positive pulse are shown in the Figure 8.12. The transmission coefficient r defined by

$$\frac{|K(-\infty)|}{|K(+\infty)|} = \frac{1}{r} \quad (8.21)$$

as a function of wavenunber α and δ is shown in Figures 8.13 At $\alpha = 0$ and at $\alpha \to \infty$, $r = 1$ and it has a maximum near the neutral wavenumber $\alpha/\alpha_0 = 1$.

With increasing δ, r grows rapidly until it reaches extremely large values on the order of hundreds.

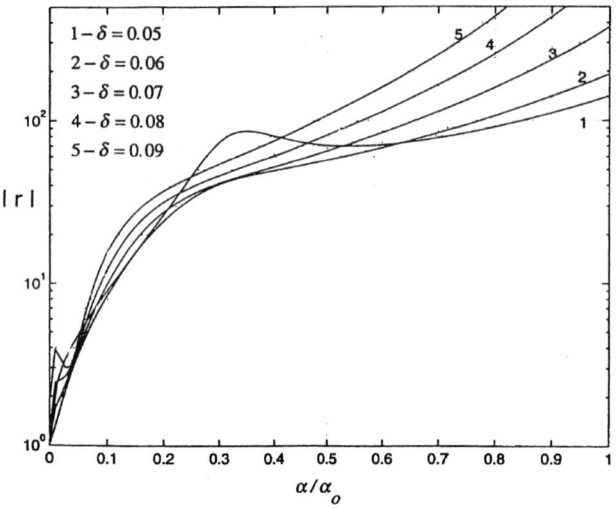

Figure 8.13: Translation coefficient r.

Such large transmission coefficients imply that, more than the gKS pulses, wave packets can be drastically attenuated when they pass through realistic pulses on a falling film - a common experimental observation best reported by Liu and Gollub (1994). In our simulations in Figure 8.14, wave packages placed before a pulse would grow originally according to linear stability theory. We can even spot a new pulse nearly created at $t = 24$ as seen in Figure 8.14. However, both the new pulse and wave packet are completely annihilated by the pulse when they meet at $t = 33$. If our packet is sufficiently long, the new pulse can escape this annihilation. If we have a lattice of pulses with some separation distance L between them, for sufficiently large L the flow will be chaotic, and for sufficiently small – regular, see example of calculation for the Kawahara equation, Figure 7.24.

Critical or neutral distance L as a function of δ is nearly constant for $\delta > 0.05$ and is about 170. It is intructive to present it for water, $\gamma = 2850$, in dimensional form, as a function of inlet Reynolds number. This distance approaches 18 cm for $R > 5$. This critical L is connected with the stability windows of the γ_2 second family of periodic waves in Figure 6.8.

Let us evaluate the influence of the essential spectrum to pulse evolution by analyzing the asymptotic behavior of the integral part of the perturbation in (8.18). Stability of a pulse with respect to the essential spectrum is more sophisticated than to the discrete one. Usually, perturbations are localized and so are a superposition of an infinite number of essential modes. We first evaluate the integral to examine whether it is growing or decaying in time. It is also convevient to replace the $\lambda-$ dependence of the contour Γ in the complex plane by α dependence for real α, $\alpha \in (-\infty, +\infty)$,

$$J = \lim_{t\to\infty} \int_{-\infty}^{+\infty} A(\alpha)K(\alpha,x)e^{i\alpha x+\lambda t}d\alpha = \lim_{t\to\infty} \int_{-\infty}^{+\infty} A(\alpha)K(\alpha,x)e^{(i\alpha\frac{x}{t}+\lambda)t}d\alpha \qquad (8.22)$$

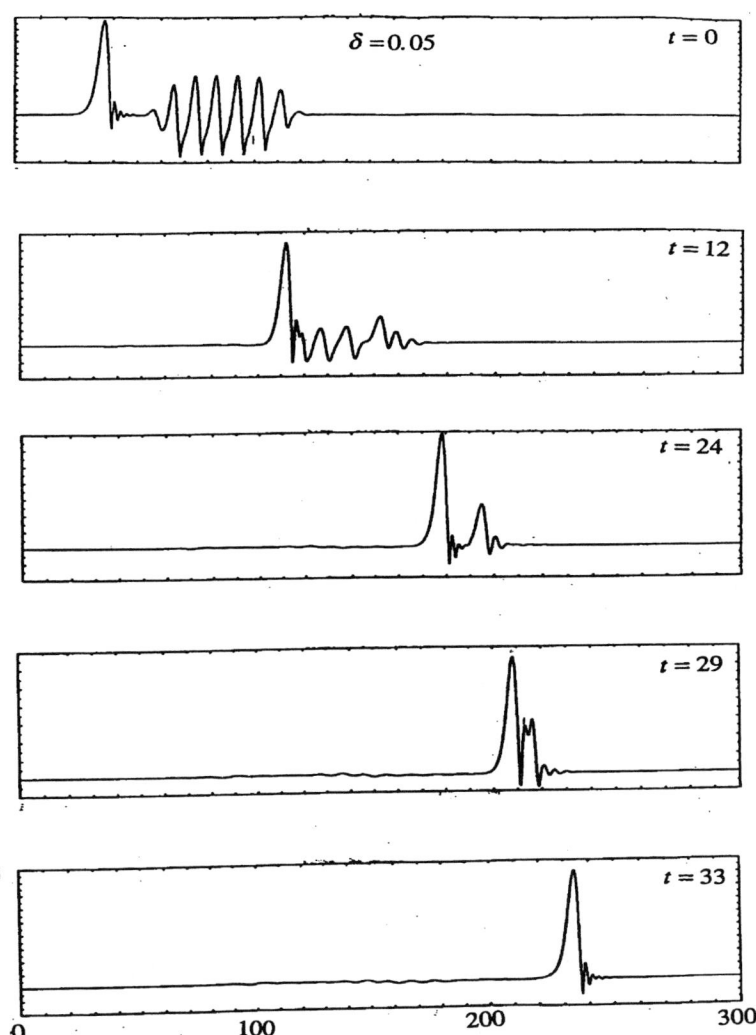

Figure 8.14: Suppression by a pulse of a large amplitude wave package.

$\lambda(\alpha)$ here is the dispersive relation in the frame moving with pulse speed frame. The large time behavior is determined by its exponential part, namely, by its dispersion relationship and can be described in terms of usual convective and absolute stability theories of a trivial solution in a moving frame. The stability can be evaluated by the steepest descent method and depends on the position of a saddle point of the dispersion relation.

$$\frac{\partial}{\partial \alpha}\left[i\alpha\frac{x}{t} + \lambda(\alpha)\right] = 0 \tag{8.23}$$

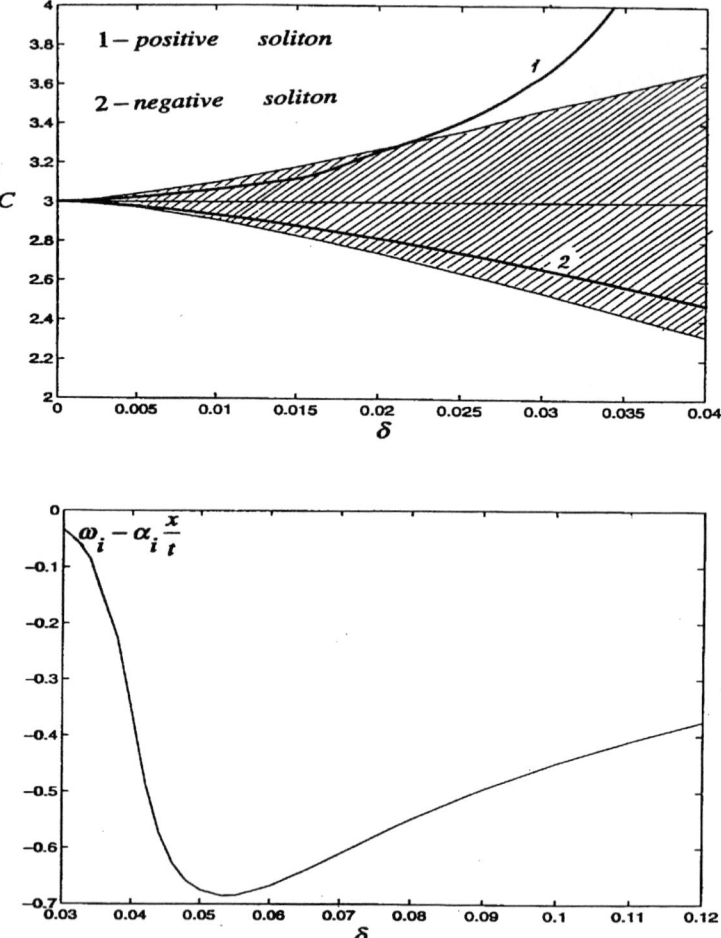

Figure 8.15: a) Interval of instability is shaded; pulse is unstable at $\delta < \delta_* \simeq 0.022$ b) growthrate taken at x/t is equal to pulse velocity.

In other words, we have to lift the integration path in the complex α-plane so that α is a complex number, $\alpha \in (-\infty + ib, +\infty + ib)$ and so that the new path crosses the saddle point α_*. Another issue which might be important is the possibility of singularities for the function $K(\alpha, x)$. We shall consider this interesting possibility later on.

If α_* as a function of x/t is known, $\lambda(\alpha_*)$ gives us the growth rate. If x/t is equal to the pulse velocity c, $\gamma = Real\lambda > 0$ means pulse instability with respect to the essential spectrum. Another way to obtain pulse instability is to obtain $(x/t)_-$ and $(x/t)_+$, so that if

$$\left(\frac{x}{t}\right)_- < \frac{x}{t} < \left(\frac{x}{t}\right)_+$$

$\gamma > 0$ and the pulse is in the instability region. In Figure 8.15a), the interval of instability is shaded and the positive pulse is unstable if $\delta > \delta_* \simeq 0.022$ while the negative pulse is always unstable. The last fact is interesting because negative pulses have never been observed in experiments. In Figure 8.15b), the growthrate γ is presented for $x/t = c$ for a positive pulse, which is negative for $\delta > \delta_* \simeq 0.022$.

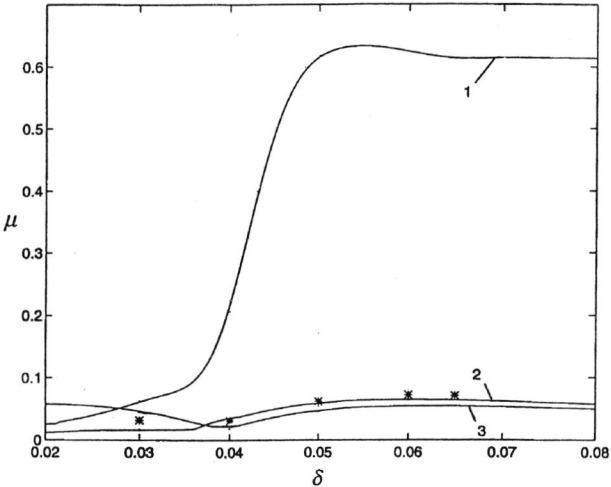

Figure 8.16: Dependence of growthrate on δ for 1 – eigenvalue λ_2, 2 – simlified semi-empirical theory of (8.15), 3 – resonance pole $\lambda_\#$. Stars corresponds to our numerical simulations.

Let us return to the singulariries of the function $K(x, \alpha)$ in the comlex α-plane. If we are moving our contour of integration in the comlex plane, the integral does not depend on the contour deformation until it crosses a singular point, a pole in the complex plane. As this happens, the residual theorem stipulates that we have to add some constant, a residue, to our integral. Taking into consideration dependence on the spatial variable x, this constant becomes a function of x, and our expansion (8.18) will now have an additional discrete eigenfunction, so called resonance pole, see Reed and Simon (1978). This eigenfunction has a very specific profile, its tail is growing on the infinity. It is the resonance pole we have seen in Chapter 7. Using an exponentially weighted space, we can convert this eigenfunction into one that decays at the infinities. The procedure of contour shifting is equivalent to increasing a in the weighted space norm, $a = Im(\alpha)$,

$$(f, g)_a = \int_{-\infty}^{+\infty} e^{-ax} fg \, dx \tag{8.24}$$

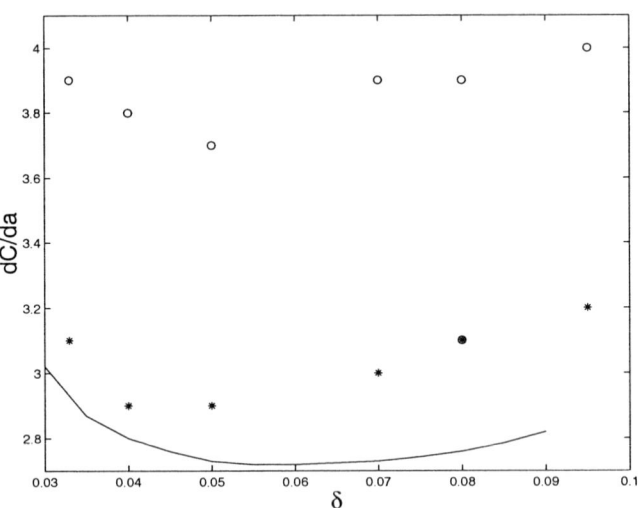

Figure 8.17: Coefficient of linear dependence speed of an excited pulse on its amplitude. Solid line - resonance pole theory, stars - numerical simutations for small amplitude of excitation and empty circles - numerical simulation for large amplitude of excitation

It is possible to show that for the discrete eigenfunction ψ_k

$$\int_{-\infty}^{+\infty} \psi_k \, dx = 0 \tag{8.25}$$

and so the dynamics of the pulse, connected with mass exchange must be connected with the resonance pole and essential spectrum.

In Figure 8.16, we present dependence of λ_2, resonance pole $\lambda_\#$ and semi-empirical growth rate γ. Stars represent decay rate from numerical simulations.

We can also apply the resonance pole theory of section 7.6 to the Shkadov model. Exploiting the fact that resonance pole function is close to the null eigenfunction h' everywhere except for negative x, where it has a growing tail, and using the same calculation in (3.88)-(3.89) for the gKS equation, we shall show that the amplitude and speed of an excited pulse are linearly correlated,

$$\hat{c} = \frac{dc}{da}\hat{a}$$

The coefficient dc/da can be obtained using the procedure for the gKS equation and is shown in Figure 8.17 as a solid line. We can see that this coefficient is close to the one obtained from the previous theory. For example, $dc/da = 0.27$ for $\delta = 0.05$ while the previous theory yields 0.24 in Figure 8.9.

Numerical simulation of drainage dynamics indicates that there is a deviation from the resonance pole theory for large amplitude perturbations. For small amplitude of excitation, $\hat{a}$ between 0.25 to 0.5 of the original pulse amplitude a, the numerical dependence is slightly above the theoretical prediction; see also Figure 7.19 for the gKS equation. These data are shown by stars in Figure 8.17. The coefficient of proportionality is nearly the same for all δ and is about 3.

We also carried out numerical experiments with a large amplitude of excitation with $\hat{a}$ about 2 to 3 times of a, see empty circles in Figure 8.17. The proportionality coefficient now increases from 3 to 4.

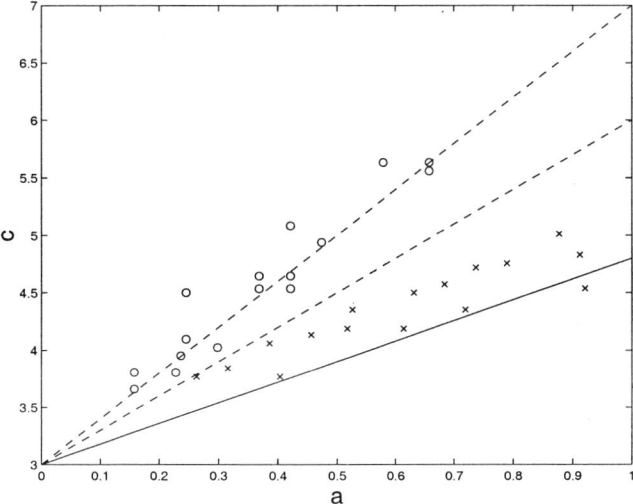

Figure 8.18: Speed of excited localized perturbations. Empty circles (triangle waves) and crosses (localized perturbations) are taken from experiments by Alekseenko et al.(1994). Solid line represents formula (5.86) for stationary pulses; lower dashed line is taken from resonance pole theory and upper dashed line is taken from our numerical experiments

Let us now compare our theoretical results with the ones from experiments of Alekseenko et al. (1994). In the experiments two kinds of solitary patterns were used, excited triangle waves with relatively large amplitudes and localized perturbations close to saturated pulses, see Figure 9.2 of their book. We present these data in Figure 8.18 as empty circles and crosses correspondingly. Solid line represents the dependence (5.86) from the earlier theory with the proportionality coefficient 1.8. The lower dashed line is taken from the resonance pole theory with a coefficient 3. The upper dashed line is taken from our numerical experiments with a proportionality coefficient 4. We can see that the experimental results are in a rather good agreement with our theory. Indeed, for large triangle waves the experimental dependence is close to the upper dashed line, while for small localized perturbations the experimental data lie between the resonance pole dependence and the dependence for stationary travelling pulses.

CHAPTER 9

Pulse Interaction Theory

The drainage dynamics of the large excited pulse determines its speed during the drainage. As it approaches its front neighbor prior ro coalescence, however, the drainage dynamics are not the only perinent dynamics. The interaction between the two pulses begin to repel or attract the pulses, resisting or accelerating the coalescence in the process. This interaction dynamics can be captured with the translational zero mode, as we have done in Chapter 6 to determine the stability of periodic waves. Although we shall mostly focus between the interaction between two pulses, such binary pulse interaction theory can be easily extended to examine a train of waves, as we did in Chapter 6. In this manner, irregularly spaced periodic wave trains that propagate in unison at constant speed, a pattern with spatial chaos but temporal order like "frozen turbulence", can be constructed using Melnikov and Shilnikov bifurcation analsyses of homoclinic orbits, see Gorshkov, Ostrovsky and Papko (1976), Collet and Elphick (1987), Eliphick et al. (1988,1990), Balmforth et al. (1995), Hecke, Wit and Saarloos (1995) and Balmforth (1997). Such construction is obviously related to the Shilnikov mechanism of generating multi-hump pulses in Chapter 5. Here, however, we shall construct the Poincare map in a different manner. It hence represents another example of how dynamical systems tools can be used to understand the complex spatio-temporal wave dynamics on a falling film.

9.1 Coherent Structure theory due to translational zero mode

We see from the numerical simulations of Chapter 8 that the drainage dynamics of the back excited pulse will eventually be affected by the front equilibrium pulse, when the separation between the two becomes sufficiently small. This binary interaction is weak and often has negligible contribution to the coalescence dynamics, compared to the drainage dynamics. Nevertheless, the steady states of such interaction dynamics give rise to the multi-hump pulses of Chapter 5 and Section 6.2 and hence understanding binary interaction can shed light on the pulse solution braches. Moreover, in other coherent structure dynamics, where excited structures do not exist and the dominant dynamics involve interaction between equilibrium structures, such interaction dynamics are the dominant dynamics, Kawahara and Toh (1988), Kawasaki (1982). This interaction theory can be constructed from the spectral theory of Chapter 7.

We shall demonstrate the concept by studing the interaction between two equilibrium pulses of the KS equation. We assume they are sufficiently separated by a distance L such that the interfacial height $h(x,t)$ can be represented by

$$h = h_0(x) + h_1(x+L) + H(x,t)$$

h_0 and h_1 are equilibrium pulse solutions in the frame x moving with their speed and H is a perturbation or correction function to account for the error introduced by the first two terms without including interaction. The overlap $h_0 h_1$ will be small for large separation distance L. If this is true, g below is also small and we can use perturbation theory. Substituting h into the KS-equation and neglecting the nonlinear term H^2, we get the equation

$$\frac{\partial H}{\partial t} + LH = -2\frac{\partial}{\partial x}(h_0 h_1) = g(x,t)$$

The interaction function $g(x,t)$ is called the Melnikov function.

For 3-pulse interaction the Melnikov function will be

$$g(x,t) = -2\frac{\partial}{\partial x} h_0 (h_{-1} + h_{+1})$$

It is easy to construct the function for n-pulse interaction or for infinite lattice of them or for any other equation. We assume that only neighboring pulses interact, for example, and neglect interaction between pulse -1 and +1, because the tails decay fast, usually exponentially.

However, the equation

$$\frac{\partial H}{\partial t} + LH = 0$$

is a linearized equation near a pulse solution and g hence represents a perturbation to the spectrum of an isolated pulse. We hence expand the forcing-term g by both the essential and discrete eigenfunction of Chapter 7.

For the inhomogeneous equation, with right hand side g, the solution is a combination of general solution of the homogeneous equation (our old sum and integral for a spectral expension) and a particular solution of the nonhomogeneous equation,

$$H(x,t) = A_1 h'(x) + A_2 \psi_2(x) e^{\lambda_2 t} + \int_\Gamma A(\lambda)\psi(\lambda,x) e^{\lambda t} d\lambda$$

$$+ B_1(t) h'(x) + B_2(t)\psi_2(x) + \int_\Gamma B(\lambda,t)\psi(\lambda,x) d\lambda$$

where A_k and $A(\lambda)$ can be calculated from the initial condition, by projecting it with the adjoint eigenfunctions (for every pulse). For binary interaction one has two such expensions

$$A_k = (f(x), \varphi_k(x)), k = 1, 2$$

$$A(\lambda) = (f(x), \varphi(\lambda, x))$$

This part of solution is the usual reaction to initial conditions $f(x)$. If the initial conditions are zero, or the system forgets its initial conditions (which is usually true for an active medium), then $A = 0$.

Equations for $B_1(t), B_2(t)$ and $B(\lambda, t)$ can be found by projecting the equation with the adjoint eigenfunctions(for both pulses):

$$\frac{dB_k}{dt} - \lambda_k B_k = (g(x,t), \varphi_k(x)), \quad B_k|_{t=0} = 0$$

$$\frac{dB(\lambda, t)}{dt} - \lambda B(\lambda, t) = (g(x,t), \varphi(\lambda, x)) \quad , B|_{t=0} = 0$$

The pulse interaction hence triggers both the discrete and continuous part of the spectrum. The question about influence of the continuous part of spectrum is a difficult problem. Indeed, we have a continuous number of equations to solve, and then one has to sum all $B(\lambda)$ for all moments of time t.

However, sometimes long time behavior can be evaluated. We shall ignore the excitation of continuous spectrum due to binary interaction here. In effect, we assume al pulses are convectively stable with a resonance pole far to the left of the imaginary axis.

It is much easier to consider the discret spectrum. We shall assume there are unstable discrete pulses — otherwise the pulses would disintegrate as shown in Chapter 7. Hence, the dominant discreta model have zero eigenvalues and the main effect will be connected with their null eigenfunctions. All null-eigenfunctions are connected with some invariance of the governing equations. For KS-equation we have a simple zero eigenvalue imbedded in the essantial spectrum; for KdV-equation - double zero eigenvalue. For KS-equation the null eigenfunction is connected with the invariance under spatial translation, if $h(x)$ is a solution, $h(x + const)$ is also a solution. Hence, the amplitude B_1 for the null eigenfunction is just the location of the pulse maximum.

After trivial calculations for the Melnikov functions, taking into account the functions decay exponentially at the infinities and keeping the main terms of the expansion, we obtain an ODE each for the rear pulse and for the front one. It is interesting that the influences of the pulses to each other for the KS-equation are not reversible. For binary pulse interaction, the back pulse position depends monotonically on the separation distance L,

$$\frac{dx_0}{dt} = Re^{-2mL}$$

and it accelerates or decelerates depending upon the sign of R. The front pulse has oscillatory behavior with respect to the distance L.

$$\frac{dx_{+1}}{dt} = Se^{-mL}\cos(\beta L + \theta)$$

Subtracting the first equation from the second one, we get the equation describing the pulse separation containing only three parameters, R, S and θ listed in Table ,

$$\frac{dL}{dt} = Re^{-2mL} - Se^{-mL}\cos(\beta L + \theta)$$

For a stationary multi-hump pulse, the pulse velocities are equal to each other

$$\frac{dx_0}{dt} = \frac{dx_{+1}}{dt}$$

we obtain the condition for stationary points $dL/dt = 0$ which corresponds to two-hump pulses

$$f(L) = Re^{-2mL} - Se^{-mL}\cos(\beta L + \theta) = 0$$

The equation has a countable set of roots, that are alternately stable and unstable. During nonstationary evolution the distance between two pulses tends to the nearest stable root. The solitary pulse with a minimal possible distance between the humps is unstable, after that there is an alternation between stable and unstable two-hump solitary pulses. This counteble infinite-number of 2-hump pulses is direct result of the Shil'nikov theorem of Chapter 5.

9.2 Repulsive pulse interaction of the gKS pulses

The stationary 2-hump pulses of the previous section often destabilizes when an additional pulse is placed near it. In fact, since a pulse in an arbitrary pulse train interact with its immediate neigbors, the pulse in front and the pulse in the back, we shall extend the theory from the 2-hump stationary pulse of previous section to a general dynamic interaction theory for multiple pulses with nearest-neigbor interaction.

Since the KdV pulses are repulsive and since the gKS equation reduces to the KdV equation as δ approaches infinity, we also expect the gKS pulses to be repulsive. However, this is a singular limit as the KdV pulse has two zero modes and its binary interaction dynamics is inertia dominated with a second order dynamics equation. The gKS pulse, however, has only one zero mode and its binary interaction is without inertia. We hence expect the gKS pulse interaction dynamics to be much less chaotic and more amenable to analysis. In fact, in the limit of large δ, the gKS pulses and their spectra can be estimated analytically. As such, its pulse binary interaction dynamics can be completely discerned analytivally! This is a welcomed result as the inverse scattering technique and other exact analytical transform theories for integrable systems fail for the gKS equation and other active/dissipative systems.

Although one does not need excessive dispersion to observe repulsive pulses, it is instructive to examine the solitary pulses in the limit of infinite δ. In this limit, the solitary pulse approaches the sech solution of the KdV equation and one can actually estimate the speed $c(\delta)$ of the large-δ solitary pulses by carrying out an expansion about the KdV solution. We do this here by Center Manifold projection and normal from techniques. We first transform the gKS equation of (5.1) in the moving frame $y = x - c\tau$ to a form suitable for expansion at infinite δ by invoking

$$H \to \frac{\delta^3}{2} u \quad y \to \frac{z}{\delta}$$

to yield

$$-\mu_1 u + \frac{u^2}{2} + \mu_2 u_z + u_{zz} + u_{zzz} = 0 \quad (9.1)$$

where

$$\mu_1 = c\delta^{-3} \quad \text{and} \quad \mu_2 = \delta^{-2} \quad (9.2)$$

are small parameters. Writing (9.1) as a dynamical system, we obtain

$$u_z = \begin{pmatrix} 0 & 1 & 0 \\ 0 & 0 & 1 \\ \mu_1 & -\mu_2 & -1 \end{pmatrix} u + \begin{pmatrix} 0 \\ 0 \\ -u_1^2/2 \end{pmatrix} \quad (9.3)$$

where $u = (u_1, u_2, u_3) = (u, u_z, u_{zz})$. With a similarity transform $u = Tv$ where

$$T = \begin{pmatrix} 1 & 0 & 1 \\ 0 & 1 & -1 \\ 0 & 0 & 1 \end{pmatrix}$$

is the diagonalizing matrix for the Jacobian when $\mu_1 = \mu_2 = 0$, (9.3) becomes

$$u_z = \begin{pmatrix} 0 & 1 & 0 \\ 0 & 0 & 0 \\ 0 & 0 & -1 \end{pmatrix} v + \begin{pmatrix} \mu_1 & -\mu_2 & 0 \\ -\mu_1 & -\mu_2 & 0 \\ \mu_1 & -\mu_2 & 0 \end{pmatrix} v + \frac{(v_1 + v_3)^2}{2} \begin{pmatrix} 1 \\ -1 \\ -1 \end{pmatrix} \quad (9.4)$$

Little can be said about the homoclinic trajectories of (9.4). However, in the limit of small μ_1 and μ_2, the Jacobian yield a double-zero eigenvalue of unit geometric multiplicity. The normal from theory then stipulates that there are smooth near-identity transformations that can transform (9.4) to the Bogdanov-Arnold normal form

$$v_z = \begin{pmatrix} 0 & 1 \\ 0 & 0 \end{pmatrix} v + \begin{pmatrix} 0 & 0 \\ \nu_1 & \nu_2 \end{pmatrix} v + \begin{pmatrix} 0 \\ av_1^2 + bv_1 v_2 \end{pmatrix} \quad (9.5)$$

The existence of homoclinic orbit to this system has been analyzed by Carr who showed that a homoclinic orbit exists when

$$\nu_2 = \frac{\nu_1 b}{7a} \quad (9.6)$$

We shall carry out the necessary transformations to convert (9.4) to (9.5) and express (9.6) in terms of the known quantities μ_1 and μ_2.

We begin by involving the "tangent space" approximation by omitting the stable v_3 mode with a -1 eigenvalue,

$$v_z = \begin{pmatrix} 0 & 1 \\ 0 & 0 \end{pmatrix} v + \begin{pmatrix} \mu_1 & -\mu_2 \\ -\mu_1 & -\mu_2 \end{pmatrix} v + \begin{pmatrix} v_1^2/2 \\ -v_1^2/2 \end{pmatrix} \quad (9.7)$$

We then remove the first component of the nonlinear portion by a near-identity nonlinear transformation

$$v \to v + \begin{pmatrix} 0 \\ -v_1^2/2 \end{pmatrix} \quad (9.8)$$

to yield

$$v_z = \begin{pmatrix} 0 & 1 \\ 0 & 0 \end{pmatrix} v + \begin{pmatrix} \mu_1 & -\mu_2 \\ -\mu_1 & -\mu_2 \end{pmatrix} v + \begin{pmatrix} 0 \\ v_1 v_2 - v_1^2/2 \end{pmatrix} \quad (9.9)$$

In removing $v_1^2/2$ in the first component, we have added a $v_1 v_2$ term in the second component. This and the original $-v_1^2/2$ term of the second component are nonresonant terms for the double-zero singularity and can never be transformed away. They appear in the irreducible Bogdanov-Arnold normal form of (9.5). We hence abandon the nonlinear terms and get to work on the linear perturbation term. From (9.5), it is clear we need to transform the entire first row to zero. This can again be done with a near-identity linear transformation

$$v \to v + \begin{pmatrix} 0 & 0 \\ -\mu_1 & 0 \end{pmatrix} v \quad (9.10)$$

to yield

$$\nu_1 = -\mu_1 \quad \nu_2 = \mu_1 - \mu_2 \quad (9.11)$$
$$a = -1/2 \quad b = 1$$

and the homoclinic condition (9.6) becomes

$$\mu_1 - \mu_2 = \frac{2}{7} \mu_2$$

or in terms of the original parameters,

$$c = \frac{7}{5} \delta \quad (9.12)$$

which is clearly in agreement with our numerical results in Table 8.1. The approximations made in transforming (9.4) to (9.5) are the tangent space approximation and the two near-identity transformations of (9.8) and (9.10). In essence, the first approximation reduces the order of the defining ordinary differential equation such that a regular perturbation about the KdV equation can be carried out. The two near-identity transformations correspond to the subsequent perturbations. The simplicity of these Center Manifold projection and normal form tramsformation techniques, compared to the classical multi-scale solution perturbation techniques is quite evident here.

Finally, substituting (9.12) into (9.3) and (9.9), we see that the two eigenvalues closest to zero approach the limit $\pm\sqrt{7/5}$, which dictate the decay rate of the pulse $H(y)$ in the limits of $\pm\infty$. Hence, the solitary pulse becomes increasingly symmetric and approaches the KdV soliton solution, $H = 3c\,\text{sech}^2\left[\frac{1}{2}\sqrt{\frac{c}{\delta}}(x-ct)\right]$. However, there is a one-parameter family of KdV solutions for any given δ, the amplitude and width of each member of the family scale as c and $c^{1/2}$, respectively. For each δ, there is a unique solitary wave for the Kawahara equation whose width is constant at about 5 and whose amplitude scales as $\frac{21}{5}\delta$ at large δ and whose speed scales as $\frac{7}{5}\delta$. The degeneracy of the integrable KdV soliton family has been broken by the growth term (H_{xx} in (5.1) and the $\nu_1\nu_2$ term in (9.5)) and the dissipation term (H_{xxxx} in (5.1) and $b\nu_1\nu_2$ in (9.5) which arises from the tangent space approximation).

For a given δ, we shall transform the gKS equation to a comoving frame with the speed $c(\delta)$ of the solitary pulse at the δ value,

$$\frac{\partial h}{\partial t} - c(\delta)\frac{\partial h}{\partial y} + 4h\frac{\partial h}{\partial y} + \frac{\partial^2 h}{\partial y^2} + \delta\frac{\partial^3 h}{\partial y^3} + \frac{\partial^4 h}{\partial y^4} = 0 \qquad (9.13)$$

It is clear that the solitary pulse $H(y)$ of (9.1) is a solution of (9.13) by definition. Linearizing (9.13) about $H(y)$ in the comoving y frame, we find the stability of $H(y)$ is determined by the spectrum of an operator L

$$L\psi_n = \lambda_n\psi_n \qquad (9.14)$$

where

$$L = \frac{d^4}{dy^4}\cdot + \delta\frac{d^3}{dy^3}\cdot + \frac{d^2}{dy^2}\cdot - c(\delta)\frac{dh}{dy}\cdot + 4\frac{d}{dy}(H\cdot)$$

For localized disturbances, the appropriate boundary conditions are

$$\psi(y \to \pm\infty) = 0$$

We hence need only to be concerved with the discret spectrum. Defining the inner product $(\cdot,\cdot) = \int_{-\infty}^{\infty}\cdot\cdot\,dy$, an adjoint operator L^+ can also be defined for the adjoint eigenvalue problem

$$L^+\varphi_n = \lambda_n\varphi_n \quad \varphi_n(\pm\infty) = 0 \qquad (9.15)$$

$$L^+ = \frac{d^4}{dy^4}\cdot - \delta\frac{d^3}{dy^3}\cdot + \frac{d^2}{dy^2}\cdot + c(\delta)\frac{dh}{dy}\cdot - 4H(y)\frac{d}{dy}\cdot$$

such that orthonormality holds

$$(\psi_m,\varphi_n) = \delta_{mn} \qquad (9.16)$$

We have found from a shooting method numerical construction of the discret spectrum $\{\lambda_n\}$ that for δ in excess of 1, all but one eigenvalue are stable and bounded well away from the imaginary axis. The only exception λ_1 is a zero eigenvalue. It is a simple matter to show from (9.14) and (2) that the null eigenfunction as

$$\psi_1 = \frac{dH}{dy} \tag{9.17}$$

The physical origin of this null eigenfunction arises from the translational invariance of the system. If $H(y)$ is a solution, so much $H(y - y_0)$. There is hence an entire family of solution parametrized by y_0 and the "tangent" along this family must correspond to a null eigenfunction of the operator L, viz. $H(y - y_0) \sim h(y) - y_0 H_y$ and H_y must be a null eigenfunction whose amplitude corresponds to the shift.

There is a corresponding null adjoint eigenfunction φ_1 for L^+ which can only be computed numerically. A constructed eigenfunction for $\delta = 5$ is shown in Figure 9.1, along with $H(y)$ and the corresponding homoclinic orbit. We can now expand any solution to the Kawahara gKS equation in the comoving y frame about a reference solitary pulse solution,

$$h(y, t) - H(y) \sim -y_0(t)\psi_1(y) \tag{9.18}$$

Since all the other eigenvalues of the solitary pulse are stable, any perturbation will only excite the shift mode ψ_1. Substituting (9.18) into (9.13) in the y frame and taking inner product with respect to φ_1, one obtains

$$\frac{dy_0}{dt} = 0$$

as expected, representing the absence of disturbance and the degeneracy due to the translational invariance. However, if disturbances in the form of two additional solitary pulses in the back and front of $H(y)$ are present, the shift dynamics of $H(y)$ is excited as it feels the presence of the two neighboring pulses. Let

$$h \sim H(y) - y_0(t)\psi_1(y) + H^+(y) + H^-(y) \tag{9.19}$$

when

$$H^+(y) = H(y - y_+)$$
$$H^-(y) = H(y - y_-)$$

represent two identical pulses located at $y_\pm$ such that $y_- < y_0 < y_+$. We shall neglect the effect of pulses beyond the two adjacent ones. Substituting (9.19) into (9.13), taking inner product with respect to φ_1, one obtains

$$\frac{dy_0}{dt} = U_+ + U_- \tag{9.20}$$

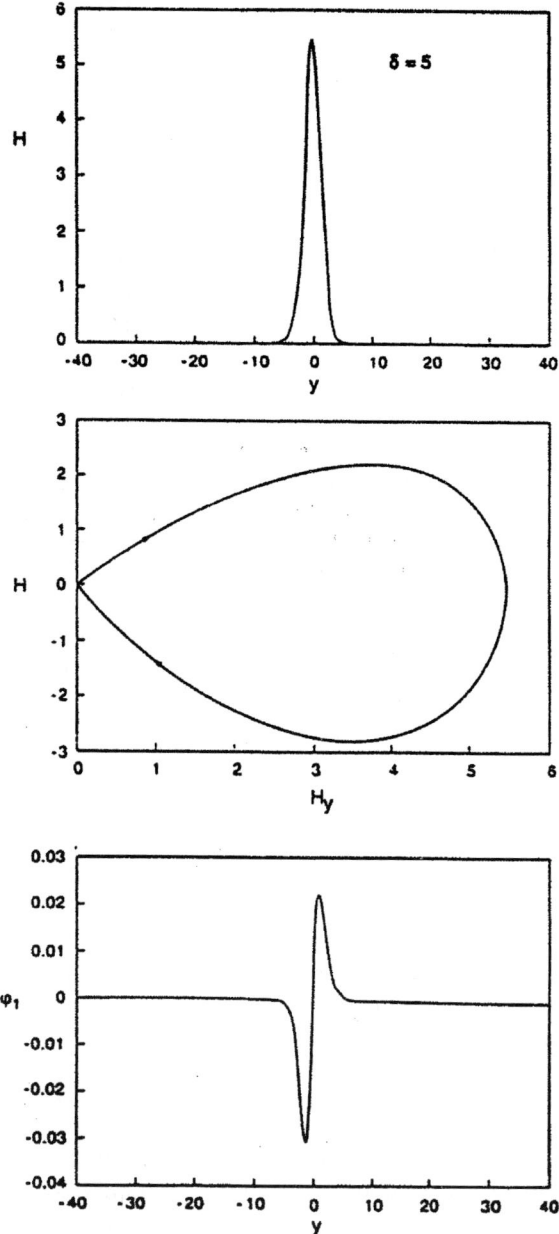

Figure 9.1: Using a shooting scheme, the solitary pulse $H(y)$, its corresponding homoclinic trajectory and its null adjoint eigenfunction $\phi(y)$ are constructed fot $\delta = 5$.

where we have chosen the maximum of the unperturbed solitary pulse toobe located at $y_0 = 0$. The interaction terms sre given by the Melnikov functions

$$U_+ = U_+(y_+) = 4 \int_{-\infty}^{\infty} H(y)H(y-y_+)\frac{d\varphi_1}{dy}dy \qquad (9.21)$$

$$U_- = U_-(y_-) = 4 \int_{-\infty}^{\infty} H(y)H(y-y_-)\frac{d\varphi_1}{dy}dy \qquad (9.22)$$

$$(9.23)$$

and they represent the effect of the front and back pulses on the position of the middle reference pulse.

Simplifications can be made of the interaction terms $U_\pm$ if the pulses are well separated such that the interaction is weak. In this limit, only the tails overlap and the tails of $H(y)$ can be approximated by

$$H(y) \sim Ae^{\sigma_1 y} \quad y \to -\infty$$
$$H(y) \sim Be^{\sigma_2 y} \quad y \to +\infty$$

where A and B are functions of δ and the spatial eigenvalues $\sigma_{1,2}$ are given in (5) and (6). The front and back pulses can likewise be approximated. The tail end of the front pulse H^+ interacts with the front of the reference pulse H in the inner product for U_+ while the front end of the back pulse H_- interacts with the tail of the reference pulse in U_-. After evaluatng the inner products, the approximate equation of motion becomes

$$\frac{dy_0}{dt} = \alpha e^{\sigma_2 y_-} - \beta e^{-\sigma_1 y_+} \qquad (9.24)$$

where the positive parameters α and β are listed in Table 9.1 and the negative spatial eigenvalue σ_2 and the positive σ_1 can be accurately estimated from (6) for δ in excess of 5. Compare with a case of dispersive medium where the order of the equation is two, Gorshkov et al. (1976). Note that y_- is negative while y_+ is positive such that both interaction terms decay exponentially with pulse separation. Also, the back pulse pushes the reference pulse forward while the front one pushes it backwards — the interaction is repulsive. To examine the accuracy of this weak interaction approximation, we plot $U_- e^{\sigma_2 y_-}$ and $U_+ e^{\sigma_2 y_+}$ as a function of y_- and y_+ in Figure 9.2 where $U_\pm$ are evaluated exactly from (9.21) and (9.22). It is clear that both approach the limits α and β when the separations $y_\pm$ are beyond 5, roughly the characteristic pulse width of the solitary pulse. Hence, the weak interaction theory is valid for separations beyond the pulse width.

Assuming only interactions with the nearest neighbors, (9.24) can be generalized to a train of pulses whose maxima lie at y_i,

$$\frac{dy_i}{dt} = \alpha e^{\sigma_2(y_i - y_{i-1})} - \beta e^{-\sigma_1(y_{i+1} - y_i)} \qquad (9.25)$$

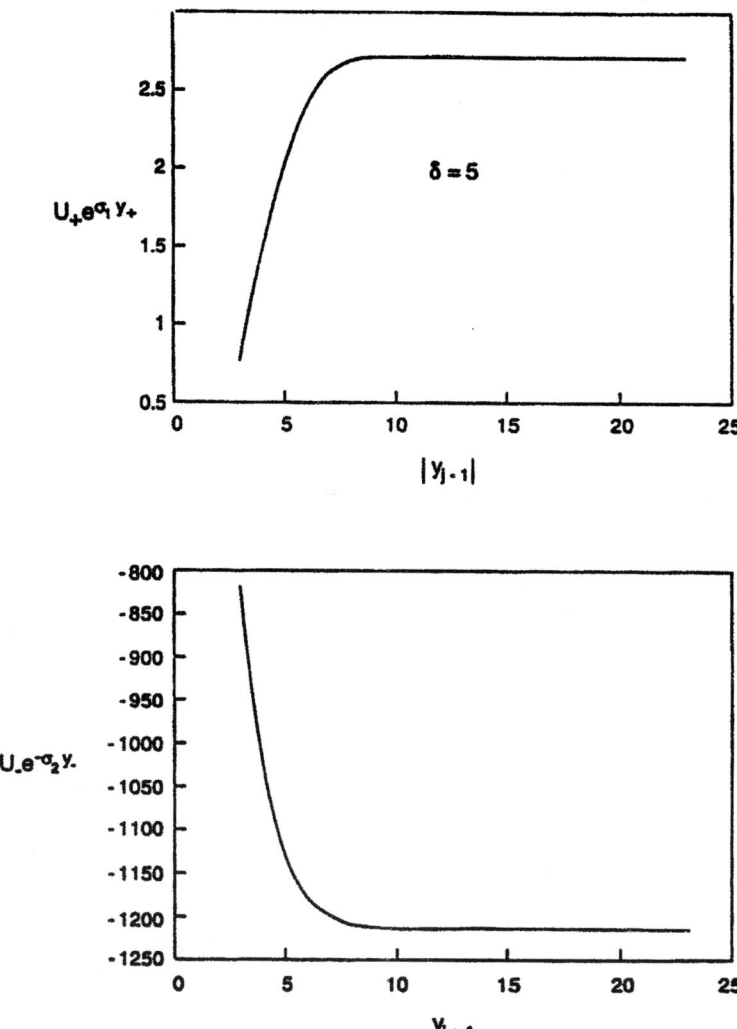

Figure 9.2: The approximation of weak interaction through tails is valid for separations in excess of 5, roughly the pulse width.

Consider first the repulsive dynamics of two pulses depicted in Figure 5. The position dynamics is governed by

$$\frac{dy_1}{dt} = -\beta e^{-\sigma_1(y_2-y_1)} \tag{9.26}$$

$$\frac{dy_2}{dt} = -\alpha e^{\sigma_2(y_2-y_1)} \tag{9.27}$$

such that the separation S is determined by

$$\frac{dS}{dt} = \alpha e^{\sigma_2 S} - \beta e^{-\sigma_1 S} \tag{9.28}$$

From (6), it is clear that σ_2 is larger in magnitude than σ_1. This implies that, due to the assymetry of the pulses, the force on the front pulse is much weaker and the dynamics is dominated by the motion of the back pulse such that

$$S \sim \frac{1}{\sigma_1} \log[\beta \sigma_1 t + e^{\sigma_1 S_0}] \tag{9.29}$$

This is favorably compared to our numerical solution of the two-pulse dynamics for the Kawahara equation in Figure 9.3. At large $t, t \gg \beta^{-1} \sim 10^{-4}$, the separation is seen to increase logarithmatically as $\frac{1}{\sigma_1} \log[\beta \sigma_1 t]$.

For a packet with many pulses, the separation S_i between pulse $i-1$ and i is governed by

$$\frac{dS_i}{dt} = \alpha e^{\sigma_2 S_i} + \beta e^{-\sigma_1 S_i} - \alpha e^{\sigma_2 S_{i-1}} - \beta e^{-\sigma_1 S_{i+1}} \tag{9.30}$$

An infinite train of periodically spaced pulse would then be defined by

$$\alpha e^{\sigma_2 S^*} = \beta e^{-\sigma_1 S^*} \tag{9.31}$$

where S^* is the separation,

$$S^* = \frac{\log(\alpha/\beta)}{|\sigma_2 + \sigma_1|} \sim \frac{6}{5\delta} \log(\alpha/\beta) \tag{9.32}$$

A simple stability analysis shows that these periodic trains are stable and they represent the asymptotic periodic trains observed in Kawahara and Toh's numerical experiment in a periodic domain.

However, periodic domains are rare in real life and it is of interest to study the repulsive dispersion of a packet of initially equal-separated pulses in an infinite domain as shown in Figure 9.4. The coupled equations of motion (9.30) cannot be solved easily. However, if the separations are not very different, a continuum approximation can be invoked. Let N be the total number of pulses in the packet, we shall use the approximation

$$\frac{S_{i+1} - S_i}{1/N} \sim \frac{\partial S}{\partial n}$$

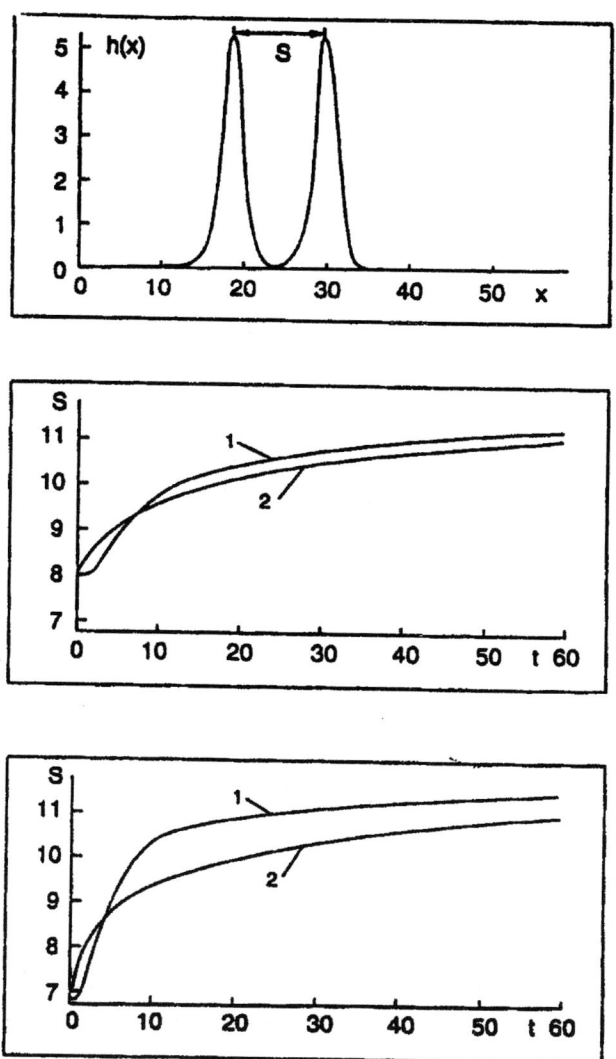

Figure 9.3: Comparison of computed pulse separation for binary interaction labelled 1 to the analitical prediction (9.29) labelled 2 for $S_0 = 8$ and 7.

$$\frac{S_{i+1} - 2S_i + S_{i-1}}{(1/N)^2} \sim \frac{\partial^2 S}{\partial n^2}$$

where $n = i/N \in [0, 1]$. Writing (9.30) as

$$\frac{dS_i}{dt} = \alpha e^{\sigma_2 S_i}[1 - e^{-\sigma_2(S_i - S_{i-1})}] + \beta e^{-\sigma_1 S_i}[1 - e^{-\sigma_1(S_{i+1} - S_i)}]$$

and since the separations are very close, we Taylor expand the terms in the square bracket about $S_i = S_{i+1} = S_{i-1}$ to get

$$\frac{dS_i}{dt} = \alpha \sigma_2 e^{\sigma_2 S_i}(S_i - S_{i-1}) + \sigma_1 \beta e^{-\sigma_1 S_i}(S_{i+1} - S_i)$$

Replacing the backward difference operator by

$$S_i - S_{i-1} = (S_{i+1} - S_i)(S_{i-1} - 2S_i + S_{i+1})$$

and invoking the continuum approximation, we obtain the following nonlinear convection-diffusion equation

$$\frac{\partial S}{\partial \tau} - f(S)\frac{\partial S}{\partial n} = g(S)\frac{\partial^2 S}{\partial n^2} \qquad (9.33)$$

where

$$f(S) = -\alpha\sigma_2 e^{\sigma_2 S} + \beta\sigma_1 e^{-\sigma_1 S} > 0 \qquad (9.34)$$
$$g(S) = -\alpha\sigma_2 e^{\sigma_2 S}/N > 0$$
$$\tau = t/N$$

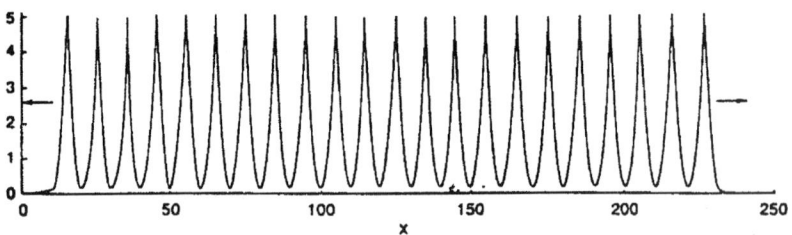

Figure 9.4: Dispertion of a packet of identical pulses separated by the same distance S_0 initially.

As $N \to \infty$, the diffusion term becomes negligible and we obtain a nonlinear hyperbolic equation whose characterics have a negative slope in the $\tau - n$ space. The initial condition is

$$S = S_0 \quad 0 < n < 1$$

$$S = \infty \quad n > 1 \quad \text{and} \quad n < 0$$

since the separation is formally infinite beyond the wave packet. We shall focus on the front edge ay $n = 1$. The slope of the characteristic line at $n = 1 - \epsilon$ is

$$\frac{dn}{d\tau} = -f(S_0) < 0$$

whereas the characteristic line for $n = 1 + \epsilon$ is

$$\frac{dn}{d\tau} = -f(\infty) = 0$$

Hence, the step change at $n = 1$ fans out in time and a smooth segment $S(n)$ appears joining the $S = S_0$ plateau with the $S = \infty$ adge (see Figure 9.5). This segment forms an expanding sector in the $n - \tau$ space and each characteristic line within the sector is described by

$$n - U\tau = 1$$

since at $\tau = 0, n$ must reduce to 1 at the point of the sector. The speed U is bounded between $-f(S_0)$ and $-f(\infty) = 0$, viz.

$$\frac{1-n}{\tau} = -U = f(S) \sim -\beta\sigma_1 e^{\sigma_1 S} \quad \text{for large } S$$

Inverting, one obtains the separation near the edge as

$$S \sim -\frac{1}{\sigma_1} \log\left[\frac{1}{\beta\sigma_1}\left(\frac{1-n}{\tau}\right)\right]$$

Defining a new index j which begins from the leading edge

$$\frac{j}{N} = 1 - n = \left(\frac{N-j}{N}\right)$$

one obtains

$$S_j \sim -\frac{1}{\sigma_1} \log[j/\beta\sigma_1 t] \qquad (9.35)$$

The logarithmic repulsion in time of the two-pulse problem in (9.29) reappears here but, in additional, we see that at any given instant, the separation increases as $\log j$ as j approaches zero towards the leading edge of the packet. This predicted behavior is confirmed by our numerical simulation of the Kawahara equation for $N = 50$ pulses and $S_0 = 8$, as seen in Figure 9.5. Repelled by the pulse behind it, the leading pulse shoots forward very rapidly since there is no other pulse in front to impede its progress. This then allows the next pulse to move forward but due to the presence of the leading pulse, its motion is not as rapid as the leading one and so forth. By symmetry, the dispersion at the back edge also follow the sane dynamics. Since this dispersion process smoothes out the jumps in S, shocks are not formed and the diffusion term never becomes significant.

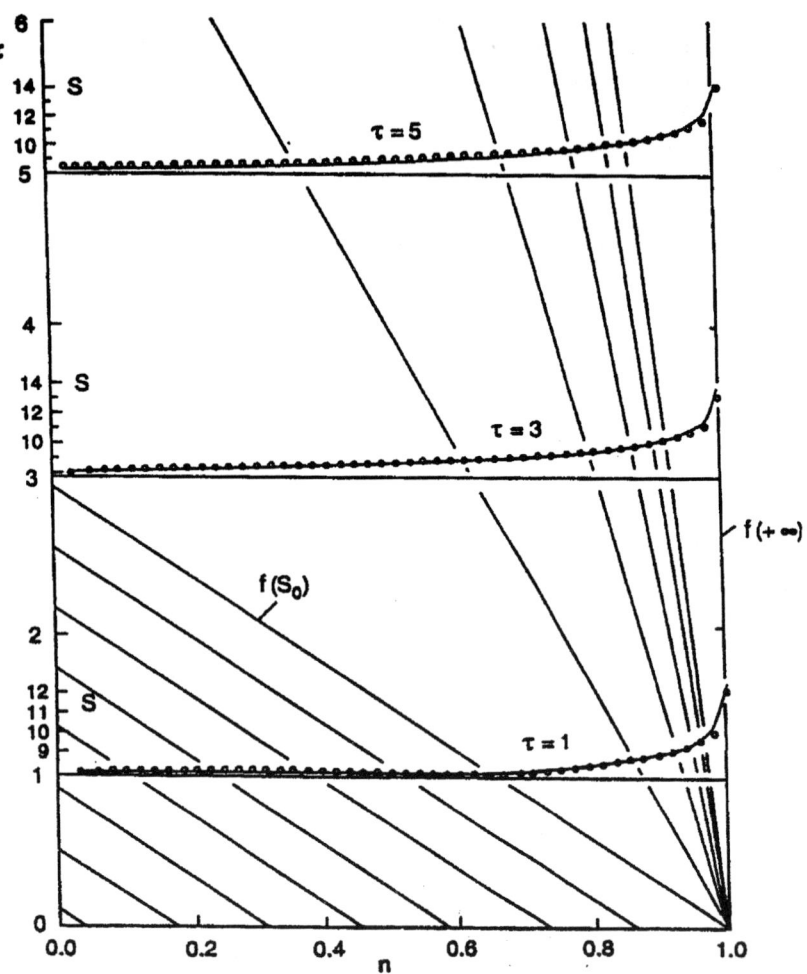

Figure 9.5: Numerical simulation in circles for $N = 50$ and $S_0 = 8$ and the predicted evolution of $S(n, \tau)$ in solid curves. The characteristic lines are also drawn, showing the sector which emanates from the jump discontinuity at $(n, \tau) = (1, 0)$.

9.3 Coupled drainage and binary interaction dynamics of pulses for the Shkadov model

If the pulses are well separated such that the pulse drainage dynamics of Section 8.3 for the Shkadov model is of the same order as the binary pulse interaction dynamics, the two dynamics can be scrutinized independently and then coupled together. This is equivalent to saying that the initial separation between the two pulses $l(0)$ is at least the order of the characteristic decay length x_∞ of an excited pulse, derived in Section 8.3.

With two variables, the expansion becomes

$$\begin{pmatrix} g \\ h \end{pmatrix}(x, y) = \begin{pmatrix} g^* \\ h^* \end{pmatrix}(x) + \begin{pmatrix} g^* \\ h^* \end{pmatrix}(x - l) - x_0(t)\psi_1(x) \quad (9.36)$$

in the moving frame. The unction $\psi_1(x)$ is the translational discrete mode of the back pulse located at $x = 0$. Substituting this expansion into the Shkadov model (3.88)-(3.89), expanding to quadratic order for the mixed term between the two pulses and taking the inner product with respect to $\varphi_1(x)$ of the back pulse, we obtain

$$\frac{dx_0}{dt} = g_f(l) = (\mathbf{D}, \varphi_1) \quad (9.37)$$

where $\mathbf{D}$ denotes the quadratic interaction terms between the two pulses. Since the only nonlinearities of the averaged equation in the moving frame appear in the $\hat{q}$ component and not the $\hat{h}$ kinematic component, we need only be concerned with the $\hat{q}$ component of φ_1. Since the two pulses are well separated, interaction is through the tails of the two pulses and is weak. Both solitary pulses also satisfy the full steady nonlinear equation in the moving frame individually. Hence, the only interaction terms come from $\hat{c}/\hat{c}x(q^2/h), hh_{xx}$ and q/h^2 in the averaged equation. Expanding to mixed terms for $q = q^*(x) + q^*(x - l)$ and $h = h^*(x) + h^*(x - l)$ and using (4) to eliminate $q^*(x)$ and $q^*(x + l)$, we obtain

$$\mathbf{D} = \begin{pmatrix} f \\ 0 \end{pmatrix}$$

where the scalar interaction term is a complex function of $\hat{h} = h^* - 1$:

$$f(x) = \frac{12}{5}(c^* - 1)^2 \frac{d}{dx}[\hat{h}(x)\hat{h}(x - l)]$$
$$- \frac{1}{5\delta}\left[\hat{h}(x)\frac{d^3}{dx^3}\hat{h}(x - l) + \hat{h}(x - l)\frac{d^3}{dx^3}\hat{h}(x) + 2(2c^* + 3)\hat{h}(x)\hat{h}(x - l)\right]. \quad (9.38)$$

The tails described by linearized version of (3.88)-(3.89) can be used with good accuracy to represent $h^*(x)$ and $h^*(x - l)$ in the interaction term f. In evaluating the inner product (9.37), it is important to line up the origins of φ_1 and $h^*(x)$ defined to be at the maximum of the reference pulse $h^*(x)$. Since

$h^*(x-l)$ appears linearly in d and $\mathbf{D}$ and since the back tail of this front pulse behaves as e^{-2ml}, it is clear that the interaction force must be of the form

$$g_f(l) = -Ee^{-2ml} \qquad (9.39)$$

where R is only a function of the reference pulse and hence a function of δ, and $2m$ is the decay rate of the back tail of the front pulse. The explicit and only dependence on the separation l, however, is contained in the exponent. Both m and R are tabulated in Table 8.1. Since R is positive, (9.39) indicates that the back pulse is repelled by the front one and its position is pushed to the left. Since h^* is a homoclinic orbit in the phase space of (h^*, h_x^*, h_{xx}^*), $g_f(l)$ is essentially the Melnikov function in homoclinic theory which measured how the front pulse perturbs the trajectory of the homoclinic orbit C (see Balmforth et al. (1994), for a more detailed discussion).

The back pulse also applies a force on the front pulse and unlike $g_f(l)$ this force can be either attractive or repulsive. Let l be negative in (9.39) and let $x_1(t)\psi_1(x)$ denote the position mode of the front pulse now situated at $x = 0$. One obtain

$$\frac{dx_1}{dt} = g_b(l) = (\mathbf{D}, \varphi_1) \qquad (9.40)$$

where the front end of $h^*(x)$ with the decaying oscillation is now used for $h(x-l) - h(x+|l|)$ in (9.38). The result is that the back Melnikov function is

$$g_b(l) = Se^{-ml}\cos(\beta l + \theta) \qquad (9.41)$$

where S and θ are listed in Table 8.1. We note that, owing to the oscillatory front tail of the back pulse, viz. the capillary ripples or bow waves, the interaction can push the front pulse both forward and backward, depending on the value of l. This is consistent with earlier theories on pulse interaction (Kawahara and Toh, 1988; Elphick et al., 1991; Chang et al., 1993; Balmforth et al., 1994).

The physical mechanism behind the attractive and repulsive forces lies in the capillary force within the ripples, which is also responsible for the unique shape of the solitary pulse. As the pulse develops and steepens owing to the pull of the gravity, a stationary pulse can only result if liquid is drained out of the pulse crest to relive the steepening. This is provided by the first "dimple" in front of the pulse which generates a sufficient negative capillary pressure to suck liquid out of the crest (see Wilson and Jones, 1983 for a lubrication analysis of this capillary drainage machainsm). The smaller capillary waves in front of the first dimple also possess these capillary forces although their strength decays further away from the pulse as the amplitude (and curvature) diminishes. As two pulses approach each other, the back slope of the front pulse experiences a positive (negative) differential pressure in the liquid as it overlaps with a maximum (minimum) of one of the capillary waves. this elevates (depresses) the back slope of the front pulse and decreases (increases) its curvature. The genenrateed differential capillary pressure then drains (sucks) liquid into (out of) the back half of the front pulse. This, in turn, causes the front pulse to move backwards (forwards) in a quasi-steady manner. This is why, depending on the separation between the two pulses, the front pulse can be attracted to or repelled by its neighbour.

Since $l = x_1 - x_0$, we can combine (9.37) and (9.40) to yield a dynamical equation for the separation between two pulses. If we now combine this with the linear decay dynamics of the back excited pulse, a self-contained dynamical system that describes the binary interaction between an excited back pulse and a stationary front pulse results:

$$\frac{d}{dt}\begin{pmatrix} l \\ \hat{\chi} \end{pmatrix} = \begin{pmatrix} o & -\alpha \\ 0 & -\gamma \end{pmatrix}\begin{pmatrix} l \\ \hat{\chi} \end{pmatrix} + \begin{pmatrix} g(l) \\ 0 \end{pmatrix} \qquad (9.42)$$

where $g(l) = g_b(l) - g_f(l)$. This is the simple dynamical system we seek that elucidates binary interaction. The decay rate γ is from the resonance pole $\lambda_\#$ of Figure 8.21.

The attractive binary force generated by the oscillatory tail immediately suggests the possibility of bounded pairs and periodic pulse trains. The former corresponds to two identical pulses $\hat{\chi} = 0$ traveling in synchrony such that the separation remains constant, as seen in Figure 9.2, while the latter corresponds to a train of equally spaced pulses. Kawahara and Toh (1988) and Elphick et al. (1991) have shown the existence of the former while Chang et al. (1993) constructed the latter for falling films. Pulses traveling with identical speeds but random spacing have also been shown to be possible (Balmforth et al., 1994). Most of these earlier theories focused on the KS equation and its generalization which are not suitable for the coherent structure theory as we have shown. More importantly, the additional speed mode ψ_2 with amplitude $\hat{\chi}$ or $\hat{c}$ in (9.42) changes the dynamics significantly.

Coalescence is the key difference and the bounded pairs play an important role in this event. Compare for example the decay length $x_\infty (\sim 50)$ of an excited pulse after coalescence to the characteristic binary interaction length $m^{-1} \sim 5$. without only binary interaction and bounded pairs are important. The bounded pairs have separations l_i defined by $g(l_i) = 0$. As shown in the sketch of Figure 9.6, a countable infinite number of possible l_i exists due to the particular form of the Melnikov functions g_f and g_b. This actually corresponds to a Shilnikov bifurcation from a single-loop homoclinic orbit to a double-loop one (Glendinning and Sparrow, 1984). We have numerically constructed the actual two-hump solitary waves, corresponding to the double-loop homoclinic orbit, for the averaged equations with separation l_2 and speed c_2^*. (See Chapter 5, for a clarification of how the two-hump solitary pulse family bifurcates from the one-hump family.) These two-hump solitary waves hence correspond to the shortest unstable bounded pairs estimate from the coherent structure theory here. They can be approximated by a root l^* of $g(l)$ which is unstable, $g'(l^*) > 0$. We choose l^* to be the closest unstable root of $g(l)$ to l_2. At low δ, there are typically another one or two smaller unstable roots but they are much smaller than the pulse width w and are hence meaningless. Our model then predicts that any two identical pulses on the same substrate separated by a distance less (larger) than l^* will attract (repel) each other irreversibly. The two-hump solitary pulse hence becomes a transition state for coalescence. At exactly l^*, more precisely l_2, the unstable transition state can be sustained indefinitely as in Figure 8.3b. Note the resemblance of the asymptotic state in Figure 8.3b to the two-hump bounded pair in Figure 9.6. Note also that the values of l_2 and l^* are of the order of the pulse width w in table 1. The accurate estimate of l_2 by l^* is then rather surprising since than l^*. We are unable to explain this coincidence. Our numerical experiments with the actual two-hump solitary waves are also consistent with those of Figure 8.1 and 8.2 — slight perturbations leads to either coalescence or repulsion. The passage through the transition state represented by this unstable bounded pair with separation l^* is evident in Figures 8.1 and 8.2. Every close encounter actually takes place. It is also observed in Liu and Gollub's experimental tracings for binary interaction on an inclined film. We also note that in the repulsion case of Figure 8.2b and 8.3a, the two pulses approach a stable bounded pair whose separation is larger than l^*.

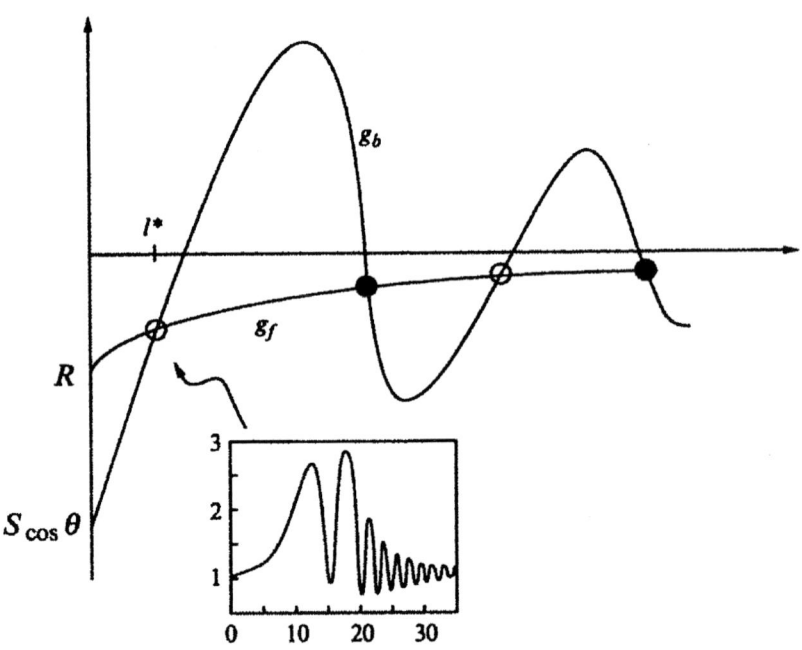

Figure 9.6: Schematic of the Melnikov functions for binary interactions. There is an infinite number of possible bounded pairs represented by the intersections. The one with the shortest separation l^* is unstable (open circles). The constructed two-hump solitary pulse corresponding to this transition state is shown ion the inset.

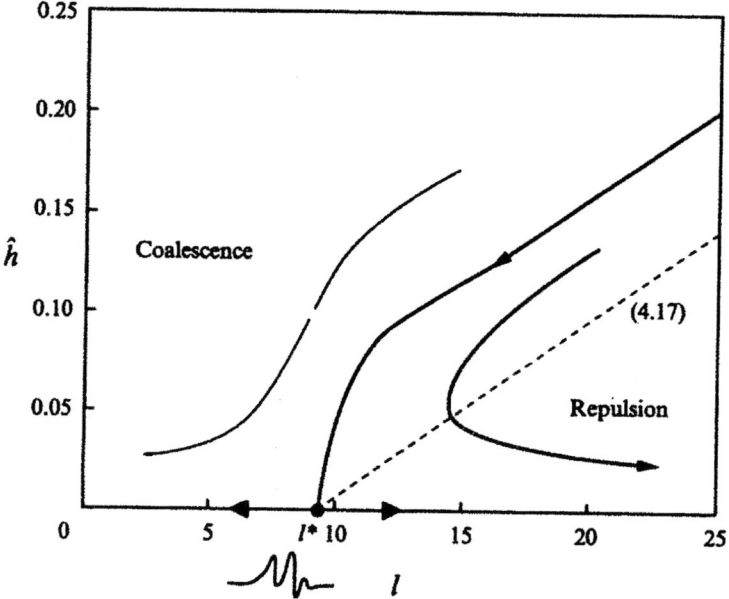

Figure 9.7: Phase-plane trajectories of dynamical system (9.43) for $\delta = 0.03$ showing the bounded-pair transition state as a saddle point and its stable manifold as a boundary for the domain of attraction of the coalescence phenomenon. Coalescence experiment with $l(0) = 25$ yields $\hat{h}_c(0) = 0.26$ which is close to the 0.21 value on the stable manifold. Approximation (9.46) which yields $\hat{h}_c(0) = 0.14$ is also shown. The deviation occurs at the inner region close to l^* where the interaction dynamics is as important as the outer decay dynamics.

It is convenient to replace $\hat{\chi}$ by $\hat{c}$ in (9.42) to yield

$$\frac{d}{dt}\begin{pmatrix} l \\ \hat{c} \end{pmatrix} = \begin{pmatrix} 0 & -1 \\ 0 & -\gamma \end{pmatrix}\begin{pmatrix} l \\ \hat{c} \end{pmatrix} + \begin{pmatrix} g(l) \\ 0 \end{pmatrix}. \tag{9.43}$$

The phase-space trajectory of (9.43) is shown in Figure 9.7. It is clear that the transition state is actually a saddle point whose stable manifold demarcates initial conditions which lead to coalescence from ones that do not. These initial conditions are determined only by the initial excitation of the back pulse, as measured by $\hat{c}(0)$, and the initial separation $l(0)$. The degree of excitation of the back pulse can also be measured by $\hat{h}(0)$ and $\hat{\chi}(0)$. The numerical experiments of Figure 8.1 and 8.2 then correspond to the initial condition to either side of the stable manifold and right on it. This stable manifold can be estimated by exploiting the different length scale of the long-range decay as in (8.16) and

(8.17), which can be combined to yield

$$\hat{c} = \hat{c}(0) + \gamma(l - l(0)). \tag{9.44}$$

This is a good approximation of the long-range dynamics except within a small neighborhood of the saddle point when the short-range forces in $g(l)$ become important. However, since $l(0) \gg l^*$ and $\hat{c}(0) \gg 0$, it is a good approximation to assume that (9.44) goes through the saddle point to obtain an estimate of the stable manifold and hence a condition for coalesence

$$\hat{c}(0) = \gamma(l(0) - l^*). \tag{9.45}$$

Converting to $\hat{h}(0)$ by the correlation $\hat{c} = 3.0\hat{h}$ of section 8.4, we get

$$\hat{h}(0) = \frac{\gamma}{3.0}(l(0) - l^*). \tag{9.46}$$

This then represents an estimate of the critical separation $l(0)$ for coalescence for a binary pair whose back excited pulse has a degree of excitation of $\hat{h}(0)$. The decay rate can be obtained from the resonance pole $\lambda_\#$ of Figure 8.21 and is roughly 0.4 to 0.8.

CHAPTER 10

Coarsening Theory for Naturally Excited Waves

We have now accumulated sufficient tools to analyze the pulse formation dynamics from inlet noise revealed in Chapter 4. In particular, we will modify the wave transition scenario of Chapter 6 towards soliary pulses. Instead, we shall show that, in naturally excited waves, the primary spectrum at the inlet is remembered through out the evolution downstream, including the final formation of solitary waves or pulses. In fact, this residue primary spectrum plays an important role in deciding the initial pulse density at formation and the initial fraction of excited pulses. Knowing these from fundamental theories, we can then use the drainage and interaction theories to capture the coarsening dynamics downstream. As stated earlier, such coarsening involving pulses is the dominant dynamics on a falling film–it begins as early as 20 cm from the inlet and persist for more than one meter. We shall utilize the Shkadov model for our theoretical derivation. Of particular interest is the creation of "excited" pulses from a modulation instability of the primary waves described in section 4.4. There seems to be a dominant secondary modulation frequency Δ that contradicts the Floquet growth rate of Chapter 6, which predicts a growth rate that increases monotonically with the wavenumber and wave frequency. The dominant modulation frequency is hence not just triggered by the instability of a saturated periodic wave, as is true in the scenario for periodically forced waves in Chapter 6, but also by some other mechanism prior to saturation–the residue primary spectrum. Once we decipher this mechanism, the density of excited pulse rlative to that of the equilibrium pulse is then known. A simple application of the binary interaction theory of Chapter 9 then readily provides the coarsening rate as a function of the normalized Reynolds number δ.

10.1 Spatial evolution, linear filtering and excitation of low-frequency band

We begin with the inception region and resolve the spatial amplification of certain frequency harmonic and filtering of thers from the white inlet noise of (4.18). The Shkadov model yields the dispersion relationships (3.121),

$$c^2 - c\left(\frac{12}{5} - \frac{i}{5\alpha\delta}\right) + \frac{6}{5} - \frac{\alpha^2}{5\delta} - \frac{3i}{5\alpha\delta} = 0 \quad (10.1)$$

$$\alpha^4 - (\omega^2 - 12\alpha\omega + 6\alpha^2)\delta + i(3\alpha - \omega) = 0 \quad (10.2)$$

The first dispersion relationship is for temporal evolution with a real wavenumber α and two complex wave speeds c. The second one corresponds to spatial evolution with a real wave frequency ω and four complex wavenumbers α. We shall focus on spatial evolution and will hence utilize (10.2). There are, however, 4 spatial roots α to (10.2). As shown in section 3.6, only those two spatial modes that propagate downstream are pertinent. Each spatial mode is also associated with a distinct wave speed $c = \omega/\alpha_r$. We assign the two modes as $c_{1,2}$ and $\alpha^{1,2}$ with mode 1 unstable ($-\alpha_i^1 > 0$) and mode 2 stable ($-\alpha_i^2 < 0$), as in section 3.6. In fact, given the single-harmonic disturbance quantities

$$h - 1 = \hat{h}(x)e^{i\alpha_r x - i\omega t} \quad (10.3)$$

$$q - 1 = \hat{q}(x)e^{i\alpha_r x - i\omega t} \quad (10.4)$$

for a given frequency ω, since $\hat{h}(0) = 0$ and $\hat{q}(0) = \epsilon = F_0$ from (4.16), solution of the linearized Shkadov model readily yields

$$\hat{h}(x) = \frac{\epsilon}{c_1 - c_2} exp(-\alpha_i^1 x) - \frac{\epsilon}{c_1 - c_2} exp(-\alpha_i^2 x) \quad (10.5)$$

As $x \to \infty$, the stable mode contribution vanishes and

$$\hat{h}(x) = \frac{\epsilon}{c_1 - c_2} exp(-\alpha_i^1 x) \quad (10.6)$$

Hence, the inlet flow rate disturbance of amplitude ϵ excites an interfacial mode that extrapolates to the amplitude $\frac{\epsilon}{|c_1-c_2|} = \epsilon'$ at the inlet –eventhough there is no interfacial disturbance at the inlet. Typical values of $-\alpha_i^1(\omega)$ and $|c_1 - c_2|$ at ω_m, the frequency of the fastest-growing mode, are shown in Figure 10.1a. The damping factor $\frac{1}{|c_1-c_2|}$ is small for small δ but approaches $1/4$ for δ in excess of 0.2.

The downstream linear evolution of the entire spectrum of harmonics is then

$$\hat{h} = \epsilon' \int_{-\infty}^{\infty} e^{-i(\omega t - \alpha_r x) - \alpha_i x} d\omega \quad (10.7)$$

where only the unstable first mode is considered now. The interfacial Fourier spectrum of this linear evolution is then

$$F(\omega) = \epsilon' e^{-\alpha_i(\omega)x} \quad (10.8)$$

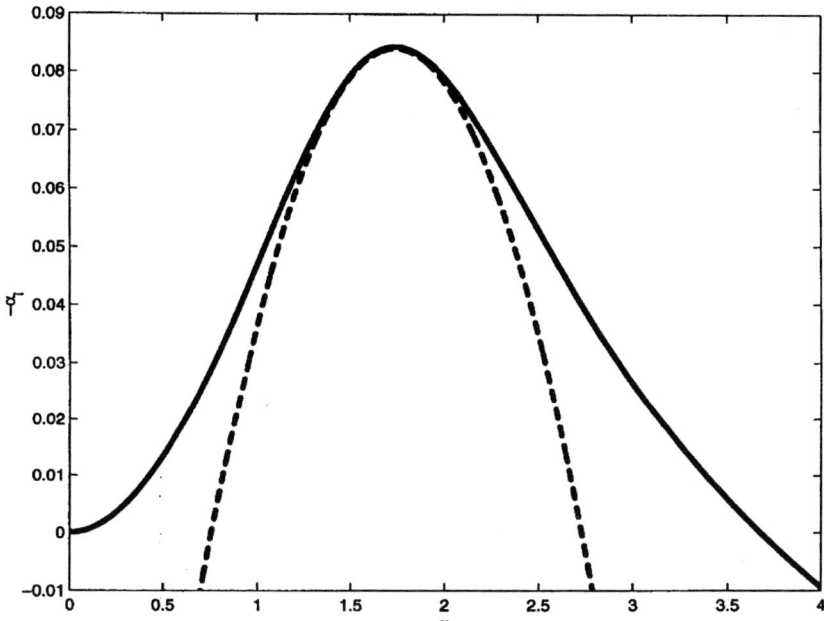

Figure 10.1: Primary spatial growth rate a function of wave frequancy ω for $\delta = 0.1$ and a parabolic approximation at ω_m.

Given the parabolic spatial growth rate of Figure 10.1a, this linear evolution will soon filter the white-noise spectrum at $x = 0$ into a sharp Gaussian band

$$F(\omega) \sim \epsilon' e^{ax - bx(\omega - \omega_m)^2} \tag{10.9}$$

where $a = -\alpha_i(\omega_m)$, $b = \frac{1}{2}\frac{\partial^2 \alpha_i}{\partial \omega^2}(\omega_m)$ and ω_m corresponds to the maximum of the growth rate, $\frac{\partial \alpha_i}{\partial \omega}(\omega_m) = 0$. This parabolic approximation is sketched in Figure 10.1b. Dependence of $\omega_m, -\alpha_i(\omega_m)$, $2b = \frac{\partial^2 \alpha_i}{\partial \omega^2}(\omega_m)$ and $\frac{\partial^2 \alpha_r}{\partial \omega^2}(\omega_m)$ of the unstable first mode on δ is shown in Figure 10.2 and in Table 10.1, along with its group velocity $c_g = 1/\left(\frac{\partial \alpha_r}{\partial \omega}\right)(\omega_m)$.

With proper normalization, this evolving primary band after filtration by the primary linear instability has a probability density

$$p(\omega) = \sqrt{\frac{bx}{\pi}} e^{-bx(\omega - \omega_m)^2} \tag{10.10}$$

with a standard deviation $\sigma = \frac{1}{\sqrt{2bx}}$ that decays as $x^{-1/2}$ downstream.

Table 10.1 Dispersion coefficients of the primary mode

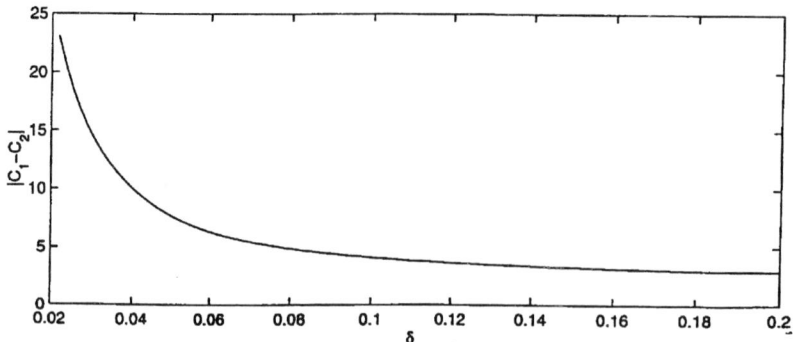

Figure 10.1: The difference in phase speeds of the two spatial modes as a function of δ-this difference corrects the spatial evolution.

δ	$c_g(\omega_m)$	$\alpha_r''(\omega_m)$	$2b = \alpha_i''(\omega_m)$	ω_m
0.02	2.983	-0.004	0.044	1.151
0.04	2.889	-0.017	0.085	1.539
0.06	2.730	-0.035	0.120	1.703
0.08	2.569	-0.053	0.149	1.748
0.10	2.432	-0.068	0.172	1.744
0.12	2.324	-0.081	0.190	1.720
0.14	2.238	-0.091	0.204	1.689
0.16	2.170	-0.100	0.213	1.658
0.18	2.116	-0.106	0.219	1.628
0.20	2.071	-0.112	0.223	1.599

The primary band hence focuses due to linear filtering. Both the theoretical probability density (10.10) and the predicted standard deviation evolution are in quantitative agreement with the simulated spectra for $x < 150$ at $\delta = 0.1$ and $\epsilon = 10^{-7}$, as shown in Figure 10.3. The Gaussian band becomes increasingly accurate downstream as the effect of the stable mode diminishes. Beyond $x = 170$, however, nonlinear effects begin to corrupt this linear filtering. Overtone and zero-frequency bands appear, as seen in Figure 4.11. The exponential growth downstream is also saturated by this nonlinear interaction.

However, at the initial stage of this transition region, the primary band can still be described by $F(\omega)$ of (10.10) and we use it to estimate the secondary band $R(\omega)$ at low frequencies. In Figure 10.4, we plot the Fourier amplitudes of the power spectrum in Figure 4.11 at $\omega = \omega_m$, the fundamental fastest-growing frequency, and at $\omega = 0$ as a function of x. Exponential growth as oredicted by (9.5) is ibserved. More important, however, is the quadratic scaling of $F(0) \sim F^2(\omega_m)$. This suggests that the secondary band $R(\omega)$ near zero frequency, as observed in Figure 4.11, is initially trigged by quadratic interaction of modes in the primary band in (9.10). In fact, the quadratic slaving seen in

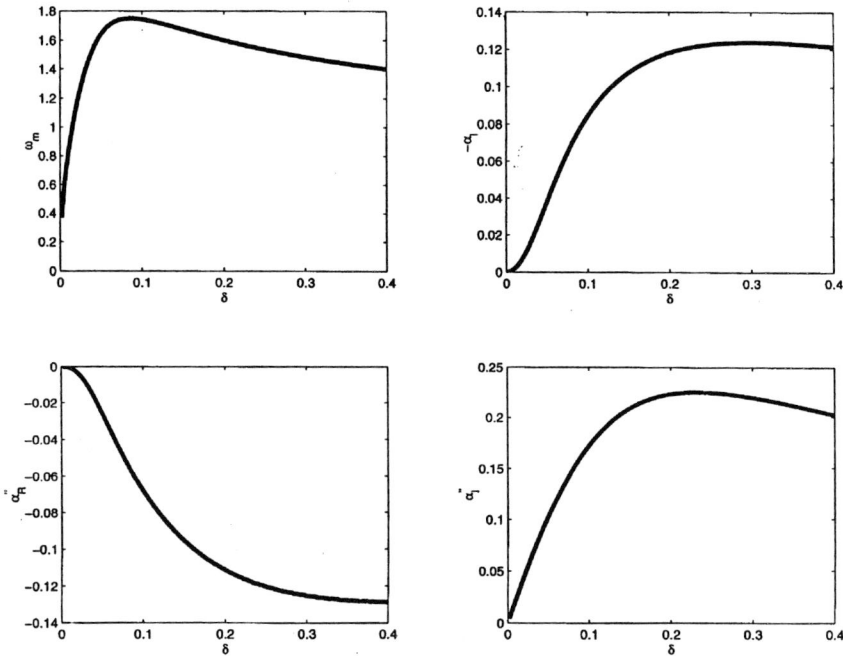

Figure 10.2: The dominant primary frequancy ω_m, spatial growth rate α_i and second derivatives of α_r and α_i at ω_m as functions of δ.

Figure 10.1 suggests that the secondary band can be approximated by

$$R(\omega) = \int_{-\infty}^{\infty} F(\beta)F(\omega - \beta)d\beta \qquad (10.11)$$

where we have omitted the phase between the harmonics β and $\omega - \beta$. A specific phase relationship exists between the two but we assume that it is independent of ω and β. As such, (10.11) is off by a constant independent of ω. Carrying out the Gaussian integral, we obtain an estimate of the secondary band

$$R(\omega) = (\epsilon')^2 \sqrt{\frac{\pi}{2bx}} exp(2\alpha x - bx\omega^2/2) \qquad (10.12)$$

An important prediction of (10.12) is that the secondary band near zero frequency with a standard deviation that is $\sqrt{2}$ higher than the primary band. This is consistent with our simulated spectra. In fact, by determining the missing constant (0.03) from the height of the secondary spectra at $x = 140$, our estimate (10.12) is able to quantitatively capture all zero-frequency secondary bands downstream, as seen in Figure 10.5. The agreement improves downstream as the contribution of the stable linear mode to the primary band becomes negligible and as the nonlinear excitation becomes more pronounced with larger

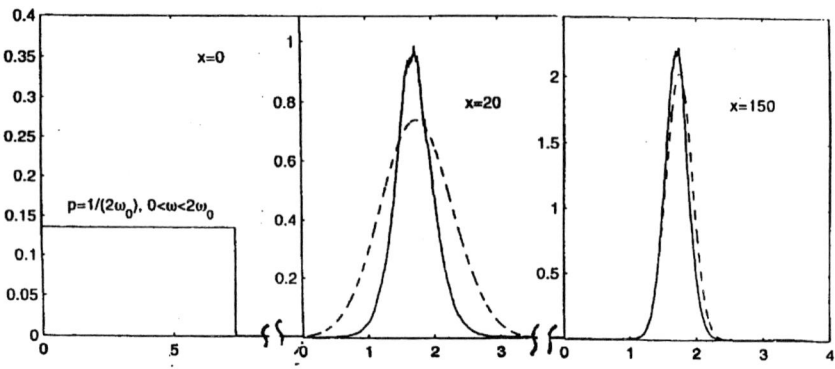

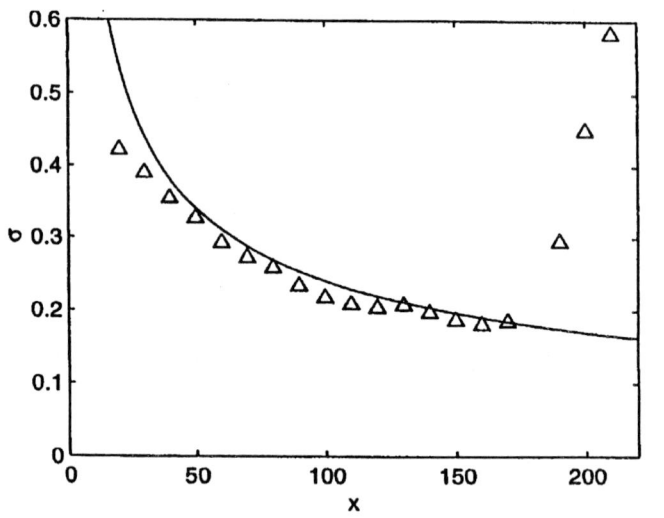

Figure 10.3: Simulated and predicted primary band of waves showing filtering of the initial white spectrum to a narrowing Gaussian with a standard deviation that decreases monotonically as $x^{-1/2}$ downstream et $\delta = 0.1$ and $\varepsilon = 10^{-7}$.

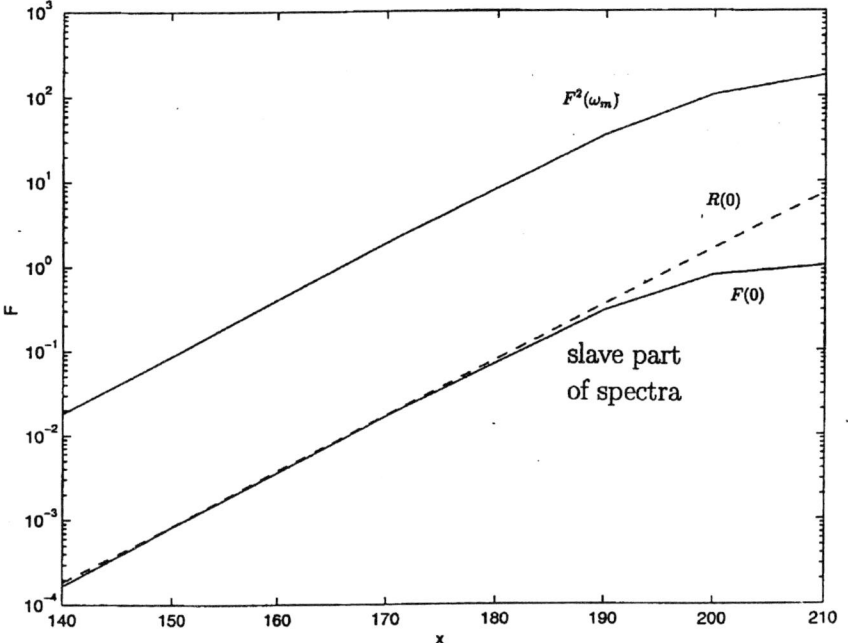

Figure 10.4: The Fourie amplitude at frecuancies ω_m and zero noise-driven wave dynamics evolve exponentially downstream with a quadratioc slaving relationship.

amplitudes in the primary band. This agreement empirically verifies our phase approximations.

10.2 A theory for the characteristic modulation frequency

As soon as the fundamental mode with frequency ω_m and the small band of waves around it saturate in amplitude, the excited secondary band of modes near zero frequency in (10.12) begins to destabilize the primary band by triggering pronounced modulations (see Figure 4.32). This saturation occurs at a distance x_0 from the inlet where

$$x_0 = -\frac{1}{\alpha_i(\omega_m)} \log(r_1/\epsilon') \qquad (10.13)$$

where $r_1 = |A_1|$ is the amplitude of the saturated fundamental that will be estimated below.

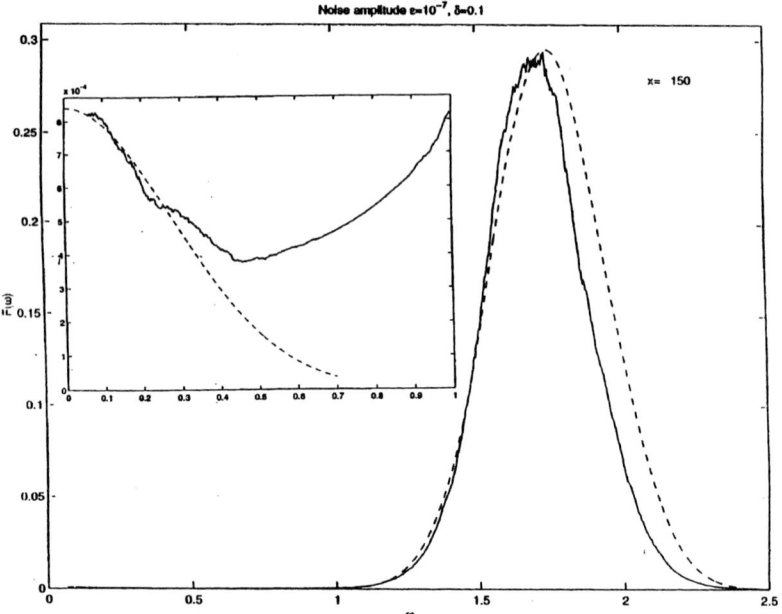

Figure 10.5: The noise-driven primary and slaved secondary bands and our predicted spectra for $\varepsilon = 10^{-7}$ and $\delta = 0.1$ at x=150

We utilize the technique of Frisch et al. (1986) to construct the periodic solutions of a longwave instability, including the fundamental mode with r_1, and to examine their modulation instability (See Chapter 6). As primary waves saturate and become stationary travelling waves, they can be described by the Shkadov model after the moving coordinate transformation $\partial/\partial t = -c\partial/\partial x$,

$$-c\frac{\partial q_0}{\partial x} + \frac{6}{5}\frac{\partial}{\partial x}\left(\frac{q_0^2}{h_0}\right) = \frac{1}{5\delta}\left(h_0\frac{\partial^3 h_0}{\partial x^3} + h_0 - \frac{q_0}{h_0^2}\right) \quad (10.14)$$

$$-c\frac{\partial h_0}{\partial x} + \frac{\partial q_0}{\partial x} = 0 \quad (10.15)$$

Integrating the kinematic equation (10.15) over a patch of periodic travelling waves and assigning the mean thickness for this wave train to be χ, one obtains

$$q_0(x) = ch_0(x) - c\chi + <q> \quad (10.16)$$

Due to the saturated periodic wave train, the mean thickness χ is different from

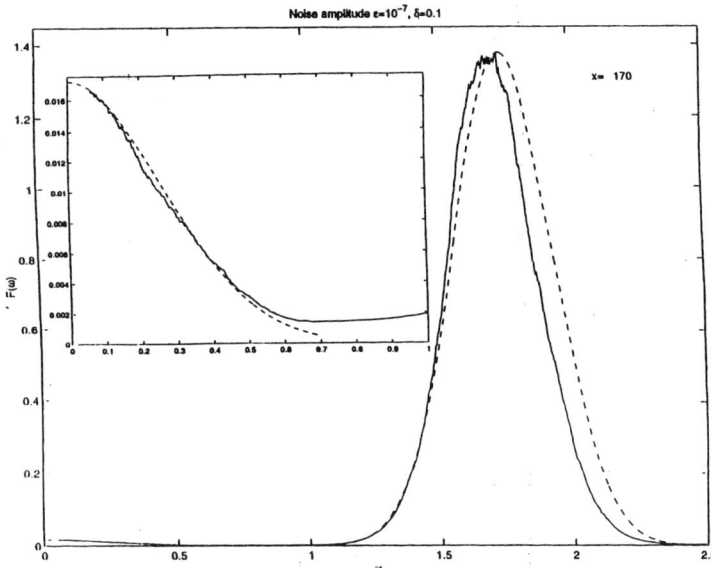

Figure 10.5: The noise-driven primary and slaved secondary bands and our predicted spectra for $\varepsilon = 10^{-7}$ and $\delta = 0.1$ at x=170

the flat-film Nusselt thickness ($h = 1$ from our scaling) and the average flow rate $<q>$ does not scale as χ^3 as would be the case for a flat film.

Substituting (10.16) into (10.14), we obtain the governing equation for the periodic travelling wave,

$$h_0 \frac{\partial^2 h_0}{\partial x^3} + \delta \left[6(<q> -c\chi)^2 - c^2 h_0^2 \right] \frac{dh_0}{dx} + h_0^3 - ch_0 - <q> +c\chi = 0 \quad (10.17)$$

with unknowns $h_0(x)$, c and $<q>$ at fixed χ, δ and the wavenumber α of the periodic wave.

A closed-form estimate of the solution to (10.17) can be obtained with a two-harmonic approximation for the profile $h(x)$, where the wavenumber α is within the unstable band,

$$h_0 = \chi + r_1 \cos \alpha x + r_1^2 (A \cos 2\alpha x + B \sin 2\alpha x) \quad (10.18)$$

Balancing terms for each harmonic, one obtains

$$A = \frac{3}{15\alpha\delta\chi^5 - 4\alpha^3\chi^2}, \quad B = -\frac{3\alpha^2\chi + 18\delta\chi^4}{15\delta\chi^5 - 4\alpha^2\chi^2}$$

$$\omega = 3\alpha\chi^2 - \frac{123}{10} r_1^2 \alpha, \quad <q> = \chi^3 + 6r_1^2\chi \quad (10.19)$$

$$r_1^2 = \frac{\alpha^3\chi^3 - 15\alpha\delta\chi^6}{\kappa}, \quad \kappa = -\frac{621}{50}\chi\alpha^3 + \frac{18}{30\alpha\delta\chi^4 - 8\alpha^3\chi}$$

This estimate includes and estimate of the amplitude r_1 for the fundamental when the average thickness is unity ($\chi = 1$) or when the average flow rate is unity ($q = 1$). The former condition is closer to the saturated primary waves of our system. However, the full χ dependence must be retained in (10.19). Unlike the classical modulation instability of short-wave instabilities, the χ dynamics are crucial in the modulation of long-wave instabilities (Chapter 6).

Hence, for the primary wave, (10.16), (10.18) and (10.19) yield $h_0(x)$ and $q_0(x)$ as functions of χ at fixed wavenumber α and normalized Reynolds number δ.

Consider small perturbations of the basic state, which is any periodic wave within the unstable band,

$$h = h_0(x) + \hat{H}(x,t) \tag{10.20}$$

$$q = q_0(x) + \hat{Q}(x,t) \tag{10.21}$$

After substituting into the Shkadov model and linearizing, one obtains the system

$$\frac{\partial \hat{Q}}{\partial t} = b_1 \frac{\partial^3 \hat{H}}{\partial x^3} + b_2 \frac{\partial \hat{Q}}{\partial x} + b_3 \frac{\partial \hat{H}}{\partial x} + b_4 \hat{Q} + b_5 \hat{H} \tag{10.22}$$

$$\frac{\partial \hat{H}}{\partial t} = c \frac{\partial \hat{H}}{\partial x} - \frac{\partial \hat{Q}}{\partial x} \tag{10.23}$$

where

$$b_1 = \frac{h_0}{5\delta}, \quad b_2 = c - \frac{12}{5}\frac{q_0}{h_0}, \quad b_3 = \frac{6}{5}\frac{q_0^2}{h_0^2}$$

$$b_4 = -\frac{12}{5}\left(\frac{q_0^2}{h_0^2}\right)' - \frac{1}{5\delta h_0^2}, \quad b_5 = \frac{6}{5}\left(\frac{q_0^2}{h_0^2}\right)' + \frac{1}{5\delta}\left(h_0''' + 1 + \frac{2q_0}{h_0^3}\right)$$

and prime denotes derivative with respect to x.

These equations are more compactly written in an operator form as

$$\frac{\partial}{\partial t}\begin{pmatrix} \hat{Q} \\ \hat{H} \end{pmatrix} = L\begin{pmatrix} \hat{Q} \\ \hat{H} \end{pmatrix} = L\boldsymbol{w} \tag{10.24}$$

An adjoint operator can be defined with respect to the inner product

$$\left\{\begin{pmatrix} \hat{Q} \\ \hat{H} \end{pmatrix}, \begin{pmatrix} \hat{\psi} \\ \hat{\varphi} \end{pmatrix}\right\} = \{\boldsymbol{w}, \boldsymbol{v}\} = \int_0^{2\pi/\alpha}(\hat{Q}\hat{\psi} + \hat{H}\hat{\varphi})dx \tag{10.25}$$

such that the adjoint operator L^* is

$$L^*\begin{pmatrix} \hat{\psi} \\ \hat{\varphi} \end{pmatrix} = \begin{pmatrix} (\frac{12}{5}\frac{q_0}{h_0} - c)\frac{\partial \hat{\psi}}{\partial x} + \frac{\partial \hat{\varphi}}{\partial x} - \frac{1}{5\delta}\frac{1}{h_0^2}\hat{\psi} \\ -\frac{6}{5}\frac{q_0^2}{h_0^2}\frac{\partial \hat{\psi}}{\partial x} - c\frac{\partial \hat{\varphi}}{\partial x} + \frac{1}{5\delta}(h_0''' + 1 + \frac{2q_0}{h_0^3})\hat{\psi} - \frac{1}{5\delta}\frac{\partial^3}{\partial x^3}(h_0\hat{\psi}) \end{pmatrix} \tag{10.26}$$

Both operators are singular with a null eigenfunction $\boldsymbol{w}_0$ for L and $\boldsymbol{v}_0$ for L^*, corresponding to translational symmetry,

$$Lw_0 = L\begin{pmatrix} q'_0 \\ h'_0 \end{pmatrix} = 0$$
$$L^*v_0 = L^*\begin{pmatrix} 0 \\ 1 \end{pmatrix} = 0 \quad (10.27)$$

There exists another generalized null mode w_1 to L defined by

$$Lw_1 = w_0 \quad (10.28)$$

and can be easily shown to be the Galilean mode of (7.60) and (8.8) due to mass conservation, but now for periodic waves instead of pulses,

$$w_1 = \begin{pmatrix} \frac{\partial q_0}{\partial c} \\ \frac{\partial h_0}{\partial c} \end{pmatrix} \quad (10.29)$$

so that

$$L\begin{pmatrix} q'_0 \\ h'_0 \end{pmatrix} = 0, \quad L\begin{pmatrix} \partial q_0/\partial c \\ \partial h_0/\partial c \end{pmatrix} = \begin{pmatrix} q'_0 \\ h'_0 \end{pmatrix} \quad (10.30)$$

by differentiating (10.16) and (10.17) with respect to c. Hence, the algebraic multiplicity of zero eigenvalue is two but the geometrical multiplicity is one, corresponding to a double zero singularity. This degeneracy has a physical origin. If one perturbs the velocity, $c \to \Delta c$, and hence the generalized eigenfunction u_1, the translation mode will also be triggered as any speed change necessarily involves an incremental translation. Instead of the $x \to x + const$ invariance (in the moving frame), the degeneracy corresponds to the invariance $x \to x - \Delta ct + const$. For the KS equation, this symmetry is expressed explicitly in (5.4). The speed-related generalized invariance eigenfunction w_1 will affect the translation mode w_0, but no vice versa, and one has a Jordan block projection of the operator L, as in (8.12) for pulses,

$$\begin{pmatrix} 0 & 1 \\ 0 & 0 \end{pmatrix}$$

For the zero modes involving the basic state, perturbation of the velocity c is only possible if one perturbs the layer thickness χ, at fixed α and δ. Hence, it is convenient to carry out the transformation

$$\frac{\partial q_0}{\partial c} = \frac{\partial q_0}{\partial \chi} / \frac{\partial c}{\partial \chi} \quad (10.31)$$

$$\frac{\partial h_0}{\partial c} = \frac{\partial h_0}{\partial \chi} / \frac{\partial c}{\partial \chi} \quad (10.32)$$

It is clear from the expressions for w_0 and w_1 that these two zero modes have the same wavelengths as the periodic travelling wave. We seek now the eigenvalues of disturbances with wavelengths much longer than the travelling wave basic state – the modulation disturbances. We exploit the separation in scales between the disturbance and the travelling wave by using the two-scale expansion in space and time

$$\frac{\partial}{\partial x} \to \frac{\partial}{\partial x} + \nu \frac{\partial}{\partial x_1}$$
$$\frac{\partial}{\partial t} \to \nu \frac{\partial}{\partial \tau_1} + \nu^2 \frac{\partial}{\partial \tau_2}$$
(10.33)

where $\nu \ll 1$ corresponds to a small wavenumber for the long modulations.

With this two-scale expansion, the operator L becomes

$$L \to L + \nu L_1 + \nu^2 L_2 \qquad (10.34)$$

We likewise expand $\boldsymbol{w} = \begin{pmatrix} \hat{Q} \\ \hat{H} \end{pmatrix}$ as

$$\begin{aligned} \boldsymbol{w} &= \boldsymbol{w}_1 + \nu \boldsymbol{w}_2 + \nu^2 \boldsymbol{w}_3 \\ &= \begin{pmatrix} \hat{Q}_1 \\ \hat{H}_1 \end{pmatrix} + \nu \begin{pmatrix} \hat{Q}_2 \\ \hat{H}_2 \end{pmatrix} + \nu^2 \begin{pmatrix} \hat{Q}_3 \\ \hat{H}_3 \end{pmatrix} \end{aligned} \qquad (10.35)$$

With these expansions, (10.33) becomes

$$\begin{aligned} L\boldsymbol{w}_1 &= \nu \left(\frac{\partial \boldsymbol{w}_1}{\partial \tau_1} - L\boldsymbol{w}_2 - L_1 \boldsymbol{w}_1 \right) \\ &+ \nu^2 \left(\frac{\partial \boldsymbol{w}_2}{\partial \tau_1} + \frac{\partial \boldsymbol{w}_1}{\partial \tau_2} - L_2 \boldsymbol{w}_1 - L_1 \boldsymbol{w}_2 \right) \end{aligned} \qquad (10.36)$$

Explicitly, we have for each order,

$$b_1 \hat{H}_1''' + b_2 \hat{Q}_1' + b_3 \hat{H}_1' + b_4 \hat{Q}_1 + b_5 \hat{H}_1 = 0 \qquad (10.37)$$

$$c\hat{H}_1' - \hat{Q}_1' = 0 \qquad (10.38)$$

$$b_1 \hat{H}_2''' + b_2 \hat{Q}_2' + b_3 \hat{H}_2' + b_4 \hat{Q}_2 + b_5 \hat{H}_2 = \frac{\partial \hat{Q}_1}{\partial \tau_1} - 3b_1 \frac{\partial \hat{H}_1''}{\partial x_1} - b_2 \frac{\partial \hat{Q}_1}{\partial x_1} - b_3 \frac{\partial \hat{H}_1}{\partial x_1} \qquad (10.39)$$

$$c\hat{H}_2' - \hat{Q}_2' = \frac{\partial \hat{H}_1}{\partial \tau_1} - c\frac{\partial \hat{H}_1}{\partial x_1} + \frac{\partial \hat{Q}_1}{\partial x_1} \qquad (10.40)$$

$$\begin{aligned} & b_1 \hat{H}_3''' + b_2 \hat{Q}_3' + b_3 \hat{H}_3' + b_4 \hat{Q}_3 + b_5 \hat{H}_3 \\ &= \frac{\partial \hat{Q}_2}{\partial \tau_1} + \frac{\partial \hat{Q}_1}{\partial \tau_2} - 3b_1 \left(\frac{\partial \hat{H}_2''}{\partial x_1} + \frac{\partial^2 \hat{H}_1'}{\partial x_1^2} \right) - b_2 \frac{\partial \hat{Q}_2}{\partial x_1} - b_3 \frac{\partial \hat{H}_2}{\partial x_1} \end{aligned} \qquad (10.41)$$

$$c\hat{H}_3' - \hat{Q}_3' = \frac{\partial \hat{H}_2}{\partial \tau_1} + \frac{\partial \hat{H}_1}{\partial \tau_2} - c\frac{\partial \hat{H}_2}{\partial x_1} + \frac{\partial \hat{Q}_2}{\partial x_1} \qquad (10.42)$$

where the primes still denote derivatives with respect to the original coordinate x. We note that the inhomogeneous parts of the first two orders are orthogonal to $\boldsymbol{v}_0$ and hence satisfy the solvability condition.

The first-order solution is spanned by the null mode $\boldsymbol{w}_0$

$$\hat{H}_1 = \alpha A(x_1, \tau_1, \tau_2) h_0'(z), \quad \hat{Q}_1 = \alpha A(x_1, \tau_1, \tau_2) q_0'(z), \qquad (10.43)$$

where α is inserted into the amplitude to simplify the algebra and $z = x\alpha$ with α the wavenumber of the periodic wave. The basic state $h_0(z)$ and $q_0(z)$ are then 2π-periodic in z. With the assistance of the generalized zero mode $\boldsymbol{w}_1$, the second-order solution can also be found

$$\hat{H}_2 = -\frac{\partial A}{\partial \tau_1} \frac{\partial h_0}{\partial \chi} / \frac{\partial c}{\partial \chi} + \alpha \frac{\partial A}{\partial x_1} \left(\frac{\partial h_0}{\partial \alpha} - \frac{\partial c}{\partial \alpha} \frac{\partial h_0}{\partial \chi} / \frac{\partial c}{\partial \chi} \right)$$

$$\hat{Q}_2 = -\frac{\partial A}{\partial \tau_1} \frac{\partial q_0}{\partial \chi} / \frac{\partial c}{\partial \chi} + \alpha \frac{\partial A}{\partial x_1} \left(\frac{\partial q_0}{\partial \alpha} - \frac{\partial c}{\partial \alpha} \frac{\partial q_0}{\partial \chi} / \frac{\partial c}{\partial \chi} \right)$$

(10.44)

Instead of solving the third-order solution explicitly, we need only to invoke the solvability condition that the inhomogeneous term must be orthogonal to v_0,

$$\int_0^{2\pi} \left(\frac{\partial \hat{H}_2}{\partial \tau_1} + \frac{\partial \hat{H}_1}{\partial \tau_2} - c\frac{\partial \hat{H}_2}{\partial x_1} + \frac{\partial \hat{Q}_2}{\partial x_1} \right) dz = 0 \qquad (10.45)$$

The solvability condition yields the dominant modulation equation

$$a_{11} \frac{\partial^2 A}{\partial \tau_1^2} + 2a_{12} \frac{\partial^2 A}{\partial \tau_1 \partial x_1} + a_{22} \frac{\partial^2 A}{\partial x_1^2} = 0 \qquad (10.46)$$

where

$$a_{11} = -\frac{\partial <h_0>}{\partial \chi} / \frac{\partial c}{\partial \chi},$$

$$2a_{12} = \alpha \frac{\partial <h_0>}{\partial \alpha} + \left[-\alpha \frac{\partial c}{\partial \alpha} \frac{\partial <h_0>}{\partial \chi} + c\frac{\partial <h_0>}{\partial \chi} - \frac{\partial <q_0>}{\partial \chi} \right] / \frac{\partial c}{\partial \chi}$$

$$a_{22} = \alpha \left(\frac{\partial <q_0>}{\partial \alpha} - c\frac{\partial <h_0>}{\partial \alpha} \right) - \alpha \frac{\partial c}{\partial \alpha} \left(\frac{\partial <q_0>}{\partial \chi} - c\frac{\partial <h_0>}{\partial \chi} \right) / \frac{\partial c}{\partial \chi}$$

and $<\cdots>$ denotes $\frac{1}{2\pi} \int_0^{2\pi} \ldots dz$, the average over 2π.

We are particularly interested in the modulation frequency triggered by the dominant primary wave within the primary band. This wave has a frequency ω_m and a wavenumber $\alpha_m = (\omega_m/c)$. These values are inserted into (10.46). Using a normal mode $e^{\mu\tau_1 + ix_1}$ solution of (10.46), the Floquet eigenvalues μ specifying the growth rates of the modulation modes are determined by the characteristic polynomial

$$a_{11}\mu^2 + 2a_{12}\mu + a_{22} = 0 \qquad (10.47)$$

Omitting the stable root μ and rescaling the other by the small modulation wavenumber ν to obtain the Floquet growth rate in the original unscaled time and spatial coordinates,

$$\tilde{\mu} = \nu\mu = ia_1\omega$$

where $\omega = c_g \nu$ is the modulation frequency and c_g is the group velocity of the fundamental mode with frequency ω_m. Strictly speaking, the slow modulation disturbance of a primary periodic wave only travels with a group velocity at the frequency of the primary wave if the primary wave is of small amplitude. Although our primary periodic wave is of finite amplitude, the amplitude is small and we find $c_g(\omega_m)$ to be a good approximation.

Due to the translational invariance of the periodic wave, the leading-order Floquet growth rate is neutrally stable. To obtain its higher-order dependence on the wave frequency, we extend the expansions in (10.33), (10.34) and (10.35) to the next three orders. Omitting the tedious algebra for the solvability condition at each order, we simply present the derived Floquet growth rate as

$$\tilde{\mu} = ia_1\omega + a_2\omega^2 + ia_3\omega^3 + a_4\omega^4 \qquad (10.48)$$

where ω is the low modulation frequency.

Table 10.2 Secondary Floquet coefficients of the fundamental periodic wave

δ	a_1	a_2	a_3	a_4
0.02	0.0182	0.130	0.0226	-0.190
0.04	0.1167	0.468	0.2074	-0.731
0.06	0.2076	0.857	0.5656	-1.714
0.08	0.2430	1.204	1.1311	-2.762
0.10	0.2413	1.501	1.6967	-4.135
0.15	0.1593	2.057	3.1860	-9.001
0.20	0.1110	3.154	5.1617	-14.06

As expected from the Floquet theory of Chapter 6, the real part of the Floquet growth rate grows quadratically with ω from zero frequency. In Table 10.2, we tabulate the computed coefficient a_i for the dominant (fundamental) primary wave with frequency ω_m of Table 10.1. We also favorably compare these results to the Floquet growth rates of Chapter 6.

Since the modulation packet travels with the group velocity c_g of Table 10.1, the spatial evolution of the secondary band (10.12), due to the modulation instability with Floquet growth rate (10.48), is represented by

$$R(\omega) = R_o\sqrt{\frac{\pi}{2bx}} exp\left(-bx_0\omega^2/2\right) exp\left[(a_2\omega^2 + a_4\omega^2)(x-x_0)/c_g\right] \qquad (10.49)$$

where R_0 is a normalization constant and x_0 is the location (10.13) where the primary fundamental has saturated.

This spectrum begins with a maximum at $\omega = 0$ for x below a critical value x_1,

$$\frac{x-x_0}{x_0} < \frac{c_g b}{2a_2} = \frac{x_1 - x_0}{x_0} \qquad (10.50)$$

This critical latent distance x_1 is typically less that $1/10$ of x_0. Beyond this latent distance, the maximum takes on a finite value in ω which increases downstream. This dominant modulation frequency due to a competition between the initial slaving excitation and the subsequent Floquet instability of the primary wave is then

$$\Delta = \left[\frac{bc_g(\frac{x_0}{x-x_0}) - 2a_2}{4a_4}\right]^{1/2} \tag{10.51}$$

Theoretical prediction (10.47) of the secondary spectrum and the predicted maximum modulation frequency (10.51) are seen to be in good agreement with the simulation results in Chapter 4 in Figure 10.6 for $190 < x < 210$ beyond $x_0 = 180$ for that case.

The transition to pulses occurs when the dominant modulation with frequency Δ reaches an amplitude comparable to the amplitude of the fundamental wave r_1 in (10.13) and (10.19). As discussed in section 6.3, the modulation synchronizes all the harmonics to form a pulse at its nodes and this process is quite apparent in the second frame of Figure 4.10. These first pulses are larger than those created later and are the excited pulses. After their creation, very fast Fourier phase synchronization dynamics occur at the wave field between them, eventually converting all remaining sinusoidal waves into solitary pulses. These later pulses are almost identical in amplitude and are called equilibrium pulses. The earlier excited pulses are larger and begin to overtake and absorb the equilibrium pulses irreversibly, as seen in the last few frames of Figure 4.10 and in the world lines of Figure 4.12.

The density of the equilibrium pulses is approximately ω_m, as each sinusoidal peak at inception is converted to an equilibrium pulse. The density of the excited pulse, however, is determined by the value of Δ when they are first created. We estimate this position x_2 by assuming a linear growth of the dominant modulation until its amplitude reaches the amplitude r_1 of the fundamental

$$a_0 \exp\left(\int_{x_0}^{x_2} \frac{\mu_r}{c_g} dx\right) = r_1 \tag{10.52}$$

where a_0 is the initial amplitude of the modulation. Substitution of the Floquet growth rate (10.48) into (10.52) yields the following transcendental equation for x_2

$$x_2 - x_0 = \frac{\Gamma_2}{\Gamma_1} x_0 \log\left(1 + \frac{x_2 - x_0}{x_1 - x_0}\right) + \frac{1}{\Gamma_1} \log\frac{r_1}{a_0} \tag{10.53}$$

where

$$\Gamma_1 = \left(-\frac{a_2^2}{2c_g a_4}\right) \text{ and } \Gamma_2 = \left(-\frac{ba_2}{4a_4}\right)$$

In some cases, $(x_2 - x_0)$ is small compared to $x_1 - x_0$ and an explicit estimate can be obtained

$$\frac{x_2 - x_0}{x_1 - x_0} = \sqrt{\frac{2}{\Gamma_2 x_0} \log(\frac{r_1}{a_0})} \tag{10.54}$$

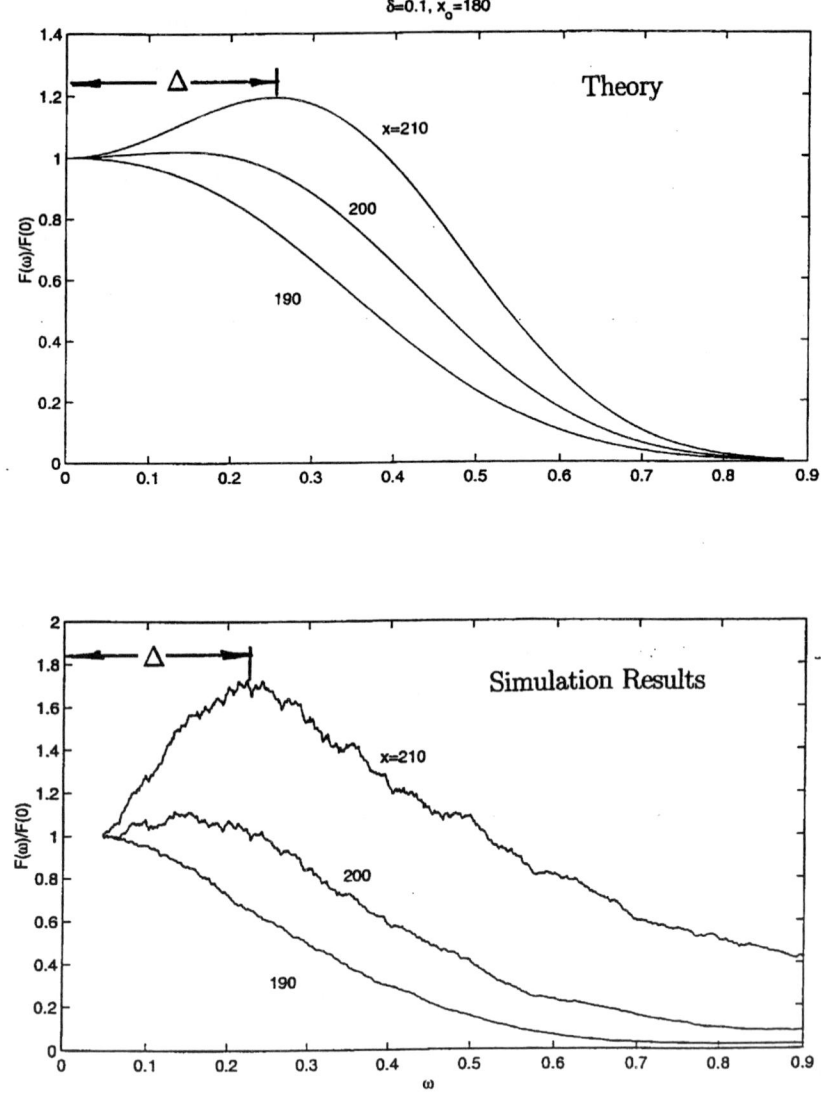

Figure 10.6: The normalized secondary band after the onset of modulation instability with the saturation of the primary band. Both simulation and theory show a dominant modulation frequancy Δ for $\delta = 0.1$ and $\varepsilon = 10^{-7}$. The onset of modulation is estimated omit to be $x_0 = 180$.

We are unable to estimate the initial modulation amplitude a_0 at x_0 – the unknown constant in (10.12) and (10.49) for the secondary band prevents an accurate estimate. However, comparison to our numerical of x_1 and x_2 suggest that a_0 ranges from $0.1\,r_1$ to $0.25\,r_1$. We use these two limits to estimate a bound for x_2 from (10.53). These bounds are then inserted into (10.51) to obtain Δ_2, the dominant modulation frequency at x_2, when the pulses are created. As seen in Figures 10.7a and 10.7b, these bounds are in good agreement with the simulated pulse frequency at inception for a range of noise amplitude ($\epsilon' = 10^{-8}$ to 10^{-5}) and a range of δ up to unity. In essence, the initial pulse

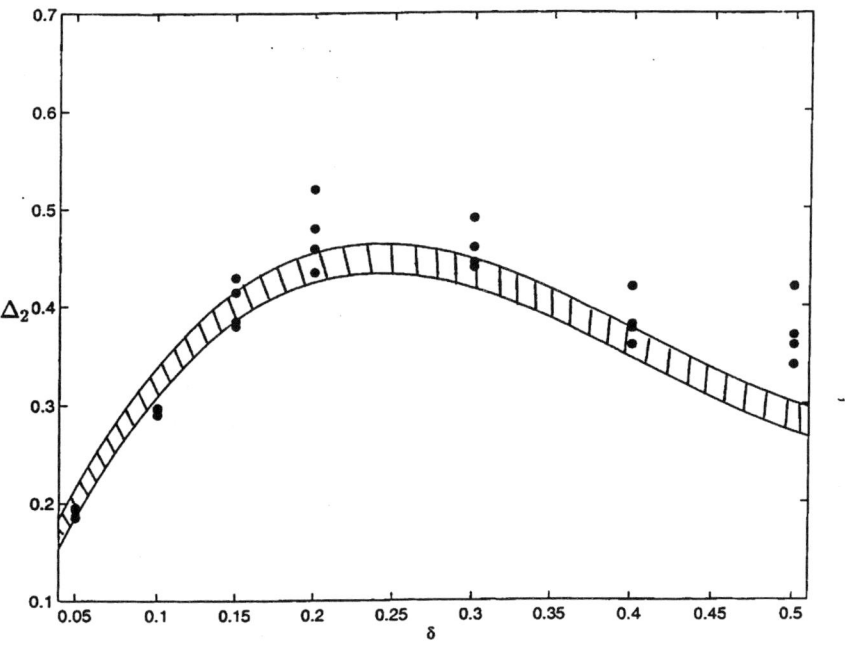

Figure 10.7: The predicted band of excited pulse frequency Δ_2 compared to simulation results.

density is almost independent of the white noise amplitude over this large range of ϵ'. This insensitivity only breaks down after ϵ' exceeds 10^{-3}. We note the Δ_2 dependence on δ in Figure 10.7a resembles that of ω_m in Figure 2. As a result, when Δ_2 is divided by ω_m to produce the excited pulse fraction at inception, Figure 10.7b and 4.36 shows an almost constant fraction beyond $\delta = 0.10$ that is in good agreement with our simulation results. About one out of every four pulses is excited for δ in excess of 0.1 and about one out of ten is excited for δ less than 0.1. These seem to be robust, universal numbers that are insensitive to noise, provided it is small, and independent of operating fluids/conditions. For water, flow conditions below $\delta = 0.1$ correspond to very

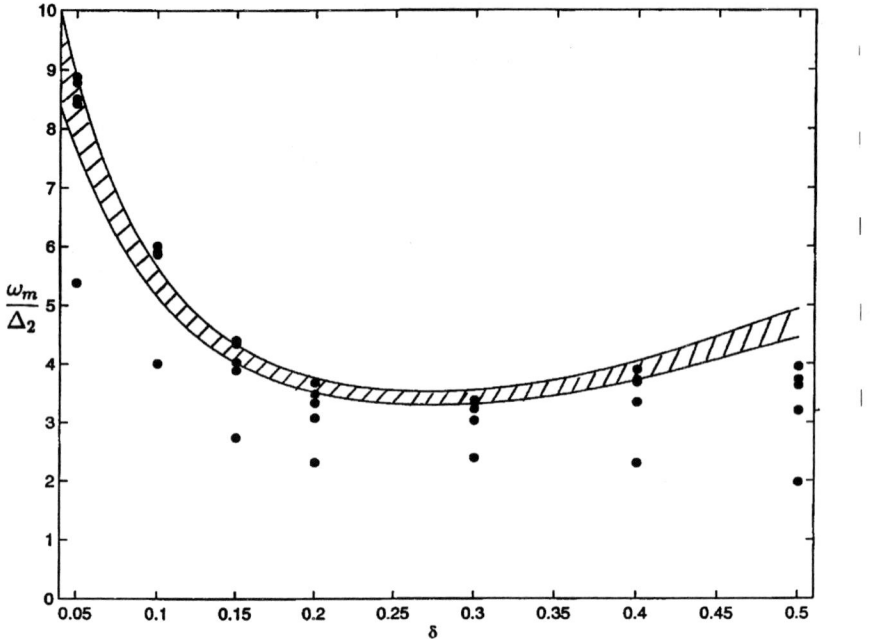

Figure 10.7: The relative density of equilibrium to excited pulses from simulations compared to the theoretical band ω_m/Δ_2. The numerical results correspond to $\varepsilon = 10^{-3}$, 10^{-4}, 10^{-5}, 10^{-6} and 10^{-7}.

thin films at Reynolds numbers less than 10. These conditions are dominated by capillary forces and seem to have a distinctively different dynamics. For the more practical conditions of $\delta > 0.1$, the constant asymptote of Δ_2/ω_m is one key to the universal coarsening rate seen in Figure 4.13.

10.3 Universal coarsening rate based on Δ

With an estimate for the excited pulse fraction from Δ_2, we are now in position to quantify the dynamics of pulse coarsening seen in Figure 10.8 — the dominant dynamics of pulses on a falling film. As seen in the schematics of Figure 8, we neglect the adjustment dynamics of the equilibrium pulses and the complex pulse coalescence and mass drainage dynamics at the excited pulses. Instead, we assume the excited and equilibrium pulses all travel at speeds that remain constant for the entire flow domain. This assumption is consistent with the world lines of Figures 4.12 and 4.14. This implies that the liquid acquired by an excited pulse during a coalescence event is roughly equal to the mass it drains out before the next event. As such, the average speed of the excited pulse between events is constant over the entire flow domain. This occurs only for very fast drainage. For drop coalescence on a thin vertical wire in Chapter 12, we find the excited drops to retain much of the mass after each coalescence. As a result, these drops accelerate downstream and its coalescence frequency also increases monotonically. This scenario seems to occur for $\delta < 0.1$ and will not be modelled here.

Let c_1 and c_2 be the speeds of the excited and equilibrium pulses, respectively. Consider a vertical line parallel to the t axis in the world lines of Figure 4.12. An excited pulse and an equilibrium one pass by this station in space at different times within an interval Δt. If these two pulses coalesce at a different spatial station Δx downstream, the following must be true from simple geometric arguments

$$\Delta t = \frac{\Delta x}{c_2} - \frac{\Delta x}{c_1} \quad (10.55)$$

or in differential form

$$\frac{dt}{dx} = -\beta = -\frac{c_1 - c_2}{c_1 c_2} \quad (10.56)$$

The negative sign of $\left(\frac{dt}{dx}\right)$ indicates all the equilibrium pulses within the time interval dt have been eliminated over a distance of dx. The differential number of equilibrium pulses within this differential period dt is

$$dn = dt \left(\frac{\omega_m}{2\pi}\right) \quad (10.57)$$

and hence

$$dn = -\left(\frac{\beta \omega_m}{2\pi}\right) dx \quad (10.58)$$

Integrating from the pulse inception location x_2, one obtains

$$\begin{aligned} n(x) &= n_0 + \tfrac{\beta \omega_m}{2\pi}(x_2 - x) \\ &= \tfrac{\omega_m}{\Delta_2} + \tfrac{\beta \omega_m}{2\pi}(x_2 - x) \end{aligned} \quad (10.59)$$

Since we have only considered one excited pulse, $n(x)$ is the number of equilibrium pulses between two excited pulses. The average pulse period is then

$$\begin{aligned} <t> &= \tfrac{2\pi/\Delta_2}{n(x)} = \frac{2\pi}{\omega_m + \left(\frac{\beta \omega_m \Delta_2}{2\pi}\right)(x_2 - x)} \\ &\sim \tfrac{2\pi}{\omega_m} + \left(\tfrac{\Delta_2}{\omega_m}\right) \beta (x - x_2) \end{aligned} \quad (10.60)$$

The coarsening rate is then

$$\frac{d<t>}{dx} \sim \left(\frac{\Delta_2}{\omega_m}\right)\left(\frac{c_1 - c_2}{c_1 c_2}\right) \quad (10.61)$$

The linear correlation between excess pulse speed and excess amplitude has been verified experimentally (Alekseenko et al., 1994) and has been captured

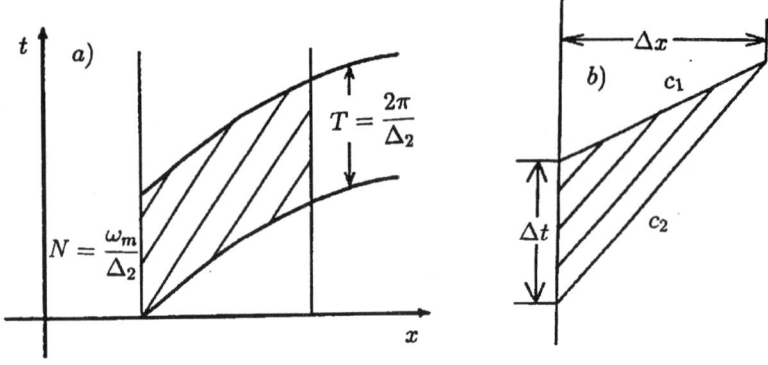
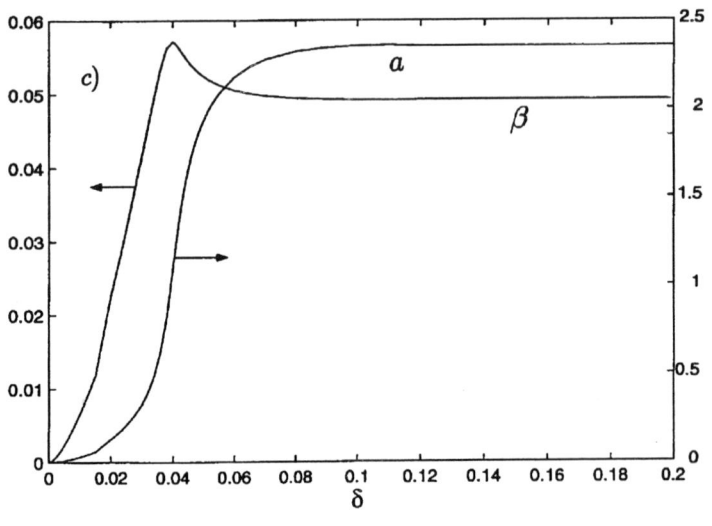

Figure 10.8: Schematic for the pulse coaleacence model. Since the equilibrium pulse amplitude a reaches a constant asymptotem, the differential speed β between excited and equilibrium pulses also approach a constant asymptote.

by our earlier version in sections 8.3 and 8.4. We shall use the resonance pole theory of section 8.4. In the present variables, it appears as

$$c \simeq 3.0 + 3.0a \qquad (10.62)$$

where a here is the total pulse amplitude instead of the excess one. At $a = 0$, one obtains the phase velocity of low-δ sinusoidal waves at inception. Equilibrium pulse amplitude a as a function of δ has been computed in Chapter 3 and 5.

We have estimated the excited pulses to have roughly twice the mass as the equilibrium ones due to the coalescence events in Chapter 7. Since the pulse width remains the same, the excited pulse should also be twice as large in amplitude. Using this assumption and correlation (10.62), the parameter β can be expressed in terms of the amplitude of the equilibrium pulse a,

$$\beta = \frac{c_1 - c_2}{c_1 c_2} = \frac{a}{(1+a)(3+5a)} \qquad (10.63)$$

As shown in Figure 10.8, for the large pulses beyond $\delta = 0.1$ (a larger than 2), this parameter approaches a rough constant of about 0.05. Combining this universal pulse differential speed with the generic excited pulse fraction of $1/4$ for δ larger than 0.15 in Figure 10.7b, we obtain the universal dimensionless coarsening rate

$$\frac{d<t>}{dx} = 0.015 \qquad (10.64)$$

This rough estimate is in agreement with all our numerical simulations of δ between 0.1 and 0.6 and a wide range of noise amplitudes, as seen in Figure 10.9. In dimensional form, are simply multiplies 0.015 by the Nusselt average flat-film velocity u_0.

In Figure 10.10, the final saturated wave period $<t>_\infty$ from our simulation, after the coalescence events have ceased, is favorably compared to the theoretical prediction $2\pi/\Delta_2$ from (10.51) and (10.54) and from Figure 10.7a. This indicates that only the excited pulses remain at the end and all the equilibrium pulses have been eliminated. This asymptotic wave period far downstream is actually the modulation frequency specified by the filtering of the inlet noise and the secondary instability of inception waves within $x < 200$ (less than 1 5cm for water) near the inlet! The dominance of pulse coherent structures and the creation of excited pulses from secondary modulation instability lead to this simple unique feature despite the complex noise-driven wave dynamics.

10.4 Noise-driven wave dynamics

The key difference between the periodically forced wave dynamics in Chapter 6 and the naturally excited wave dynamics of this chapter is the persistent residual primary spectrum. This spectrum is created when white noise at the inlet is filtered by the primary linear evolution during the inception region. It insists that the amplitude of low frequency modes decay with increasing wavelength.

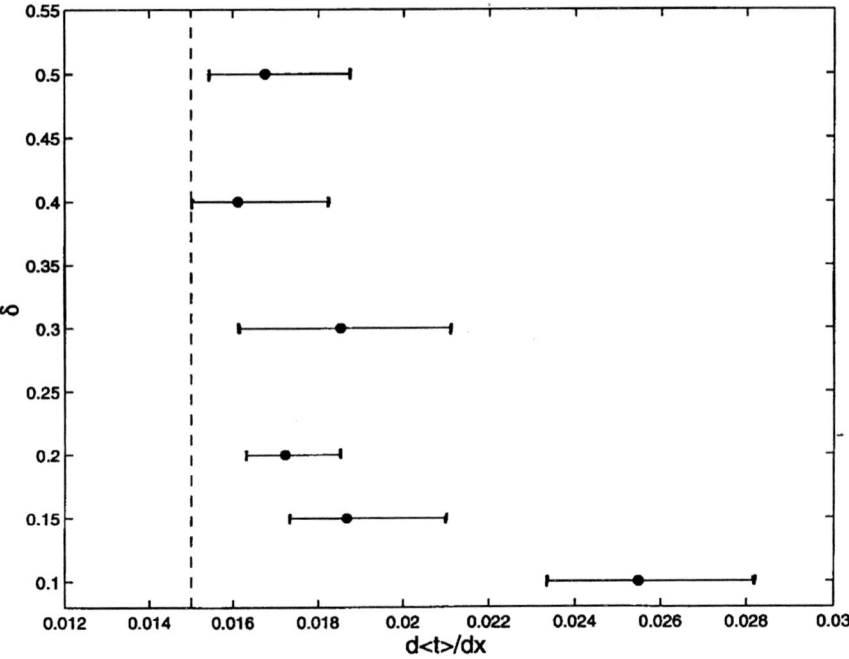

Figure 10.9: The constant asymptote of β beyond $\delta = 0.1$ produces a universal dimensionless coarsening rate $\frac{d<t>}{dx} = 0.015$ that is consistence with our simulation results. The error bars correspond to noise amplitudes $10^{-7} < \varepsilon < 10^{-3}$.

Such a spectrum cannot be created by periodic forcing or even a narrow-banded inlet noise that has no component in the long wave limit (much longer than the fastest-growing fundamental mode). Hence, as the secondary modulation instability of the fundamental triggers a second set of disturbances, the amplitude of such disturbances do not follow that predicted by the Floquet theory of Chapter 6, one that grows with wavelength. Instead, the residual primary spectrum couples with this modulation instability to produce a dominant modulation frequency that produces solitary waves immediately. The modulation frequency also yields the fraction of pulses formed that are larger than usual–a critical information that describes the subsequent coarsening. In periodic forcing, on the other hand, the primary spectrum is no where to be seen in the secondary transition and the wave dynamics hop from one member of the two periodic wave families to the next, as was seen in Chapter 6. Although the pulse limit of the γ_2 family is eventually reached, there is no excited pulse to speak of and little coalescence takes place to induce further wave evolution. There is hence a distinct difference between periodically forced system and one driven by white natural noise. We know of no other hydrodyamic instability or any other

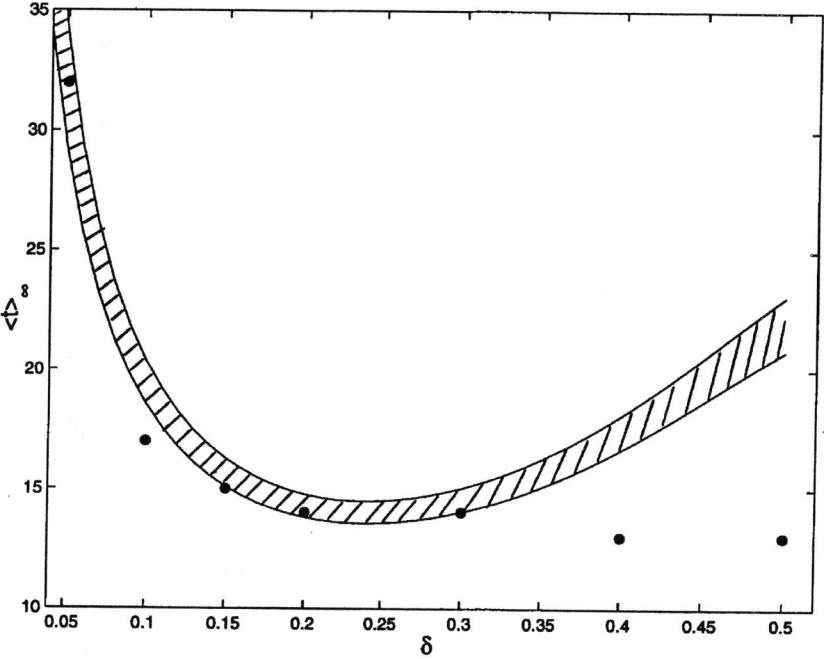

Figure 10.10: The final equilibrium pulse period of Figure 4.35 compared to the theoretical band $2\pi/\Delta_2$.

pattern formation dynamics where the role of inlet noise and the downstream dynamics triggered by the noise are so clearly understood as for the falling film.

CHAPTER 11

Transverse Instability

According to Squire's theorem in Chapter 2, the most unstable perturbations to the Nusselt flat-film basic state are two-dimensional. For sufficiently small inlet noise, linear filtering mechanism isolates a narrow band of such perturbations near the two-dimensional maximum growing frequency, as seen in Chapter 4. Hence, we see nearly two-dimensional periodic waves near the inlet region. However, some residual three-dimensional components of the inlet noise still disturb these waves, as seen in the wave image of Figure 4.4.

We shall show in this Chapter that two-dimensional saturated waves are unstable to both two- and three- dimensional perturbations. Two-dimensional perturbations are usually more dangerous and the resulting secondary instability transforms the two-dimensional periodic waves into two-dimensional solitary waves, see experiments by Liu, Schlicting and Gollub (1995), as was discussed in the previous chapter.

Two-dimensional solitary waves are shown in Chapter 7 to suppress all small radiation modes, both two-dimensional and three-dimensional, which are connected with the essential spectrum. However, while these two-dimensional solitary waves are stable to two-dimensional disturbances(see Chapter 5) ,they are unstable to three-dimensional discrete modes of the spectrum. Hence, small background three-dimensional noise eventually destroys two-dimensional pulses. Physically, it implies that the crests of the two-dimensional pulses develop pronounced transverse instability, eventually producing coupled oblique waves, localized scallop waves of Figure 4.5 or transversely modulated two-dimensional pulses. We shall delineate all these transitions involving transverse instabilities here.

Three-dimensional nonlinear wave evolution is much more complex than two-dimensional wave dynamics. Hence, we restrict our investigation to the simplest model equations.

11.1 Coupled oblique waves and triad resonance.

In a small vicinity of the critical Reynolds number, the Navier-Stokes equations for a liquid film on a vertical wall was transformed to the 3D gKS equation in (3.45)

$$\frac{\partial H}{\partial t} + 4H\frac{\partial H}{\partial x} + \frac{\partial^2 H}{\partial x^2} + \delta\frac{\partial}{\partial x}\nabla^2 H + \nabla^4 H = 0 \tag{11.1}$$

where

$$\nabla^2 = \frac{\partial^2}{\partial x^2} + \frac{\partial^2}{\partial z^2}$$

which is a generalization of two-dimensional Kawahara equation in (3.21) and we have carried out a simple transformation of $H \to \frac{4}{6}H$ to convert the coefficient 6 in (11.1) to that of 4 in the Kawahara equation. We can also generalize the invariant symmetries (5.2)–(5.4) to three dimensions

$$H(x, z, t) \to H(x + x_0, z + z_0, t + t_0) \tag{11.2}$$
$$H(x, z, t) \to -H(-x, z, t) \tag{11.3}$$
$$H(x, z, t) \to H(x, -z, t) \tag{11.4}$$
$$H(x, z, t) \to H(x - 4\chi t, z, t) + \chi \tag{11.5}$$

where x_0, z_0, t_0 and χ – are some constants.

For large capillary forces $\delta \to 0$ and (11.1) becomes a three-dimensional generalization of the KS equation,

$$\frac{\partial H}{\partial t} + 4H\frac{\partial H}{\partial x} + \frac{\partial^2 H}{\partial x^2} + \nabla^4 H = 0 \tag{11.6}$$

The equation (11.1) has a trivial solution :

$$H = 0 \tag{11.7}$$

For small sinusoidal perturbations of the trival solution

$$H = \varepsilon e^{i(\mathbf{k}\vec{x} - \omega t)} = \varepsilon e^{i(\alpha x + \beta z - \omega t)}, \quad \varepsilon \to 0 \tag{11.8}$$

we have the characteristic equation :

$$\omega = i\left[\alpha^2 - (\alpha^2 - \beta^2)^2\right] \tag{11.9}$$

which defines a circle (a neutral stability curve) in the wavenumber space for neutral waves.

When the wave vector $\mathbf{k} = (\alpha, \beta)$ is inside the circle (see Figure 11.1),

$$\left(\alpha - \frac{1}{2}\right)^2 + \beta^2 = \frac{1}{4} \tag{11.10}$$

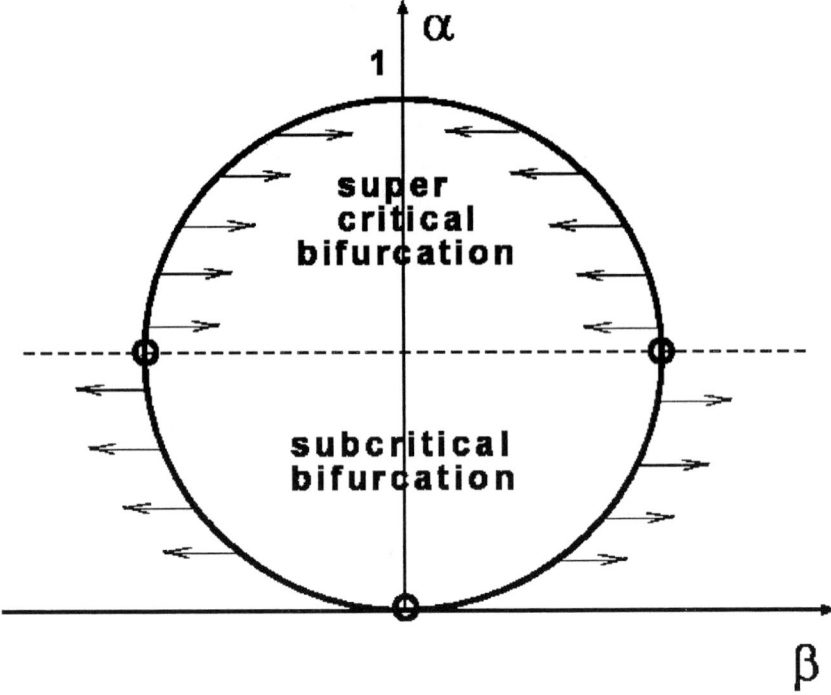

Figure 11.1: The trivial solution is unstable inside the circle for neutral stability and stable outside. Crossing the line $\beta = \frac{1}{2}$ changes the character of instability.

the trivial solution (11.7) is unstable. The most unstable disturbances, according to Squire's theorem, are two-dimensional and occur at

$$\mathbf{k} = \left(\frac{\sqrt{2}}{2}, 0\right) \tag{11.11}$$

with the growth rate

$$\omega_i^{max} = \frac{1}{4} \tag{11.12}$$

For any localized perturbation, the energy balance (5.9) can be rewritten in the generalized form :

$$\frac{d}{dt}\int_{-\infty}^{+\infty}\int_{-\infty}^{+\infty} H^2 dx dz = 2\int_{-\infty}^{+\infty}\int_{-\infty}^{+\infty}\left[\left(\frac{\partial H}{\partial x}\right)^2 - \left(\nabla^2 H\right)^2\right] dx dz \tag{11.13}$$

Energy transfer is along the gravity vector in the x-direction, while short-wave energy dissipation is isotropic in space. As for the two-dimensional case, the dispersion term does not contribute to the energy equation.

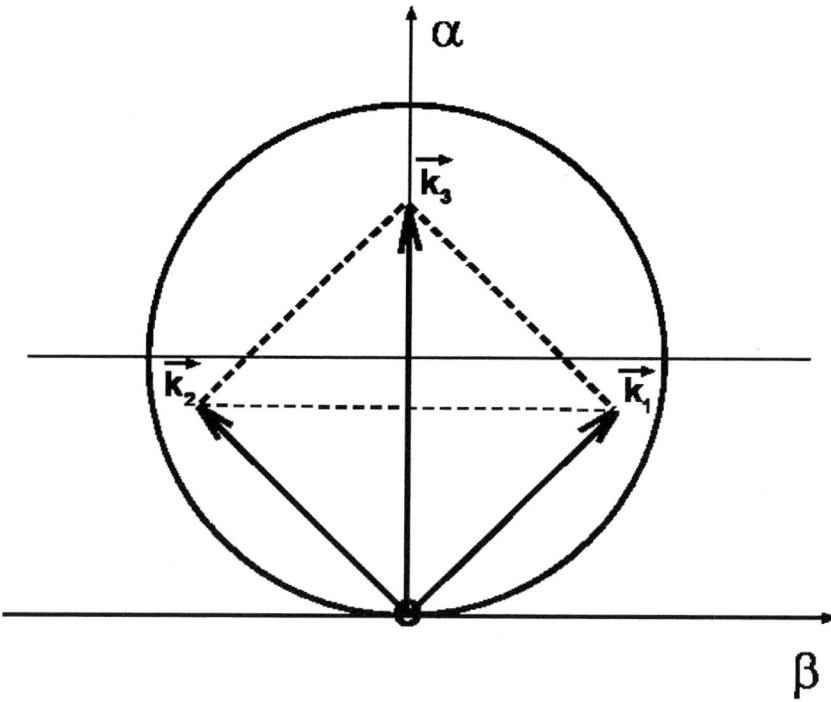

Figure 11.2: Coupled triplets of wave vectors $\mathbf{k}_{1,2} = (\alpha, \pm\beta)$ and $\mathbf{k}_3 = (2\alpha, 0)$.

Such energy transfer produces three types of three-dimensional waves. We shall first analyze stationary waves propagating at the phase velocity of near neutral waves due to oblique waves. A wave-vector $\mathbf{k} = (\alpha, \beta)$ across the neutral stability circle corresponds, in physical space to some oblique wave with its front perpendicular to $\mathbf{k}$, see Figure 11.3. A saturated oblique wave with this wave vector can be directly related to a two-dimensional wave constructed in section 5.4 after a simple rescaling of wave number. Such waves have not been observed in experiments, as Squire's theorem suggests their precursor waves are not as dominant as two-dimensional waves that are perpendicular to the direction of flow.

A more dominant saturated three-dimensional waves is presented in Figure 11.3b. Two wave-vectors $\mathbf{k}_1$ and $\mathbf{k}_2$ cross the neutral stability curve simultaneously. We shall consider the most important symmetric case, $\mathbf{k}_1 = (\alpha, \beta)$ and $\mathbf{k}_2 = (\alpha, -\beta)$. Mutual nonlinear iteraction of these oblique waves excites a two dimensional mode $\mathbf{k}_3 = (2\alpha, 0)$. Such combinations of wave vectors are known as resonant triads and they give rise to the three-dimensional waves shown in the left frame of Figure 11.4 .

We construct these coupled oblique waves by an expansion in the double

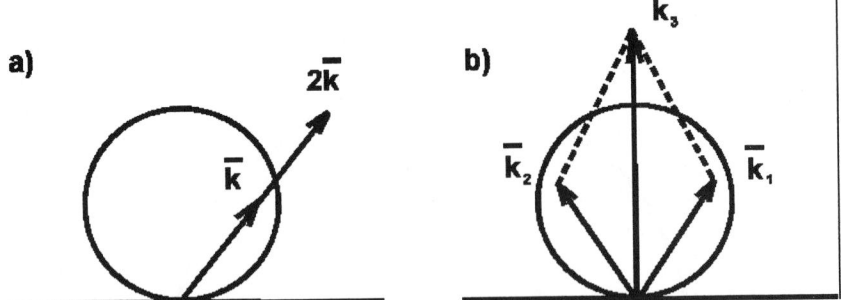

Figure 11.3: Two ways of bifurcation off the the neutral stability curve.

Fourie series
$$H = \sum_n H_n e^{in(\alpha x + \beta z)}$$

which propogates at some angle with respect to the direction of gravity.

For the resonant triads $\mathbf{k_1} = (\alpha, \beta)$, $\mathbf{k_2} = (\alpha, -\beta)$ and $\mathbf{k_3} = (2\alpha, 0)$, direct substitution into the KS equation (11.6) yields for stationary waves with zero speed, Nepomnyashchy (1974, 1976)

$$H \sim H_{11} \sin \alpha x \cos \beta z + H_{20} \sin 2\alpha x + H_{22} \sin 2\alpha x \cos 2\beta z \tag{11.14}$$

$$H_{11} = \pm \varepsilon \sqrt{a}, \quad H_{20} = -\varepsilon^2 \frac{\alpha a}{L_{20}}, \quad H_{22} = -\varepsilon^2 \frac{\alpha a}{L_{22}} \tag{11.15}$$

where

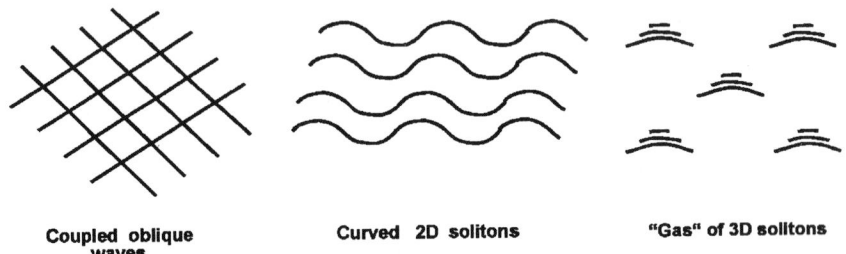

Coupled oblique waves   **Curved 2D solitons**   **"Gas" of 3D solitons**

Figure 11.4: Three possible 3D patterns.

$$a = \frac{L_{20} L_{22}}{\alpha^2 (L_{20} + 2L_{22})} \tag{11.16}$$

$$L_{nm} = (n\alpha)^4 + (m\beta)^4 - (n\alpha)^2 (1 - 2m^2 \beta^2) \tag{11.17}$$

and we have introduced the small bifurcation parameter ε, which measures the distance from the neutral curve (11.10), as

$$\varepsilon^2 = -L_{11} \tag{11.18}$$

At $\beta \to 0$, (11.14)-(11.18) collapse into the two-dimensional solution (5.28).

It is interesting that this stationary solution exists at $L_{11} > 0$ (bifurcating into the instability region) for $\beta > \frac{1}{2}$ and $L_{11} < 0$ (bifurcation out of the instability circle) for $\beta < \frac{1}{2}$, see Figure 11.1 and 11.2. In the last case, the mode $\mathbf{k}_3 = (0, 2\alpha)$ lies in the unstable region, as it is shown in Figure 11.2, and the stationaty wave results from coupling among a triad of unstable waves. Unfortunately, these checker-board patterns triggered by coupling among triads of oblique waves are unstable to long distubances (modulation instability). They are typically only observed at the base of localized scallops. Three-dimensional waves on the right two frames of Figure 11.4 are much more stable and prevalent.

11.2 Transverse breakup of equilibrium 2D-waves

The other two types of three-dimensional waves in Figure 11.4 arise from instabilities of saturated two-dimensional periodic and solitary pulses. The KS equation (5.1a)

$$\frac{\partial H}{\partial t} + 4H\frac{\partial H}{\partial x} + \frac{\partial^2 H}{\partial x^2} + \frac{\partial^4 H}{\partial x^4} = 0 \qquad (11.19)$$

possesses two-dimensional travelling periodic waves with velocity C which obey equations (5.11) and (5.13)

$$H''' + H' + 2H^2 - CH = Q \qquad (11.20)$$

where Q is an integration constant. In Figure 5.23 we depict the most important

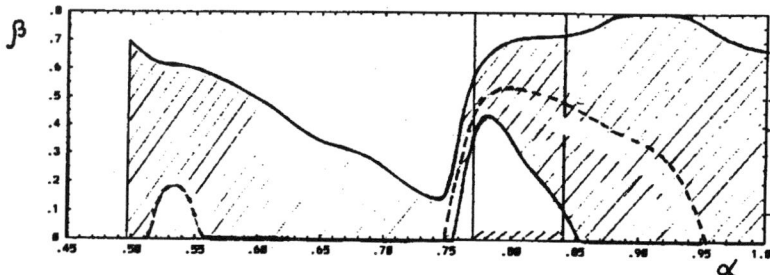

Figure 11.5: Instability region of the S-family against 3D perturbations. All the waves of the S-family are unstable. Dashed line corresponds to the most unstable β.

of these periodic wave solution branches. The primary branch S is a periodic wave solution of (11.20) with $C = 0$, which bifurcates at $\alpha = 1$ from the trivial solution $H = 0$. Its zero speed indicates that it travels at the near-critical linear phase speed - three times the average film velocity of the Nusselt flat film. At, $\alpha = 0.498$, this S-branch disappears. C-waves branch from S at $\alpha = 0.5547$ and practically the whole branch consist of periodic waves that resemble one-hump pulses. The shapes of the S-waves are shown in Figure 5.19 and C-waves in Figure 5.21.

Consider the stability of the solutions of (11.20) to three-dimensional disturbances:

$$H(x,z,t) = h(x) + \varepsilon f(x)e^{i\beta z}e^{\mu t} \qquad (11.21)$$
$$\varepsilon \to 0$$

after substitution (11.21) in (11.19) and linearizing, we can obtain

$$\mu f - Cf' + (1 - 2\beta^2)f'' + f^{IV} + \beta^4 f + \frac{d}{dx}(hf) = 0 \qquad (11.22)$$

According to the Floquet theorem, solutions of (11.22) have the form

$$f = F(x)e^{i\nu\alpha x} \qquad (11.23)$$

where $F(x)$ has the same period $2\pi/\alpha$ as $H(x)$ and ν is a real number because we seek spatially-bounded solutions with $\nu \in (0, 1/2)$. If we expand $H(x)$ and $F(x)$ in a Fourie series and substitute into (11.22), we obtain an eigenvalue problem that can be solved by the QR-algorithm. The results of stability calculation for the S-family are presented in Figure 11.5, the S - branch has a region stable to two-dimensional disturbances at $\alpha \in [0.768, 0.837]$. However, these S-solutions stable to two-dimensional disturbances are unstable to three-dimensional ones.

Consider now three-dimensional solutions which branch from the two-dimensional one-hump pulses of the C-family. The linearized equation for perturbations is still (11.22), but we now seek solutions that decay at $x = \pm\infty$, corresponding to the discrete modes of the one-hump two-dimensional pulse,

$$f(x) \to 0 \quad at \quad x \to \pm\infty \qquad (11.24)$$

The characteristic equation of the differential equation at $x \to \pm\infty$

$$f^{IV} + (1 - 2\beta^2)f'' - Cf' + (\beta^4 + \mu)f = 0$$

is

$$\sigma^4 + (1 - 2\beta^2)\sigma^2 - C\sigma + (\beta^4 + \mu) = 0 \qquad (11.25)$$

It has two solutions that decay at $+\infty$ and two that decay at $-\infty$. The process of matching of these asymptots is described in Chapter 7. The results are presented in Figure 11.6. At $\beta = 0$ the specrum is the same as for two-dimensional perturbations with a zero discrete mode due to translation invariance (11.2). With finite β, the translation mode becomes unstable.

In some cases, this implies the two-dimensional pulses in the C family develop transverse instability and evolve into localized three-dimensional scallop waves, as seen on the right frame of Figure 11.4. In many cases, they evolve instead, into a "modulated two-dimensional" wave, see the middle frame of Figure 11.4 and 11.7, without further evolution into isolated scallop waves. We shall analyse these two district transitions in the next sections.

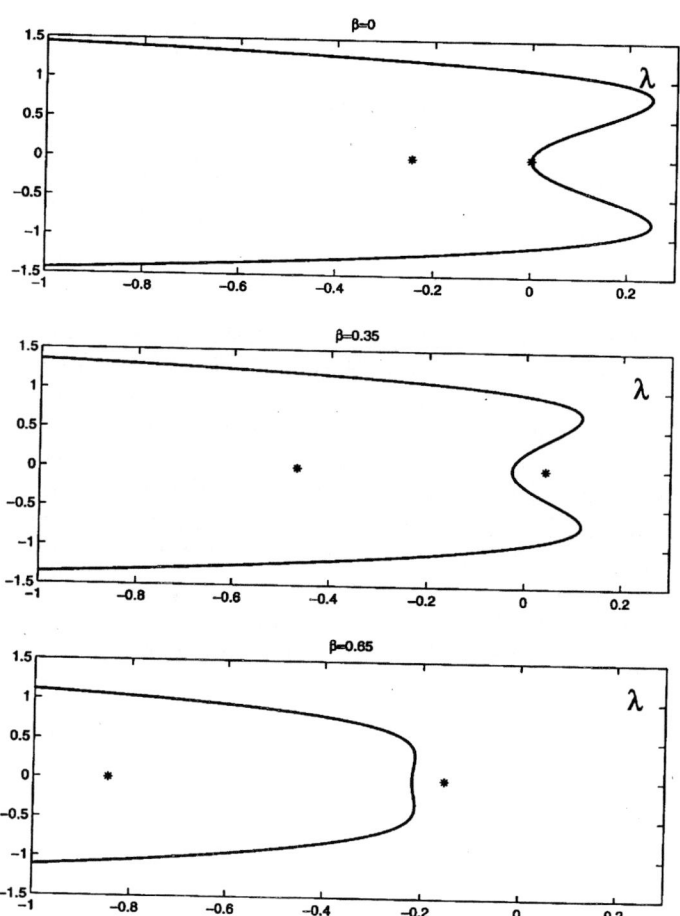

Figure 11.6: Discrete and essential spectrum of the C-pulses of the KS equation for different β; the translational zero mode becomes unstable.

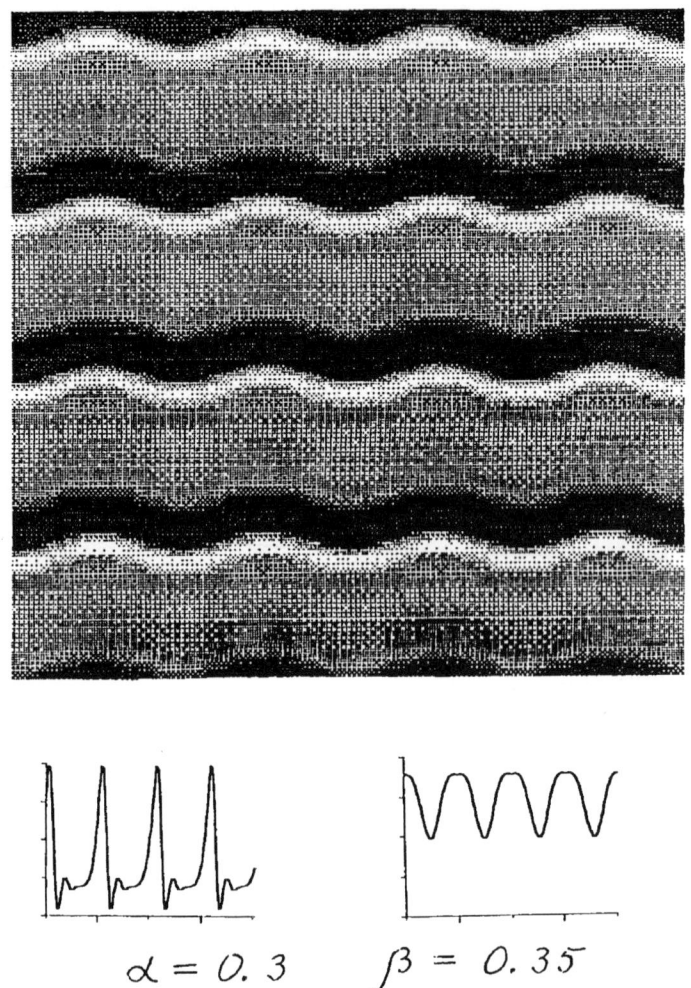

Figure 11.7: Transversely Modulated 2D pulses, $\alpha = 0.3$, $\beta = 0.35$.

11.3 scallop waves

Three-dimension travelling waves of the gKS equation (11.1) are described by the equation:

$$-c\frac{\partial H}{\partial x} + 2\frac{\partial H^2}{\partial x} + \frac{\partial^2 H}{\partial x^2} + \delta\frac{\partial}{\partial x}\nabla^2 H + \left(\frac{\partial^2}{\partial x^2} + \frac{\partial^2}{\partial z^2}\right)^2 H = 0 \qquad (11.26)$$

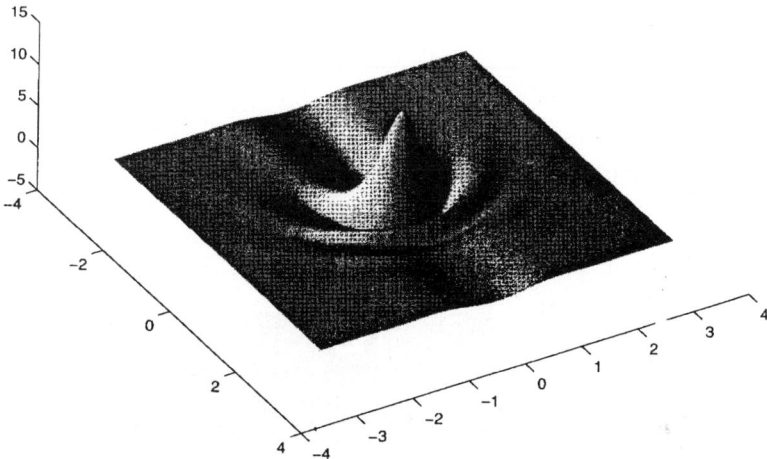

Figure 11.8: The Tsvelodub localized scallop wave of the KS equation (11.6).

For solitary waves, we have to add the decaying condition

$$H(x, z) \to 0, \quad x^2 + z^2 \to 0 \qquad (11.27)$$

The localized scallop wave of the KS equation with $\delta = 0$ was first constructed by Petviashvily and Tsvelodub (1978). This Tsvelodub soliton is shown in the Figure 11.8. It has speed of $c = 0.62$ and an amplitude of $H_{max} - H_{min} =$. Characteristic z-size of this soliton $\simeq$ half of x-size. This aspect ratio originates from the fact that the neutral wave number in the x-direction α_0 is twice that in the z-direction β_0, see Frigure 11.1. The scallop wave appears after several bifurcations from this neutral curve but it still retains its basic aspect ratio of the selected length scales during primary instability.

We have extended Tsvelodub's soliton to finite δ for the gKS equation (11.1). In Figure 11.9 we depict the scallop wave velocity C as a funstion δ; as is

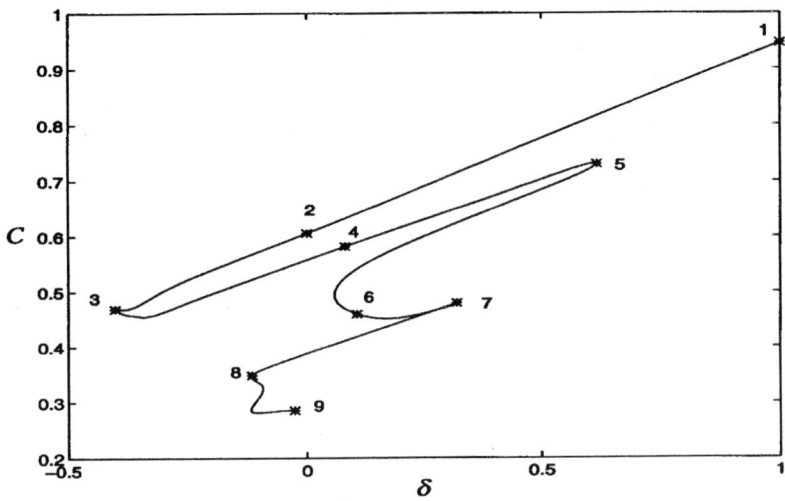

Figure 11.9: Scallop wave speed in the $\delta - C$ space. Wave profiles at points $1, 2, \ldots, 8$ are presented in Figure 11.10.

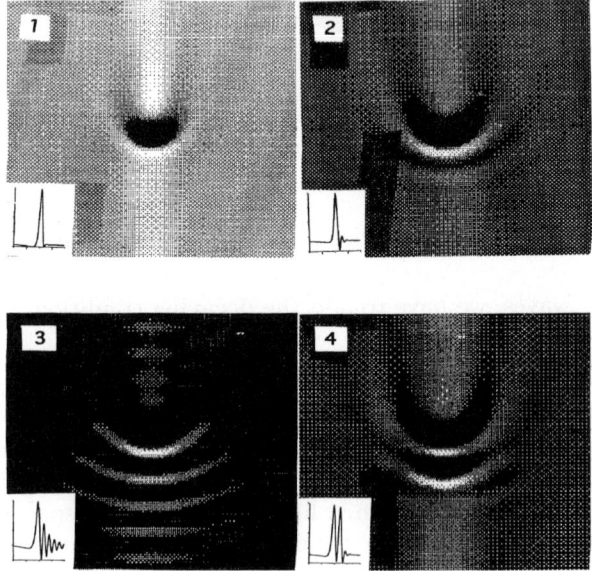

Figure 11.10: Profiles of three-dimensional waves from the bifurcation diagram, Figure 11.9.

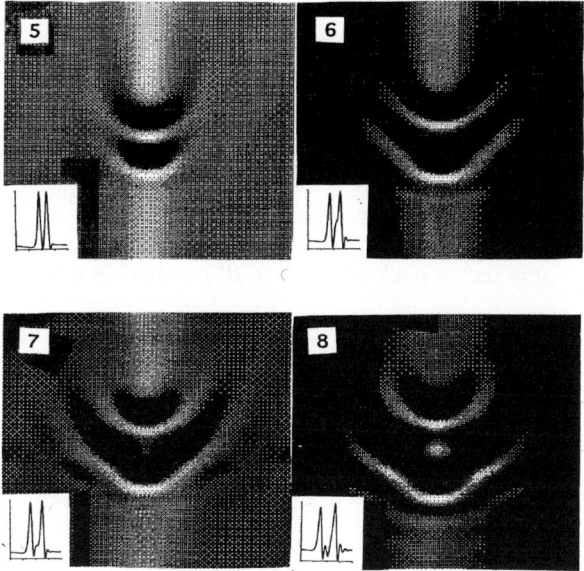

Figure 11.10: Continuation.

analogos to the two-dimensional curve in Figure 5.27. The wave profiles at points $1, 2, \ldots, 8$ are presented in Figure 11.10.

The multi-valued $C(\delta)$ curve retain the same Shilnikov multi-valued shape of Figure 5.27. The profiles in Figures 11.10 show that the similarity is even deeper as different segments also correspond to different multi-hump pulses. At point 1, the wave has a characteristic scallop profile but, at even larger δ, the oscillations dissapear. At lower δ, the profile evolves into the Tsvelodub solution at points 2. At point 3, the oscilations grows the bow wave increases in amplitude until the crests closest to the hump becomes a second hump at point 4. Such behavior is consistent with the two-dimension pulses of the gKS equation in Figure 5.27. The curve $C(\delta)$ spirals down the δ-axis with increasing number of humps and decreasing hump amlitude. The secondary humps often trigger checker-board patterns at their base, as seen in points 6-8. Unfortunatelly, our numerical code did not work beyond point 9.

11.4 Stability of nonlinear localized patterns.

Localized two- and three- dimensional nonlinear patterns can be unstable with respect to modes of essential and discrete spectrum. Let us first consider stabilty with respect to radiation modes, using the same method we applied for two-dimensional waves in Chapter 7. Small and localized perturbations away from the pulse can be represented as a Fourier integral over all the wave numbers α

and β,

$$H = \int_{-\infty}^{+\infty}\int_{-\infty}^{+\infty} F(\alpha,\beta)e^{i(\alpha x+\beta z-\omega t)}\,d\alpha d\beta = \int_{-\infty}^{+\infty}\int_{-\infty}^{+\infty} F(\alpha,\beta)e^{i(\alpha\frac{x}{t}+\beta\frac{z}{t}-\omega)t}\,d\alpha d\beta \qquad (11.28)$$

Disspersive relation $\omega(\alpha,\beta)$ has the form

$$\omega = i\alpha^2 - i(\alpha^4 + 2\alpha^2\beta^2 + \beta^4) - \delta(\alpha^3 + \alpha\beta^2) \qquad (11.29)$$

for the gKS equation. In order to evaluate the integral (11.28) at $t \to \infty$, we find the saddle points of the dispersion relation,

$$\begin{cases} \frac{x}{t} - \frac{\partial\omega}{\partial\alpha} = 0 \\ \frac{z}{t} - \frac{\partial\omega}{\partial\beta} = 0 \end{cases} \qquad (11.30)$$

where

$$\begin{cases} \frac{\partial\omega}{\partial\alpha} = 2i\alpha - i(4\alpha^3 + 4\alpha\beta^2) - \delta(\alpha^3 + \alpha\beta^2) \\ \frac{\partial\omega}{\partial\beta} = -i(4\alpha^2\beta + 4\beta^3) - 2\alpha\beta\delta \end{cases} \qquad (11.31)$$

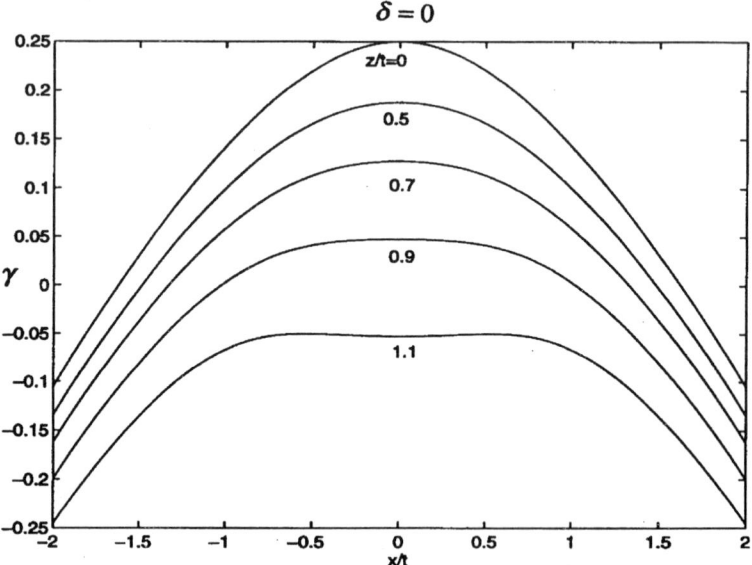

Figure 11.11: The real part γ of $\lambda(\alpha_*)$ at the saddle point α_* for the gKS scallop wave at $\delta = 0$. This value of γ is a function of $\frac{x}{t}$ and β.

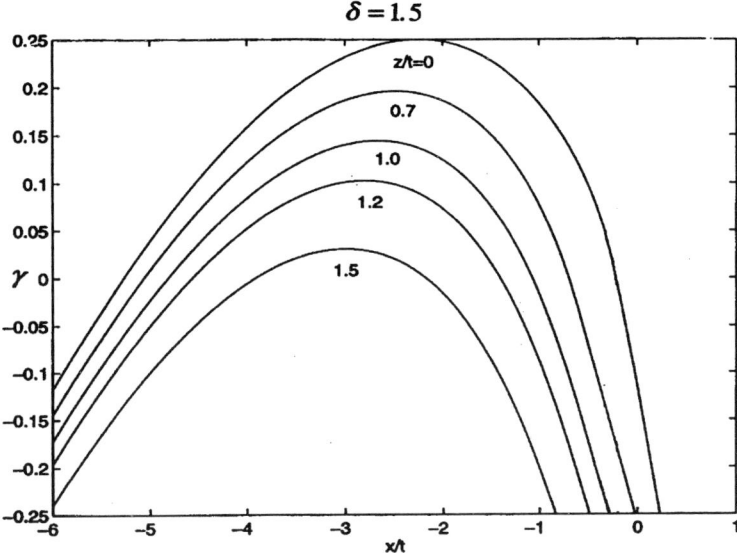

Figure 11.12: The same as for previous picture, $\delta = 1.5$.

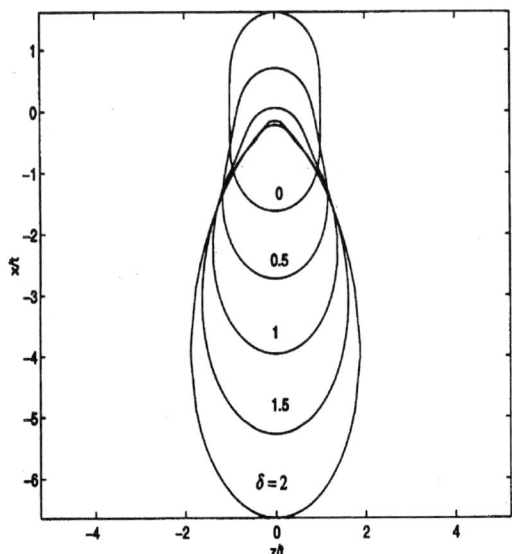

Figure 11.13: Closed figures in the picture are marginal stability curves in the $\frac{z}{t} - \frac{x}{t}$ coordinates for different δ. Inside them, the disturbances unstable.

Rearranging, we get

$$\begin{cases} \alpha^3 + \alpha^2 \left(-\frac{3}{4}i\delta\right) + \alpha\left(\beta^2 - \frac{1}{2}\right) - \frac{i}{4}\left(\frac{x}{t} + \delta\beta^2\right) = 0 \\ \beta^3 + \beta\left(\alpha^2 - \frac{1}{2}i\delta\alpha\right) = \frac{z}{t} \end{cases} \quad (11.32)$$

Solution of the system (11.32) for a given δ provides us the complex saddle point,

$$\alpha_* = \alpha_*\left(\frac{x}{t}, \frac{z}{t}\right)$$

and its complex growth-rate $\lambda(\alpha_*)$. The real part of λ determines the stability along these characteristics

$$\gamma = Real(\lambda(\alpha_*))$$

For $\delta = 0$, the dependence γ on $\frac{x}{t}$ for several values of $\frac{z}{t}$ is given in Figure 11.11 and Figure 11.12. We see that, at fixed $\frac{z}{t}$, $\gamma\left(\frac{x}{t}\right)$ has stable and unstable regions with two neutral points. At fixed $\frac{x}{t}$, we can also find two neutral points (speeds) along $\frac{z}{t}$. The generalization of these neutral points for various δ is given in Figure 11.13. Closed curves are the neutral stability curves for different δ. These curves define an expanding turbulent spot near the scallop wave due to propagation and amplification of localized disturbance. All the curves are symmetric with respect to the $\frac{z}{t}$-axis which comes from the symmetry of the governing equation $H(x, z) = H(x, -z)$ in (11.4). At $\delta = 0$, the curve is also symmetric with respect to $\frac{x}{t}$-axis, but is stretched along $\frac{x}{t}$-axis. With increasing δ, the neutral curves drop in the $\frac{x}{t}$ direction indicating that the "turbulent" wave spot slows down. The unstable region also becomes wider in $\frac{z}{t}$ with increasing δ. This is connected with the deceleration action of dispersion. The stability of the scallop wave is determined by is ability to escape the turbulent spot. If it has a larger or smaller velocity than the expanding "turbulent" spot, it will manage to escape and is convectivelly stable. These two limiting speeds are represented by the two tips of the closed neutral stability cutve (the $\frac{x}{t}$ neutral points).

A combination of this generalized picture of convective instability for two-dimensional pulses in Figure 7.4 and 7.5 and three-dimensional scallop waves is presented in Figure 11.14. For $0 < \delta < 0.18$, 2D and 3D pulses are unstable. For $\delta > 0.18$, 2D-pulses becomes stable but 3D pulses do so only for $\delta > 0.5$.

We have validated our stability analysis with numerical experiments. The results comfirmed and extanded the ones of Toh and Kawahara (1989, 1996) and Frenkel and Indeshkumar (1997) The entire δ line can be devided into four subregions. For $0 < \delta < \delta^{(1)} \approx 0.2$, no localized coherent structure is ever approached. During their evolution from initially random perturbations, irrgelar patterns consisting of fragments of quasi-two-dimensional waves to scallop waves appear intermiffenty. We fail to extract some average "generalized portrait" of the structure, and will called this region absolute chaos. All localized structures in this region are convectivelly unstable, see Figure 11.21.

We also carried out a special numerical experiment for $\delta = 0$ in Figure 11.15. At $t = 0$, the Tsvelodub solution of Figure 11.8, with a small localized

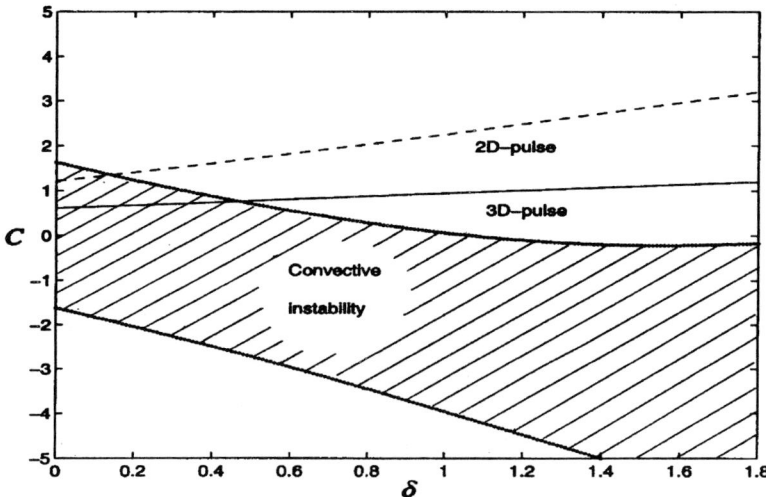

Figure 11.14: In the shaded region, both two-dimensional and three-dimensional pulses are convectivelly unstable.

pertubations imposed as initial conditions. At $t = 20$ this small perturbation evolved into a "convective turbulent spot" and at $t = 35$, this spot completely engulfs the scallop wave. The profile of the numerical spot is identical to the one obtained analytically, see Figure 11.13, $\delta = 0$.

At $\delta^{(1)} < \delta < \delta^{(2)} \approx 0.5$, another wave regime exists, which can be characterized as a regime of modulated two-dimensional waves. We can see from Figure 11.14 that from $\delta \approx \delta^{(1)}$, 2D-pulses becomes convectively stable while 3D- pulses are convectively unstable. Nearly two-dimensional pulses are hence selected.

We also performed a numerical experiment in Figure 11.15 at $\delta = 0.4$ with a scallop-wave. The result is qualitatively similiar to that of $\delta = 0$ with the convectively unstable scallop wave being swallowed by the turbulent spot.

In Figure 11.16, we studied the dynamics of two-dimensional pulses. At $t = 0$, a small noise is superimposed on the two-dimensional pulse. In this region of δ, the pulse is convectively stable, but its discrete eigenvalues are unstable, as is shown in Figure 11.6. Our numerical experiment shows that the selected wave-length in the transverse direction at onset ($t = 20$) is determined by the maximum growth frequency of the discrete mode. Further evolution modulates the 2D structure without triggering the formation of localized three-dimension scallops. For this numerical experiment, the profile oscillates in time in the z-direction, see image at $t = 210$ and $t = 470$. For $\delta \approx 0.3$, the 3D-structure is frozen and the profile reminds us of the one in Figure 11.7. These observations suggest that the unstable discrete mode gives rise to saturated standing or travelling waves in the transverse z-direction.

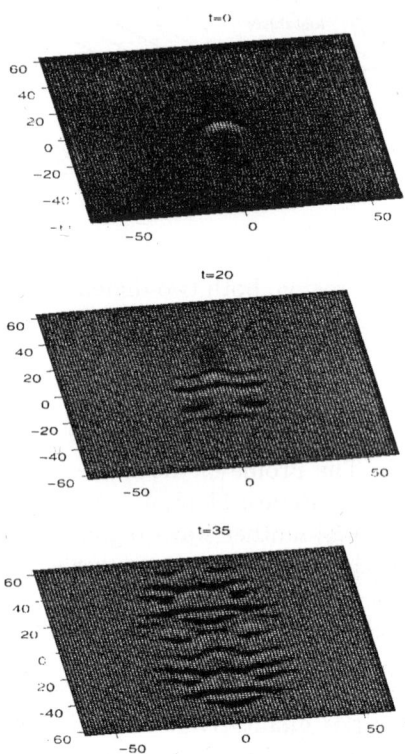

Figure 11.15:

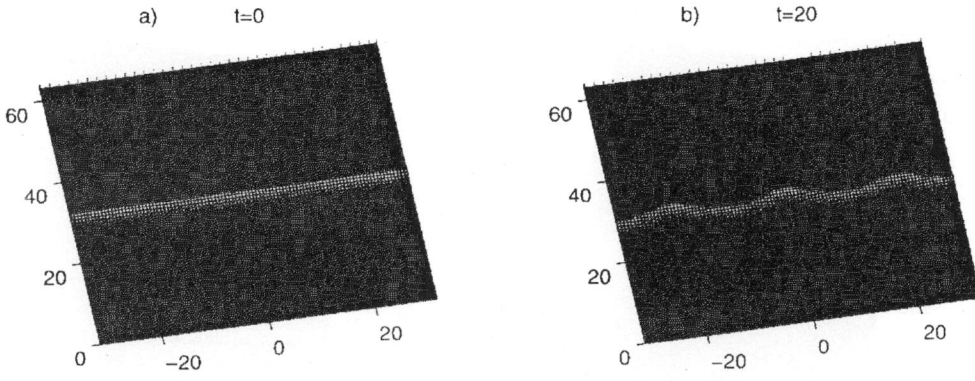

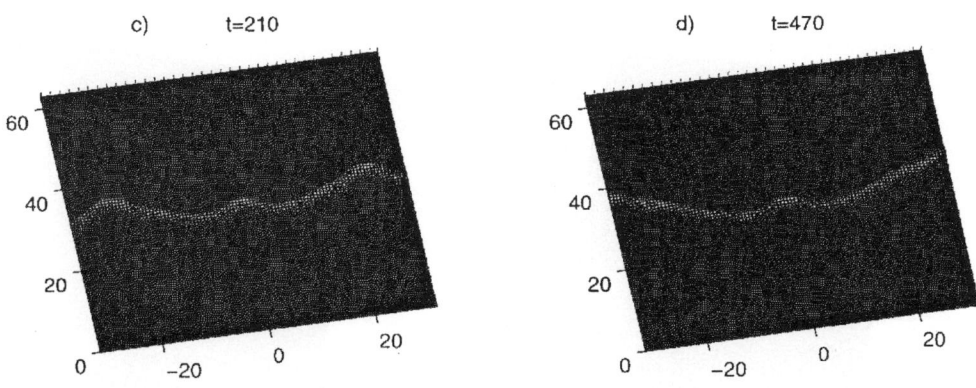

Figure 11.16: Numerical experiments — 2D pulses is modulated by 3D-perturbations but not destroyed, $\delta = 0.4$.

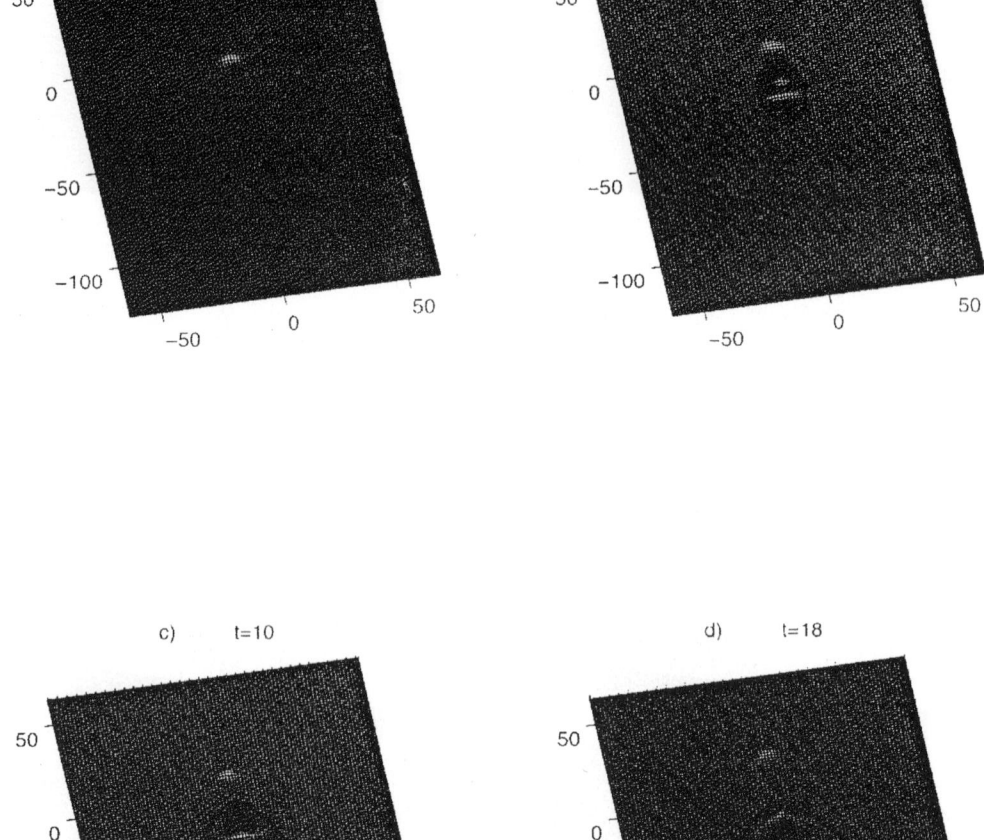

Figure 11.17: Numerical experiment— small disturbances near 3D pulse are convected away and, hence, are not able to destroy the pulse at $\delta = 2$.

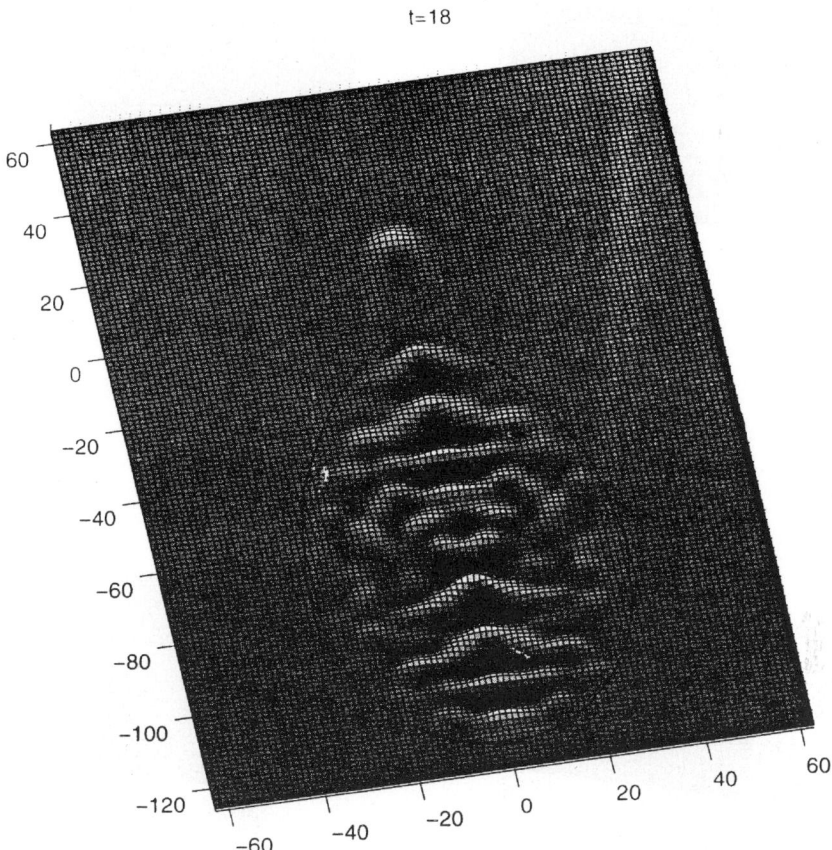

Figure 11.18: Last picture of the previous Figure amplified, $t = 18$.

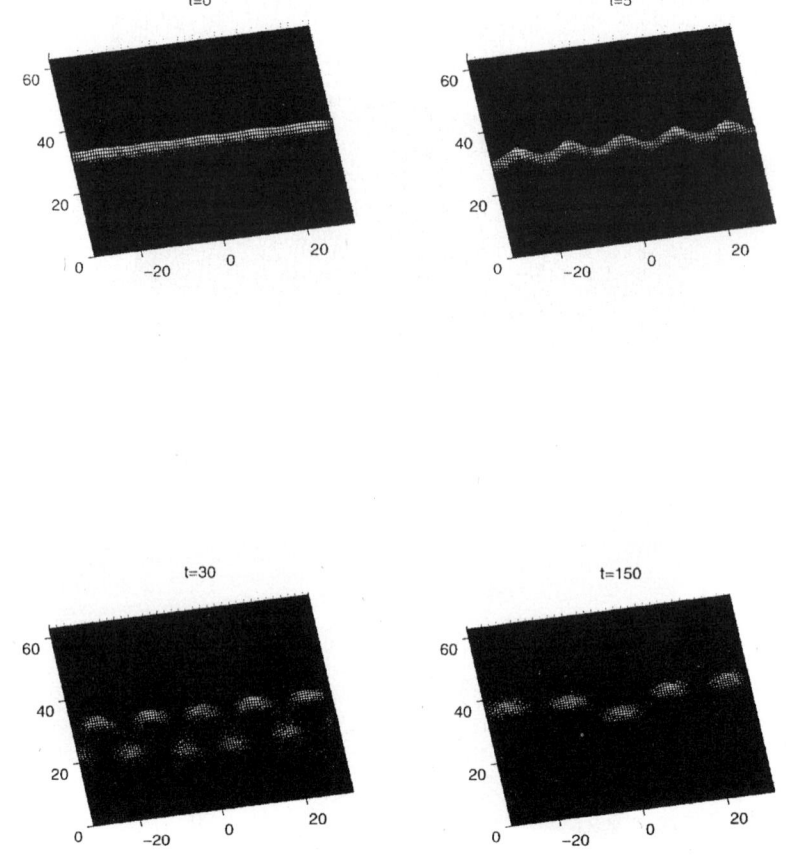

Figure 11.19: Front instability; small perturbations eventually destroy a 2D pulse and transform it into several 3D pulses. Small pulses at $t = 30$ lag behind at $t = 150$ and are out of the computation region. ($\delta = 0$)

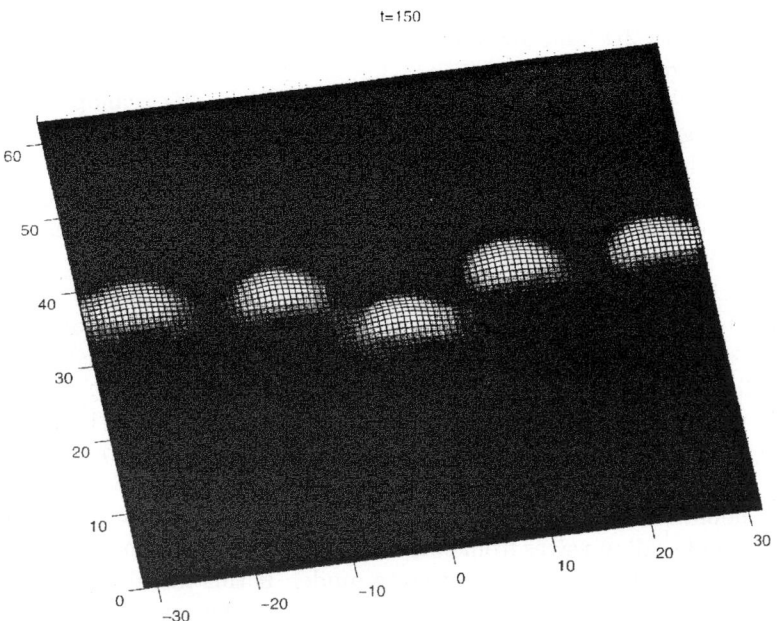

Figure 11.20: blow-up from the previous figure. Stationary localized 3d structures are clearly evident.

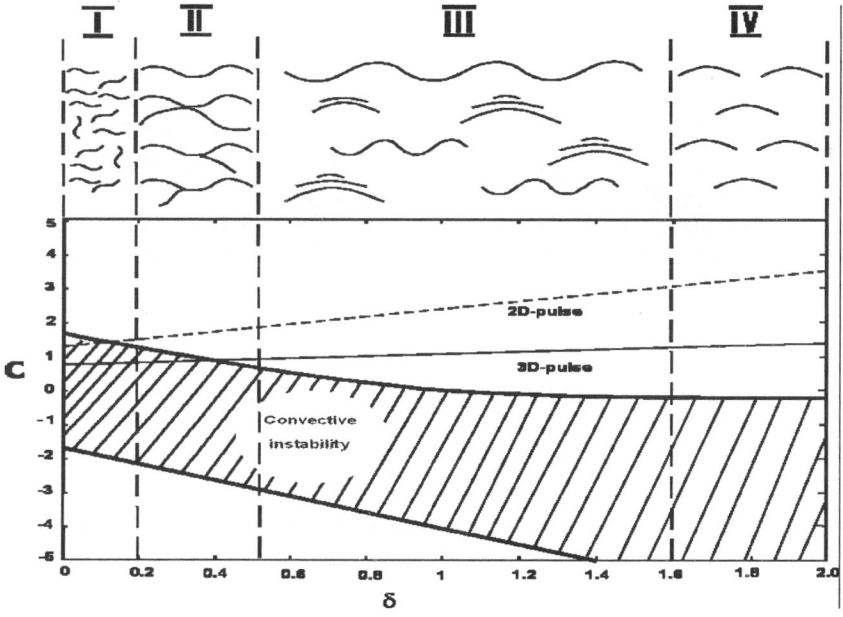

Figure 11.21: Different wave regimes at different δ.

For $\delta^{(2)} < \delta < \delta^{(3)}$, both 2D- and 3D- localized patterns can be found. In Figure 11.21, both 2D- and 3D- pulses are seen to be convectively stable in this intermediate region.

For $\delta > \delta^{(2)} \approx 1.8$, 2D-structures are replaced by 3D scallop waves. At sufficiently large $\delta > 3$ or 4, localized patterns gather into V-shaped arrays. The angle of this array decreases with increasing δ.

In the numerical experiment of Figure 11.17, with $\delta = 2$, the scallop wave is convective stable and the "turbulent spot" is convected away. The scallop wave and the trailing turbulent spot it leaves behind are shown in Figure 11.18.

Another interesting numerical experiment in Figure 11.19 shows front instability of 2D-pulses and 3D-scallop waves, at $\delta = 1$. Small perturbations at the maximum growth rate eventually destroy the two-dimensional pulse and transform it into several scallop waves which start to arrange in a staggered pattern. The blow-up in Figure 11.20 cleary shows the aligned localized 3D scallop waves. At the high value of $\delta = 1$, they do not have oscillations at the front, see Figure 11.10.

The schematic in the Figure 11.21 summarizes our analysis and observations. The creation of scallop waves from two-dimensional pulses seems to be triggered by unsaturated instability of the discrete mode. If this instability is arrested, transversely modulated waves are observed. When the two-dimensional pulse is convectively unstable, so is the scallop wave, and turbulence ensures. A non-linear bifurcation analysis of the discrete mode can differentiate the first two transitions (Ye and Chang, 1999). Despite the simplicity of the gKS equation, it has qualitatively captured the final transitions to scallop waves on a falling film, as depicted in Figure 11.22. This preliminary analysis must, of course, be

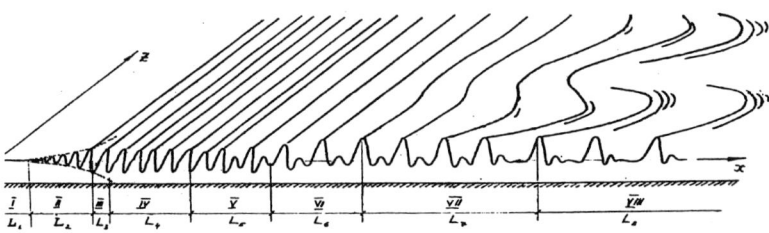

Figure 11.22: The complete wave transition regimes for the falling film.

followe up with a study with more realistic models, like the Shkadov model. We also do not know how the three-dimensional scallop waves interact or whether they coalesce to induce further wave texture coarsening. Certainly the arrangement into V-shaped arrays suggests they do interact. However, such arrays and coalescence events are not observed in the pioneer study of Tailby and Portalski (1960) shown in Figure 4.5. More careful experiments must, however, be carried out to fully decipher this final wave regime for the falling film.

Both the experimental and theoretical analyses of three-dimensional wave dynamics are extremely difficult. Both are so because of the slow wave evolution and interaction dynamics. We shall leave these efforts to future researchers and take out departure from long-wave, capillarity dominated and relatively low R falling-film waves here. In the next two chapters, we shall examine two wave evolution dynamics distinct from the falling-film wave dynamics studied thus far. However, we shall use the same mathematical tools to analyze them. All these wave dynamics are hence related in one important aspect - they all possess localized coherent wave structures.

CHAPTER 12

Hydraulic Shocks

At high flow rates (large Reynolds number R), the longwave expansion of Chapter 3 that leads to simpler equations becomes invalid. The interfacial waves still form localized pulses. However, these pulses contain a large recirculation eddy and have very steep fronts that defy the usual longwave approximation with parabolic profiles. In some extreme cases, the front profile can tip and become multi-valued. Such localized pulses are known as roll waves and they are common to high-flow rate inclined flows. The solution branch of the single-valued solitary waves exhibits a turning point at a critical Reynolds number and it is suspected that the roll waves appear beyond that point.

It is numerically difficult to construct multi-valued stationary roll waves from the Navier-stockes equation — all other equations derived from a longwave expansion would be in appropriate. Besides, it is questionable if some of their finer wave dynamics are even stationary waves as trasient events like air entrainment and atomization often occur at the roll waves. Due to these considerations, the steep front of a roll wave is often modeled as a discontinuous hydraulic jump (Brock, 1969). Moreover, at such high flow rates and with the omission of the high-curvature front, the Weber number W of (2.25) becomes negligible and capillary effects are unimportant.

Despite the difference in their appearance to capillary aolitary waves, roll waves share many features of the latter's wave dynamics. As for capillary solitary waves of Chapter 9, these roll waves are stationary — they retain their speed and shape for long durations. Also, the separation and height of these roll waves are observed to increase linearly downstream indefinitely (Brock, 1969) such that the wave density decreases monotonically by as much as one order of magnitude. The linear coarsening rate is constant at every downstream position and is hence independent of inlet noise and local wave texture. This linear coarsenining mechanism seems to be universal to all inclined layers although the coarsening rate varies as a function of fluid properties and inclination angle. Unlike capillary solitary waves, however, coarsening of roll wave seems to proceed indefinitely without reaching an equilibrium separation, as seen in Chapter 4 and 9 for capillary solitary waves.

Figure 12.1: Roll waves on a canal in the Alps.

There are other distinctions — some of them simplify the analysis for roll wave dynamics considerably.

Since capillary solitary waves are unstable to transverse disturbances due to a Rayleigh-type capillary instability, these coalescence dynamics are contaminated in real life by other dynamics involving the breakup into "scallop" waves (Chapter 9).

In fact, the latter dynamics will eventually overwhelm the coarsening dynamics. The complexity of the evolution equation and the eventual saturation of coarsening due to capillary also do not allow us to explicitly capture the symmetries and self-similarities behind the scale-invariant coarsening. In contrast, capillarity is unimportant for roll waves and the latter are experimentally observed to be stable to transverse disturbances, at least at low inclination angles (Brock, 1969). Without capillarity and with the usual hydraulic simplifications of the rolls, the simpler governing equations are also amenable to a detailed scaling analysis in conjunction with a solitary coherent wave theory. Instead of a solitary wave solution branch for the capillary waves (Chapter 3), all solitary role waves can be mapped into a single generic wave due to a subtle self-similar symmetry! Finally, the robust and dominant coarsening dynamics of roll waves over one order of magnitude in wave texture allow abundant and accurate experimental measurements for comparison. We hence tackle the downstream coarsening wave dynamics of roll waves in this chapter and capture its scale-invariant mechanism explicitly by exploiting the symmetries of the hydraulic equations and the localized coherent wave structures.

12.1 Governing equations

One difficulty in understanding and classifying wave dynamics on an inclined layer is the multitude of parameters. Standard dimensional analysis of Chpater 2 for a flat film yields three parameter, Froude number Fr, Weber number We, and the inclination angle θ. The Reynolds number R can be replaced by Fr.

Even with the omission of capillarity and thus We, two parameters remain in the equations of motion. Since localized wave structures are perched on flat liquid substrates that are consistent with this flat-film dimensional analysis, every localized wave on the film is parametrized by these two parameters. The wave coarsening dynamics hence evolve over a two-parameter family of quasi-steady wave solutions. Such evolution over a two-dimensional solution surface is extremely complex and difficult to capture. However, with a hydraulic approximation by Dressler (1949) and Brock (1970), the two parameters collapse into a single parameter. Moreover, a certain symmetry of the hydraulic equations will allow us to scale every solitary wave on the same channel into another one. This self-similarity implies that, for a given channel, the evolving wave structure at every position can be mapped into to a single generic wave - a drastic reduction in complexity! It is such simplifying symmetries that allow us to capture the coalescence dynamics explicitly.

The hydraulic equations of Dressler (1949) and Brock (1970) can be rewritten

Figure 12.2: Roll waves of the Santa Anita wash in Los Angeles.

as

$$\frac{\partial u}{\partial t} + u\frac{\partial u}{\partial x} + G\frac{\partial h}{\partial x} = 1 - \frac{u^2}{h} \qquad (12.1)$$

$$\frac{\partial h}{\partial t} + \frac{\partial}{\partial x}(uh) = 0 \qquad (12.2)$$

with a single parameter G, the modified Froude number. It is connected with Brock's Froude number $Fr = gh_N \cos\theta/u_N^2 = c_f/\tan\theta$ by the expression $G = \cos\theta/Fr^2$ where h_N and u_N are the thickness and velocity of a waveless flat film and c_f is the friction factor for the Chezy form of the frictional drag. Our independent variables x and t are stretched versions of Brock's variables x' and t'

$$\begin{aligned} x' = \kappa x, t' = \kappa t, \\ \kappa = Fr^2/\sin\theta = \cos\theta/G \end{aligned} \qquad (12.3)$$

Such a coupled system of quasi-linear first-order partial differential equations is classified by the characteristic equation

$$\begin{vmatrix} u - \lambda & G \\ h & u - \lambda \end{vmatrix} = 0 \qquad (12.4)$$

such that $\lambda = u \pm \sqrt{Gh}$. The sytem is hyperbolic for a positive G and elliptic for a negative G, while $G = 0$ is a transition value when the system is of parabolic type. A well-posed hyperbolic Cauchy problem in time only occures for an inlcined plane, $G > 0$, and hence the hydraulic equation cannot capture the wave dynamics on a vertical plane.

Evolution of our nonlinear hyperbolic system, $G > 0$, ultimately breaks down when u and h become multi-valued. In reality, a large vortex or roller is observed on the steep downstream edge of the wave. Mathematically, this relatively narrow vortex region can be modeled by a discontinuity with shock conditions (Whitham, 1974; Dressler, 1949)

$$-c[uh] + [\tfrac{1}{2}Gh^2 + u^2h] = 0$$
$$-c[h] + [uh] = 0 \tag{12.5}$$

where the brackets indicate the jump in the quantity and c is the speed of the discontinuity. The hydraulic equations (12.1) and (12.2) must be considered with (12.5) to permit a weak shock solution.

They always admit a flat-layer solution, $h = 1$ and $u = 1$. A simple stability analysis of this flat-film solution to the a normal mode disturbance, $e^{i(\alpha x - \omega t)}$, yields a dispersion relationship for the complex frequency ω, analogous to that in Chapter 2,

$$\omega^2 + 2\omega(i - \alpha) + \alpha^2(1 - G) - 3i\alpha = 0 \tag{12.6}$$

There are two wave modes, $\omega = \omega(\alpha)$ for (12.6). One branch is always stable, $\omega_i < 0$, while the second branch is unstable, $\omega_i > 0$, for $0 < G < 1/4$. Beyond criticality, $G = 0$, this branch is actually unstable for all real wavenumbers with $\omega_i \to 1/(2\sqrt{G}) - 1$ as $\alpha \to \infty$. A maximum in the growth rate is also absent beyond onset to yield a preferentially amplified wavenumber.

The hyperbolic system (12.1) and (12.2) hence cannot capture the filtering mechanism within the inception region, like that effected by viscosity and surface tension in sections 4.3 and 9.2, where a band of inlet disturbances near a specific wavelength is amplified into roll waves. This shortcoming can be remedied if an artificial friction is introduced to (12.1) and (12.2), as was done by Needham and Merkin (1984) and Hwang and Chang (1987)

$$\frac{\partial u}{\partial t} + u\frac{\partial u}{\partial x} + G\frac{\partial h}{\partial x} = 1 - \frac{u^2}{h} + \frac{\nu}{h}\frac{\partial^2 u}{\partial x^2} \tag{12.7}$$

$$\frac{\partial h}{\partial t} + \frac{\partial}{\partial x}(uh) = 0 \tag{12.8}$$

With the added ν, the corresponding dispersion relationship to (12.6) then yields stable short waves and a preferred wave frequency at ω_{max}. The disadvantage of this ad hoc approach is that ν is unknown. However, as we shall show in our numerical experiments, it can be specified by fitting the experimental inception length for roll waves. The subsequent roll wave coarsening dynanics are unaffected by the value of ν and the inception length. We shall hence analyze the downstream dynamics with the much simpler hydraulic equations although the simulation is carried out with the artificial ν, which also renders the shocks continuous but with a very steep front slope.

12.2 Numerical simulation

In order to solve (12.7) and (12.8), boundary conditions must be chosen judiciously. Periodic boundary conditions are more amenable to numerical treatement but the results would not yield the linear coarening dynamics observed in reality. However, as discussed in Chapter 4, periodicity improperly connects convective instability of open systems to absolute instability of closed systems. In essence, the upstream feedback provided by the artificial periodicity destroys the downstream evolution of the wave dynamics. Thus, we impose at $x = 0$ the open-flow conditions for thickness and flow rate, $q = uh = 1, h = 1 + f(t)$, where f(t) represents zero-mean stochastic external disturbances and is simulated using a random number generator. This stochastic noise is white above a cut-off frequency with a noise amplitude, ϵ.

"Soft" boundary conditions,like (4.6), with vanishing derivatives for h and u are taken at the end of the channel, in order to reduce upstream feedback of disturbances. Since these exit conditions are artificial, the wave dynamics within a small region at the exit are not realistic. However, the spatial-temporal wave dynamics within the bulk of the channel are faithfully captured.

In the presence of sufficiently rich noise at the inlet, as generated by our stochastic forcing, a broad spectrum of modes is excited upstream. However, within a short distance from the inlet, the linear instability mechanism and short-wave suppression by ν in (12.7) filters out most of the broad spectrum except a narrow band near the maximun growth rate which depends on ν. Beyond this inception region, this band of Fourier modes quickly synchronize to form discontinuous roll waves. The spatial evolution of the average wave period $<t>$, before and after roll wave inception, is presented in Figure 12.3a. Changing the friction coefficient ν, and thus ω_{max}, only influences the initial evolution within the inception region. We note that, in all cases, $<t>$ increases linearly beyond the inception region and its slope is independent of ν.

Changing the inlet noise amplitude results in a downstream translation of the straight, post-inception $<t>(x)$ line without affecting its slope, corresponding to a shift of the roll wave inception length. Hence, only the roll wave inception length is affected by noise but not their coarsening dynamics, as seen in Figure 12.3b. Linear coarsening of roll waves is hence a robust phenomenon insensitive to inlet noise and the selected band of Fourier modes in the inception region.

We choose the friction coefficient to fit the experimental minimum of $<t>$ vs x in Brock's data for $G = 0.04$ or $Fr = 5$ and $sin\theta = 0.084$ in Figure 12.4. It is seen that, by slightly adjusting the inlet noise amplitude ϵ, the simulated $<t>$ is in a good agreement with the measured value for the entire channel, including excellent agreement of the linear coarsening rate for the roll waves. This again confirms the roll wave dynamics are insensitive to the choice of ν or

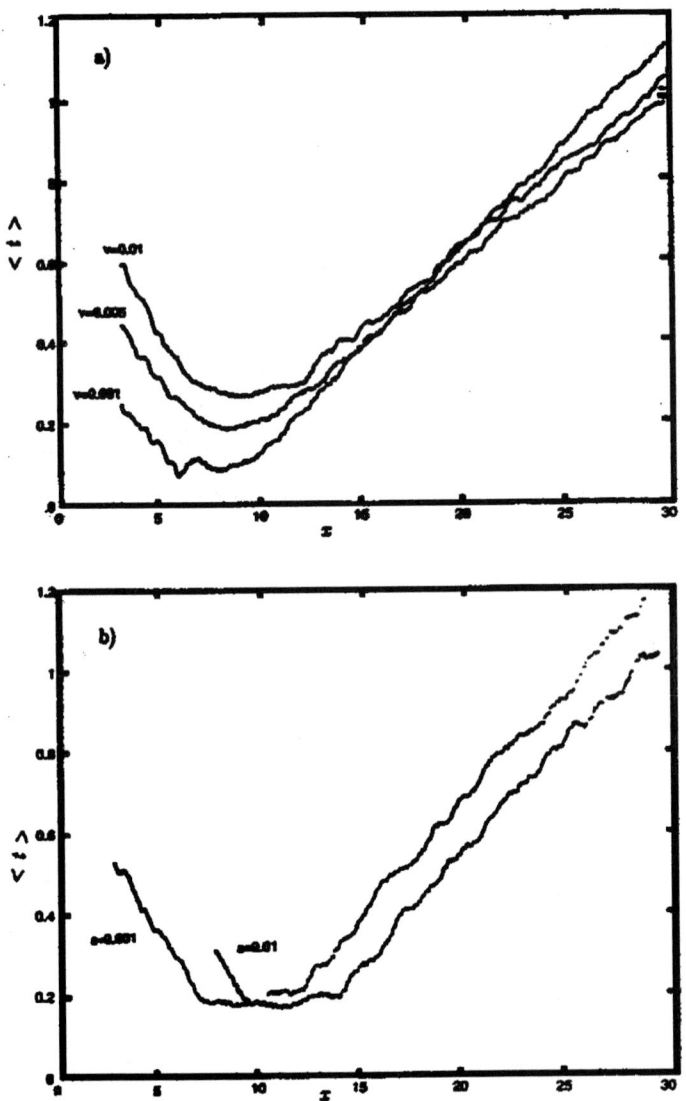

Figure 12.3: (a) Simulated average period downstream for $G = 0.04$ at $\nu = 0.001$, 0.005, and 0.01. (b) Simulated average period downstream for $G = 0.04$ at $\varepsilon = 0.001$ and 0.01.

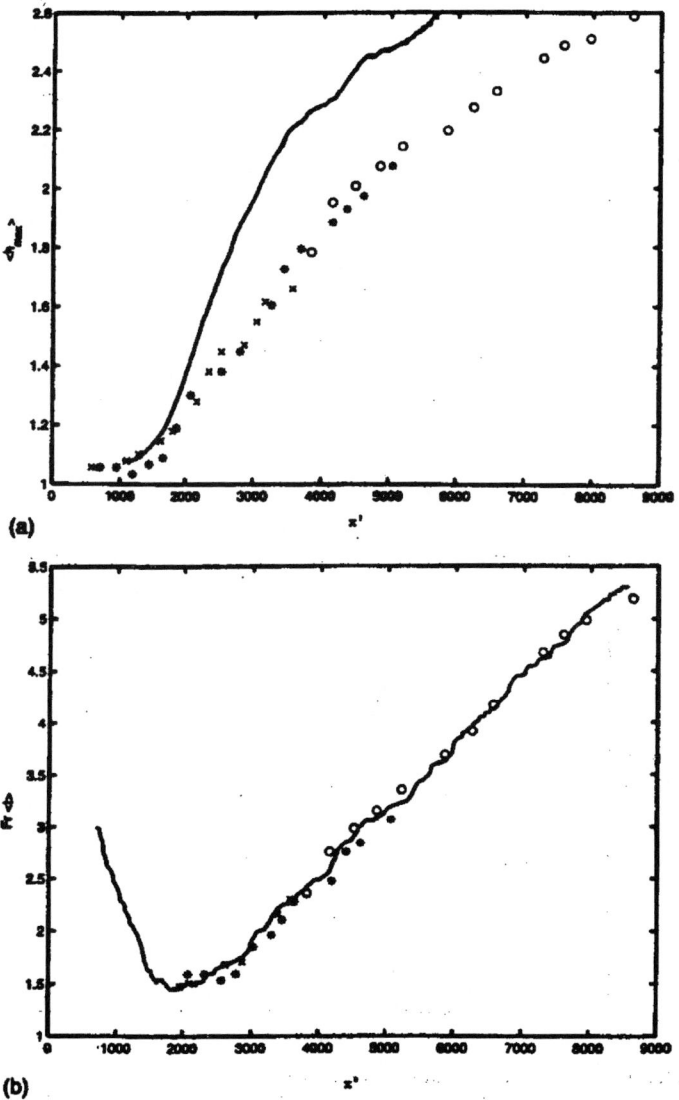

Figure 12.4: Simulated time-average quantities at various location of (a) wave amplitude and (b) wave period, $G = 0.04$. Brock's data for $G = 0.047$ (circle), 0.041 (star), and 0.039 (cross) are also shown for comparison.

the inlet noise spectrum – it seems to involve entirely deterministic dynamics governed by roll waves of the hydraulic equation (12.1) and (12.2).

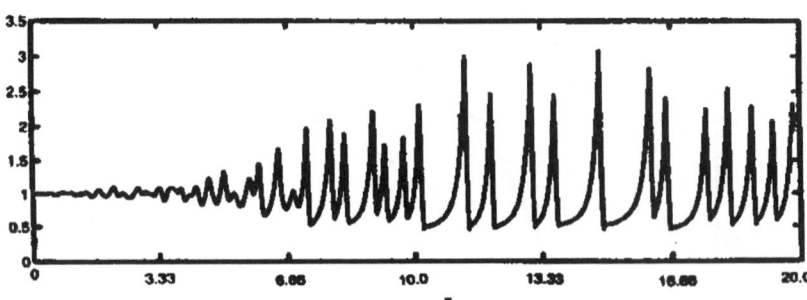

Figure 12.5: A snapshot of the spatially evolving waves due to random inlet disturbances ,$G = 0.04$.

Closer scrutiny of the simulated dynamics reveals the following stages of evolution: a) From $x = 0$ to 4, exponential growth of the surface waves preferentially excites a narrow band of frequencies near ω_{max}. b) From $x = 4$ to 7, this band of Fourier modes synchronize to form shocks. c) By $x \simeq 7$, the separation between shocks has increased sufficiently such that the waves could be considered a train of nearly independent roll waves. These waves perch on a flat substrate and propagate with stationary shape and speed downstream. However, beyond this point, coalescence events begin to alter the average roll wave characterictics over a length scale far larger than the roll wave width. This coarsening regime persists to the end of our channel at $x = 30$ or the length of calculation. All transition points depend on G; they are given here only for $G = 0.04$. The average wave amplitude and wave period at every station x are presented in Figure 12.4 and compared to Brock's data in Brock's variables x' and t' in (12.3) for accuracy. A snap-shot for the wave profile $h(x)$ is shown in Figure 12.5. Tracks of the wave crests in the $x - t$ coordinates, the "world lines", are shown in Figure 12.6.

The simulation yields an accurate estimate of the wave period $<t>$ and a less satisfactory prediction for the wave amplitude h_{max}. This is not surprising because the amplitude of a stationary running jump is always overestimated by the hydraulic approach (Brock, 1970). The difference is a consequence of the one-dimensional shock approximation of the front vortex. We have also recorded the standard deviation of h_{max} and the average substrate thickness between schocks at every station. Both are in staisfactory agreement with Brock's data. The standard deviation of h_{max} grows downstream in a similar fashion as h_{max} – a self similar characteristic that will be scrutinized subsequently. While the average wave period increases by almost an order of magnitude due to the coalescence-driven coarsening, the average substrate thickness varies by a factor less than two. We shall exploit the relative invariance of this latter quantity to

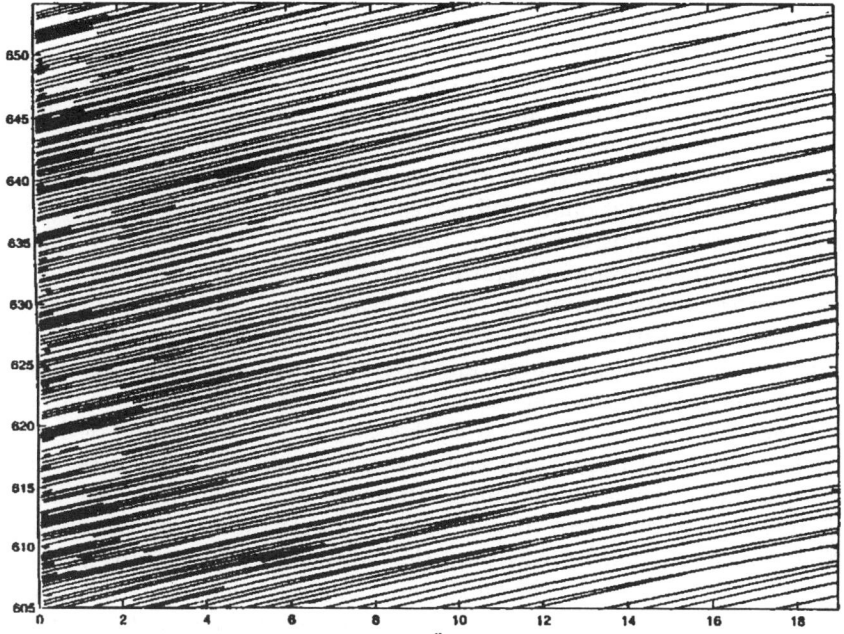

Figure 12.6: The world lines tracking the wave peaks are ploted in the space-time coordinates, $G = 0.04$.

capture the linear coarsening of the former.

12.3 Coherent wave structures and self-similarity

Beyond some station, the surface of the layer is covered by nearly independent and nearly stationarily propagating roll waves which sit on some substrate χ, as seen in Figures 12.5 and 12.6. The speed c of each stationary roll wave is deternined by a steady version of (12.1) and (12.2) in the moving frame $x \to x - ct$,

$$\frac{\partial}{\partial x}[\frac{(u-c)^2}{2} + Gh] = 1 - \frac{u^2}{h} \qquad (12.9)$$

$$\frac{\partial}{\partial x} h(u-c) = 0 \qquad (12.10)$$

which has to be solved with the jump conditions (12.5). At $x \to \pm\infty, h$ obeys the condition $h \to \chi$. As such, the stationary roll waves are to two-parameter family of solutions parameterized by G and χ. The wave speed and wave shape will be expressed in terms of these two parameters.

As is verified in our numerical experiments, the friction coefficient ν of (12.7) and (12.8) is unimportant for the roll wave dynamics beyond this inception region and hence only the stationary version of the hydraulic equations (12.1) and (12.2) are used here. The flat velocity profile in shallow-water flow, the localized roll wave structure and the turbulent friction coefficient endow the solitary wave equation with a very strong invariance symmetry with respect to power-law affine coordinate stretching. This, in turn, reflects the intriguing self-similarity of the stationary roll waves that we will exploit to simplify and categorize the coarsening dynamics.

In fact, the annihilation of localized waves by coalescence implies that there are fewer waves downstream to carry the downflowing liquid. To ensure that the time-averaged flow rate at every station remains the same as the inlet flow rate, the remaining waves must be larger and faster than their predecessors upstream. Due to the quasi-stationary and localized nature of these waves, these larger and faster waves can only survive on a thicker substrate. The complex dynamics that adjust the wave height and speed and the substrate thickness, without coalescence, are difficult to decipher. However, it is clear from Figures 12.5 and 12.6 that they do occur. It will be shown that estimating the coarsening rate does not require full knowledge of this adjustment. Most conveniently, the substrate thickness χ will be shown to parametrerize the quasi-stationary wave at each station and hence, indirectly from flow balance, the wave texture locally. We hence seek to scale χ away to obtain a scale-invariant description.

Due to the localized roll wave structure, there is no length scale in the streamwise direction and the substrate thickness χ is the lone length scale for any stationary roll wave. We hence scale h by χ. The wave speed and the liquid velocity should also be scaled by the same characteristic velocity . The mass balance kinematic equation (12.10) is invariant to this scaling due to the flat velocity profile. Even with a velocity profile that is not flat, a conversion of the velocity u to the flow rate q would still render the kinematic equation invariant if q scales as some algebraic power of h. This was the case for lubricating films with parabolic velocity profiles (q scales as h^3) as in falling-film wave dynamics driven by solitary capillary waves.

This invariance is usually lost on the force balance equation of motion (12.9).

Since the parameter G is based on tha flat substrate, rescaling the substrate thickness usually alters G, as was the case for solitary capillary waves. As a result, every solitary wave at every station downstream of the channel corresponds to a distinct member of a wave family. The wave coarsening then corresponds slow evolution along this family whose dynamics as we have shown in Figure 4.38 for the falling film and have deciphered with a wave coalescence theory in Chapter 10. Intriguingly, the equation of motion (12.9) remains invariant to affine scaling with respect to χ –G remains unchanged by the scaling. The force balance is simply equating the gradient of the total local energy, kinetic and gravitational potential, to the total force exerted on it, the tangential gravitational force and the wall friction. Due to the flat-velocity shallow water formulation, the local kinetic energy is simply u^2 while the gravitational potential is Gh. This yields a scaling of u^2/Gh for these two energy terms. On the other hand, the scaling for the friction force to tangential gravitational force is identical for the Chezy formula, used in (12.9). These two identical relative scalings for the two terms on each side of the force balance (12.9) and the scaling of x by χ yield the invariance of the force equation (12.9) and the kinematic equation (12.10) to the following affine scaling:

$$h(x) = \chi H(\xi), x = \chi\xi$$
$$u(x) = U(\xi)\sqrt{\chi}, c = C\sqrt{\chi} \qquad (12.11)$$
$$G = G$$

which converts the stationary wave equations into

$$\xi \to -\infty : \quad H \to 1$$
$$-\infty < \xi < 0 :$$
$$\frac{dH}{d\xi} = \frac{H^3 - (CH + 1 - C)^2}{GH^3 - (C-1)^2} \qquad (12.12)$$
$$\xi = 0 : \quad \frac{1}{2}GA(A+1) = (C-1)^2$$
$$0 < \xi < +\infty : \quad H = 1$$

Here $A = H_{max} = H(\xi = 0)$ is the amplitude of the normalized roll wave. Since χ has been normalized to unity by this scaling, the two-parameter family of stationary roll waves is reduced to one-parameter one paramaterized by G. By this transformation, every roll wave on the channel with different χ but the same G can then be mapped into a generic one with a unit substrate. The transient version (12.1) and (12.2) can likewise be transformed to eliminate χ dependence but the scaling in time $t = T\sqrt{\chi}$ must be used.

A version of (12.12) for small inclination angles was solved by Dressler (1949) and for periodic solutions by Brock (1970). The solution can be found in closed-form

$$G = 4\left(C\left(C + 1 + \sqrt{(C+3)(C-1)}\right) - 2\right)^{-2}$$
$$A = \frac{-G + \sqrt{G(G + 8(C-1)^2)}}{2G} \qquad (12.13)$$
$$\xi - \xi_0 = G(h - S_1 \ln(h-1) + S_2 \ln(h-a))$$

where

$$S_1 = \frac{b^2 + b + 1}{a - 1}, \quad S_2 = \frac{b^2 + ab + a^2}{a - 1}$$
$$a = 0.5(C-1)(C+1-\sqrt{C^2 + 2C - 3}), b = \frac{(C-1)^{2/3}}{G^{1/3}}$$

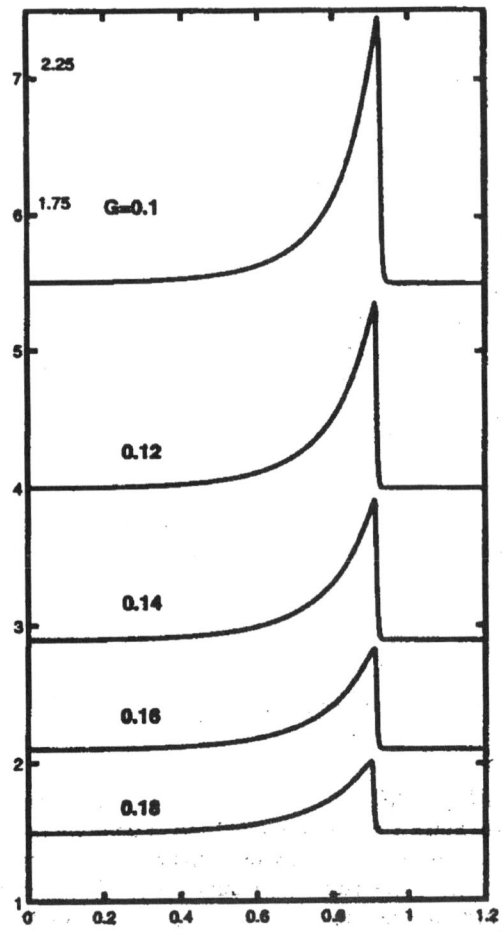

Figure 12.7: The roll wave family as a function of the modified Froude number G. All substrate are of unit thickness.

ξ_0 is such a constant which places the maximum of $H(\xi) = A$ at the origin of the coordinate $\xi = 0$. These roll waves and their amplitude as functions of G are shown in Figures 12.7 and 12.8.

The most important feature of transformation (12.11) is that the modified Froude number G is independent of the substrate layer χ. Thus, G is constant everywhere downstream, and without loss of generality, could be taken to be the value at the inlet where $\chi = 1$. Moreover, every roll wave at every station is self-similar to every other one through the similarity variable χ. This invariance is unique to hydraulic roll waves. This universal dependence of $A = H_{max}$ on the modified Froude number is shown in Figure 12.8 for the experimental data of Brock and our numerical simulation data. Due to the inaccuracy of the estimating χ from h_{min}, we estimate the former from the solitary wave speed, $\chi = (c/C)^2$. By virtue of the self-similarity, all the experimental points of Brock for the same run (for x' ranging from 1000 to 8000, wave amplitude varying over a factor of 5, χ ranging from 1 to 0.5 and wave period by a factor of 8 for $3° < \theta < 10°$ and $3.5 < Fr < 5.6$) collapse into a lone point, albeit the experimental point is slightly lower than the theoretical prediction.

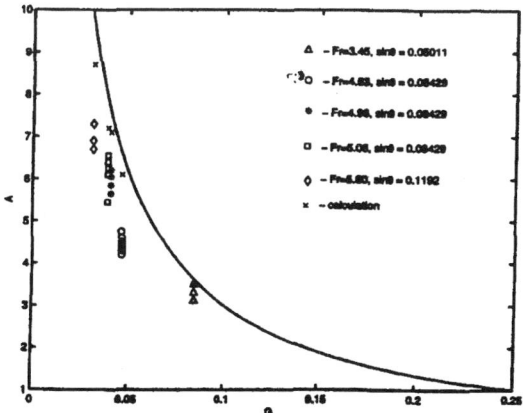

Figure 12.8: The universal dependence of normalized amplitude on the modified Froude number. Comparison of the theoretical prediction (12.13), numerical simulations, and experiments, of Block. The simulation results collapse so well that only one point (the cross) is shown. Instead of using the measured h_{min} for χ, we find it more accurate to determine χ from the speed measurement $\chi = (c/C)^2$, where c is the measured wave speed and C is the predicted normalized speed of (12.13)

As seen in the typical case of Figure 12.5, χ reaches a minimum of about 0.5 along the channel. By the above scaling, (12.11) stipulates that its local wave amplitude h_{max} and local wave speed c are correlated by $(h_{max}/A) = (c/C)^2$ where A and C are the wave amplitude and speed of the normalized wave with unit substrate of the same G. Expanding about $\chi = 1$ of this χ-family, we get

$$\frac{dh_{max}}{A(G)} \sim 2\frac{dc}{C(G)} \qquad (12.14)$$

Hence, for every G, the deviation of the wave amplitude at every station (with $\chi \neq 1$) from the normalized amplitude A scales linearly with repsect to the analogous deviation speed at the same station. Since χ does not change much from the inlet unity value along the entire channel, this linear scaling is valid for all waves on the channel at a given G. However, since A is in excess of 2 for small G, as seen in Figure 12.8 and since h_{max} scales as χ by (12.11), a small differentail change in χ can lead to a large change in h_{max}. Hence, an expansion at $\chi = 1$ still captures large variations in h_{max} when χ changes. This is a basic scale-invariant approximation that allows us to caputre the generic linear scaling (12.14) for the χ-family of stationary roll waves with G fixed. We shall use this linear scaling, based on χ expansion about unity, to caputre the pertinent scale-invariant dynamics of an excited wave that drive coalescence.

12.4 Self-similar coarsening dynamics

Consider now the coarsening dynamics of the roll waves downstream depicted in Figure 12.6. The numerical experiments demonstrate a slow self-similar coarsening of nearly stationary and idependent roll waves with increasing separation and amplitude downstream. The key mechanism behind this coarsening dynamics is an irreversible wave coalescence at every convergence of the world lines in Figure 12.6. The coalescence is irreversible as one world line is eliminated after every convergence showing the disappearance of one wave.

Like capillary solitary waves, the salient feature of this coalescence mechanism is that it is self-perpetuating. The coalescence is triggered by an "excited" roll wave with larger amplitude and speed than the stationary equilibrium wave of (12.13). This excited wave chases down its equilibrium neighbor in front at a convergence point of the world lines in Figure 12.6. The resulting coalescence reduces the number of waves by one. Moreover, the wave resulting from the coalescence has the liquid mass of its two parent waves and is itself another excited wave. It would then perpetuate the coalescence cascade.

Statistical evidence of this presence of two types of waves, an equilibrium one and an excited one, is shown in Figure 12.9. The measured distributions of wave amplitude at several locations down the channel are self-similar as they collapse into one universal distribution by scaling with respect to the mean amplitude $< h_{max} >$ with a long decaying tail to the right. The sharp distribution around $< h_{max} >$ corresponds to the equilibrium stationary wave while the long tail corresponds to an excited wave. The reason for the latter's broad distribution is that every excited wave decays in amplitude and approaches an equilibrium wave. The tail distribution reflects the probability of finding an excited wave at different stages of its decay. The decay slows as the equilibrium wave is approached and hence the slowly decaying tail to the distribution. In

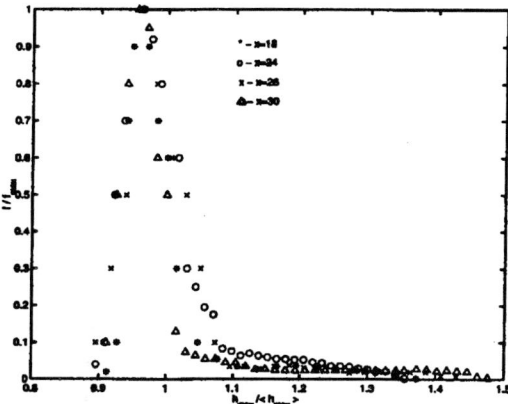

Figure 12.9: The wave amplitude distribution at various downstream locations normalized by scaling with respect to the mean and maximum value at the mean.

contrast, the measured distribution in wave period t at a given station does not exhibit this long tail. The equillibrium and excited waves seen to obey the same distribution in separation. To estimate the relative density of excited and equilibrium waves, we use the criterion $h_{max} > 1.1 < h_{max} >$, from the cutoff of the long tail in Figure 12.9, for an excited wave and record in Figure 12.10 the ratio between the average wave separation in time $< t >$ and the average excited wave separation in time $< \tau >$ at all stations. About 13% of the waves are excited and, most intriguingly, $< t > / < \tau >$ remains constant throughout the channel.

The picture that emerges from the above statistics is that there exists a lattice of localized roll waves with roughly equal separation in time at every position. About 13% of them are excited waves that precipitate coalescence. After each coalescence, the waves adjust (increase) their separation $< t >$ in time with identical distribution. Since the lattice separation does not distinguish between equalibrium and excited waves, $< t >$ and $< \tau >$ increase in the same self-similar manner to yield a constant $< t > / < \tau >$.

To proceed further, we must identify the excited wave and the wave adjustment dynamics after coalescence. For the coarsening dynamics to be self-similar at every station, the excited wave, like the equilibrium one, must be self-similar to transformation (12.11). We assume that, on the average and at equilibrium, the excited wave created by coalescence has an amplitude that is twice the equilibrium one. This assumption is based on the observation that the excited wave decays after it is generated by the coalescence and, immediately before the next coalescence, its amplitude has approached that of the equilibrium value A in the normalized version of (12.13). The width of the wave remains constant during the coalescence but the amplitude and the total area (liquid) have roughly

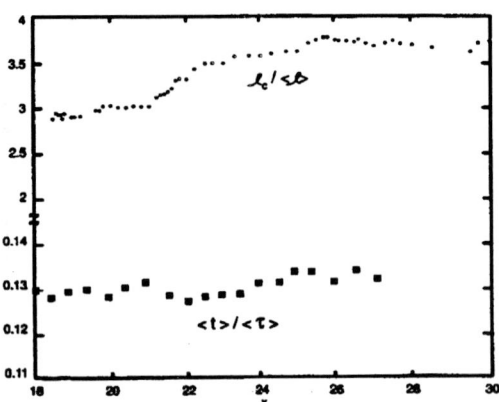

Figure 12.10: The invariance of the fraction of excited waves in time and $l_c/<l>$ with respect to x. The statistic for the two are average over a window of lenth 3 for $G = 0.04$.

doubled. Although this factor of 2 is somewhat arbitrary, one could assign a different invariant constant to preserve self-similarity and would not change the scalings of the final result.

The adjustment in $<\ell>$ after the coalescence to yield an average period $<t>$ that varies downstream is due to global mass balance –the time average flow rate $<q> = <uh>$ at every station must be equal to the one at the intet, $<q> = 1$. For a periodic train of equilibrium roll waves with a separation distance $<l>$, or $<L>$ in the normalized variables, one acquires from (12.10)

$$<Q> = CJ/<L> + 1,$$

where

$$J = \int_0^{<L>} (H-1)d\xi \simeq \int_{-\infty}^{\infty} (H-1)d\xi$$

is the excess liquid mass contained in the normalized roll wave. This integral is practically independent of the length of integration $<L>$ if it is larger than the characteristic wave width, because $(H-1)$ is exactly zero downstream of the jump, and $(H-1)$ decays exponentially fast upstream. Thus, J is a function of G only. Then, $<q> = \chi^{3/2} <Q>$, and eventually one obtains the separation distance and the wave period as functions of the similarity variable χ – the wave texture is also parameterized by χ,

$$<l> = \frac{CJ\chi^{5/2}}{1-\chi^{3/2}}$$

$$<t> = <l>/c = \frac{J\chi^2}{1-\chi^{3/2}} \qquad (12.15)$$

In essence, the global mass balance and the self-similarity of the roll waves allow us to relate the average separation of a periodic train to the local substrate thickness χ for a given inlet value of G. Since χ varies from unity to 0.5 over the channel, $<l>$ and $<t>$ vary quite significantly across the entire channel. It is amazing that their rate of variation is constant at every station. This universal linear coarsening rate is actually not dependent on the global mass blance (12.15) although the exact value of $<l>$ or $<t>$ is determined by it! More specifically, the texture is a strong function of χ but the rate of change of the texture, as driven by the cascaded coalescence events, is a weak function of χ. This is the origin of the scale-invariant linear coarsening rate.

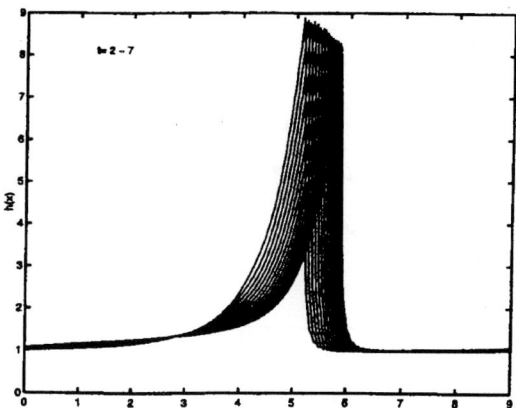

Figure 12.11: The decay of an excited roll wave over an interval of $t = 5$ for $G = 0.04$. Note that the decaying wave resembles the equilibrium family of Figure 12.7 and that the substrate layer is slightly elevated in the back due to liquid drained from the wave.

The final required information is an estimate of the coalescence rate. Consider the distance travelled by an equilibrium wave in the train before it is captured by a trailing excited wave some specified distance behind. This distance requires only the speed of the excited roll wave since the speed of the equilibrium roll wave is known. The self-similarity of the roll-waves again allows us to examine a generic excited wave with a specific series of simulations in the normalized coordinates of (12.11). An equilibrium roll wave solution (12.13) is taken and excited with the addition of an excess mass such that the wave amplitude is as much as twice of the equilibrium amplitude, as seen in Figure 12.11. The exact manner the extra mass is distributed is important only at small times. Beyond this fast transient, the excess mass drains out of the roll wave in the form of a wave packet and there is a corresponding decay of the roll wave amplitude. Note that the wave packet forms a shelf behind the wave such that the excited wave resembles a transient hydraulic jump whose ampliitude decays in time. Both the amplitude and wave speed decay exponentially with the same exponent $\lambda_{\#}$ towards the equilibrium values, as seen in Figure 12.12. This correlation between the decay dynamics of the amplitude and speed can be explained by the weighted spectral theory of Chapter 7. This coupled decay yields the linear correlation in Figure 12.12 between the speed and the amplitude during the decay towards the equilibrium pulse. Hence, in the normalized coordinate of (12.11), a generic excited wave will shed its extra mass to become an equilibrium one. During this decay towards equilibrium, its instantaneous speed reduced with respect to the equilibrium one is linearly related to the analogously reduced roll wave amplitude,

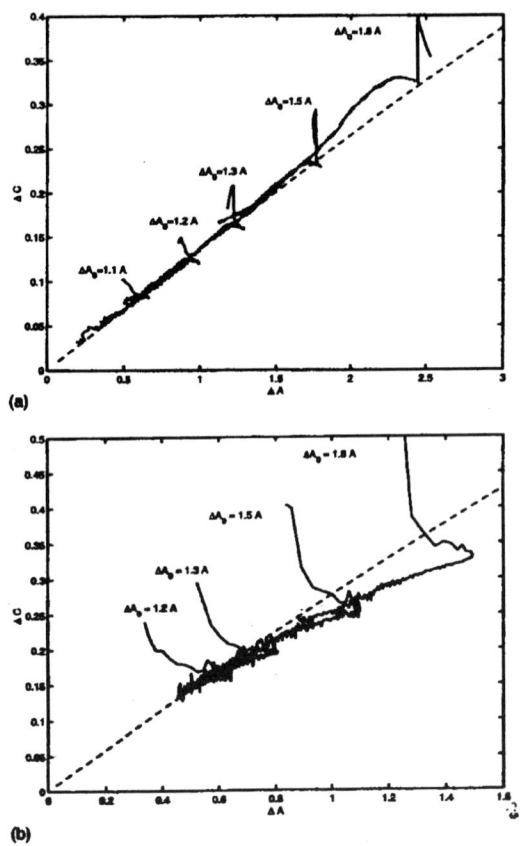

Figure 12.12: The linear correlation between the speed ΔC and amplitude ΔA as compared to (12.16) and (12.17) for various degrees of excitation for $G =$(a) 0.04 and (b) = 0.1 . All correlations are linear but (12.17) breaks down beyond $G = 0.15$.

$$\Delta C = \gamma \Delta A \qquad (12.16)$$

That a decaying wave pulse or wave front exhibits a linear scaling between

its instantaneous deviation speed and deviation amplitude, both reduced with respect to equilibrium, has been observed in many experiments on thin-film flows (Alekseenko et al.,1985). It is also in agreement with scaling (12.14) for the χ-family equilibrium waves on different substrates but the same G, if one equates dh_{max} and dc to ΔC and ΔA, recalling that A and C are only functions of G. This analogy suggests a quasi-equilibrium approximation of the transient, decaying excited wave in Figure 12.11, as we have successfully done in Chapter 7.

In front of the shock of the normalized excited wave, the substrate thickness remains at unity. Behind the shock, however, it is modelled by an equilibrium wave of Figure 12.7 on a thicker substrate $\chi > 1$. Due to the jump in χ, the amplitude and speed both decay as more liquid is drained out of the localized excited wave than is injected in front. However, at any stage of the decay, the excited wave corresponds to an equilibrium one behind the shock and the instantaneous amplitude and speed are correlated by the self-similar scalings of (12.11) for equilibrium waves. According to this scaling, the wave amplitdue scales as χ and hence the difference in amplitude between the excited and equilibrium waves is $\Delta A = A(\chi - 1)$. The speed, on the other hand, scales as $\sqrt{\chi}$ and hence $\Delta C = C(\sqrt{\chi} - 1)$. As a result,

$$\gamma = \frac{\Delta C}{\Delta A} = \frac{C(\sqrt{\chi} - 1)}{A(\chi - 1)} = \frac{C}{A(\sqrt{\chi} + 1)}$$

where $C(G)$ and $A(G)$ are given by the equilibrium values in (12.13). During the decay of the excited pulse with initially twice the amplitude, χ decays from 2 to 1. Hence, γ ranges from $\frac{C(G)}{A(G)}(\frac{1}{1+\sqrt{2}})$ to $\frac{C(G)}{2A(G)}$, a very small variation. For convenience, we take

$$\gamma(G) = \frac{C(G)}{2A(G)} \qquad (12.17)$$

which is equivalent to an expansion at $\chi = 1$ as we have done in (12.14) for equilibrium waves. This linear scaling is identical to (12.14) for equilibrium roll waves and is seen to be in good agreement with the numerical simulation of the decaying excited wave in Figure 12.12 for various degree of excitation. The proportionality constant breaks down beyond $G = 0.15$ although ΔC is still linear with respect to ΔA. We note from Figure 12.8 that, beyond 0.15, A and χ are of the same magnitude and a χ expansion is not expected to capture the large change in A during the decay.

Hence, in the normalized coordinates, the amplitude of the excited wave decays exponentially in the normalized time $T = t/\sqrt{\chi}$ and its differential speed decays as

$$\Delta C = \gamma \Delta A(T = 0) e^{-\lambda_\# T}$$

where $\Delta A(T = 0)$ corresponds to the amplitude of the excited wave after the last coalescence. Hence, the time T_c it takes for the excited wave to capture an equilibrium wave a distance of $<L>$ in front is then determined by

$$<L> = \int_0^{T_c} \Delta C \, dT = \frac{\gamma \Delta A}{\lambda_\#}(1 - e^{-\lambda_\# T_c})$$

or
$$T_c = -\frac{1}{\lambda_\#} \ln(1 - \frac{\lambda_\# <L>}{\gamma \Delta A})$$
and the characteristic distance travelled by the equilibrium wave over T_c is
$$L_c = CT_c$$
We find from our numerical experiments that $\frac{\lambda_\# <L>}{\gamma \Delta A} << 1$ and the decay of the excited wave is negligible to leading order. Hence T_c and L_c can be accurately approximated by
$$L_c = CT_c \sim \frac{C<L>}{\gamma \Delta A} = \left(\frac{C}{\Delta C}\right) <L>$$
Based on the scalings of the excited wave $\Delta C/C = (\sqrt{\chi} - 1)$ where χ is the substrate thickness of the excited wave immediately after the previous coalescnce event. For the coarsening to be scale-invariant, this number has to be constant at every loction in the normalized coordinates. Since the waves are nearly identical such that the amplitude of the excited wave is nearly twice as the equilibrium one and since the amplitude scales as χ, a reasonable universal value is $\chi = \sqrt{2}$ such that
$$\frac{<\ell>}{<\ell_c>} = \frac{<L>}{<L_c>} = \frac{\Delta C(T=0)}{C} = \sqrt{2} - 1 = 0.414 \quad (12.18)$$
The coalescence length ℓ_c is roughly 2.4 times $<\ell>$ at every position independent of the local texture $<\ell>$.

Note that the ommission of wave decay is not inconsistent with a universal size for the normalized excited wave. After each coalescence, significant increase in the height of equilibrium wave occurs, as seen in Figure 12.5, to ensure the average flow rate remains constant, as dictated by (12.15). It is hence feasible that the excited wave remains twice the size of the equilibrium one eventhough it absorbs successive equilibrium waves without significant decay. In fact, the scale-invariant coarsening rate requires this to be true.

The self-similarity in (12.18) results because the excited wave can be modelled as an equilibrium one with a larger substrate. That $l_c/<l>$ is a constant invariant to G and the spatial location is verified from our numerical simulation in Figure 12.10 . The estimate of l_c is from all simulated coalescence events within a window around every spatial station. A larger value between 3.0 and 3.5 instead of the predicted 2.4 is observed but the invariance is clearly established. With the local wave period or separation (12.14) and the coalescence rate given by l_c or $t_c = <l>/c$ in (12.18), the universal linear coarsening rate is immediately within reach. At a given location x, consider an average train of waves over a time segment $<\tau>$ between two excited waves. By $x + l_c$, one equilibrium wave between the two excited waves will be annihilated by coalescence. This increases the average wave period within the train by $\frac{<t>^2}{<\tau>}$ at x. With a continuum coarse-graining for l_c, this is simply
$$\frac{d}{dx}<t> = \frac{<t>^2}{<\tau> l_c} \quad (12.19)$$

Since $<l> = c<t>$, the corresponding increase in the average wave separation is

$$\frac{d<l>}{dx} = \frac{<t>}{<\tau>} \cdot \frac{<l>}{l_c} \qquad (12.20)$$

if one converts the wave train in time to wave train in space while holding the local c constant. The wave speed c scales as $\chi^{1/2}$ as seen in (12.11) as compared to the much more sensitive χ scalings of $<l>$ and $<t>$ in (12.15). One can hence neglect the local speed variation in x as χ decreases from 1 to 0.5 over the entire channel, as seen in Figure 12.6. Since $<t>/<\tau>$ and $<l>/l_c$ are invariant constants with respect to x and wave texture, we obtain $<l>$ from (12.18),

$$\frac{d<l>}{dx} \sim \frac{<t>}{<\tau>}(\sqrt{2} - 1) \qquad (12.21)$$

is constant for every downstream station independent of the local $<l>$. The coarsening rate is linear, indepndnet of downstream position, the local wave texture and length scale.

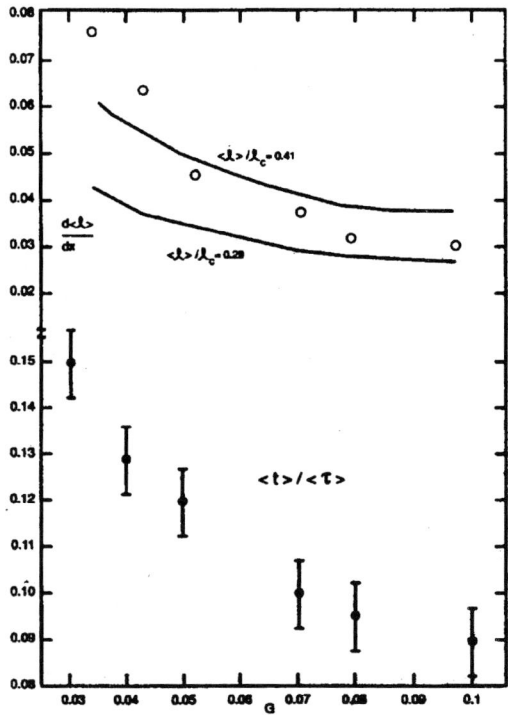

Figure 12.13: The function of excited wave $<t>/<\tau>$ as a function of G and the predicted linear coarsening rate using $<t>/<\tau>$ compared to simulated values with $<l>/l_c = \sqrt{2} - 1$ and $\frac{1}{3.5}$.

The only missing information is how the fraction of excited waves $<t>/<\tau>$ depends on G. We detect a weak dependence on G from our numerical simulation as seen in Figure 12.13. There is, however, little dependence on noise amplitude ϵ provided the noise is sufficiently white, as seen in Figure 12.13. This is most likely determined by the modulation of the band near ω_{max} at the end of the roll wave inception region, as for capillary waves in Chapter 4 and 9. However, such modulations are critically dependent on viscosity and srface tension. The empiricism of our artificial friction ν implies such an analysis is inappropriate for all waves.

Nevertheless, for every G, $<t>/<\tau>$ remains constant downstream. Using the measured $<t>/<\tau>$ and the theoretical value $<\ell>/\ell_c = \sqrt{2}-1$ or the empirical value $<\ell>/\ell_c = 1/3.5$ in Figure 8, two narrow bounds for $d<\ell>/dx$ are obtained for $0.02 < G < 0.1$. They tightly bound most of our simulated coarsening rate in Figure 11.

12.5 Summary and discussion

Two main invariances are responsible for the constant scale-invariant rate: $<t>/<\tau>$ and $<l>/l_c$ are independent of position or χ to leading order. The former results because the temporal spacing of waves is independent of the equilirium or excited status of two neighboring waves. As a results, the spacing between two excited waves $<\tau>$ scales and increases the same way as the average wave spacing $<t>$ downstream in this "mean-field" model. The invariance of $<l>/l_c$, on the other hand, results from the invariance of $\Delta C/C$ for every stage of a decaying excited wave. Universal decaying dynamics (derived by an expansion about $\chi = 1$) of an excited wave is, in turn, allowed by the self-similarity and sensitivity of the large localized equilibrium waves with respect to χ scaling in (12.11), used to model the decaying excited pulse in (12.16) as a substrate jump. It is only because of the simplicity and symmetries of the hydraulic equations that such unique features of complex spatio-temporal wave dynamics can be deciphered explicitly for all conditions.

CHAPTER 13

Drop Formation on a Coated Vertical Fiber

We use the same mathematical tools to address another kind of wave dynamics here. Localized pulses and coalescence cascades are also present here. However, there is a new feature — there exists a localized wave structure (a shock) that grows in time instead of translating stationarily. This shock also drives the coalescence events. Such wave dynamics appear in a unique hydrodynamic instability, about which our understanding has advanced beyond linear theory for wave inception to strongly nonlinear largetime dynamics. It is the Rayleigh instability for an annular film on a vertical fiber. If the thickness of the film is small compared to the fiber radius, a leading-order long-wave evolution equation can be derived to replace the far more complex equations of motion (Trifonov, 1992; Frenkel, 1992). Theoretical and numerical analyses of this equation (Kalliadasis and Chang, 1994a and Kerchman and Frenkel, 1994) have focused on a curious experimental observation by Quere (1990) on the large-time asymptotic dynamics of this instability. Quere observes that, if the initial film thickness h_0 exceeds a critical value, small-amplitude waves on the film around a thin fiber can form large capillary drops of the dimension of the capillary length $H = \sqrt{\sigma/\rho g}$ which are at least one order of magnitude larger than the former waves.

Kalliadasis and Chang (1994a) have constructed lone stationary pulses that travel steadily at constant speeds on a substrate of thickness h. They correspond to equilibrium states that isolated pulses can evolve into. They find that such equilibrium pulses can only exist for h less than

$$h_c = 1.68 R^3/H^2 \qquad (13.1)$$

where R is the fiber radius.

Both Kerchman and Frenkel (1994) and Kalliadasis and Chang (1994a) have carried out simulations on an extended domain with random initial conditions on an initial flat film of thickness h_0. The wave dynamics are also observed to be fundamentally different depending on the relative magnitude of h_0 to h_c. In both cases, the non-stationary, small-amplitude waves at inception evolve into well-separated pulses on a thin substrate film. This pulse formation mechanism should be similar to that of Chapter 4 and 9. Once formed, however, the dynamics of each pulse seem to be determined only by the local substrate thickness h. A subcritical pulse ($h < h_c$) would equilibrate into one of the stationary pulses constructed by Kalliadasis and Chang. This equilibration can involve collecting fluid from the substrate or draining fluid from the pulse but the amplitude and speed of an isolate pulse would eventually become stationary. The pulses may continue to interact weakly with their neighbors and adjust their spacing. However, extensive simulations show that coalescence rarely occurs after the subcritical pulses are formed. In the absence of coalescence and individual pulse growth, these subcritical pulses do not grow into drops.

This scenario is dramatically different for supercritical pulses with $h > h_c$. These pulses grow individually by collecting fluid from the substrate. Moreover, their growth rate is very different such that the larger pulses are much faster than the smaller ones. As a result, a large pulse eventually overtakes and captures its smaller and slower front neighbor in a coalescence event. The former pulse gains more fluid in the process and becomes even larger and faster. An entire train of smaller pulses can then be captured successively in a coalescence cascade by a trailing large pulse. This large pulse grows with each coalescence and continues to collect fluid from the substrate between coalescence events. Although the simulations cannot be carried out indefinitely and the model equation breaks down eventually when the growing pulses become too large, this combination of coalescence and individual growth is expected to drive the large supercritical pulses into drops.

13.1 Pulse coalescence dynamics

We shall utilize the leading order evolution equation derived by Trifonov (1992) and Frenkel (1992) for $(h_0/R) \ll 1$

$$\frac{\partial h}{\partial t} + \frac{\partial}{\partial x}[\delta h^3(\frac{\partial^3 h}{\partial x^3} + \frac{\partial h}{\partial x}) + \frac{2}{3}h^3] = 0 \tag{13.2}$$

where

$$\delta = (2\sigma h_0/3\rho\, g\, R^3) = (2H^2 h_0/3R^3)$$

measures the ratio of curvature-driven flow of the Rayleigh instability to the gravity-driven mean flow. The film thickness h_0 is taken to be that of the initial waveless film and it has been used to scale the interfacial height. The fiber radius R is used to scale the axial coordinate x and the characteristic time used is R/U where $U = (gh_0^2/2\nu)$ is the interfacial velocity of the film. As a result of the scaling, the thickness of the initial waveless film is always unity. The critical condition $h_0 = h_c$ now corresponds to a critical $\delta_* = 1.12$. It is sometimes convenient to present the graphics in a frame moving with speed c

and the equation in that frame becomes

$$\frac{\partial h}{\partial t} + \frac{\partial}{\partial x}[\delta h^3(\frac{\partial^3 h}{\partial x^3} + \frac{\partial h}{\partial x}) + \frac{2}{3}h^3 - ch] = 0 \qquad (13.3)$$

Typically, c is chosen to be the pulse speed of an equilibrium subcritical pulse at the particular value of δ. It is hence well-defined only for subcritical values of δ, $\delta < \delta_*$.

We have developed a high-order finite-difference scheme similar to the one described in Chapter 4 for falling film waves. For the current problem, (13.2) and (13.3) are integrated in time using 2000 spatial grid-points. Unlike the falling-film problem, however, we shall not pursue the coarsening statistics. Instead, only the formation of the excited pulse will be scrutinized. Hence, periodic computational domains will suffice.

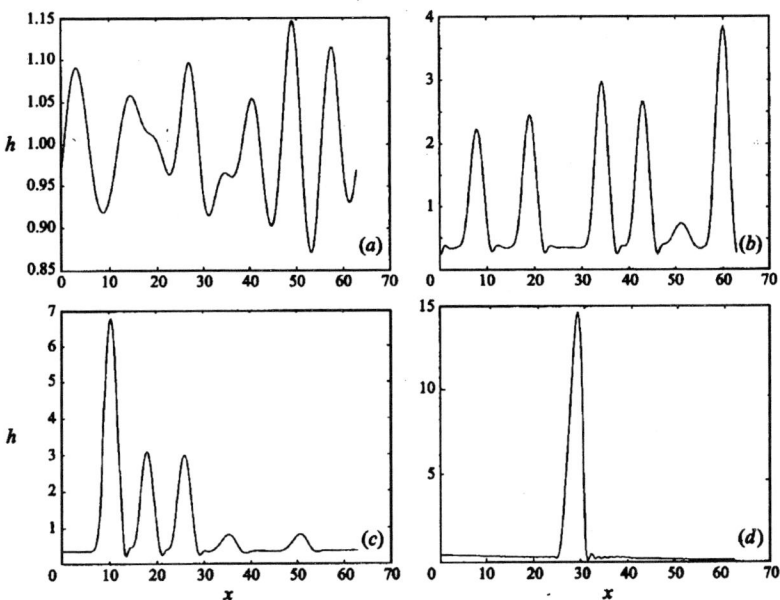

Figure 13.1: Simulated coalescence cascade with random initial condition at $\delta = 3$ and (a) $t = 2$, (b) 15, (c) 23 and (d) 30.

In Figure 13.1, we depict the simulated interfacial profiles from (13.3) in a periodic domain of length 20π beginning with a small-amplitude ($< 10^{-3}$), zero-mean, random disturbance to $h_0 = 1$ at the supercritical condition of $\delta = 3$. As is evident, the evolution evolves through the linear filtering stage and forms heavily modulated sinuous waves roughly the wavelength λ of the fastest-growing mode from linear theory, $\lambda = 2\sqrt{2}\pi$, as is analogous to falling-film waves. The modulation annihilates certain wave peaks at this stage to yield about 6 distinct wave crests. These fluctuating crests seem to phase lock and saturate for some time at about $h = 1.2$ before they slowly blossom individually into distinct pulses by $t = 10$ with h in excess of 2. These pulses possess the signature front dimple of capillary film flows (Wilson, 1982; Wilson and Jones, 1983; Hammond, 1983). They are also separated by thin flat substrate films. Although the pulses are similar in shape, there are variations in their speed, height and separation which are legacies of the earlier modulations. As seen by comparing figures 1b and 1c, isolated pulses continue to grow by accumulating fluid from the substrate. Well-packed pulses do not grow appreciably, presumably because of the limited reservoir of fluid in the substrate between pulses. The larger pulses travel faster and begin to encroach the smaller ones in front. By $t = 15$, the largest one has coalesced with its front neighbor to induce a jump in its amplitude.

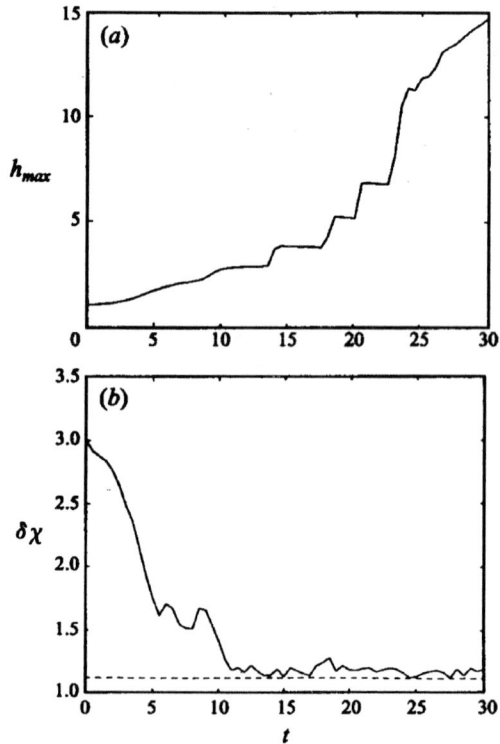

Figure 13.2: (a) Amplitude and trailing substrate thickness χ of the large driving pulse if Figure 13.1.

We follow the amplitude of this large pulse in Figure 13.2a. Its solitary growth prior to the first coalescence is apparent and we see that it successively captures all five other pulses in due time, the last two in one violent gulp at $t = 24$, to form a monster pulse of amplitude 10. We also note that the large pulse either retains its amplitude or actually grows between coalescence events.

The final lone pulse smoothes the substrate and continues to grow, albeit more slowly, by collecting liquid from the substrate, as is evident from the amplitude growth beyond $t = 24$ in Figure 13.2a. Although it is not evident in Figure 13.2a, its amplitude eventually saturates at about 15, a full order of magnitude larger than the initial quasi-saturated waves at $t = 5$. This monster pulse stops growing when the substrate thickness reaches a critical value. The saturation occurs because we use a periodic computation domain with a finite amount of liquid. In an extended domain, the substrate beneath a supercritical pulse does not thin appreciably and the practically infinite reservoir of fluid allows the pulse to evolve into a drop. We shall demonstrate this in a later section. To demonstrate the large supercritical pulse continually accumulates fluid from the substrate film during the coalescence cascade, we track the substrate thickness χ behind the large pulse as a function of time in Figure 13.2b. It is clear that χ decreases monotonically until $\chi\delta$ approaches the critical value δ_*. Since the substrate thins and its thickness decreases from its initial value of unity to χ, the effective δ for the growing pulse, based on its own substrate thickness, is not the original δ but $\delta\chi$. Figure 13.2b then suggests that the pulse ceases to grow under subcritical conditions $\delta < \delta_*$. We also note that χ has reached its asymptotic value by $t = 15$ while the pulse is still growing at $t = 25$. This suggests that the substrate thickness in front of the growing pulse is larger than χ and the jump in the substrate thickness across the pulse fuels the growth. The scenario is

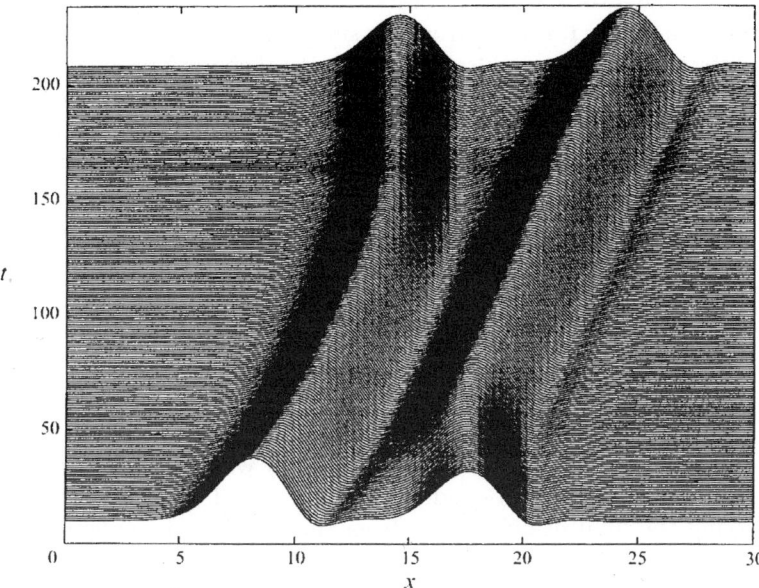

Figure 13.3: A subcritical equilibrium pulse at $\delta = 0.8$ followed by larger pulse with 1.265 the equilibrium amplitude. Thr initial separation is $\Delta x = 10$.

completely different for subcritical conditions $\delta < \delta_*$. Extensive simulations by

Kerchman and Frenkel (1994) and Kalliadasis and Chang (1994a) with random initial conditions have shown that subcritical pulses rarely coalesce. In Figure 13.3, we have placed two subcritical pulses next to each other for $\delta = 0.8$. The front one is an equilibrium stationary pulse constructed by Kalliadasis and Chang (1994a) while the back one is larger with an amplitude 1.265 times the equilibrium value. It is clear that the two pulses interact with each other but there is no coalescence at the end. The back pulse drains its excess fluid and equilibrates by slowing down to a stationary pulse in the moving frame of Figure 13.3. The front pulse gains some fluid during the interaction and moves forward. Although it is not shown in Figure 13.3, the front pulse will also shed its excess fluid and equilibrates. While the separation and amplitude variation in Figure 13.3 are typical of the "natural" conditions with random, small-amplitude initial conditions, one can force coalescence of subcritical pulses by using extremely large and unnatural pulses. We place a pulse twice as large as the stationary in front of a train of equilibrium pulses spaced about 10 units apart. We track the amplitude of the large pulse, which travels much faster than the equilibrium pulses, as a function of time in Figure 13.4 for the subcritical conditions of $\delta = 0.6$ and 0.4. As is evident, the coalescence cascade still occurs for $\delta = 0.6$ eventhough the large pulse drains significant amount of its liquid between coalescence events. As a result, its amplitude actually decreases in time despite the coalescence cascade. The drainage is so severe for $\delta = 0.4$ that the cascade stops after the first coalescence event. The large pulse has decayed into an equilibrium one after the first coalescence event and is unable to capture the next equilibrium pulse. It is clear from the simulations that coalescence is driven by speed differential between pulses. For subcritical conditions, coalescence only occurs if there are excessively large pulses compared to the equilibrium ones. This coalescence mechanism is the same as that on an inclined film. However, due to the fast drainage from the large pulses for this system, their amplitude can decay in time despite the fluid gain during coalescence and the coalescence cascade may not be sustainable. In constrast, drainage is always from the substrate to the pulse under supercritical conditions and this drainage amplifies the difference between pulses – it promotes coalescence.

13.2 Equilibrium subcritical pulses and stability

We shall demonstrate in this section that, in the absence of coalescence, a large non-equilibrium pulse (or any localized structure) will decay towards an equilibrium pulse under subcritical conditions and we shall estimate the decay rate by quantifying the fluid drainage rate from the pulse to the substrate. We do so by showing that the equilibrium pulses are linearly stable with a new spectral theory. Consider a single stationary pulse on a substrate of unit thickness (h_0 in the definition of δ is now taken to be the substrate thickness of a single pulse), these equilibrium pulses are defined by the following equations in a frame

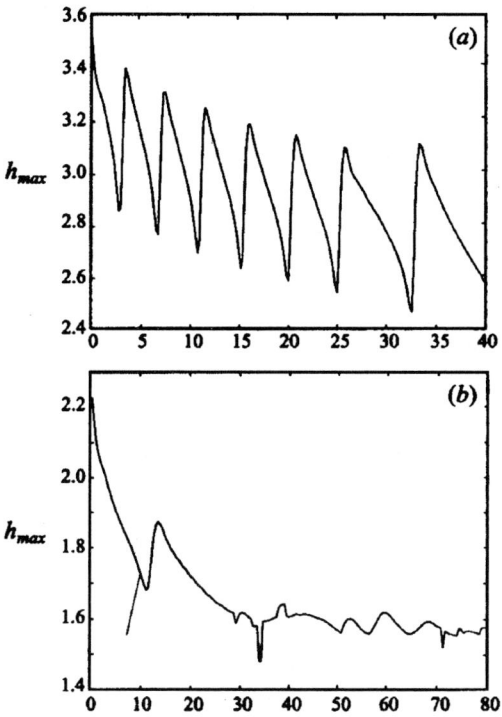

Figure 13.4: Amplitude of the large pulse (twice the equilibrium) behind a train of equilibrium pulses for (a) $\delta = 0.6$ and (b) 0.4.

moving with constant speed c, a stationary version of (13.3),

$$\delta h_s^3 \left(\frac{\partial^3 h_s}{\partial x^3} + \frac{\partial h_s}{\partial x} \right) + \frac{2}{3} \left(h_s^3 - 1 \right) - c(h_s - 1) = 0 \quad (13.4a)$$

$$h_s(x \to \pm \infty) = 1 \quad (13.4b)$$

Construction of the stationary pulse then amounts to determining $c(\delta)$. Kalliadasis and Chang (1994a) showed by matched asymptotics that $c(\delta)$ blows up to infinity at the limiting δ value of δ_* such that equilibrium pulses only exist when δ is less than

$$\delta_* = 1.12 \quad (13.5)$$

which is a dimensionless version of (13.1). This corresponds to a substrate thickness thinner than h_c of (13.1). For $\delta < \delta_*$, each equilibrium pulse has the distinct shape of a large pulse preceded by a deep dimple. As δ increases towards δ_*, the pulse becomes larger, the speed faster and the dimple curvature more pronounced until all three approach infinity at δ_*. Near δ_*, the speed c, amplitude h_s^{max} and area J of the stationary pulses can be estimated as

$$c - 2 = \frac{2\delta^{2/3}}{\delta_*^{2/3} - \delta^{2/3}} \tag{13.6a}$$

$$h_s^{max} - 1 = 0.371(c - 2) \tag{13.6b}$$

$$J = \int_{-\infty}^{\infty} (h_s - 1) dx = 1.175(c - 2) \tag{13.6c}$$

Correlations (13.6) represent slight empirical improvement of the analytical expressions derived by Kalliadasis and Chang near δ_* to extend the validity of the correlations further away from δ_*. We have also obtained the true values by constructing the stationary pulses numerically. These values are tabulated in Table 1.

As a check of their validity, we note that the decaying pulse in Figure 13.4b approaches the equilibrium height of $h_{max} = 1.5$ in Table 1 at $\delta = 0.4$. This again supports the observation that subcritical pulses decay towards the equilibrium pulses constructed from (13.4). We linearize the evolution equation (13.3) about the pulse solution (13.4) to yield the linearized equation for the disturbance $u(x,t) = h(x,t) - h_s(x)$,

$$\frac{\partial u}{\partial t} = Lu \tag{13.7}$$

where the linearized operator is

$$L = -\frac{\partial}{\partial x}[\delta h_s^3 \left(\frac{\partial^3}{\partial x^3} + \frac{\partial}{\partial x}\right) \cdot + 3\delta h_s^2 \left(\frac{d^3 h_s}{dx^3} + \frac{dh_s}{dx}\right) \cdot + (2h_s^2 - c) \cdot] \tag{13.8}$$

and the disturbances are bounded as discussed in Chapter 7

$$L\psi = \lambda \psi$$

$$\psi \text{ bounded as } x \to \pm \infty \tag{13.9}$$

The essential eigenfunctions $\psi(\lambda, x) = K(\alpha, x)e^{i\alpha x}$ which approach bounded oscillations in the infinities belong to the essential spectrum defined by

$$\lambda(\alpha) = i\alpha(c - 2) + \alpha^2 \delta(1 - \alpha^2) \tag{13.10}$$

for $\alpha \in (0, \infty)$. The reason that (13.10) is simply the dispersion relationship for a flat film of unit thickness is because the equilibrium pulse decays into such a film at both infinities and the oscillations of the eigenfunctions must also be described by the flat-film dispersion relationship. There is a continuum of such "radiation" modes since α takes on all real values. A sample spectrum for the present operator, constructed with the technique of Chapter 7, is shown in Figure 13.5. Due to translational invariance, there is always a simple zero eigenvalue λ_1 at the origin that corresponds to the eigenfunction $\psi_0 = \frac{dh_s}{dx}$. (This ever-existing neutral eigenvalue is not shown in the figure.) There is only

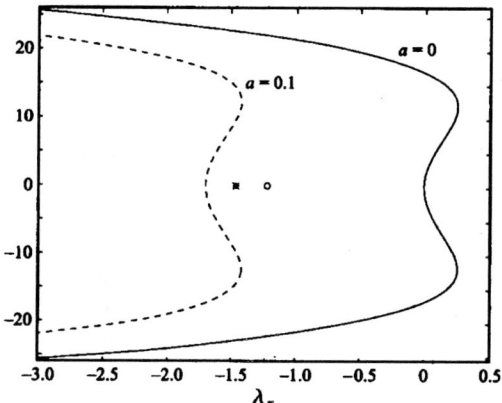

Figure 13.5: The spectrum of the equilibrium pulse at $\delta = 1$ in the complex λ-plane. The original and shifted essential spectra for indicated values of a are shown. The discrete eigenvalue is denoted by a star and the resonance pole an open circle. After the shift, the resonance pole becomes a discrete eigenvalue a resonance pole. The former dominates the decay dynamics. The neutral eigenvalue, corresponding to translation symmetry, at the origin is not shown.

one other discrete mode λ_2 that we could locate for the entire pulse family. It is always stable and is marked by an asterisk in Figure 13.5 at the position of -1.463 for $\delta = 1$. The unstable essential modes of Figure 13.5, which are the mass carrying ones, again suggest a weighter spectral formulation to capture local mass drainage near the pulse. We hence use the e^{ax} weight of Chapter 5 on the disturbance u in (7), $v = e^{ax}u$, and define a corresponding eigenvalue problem

$$L_a \phi = e^{ax} L(e^{-ax}\phi) = \lambda \phi \qquad (13.11)$$

where $\phi = e^{ax}\psi$ is the weighted eigenfunction that spans the weighted disturbance $v(x,t)$. It is clear that the essential spectrum Γ_a of L_a is related to that of L in (13.10) by the transformation $\alpha \to \alpha + ia$.

$$\lambda_a = i(\alpha + ia)(c - 2) + \delta(\alpha + ia)^2[1 - (\alpha + ia)^2] \qquad (13.12)$$

The net result is that the essential spectrum is shifted to the left in the complex plane. This is demonstrated in Figure 13.5. The fact that a particular value of a exists that can shift the entire essential spectrum to the left half plane implies that the pulse is convectively stable – the weighted disturbance decays to zero in amplitude.

Eventhough the drained mass will eventually grow on the trailing substrate, the pulse suffers no long-time perturbation locally. Unlike the essential spectrum, which is shifted by the weight, the discrete spectrum of L is also a discrete spectrum of L_a - with an important exception. The discrete eigenfunctions ψ_k of L decay to zero as $|x| \to \infty$ with an exponential rate determined by the roots

α of the spatial characteristic polynomial

$$P(\alpha) = \lambda_k - \lambda(\alpha) \qquad (13.13)$$

where λ_k is the discrete eigenvalue corresponding to ψ and $\lambda(\alpha)$ is given by the dispersion relationship (13.10). The roots with positive real part determine the decay rate of $x \to \infty$ and ones with negative real parts the rate at $x \to \infty$. However, one can, in principle, also construct functions which satisfy $L\psi_\# = \lambda_\# \psi_\#$ but do not decay to zero at one infinity. If, for example, there are two roots with negative real parts α_1 and α_2, $\psi_\#$ can decay to zero at $+\infty$ with rate α_1 but grow exponentially at $-\infty$ with rate α_2. Since the eigenfunctions $\psi_\#$ of these $\lambda_\#$ modes do not decay to zero, they are not part of the discrete spectrum but are the resonance poles of Chapter 5. We have found only one real and negative resonance pole ($\lambda_\# < 0$) which is shown as an open circle in Figure 13.5. These resonance poles and their eigenfunctions also satisfy the weighted operator $L_a \phi_\# = \lambda_\# \phi_\#$ where $\phi_\# = \psi_\# e^{ax}$ but their asymptotic behavior at $x \to \pm\infty$ can now change. Since the essential spectrum Γ_a for L_a is defined by $\lambda_a(\alpha)$ for α real, when there is a discrete eigenvalue on the essential spectrum Γ_a, one of the roots to the spatial characteristic polynomial (13.13) is purely imaginary. Consequently, as Γ_a is shifted across the resonance pole $\lambda_\#$, the negative root α_2 crosses the imaginary axis and its real part becomes positive. Hence, as Γ_a is shifted across the resonance pole, $\phi_\#$ now decays to zero at both infinities and the resonance pole becomes a discrete eigenvalue of L_a. Similarly, if Γ_a is shifted across the discrete eigenvalue λ_2, it becomes a resonance pole of L_a. The translational zero mode λ_1 remains an eigenvalue as Γ_a passes by due to a hole in the Riemann surface defined by Γ_a. Other rules on the exchange between eigenvalues and resonance poles can be found in Chapter 5. For all equilibrium pulses with $\delta < \delta_*$, we are able to shift Γ_a such that $\lambda_\#$ becomes a discrete eigenvalue of L_a and λ_2 a resonance pole as seen in Figure 13.5. Hence, $\lambda_\#$ is the dominant mode of the weighted disturbance and it determines the asymptotic decay rate towards equilibrium pulses – the asymptotic drainage rate. The value of this resonance pole $\lambda_\#$ is tabulated in Table 13.1. It becomes increasingly negative as δ approach δ_* from below. All equilibrium pulses are hence stable and all non-equilibrium pulses with a local substrate thinner than h_c are expected to decay into equilibrium pulses if they are not involved in further coalescence events. The decay rate $\lambda_\#$ increases with increasing substrate thickness. To verify the decay rate estimated by the dominant resonance pole, we carry out a sequence of numerical studies with "excited" lone pulses. The initial non-equilibrium pulse is $h(x, t = 0) = 1 + 1.2(h_s(x) - 1)$ in the frame moving at the equilibrium pulse speed c. This represents an added mass 20% of that carried by the equilibrium pulse. The drainage rate and the decay rate of the maxiumum pulse height h^{max} towards the equilibrium value should both be $\lambda_\#$ at large time, $h^{max} - h_s^{max} \sim 0.2(h_s^{max} - 1)e^{\lambda_\# t}$. We hence track $\eta(t) = t^{-1} \ln \left[\frac{h^{max}(t) - h_s^{max}}{0.2(h_s^{max} - 1)} \right]$ as the excited pulse decays. As seen in Figure 13.6a, $\eta(t)$ approaches a constant negative asymptotic value for $\delta < \delta_*$, indicating an exponential decay towards the equilibrium pulse. That large pulses decay rapidly towards equilibrium pulses under subcritical conditions explains why they do not evolve into drops. They cannot grow individually beyond the equilibrium pulse. For $\delta < 0.8$, the asymptotic value of η is in excellent agree-

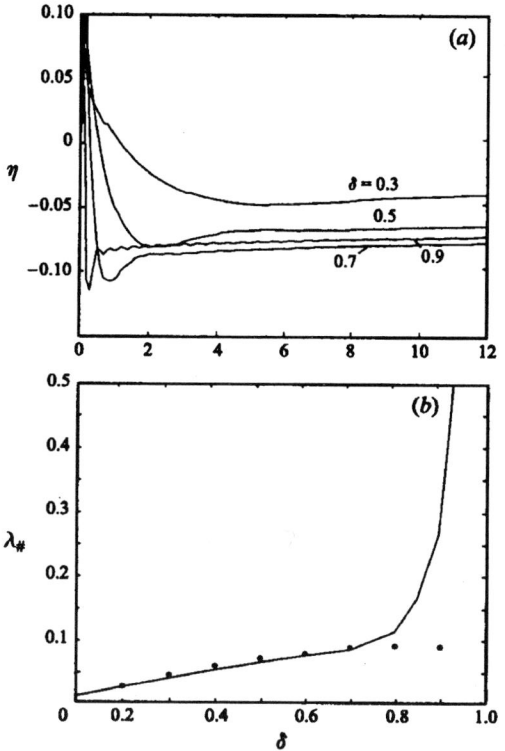

Figure 13.6: (a) The decay rate $\eta(t)$ from numerical simulation of how an 'excited' pulse with 50% added mass decays towards the equilibrium pulse for $\delta < 0.9$. (b) Comparison of the asimptotic decay rate $\eta(\infty)$ (aterisks) to the theoretical curve from the resonance poles.

ment with the resonance pole $\lambda_\#$ as seen in Figure 13.6b. For $0.8 < \delta < \delta_*$, however, oscillations are observed in $\eta(t)$ which become more pronounced as δ approaches δ_*. Some oscillations are evident in the $\delta = 0.9$ curve in Figure 13.6a. There seems to be an apparent equilibration of $\eta(t)$ as well but this asymptotic value deviates from the resonance pole $\lambda_\#$. The cause of the deviations of the resonance pole theory from the numerical results near δ_* is not presently understood.

Table 1 Properties of subcritical equilibrium pulses ($\delta < \delta_*$) with unit substrate

δ	c	J	h_s^{max}	$\lambda_\#$
0.1	2.152	0.316	1.095	-0.0137
0.2	2.304	0.634	1.191	-0.0267
0.3	2.521	1.061	1.321	-0.0388
0.4	2.809	1.603	1.488	-0.0496
0.5	3.205	2.314	1.707	-0.0592
0.6	3.781	3.289	2.012	-0.0668
0.7	4.687	4.714	2.458	-0.0738
0.8	6.279	6.997	3.175	-0.1000
0.85	7.606	8.746	3.728	-0.1510
0.9	9.645	11.240	4.516	-0.2510
0.95	13.020	14.990	5.701	-0.5670
1.0	18.900	20.790	7.544	-1.2200

13.3 Growth dynamics of supercritical pulses

When the local substrate thickness exceeds h_c, the corresponding δ is larger than δ_* of (13.5) and no stationary pulses exist. As a result, a local large structure cannot decay and approach an equilibrium pulse. To study its evolution towards a new asymptotic state, we place a large pulse of arbitrary shape on a unit substrate at $\delta = 1.5$ in Figure 13.7 . Soft boundary conditions are used instead of

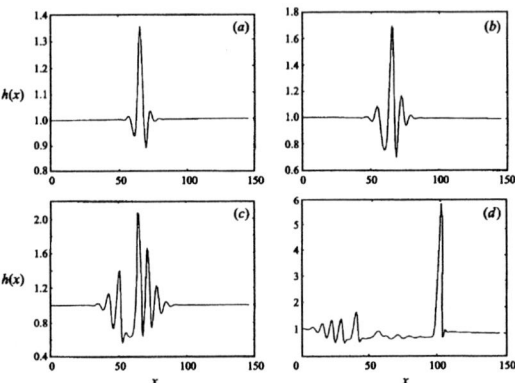

Figure 13.7: The transient for the formation of a growing pulse when a lager pulse is placed on a unit substrate for $\delta = 1.5$. The snapshots are taken at $t = 2, 5, 10,$ and 20 $(a - d)$. A quasi-steady jump in the substrate thickness is clearly evident by $t = 20$ after the transient wave packets convens away.

periodic boundary conditions for this large domain to ensure that a sufficiently large amount of liquid is available to fuel growth. A transient adjustment ensues over about 20 units of time. During this transient period, fluid is drained to the back of the pulse as it attempts to reach a non-existing equilibrium shape. It soon finds that a quasi-equilibrium position is possible only if the thickness of the back substrate layer, denoted $\chi(t)$ for convenience here, is smaller than the unit thickness in front. Due to this jump in substrate thickness across the pulse, the flow entering the pulse is larger than the exit value in the frame moving with the pulse and the pulse grows. The stationary equilibrium pulses for $\delta < \delta_*$ are hence replaced by growing pulses as the asymptotic states of localized structures for supercritical conditions, $\delta > \delta_*$. This growth by collecting liquid from the front substrate is a slow process compared to the local adjustment time for the interface. As a result, the pulse and the inner regions connecting the pulse to the substrates remain quasi-static. The matched asymptotic analysis of Kalliadasis and Chang for stationary pulses can then be extended below to study this asymptotically growing pulse with a unique self-growth mechanism. We shall divide the slowly-growing pulse into three regions and utilize matched asymptotics. The hydrostatic outer region at the peak of the growing pulse is expected to be dominated by azimuthal and axial capillary forces to leading order, with next-order corrections from gravity. To allow matched asymptotics, the quasi-steady pulse height must be large compared to the substrate thickness. For such large pulses, we expect the quasi-steady pulse speed c to also be large. We shall hence use $c^{-1/3}$ as the small parameter in the asymptotics. In the outer quasi-static outer region dominated by hydrostatics, the x scale is of unit order since the axial and azimuthal curvature terms h_x and h_{xxx} within the parenthesis in (13.3) must balance. However, the pulse height h_{max} is large. As such, the curvature near the back of the pulse must be of order h_{max}. There are then the back inner region where the pulse meets the substrate layer of unit-order thickness χ and the front inner region where it meets the unit substrate thickness. Within both inner regions, one has the dominant Bretherton scaling which balances axial-curvature driven flow $h^3 h_{xxx}$ with shear flow due to translation ch (Bretherton, 1961). As such, the vertical length scale and the horizontal length scale must have a ratio of $h_x \sim O(c^{1/3})$. We know, however, from matching curvature with the outer region that $h_{xx} \sim O(h_{max})$ in the inner region. Since h is of unit order in both inner regions, we immediately conclude that $h_{max} \sim O(c^{2/3})$ in the outer region and $x \sim O(c^{-1/3})$ in the inner regions. The dominant balances in all three regions do not involve the growth term $\frac{\partial}{\partial t}$ and this stipulates a relatively long time scale for slow growth. To estimate the growth time scale, we integrate (13.3) from $x = -\infty$ where $h = \chi$ to $x = +\infty$ where $h = 1$ to yield a global mass balance over the jump

$$\begin{aligned}\frac{\partial}{\partial t}\int_{-\infty}^{+\infty} h\,dx &= c(1-\chi) - \tfrac{2}{3}(1-\chi^3)\\ &\sim c(1-\chi)\end{aligned} \qquad (13.14)$$

which yields a growth time of $O(c^{-1/3})$. With $h \sim O(c^{2/3}), t \sim O(c^{-1/3})$ and $x \sim O(1)$ in the outer region, the evolution term $\frac{\partial h}{\partial t}$ in (13.3) is of negligible $O(c)$ compared to the dominant $O(c^{8/3}), O(c^2)$ and $O(c^{5/3})$ terms in (13.3). In fact, if we resolve the outer solution to the following order

$$h \sim c^{2/3}(h_0 + c^{-1/3}h_1 + c^{-2/3}h_2 + \cdots) \qquad (13.15)$$

the evolution term can be omitted – the outer solution is quasi-stationary. Integrating (13.3) from $-\infty$ where $h = \chi$ and from $+\infty$ where $h = 1$, we get

$$\delta h^3(h_{xxx} + h_x) - c(h - \chi) + \frac{2}{3}(h^3 - \chi^3) + \frac{\partial}{\partial t}\int_{-\infty}^{x} h\,dx = 0 \quad (13.16a)$$

$$\delta h^3(h_{xxx} + h_x) - c(h - 1) + \frac{2}{3}(h^3 - 1) + \frac{\partial}{\partial t}\int_{x}^{+\infty} h\,dx = 0 \quad (13.16b)$$

To leading orders of $O(c^{8/3})$ and $O(c^2)$, these two equations are identical,

$$\delta h^3(h_{xxx} + h_x) + \frac{2}{3}h^3 = O(c^{5/3}) \quad (13.17)$$

Substituting (13.5) into (13.7), we get

$$\frac{d^3 h_0}{dx^3} + \frac{dh_0}{dx} = 0 \quad (13.18a)$$

$$\frac{d^3 h_1}{dx^3} + \frac{dh_1}{dx} = 0 \quad (13.18b)$$

$$\frac{d^3 h_2}{dx^3} + \frac{dh_2}{dx} = -\frac{2}{3\delta} \quad (13.18c)$$

The leading-order equation (13.18a) is just the long-wave Laplace-Young equation which possesses a static pulse solution symmetric about $x = \pi$

$$h_0 = A'(t)(1 - \cos x) \quad (13.19)$$

This static solution has a constant width of 2π and makes contact with the substrate at $x = 0$ and 2π. With a non-trivial h_0, the next order term h_1 vanishes exactly and the next non-trivial correction to the outer solution is

$$h_2 = -\frac{2}{3\delta}(x - \sin x) + B \quad (13.20)$$

It elevates the substrate thickness to B at $x = 0$ and 2π and introduces an asymmetric correction to (13.19) after the baseline correction. The tilt forward is due to gravitational steepening in the $\frac{2}{3}h^3$ term in (13.16) and (13.17). Combining h_0 and h_2 and expanding about the contact points at the back $x = 0$ and at the front $z = x - 2\pi = 0$, we obtain

$$h(x \to 0) \sim A(t)\frac{x^2}{2} + B \quad (13.21a)$$

$$h(z \to 0) \sim A(t)\frac{z^2}{2} + (B - \frac{4\pi}{3\delta}) \quad (13.21b)$$

where $A(t) = A'(t)c^{2/3}$. In the two inner regions, we shall rescale the coordinates with the proper scales, $h \sim O(1)$ and $x \sim O(c^{-1/3})$. We shall also examine both sides at the same time and define the Bretherton variables

$$\begin{aligned} f &= h/\chi \\ \xi &= \frac{1}{\chi}(\frac{c}{\delta})^{1/3}\,x \end{aligned} \quad (13.22)$$

such that $\chi = 1$ in front and $\chi \neq 1$ in the back. The rescaled (13.4) does not involve the dynamic term $\frac{\partial}{\partial t}(\cdot)$ to leading three orders - the inner regions are also quasi-steady due to the small scales. In fact, only the dominant inner solution is required to match the outer expansion of (13.21). This the Bretherton equation (Bretherton, 1961)

$$f_0''' - \frac{f_0 - 1}{f_0^3} = 0 \qquad (13.23)$$

subject to the following boundary conditions at the two inner regions,

$$f_0 \to 1 \quad \text{as} \quad \xi \to \pm\infty \qquad (13.24)$$

with the plus sign corresponding to the front inner region and minus the back inner region. Since the leading order outer solution (13.19) is symmetric about $x = 0$ and makes tangential contact with the negligibly thin substrate, the two leading order inner solutions must blow up quadratically to allow matching. In general, the asymptotic solutions of the Bretherton equation blows up with a vanishing third derivative and hence

$$f_0^\pm(\xi \to \pm\infty) \sim \alpha^\pm \xi^2 + \gamma^\pm \xi + \beta^\pm \qquad (13.25)$$

where $+$ denotes the behavior of the back inner solution as ξ approaches $+\infty$ and $-$ denotes the asymptotic behavior of the front inner solution at $-\infty$. There is also a possibility of a higher order $\xi \ln \xi$ behavior (Kalliadasis and Chang, 1996) that can be matched with higher order terms of (13.15). There is an additional degree of freedom in choosing the origin of ξ and this is chosen to suppress the linear $\gamma_\pm \xi$ term to ensure quadratic blow-up. For the back inner region near $x = 0$, integrations by Bretherton (1961) and many others (see Kalliadasis and Chang, 1994a,b, for example) show that there is only a unique asymptotic behavior with

$$\alpha^+ = 0.32171 \qquad \beta^+ = 2.898 \qquad (13.26)$$

In the z coordinate of the outer region, this corresponds to a unique inner asymptote for the back

$$h^+ \sim \frac{\alpha^+}{\chi}\left(\frac{c}{\delta}\right)^{2/3} x^2 + \chi\beta^+ \qquad x \to 0 \qquad (13.27)$$

For the front inner region near $z = x - 2\pi = 0$, however, a family of inner asymptotes are now possible. One integrates the corresponding leading-order Bretherton equation in (13.23) towards $\xi = -\infty$ with the initial condition

$$f_0 \sim 1 + \epsilon e^{-m\xi} \cos(n\xi + \theta) \qquad (13.28)$$

where $m = 1/2$ and $n = 1/\sqrt{2}$. The parameters m and n correspond to the complex conjugate eigenvalue pair for the flat-film dispersion relationship that grow as $\xi \to -\infty$, viz. $m > 0$. Because it is a complex pair, a phase θ enters the initial condition. For vanishingly small ϵ, integration of the Bretherton equation with (13.28) shows that a range of θ values yield quadratic asymptotic behavior for f_0 at $\xi \to -\infty$. There is hence a family of (α^-, β^-) pairs which we shall represent in a functional form

$$\beta^- = \beta^-(\alpha^-) \tag{13.29}$$

The computed values of this function will be presented in a more convenient form later. In any case, a family of inner asymptotes exist for the front

$$h^- \sim \alpha^- \left(\frac{c}{\delta}\right)^{2/3} z^2 + \beta^- \qquad z \to 0 \tag{13.30}$$

Baseline and curvature matching of the zeroth and quadratic order terms in the outer solution in (13.21b) to the inner asymptotes (13.27) and (13.30) yield four equations,

$$\frac{\alpha^+}{\chi}\left(\frac{c}{\delta}\right)^{2/3} = \frac{A}{2} \qquad \beta^+ \chi = B \tag{13.31a}$$

from matching at the back $x = 0$ and

$$\alpha^- \left(\frac{c}{\delta}\right)^{2/3} = \frac{A}{2} \qquad \beta^- = B - \frac{4\pi}{3\delta} \tag{13.31b}$$

from matching at the front $x = 2\pi$. They allow us to eliminate the unknown B in the dominant outer terms and relate A, c, χ, δ and α^- by three relationships

$$c = \left(\frac{A}{2\alpha^-}\right)^{3/2} \delta \tag{13.32a}$$

$$\chi = \alpha^+/\alpha^- \tag{13.32b}$$

$$\beta^- + \frac{4\pi}{3\delta} = \frac{\beta^+ \alpha^+}{\alpha^-} \tag{13.32c}$$

where β^- is a funciton α^-. We note that at $\delta = \delta_* = 1.1201$, the solution to (13.32) is $\chi = 1, \alpha^- = \alpha^+ = 0.32171$ and $\beta^- = -0.8415$. This shows that the positive substrate jump vanishes at δ_* and reverses for $\delta < \delta_*$. Hence, the growing pulse solution exists only beyond δ_* where χ is less than unity. We can use (13.32c) to map the funciton $\beta^-(\alpha^-)$ into the functions $\beta^-(\delta)$ and $\alpha^-(\delta)$ in Figure 13.8. From these relationships, we obtain how χ and c depend on δ from (13.32a) and (13.32b). To get the drainage rate, we substitute the leading-order outer solution (13.19) into (13.14) and upon integrating from $x = 0$ to 2π,

$$\frac{dA}{dt} \sim \frac{c(1-\chi)}{2\pi} = \kappa A^{3/2} \tag{13.35}$$

after substituting (13.32a), where the growth constant is

$$\kappa = \frac{\delta(1-\chi)}{2\pi(2\alpha^-)^{3/2}} \tag{13.36}$$

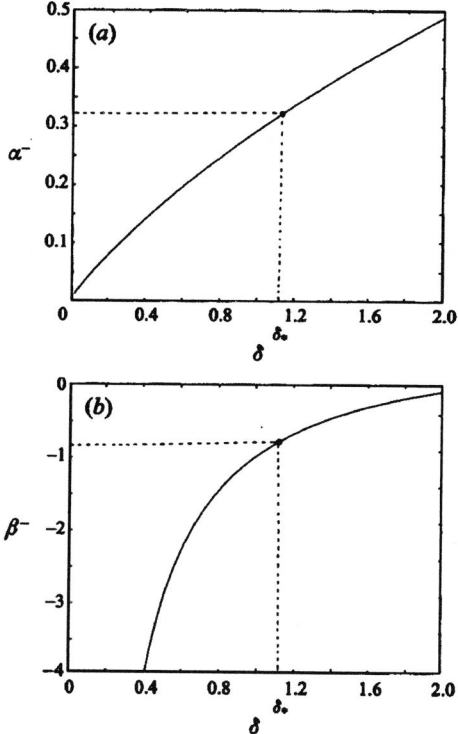

Figure 13.8: Dependence of α^- and β^- on δ from matches asymptotic. Even value below δ_* are included.

is only a function of δ through (13.32b) and Figure 13.8. This equation can be easily integrated to yield the blow-up behavior of the growing pulse

$$A(t) = \frac{4}{\left(\frac{2}{\sqrt{A(0)}} - \kappa t\right)^2} \tag{13.37}$$

The computed values of $\chi(\delta)$ and $\kappa(\delta)$ are shown in Figure 13.9. It is then clear that the self-growth mechanism by using the substrate jump to collect liquid from the front substrate is only possible for $\delta > \delta_*(\kappa > 0)$. The blow-up time decreases with increasing $\delta - \delta_*$ and the blow up evolves in the manner described by (13.33) and (13.37). This blow-up behavior is supported by the wave tracings of Figure10 where the large-time asymptotics of Figure 13.7 are overlayed. The growth of the supercritical pulse is dramatically different from the decay of a subcritical pulse in Figure 13.6a. The asymptotic blow-up behavior of (13.37) and (13.33) is confirmed in Figure 13.11a by carrying out simulation for five values of $\delta > \delta_*$. The recorded trailing substrate thickness χ of the $\delta = 1.5$ case of Figure 13.10 is shown in Figure 13.11b. The substrate thins rapidly during

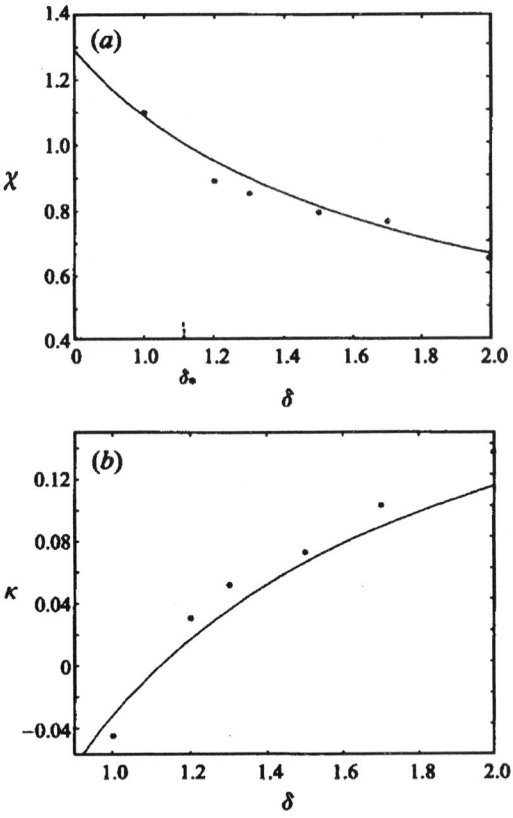

Figure 13.9: Measured χ and k during the shock-drived growth of Figure 13.10 and 13.11 compared to theoretical predictions of (13.32) and (13.36). The first data point is actually for a decaying pulse but the theory seems to still work.

the pulse formation transient and slowly approaches the equilibrium value by $t = 10$ when the growing pulse is fully developed. The measured asymptotic $\chi(\delta)$ and $\kappa(\delta)$ values are favorably compared to the theoretical values of (13.36) and (13.32) in Figure13.9. Although the theory is developed for the growth dynamics of positive substrate jumps ($\chi < 1$), it seems to also capture the nonlinear decay dynamics towards equilibrium for δ between 0.8 and δ_*. This confirms the observation that the decay dynamics towards these large equilibrium pulses captured in Figure 13.7 are actually transient nonlinear dynamics driven by finite amplitude negative substrate jumps ($\chi > 1$). Unlike the growth dynamics that can proceed indefinitely with a large reservoir of liquid, the decay dynamics will eventually evolve into exponential decay. The algebraic blow up behavior captured by (13.33) also indicates that the differential speed of two supercritical pulses blows up in time unless they are exactly the same amplitude. Hence, a

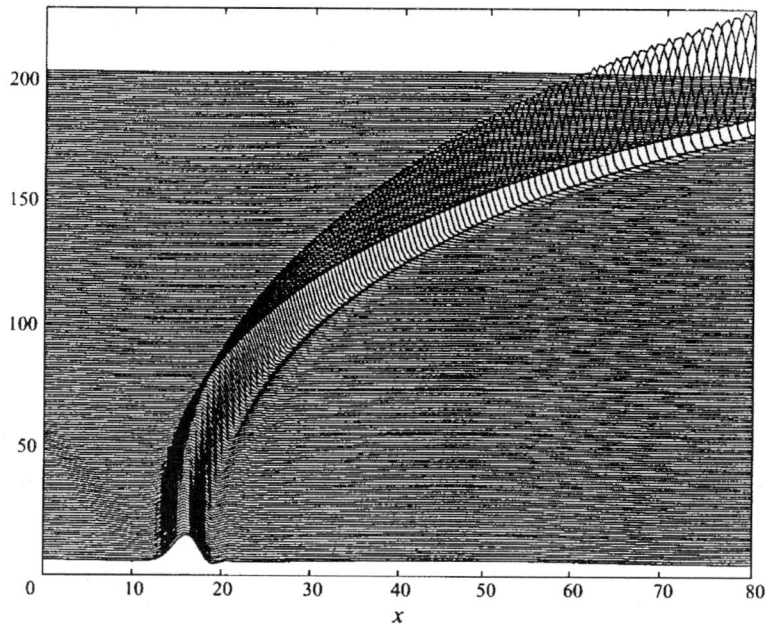

Figure 13.10: Large-time self-similar blow-up of the pulse in Figure 13.7

large supercritical pulse will quickly outrun any smaller supercritical pulses in front. This explains the propensity of supercritical pulses to coalesce and the robustness of their coalescence cascades seen in Figures 13.1 and 13.2.

13.4 Discussion

As is evident in Figures 13.1 and 13.2b, the substrate thickness thins monotonically in a finite periodic domain as liquid drains into the growing pulses. As a result, the growing pulses see a gradually thinning front substrate. The analysis of section 4 is for a "normalized" growing pulse with unit substrate thickness. We can renormalize the thinning substrate by realizing that (13.3) is invariant to

$$h \to h/\chi \quad c \to c/\chi^2 \quad \delta \to \delta\chi$$
$$x \to x \quad t \to \chi^2 t \tag{13.38}$$

where χ now refers to the thickness of the front substrate. Consequently, the effective δ for the growing pulse is $\delta\chi$. For an initially supercritical film, (13.38) indicates that as $\delta\chi$ approaches δ_*, the pulse stops growing and front and trailing substrates equilibrate to the same equilibrium value of

$$\chi_{eq} = \delta_*/\delta \tag{13.39}$$

This equilibrium substrate is clearly approached in our supercritical simulations of Figures 13.1 and 13.2 with random initial conditions. After the coalescence

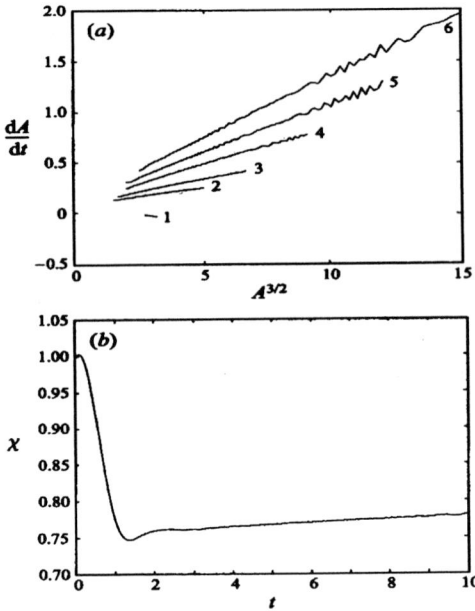

Figure 13.11: (a) The self-similar blow-up of the amplitude of a growing pulse sucking liquid from the front substrate driven by the shock. The growth rate of the pulse amplitude A relative to the mean thickness dA/dt is shown to be linear with respect to $A^{3/2}$ during the growth for $\delta = 1.0, 1.2, 1.3, 1.5, 1.7$, and 2 (curves 1 to 6). The first curve breaks the correlation since this pulse actually decays as $\delta < \delta_*$. (b) Approach to sn equilibrium trailing substrate thickness for the $\delta = 1.5$ case in Figure 13.7

cascade, only one supercritical pulse remains and it grows until the front substrate that feeds it thins to χ_{eq}, as seen in Figure 13.2b. The growth also stops at $\chi = \chi_{eq}$ when there are multiple supercritical pulses provided they have ceased to interact and coalesce. Kerchman and Frenkel have carried out extensive supercritical simulations on a periodic domain with random initial conditions. We reproduce their recorded range of substrate thickness at the end of their simulations in Figure 13.12. In some cases, there remain multiple pulses at the end that continue to interact and coalesce. Nevertheless, the upperbound of their band of equilibrium substrate thickness is closely approximated by (13.39), as seen in Figure 13.12. The recorded lower equilibrium values suggest there are patches of subcritical equilibrium pulses on thinner substrates other than the supercritical growing pulses. It must be recognized that the cessation of pulse growth at χ_{eq} is due entirely to our finite periodic computation domain. The final growing pulse in Figure1 actually returns around the domain and drains

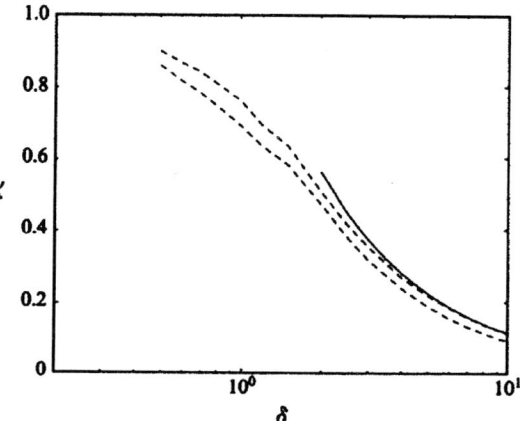

Figure 13.12: The predicted equilibrium substrate thickness with random intial noise at slightly supercritical conditions ($\delta > \delta_*$) as a function of δ (solid curve). The dotted curves bound the measured values from Kerchman & Frenkel's various simulations. The theory obviously is not appropriate for $\delta < \delta_*$.

its own substrate. On a real fiber, pulses grow down the fiber and a growing pulse would sustain the same χ as it moves down. Whether growing pulses would develop then depends on the substrate thickness h_f at the point the dimple pulses first form from the initial nonstationary waves, as in Figure1b. This thickness is smaller than the original waveless value h_0 due to drainage into the infant pulses. If h_f is smaller than h_c, the pulses will equilibrate into stationary subcritical pulses and if it exceeds h_c, growing pulses will form and develop into drops by coalescence and by collecting liquid from the substrate. Since h_f is smaller than h_0, it is clear that if h_0 is smaller than h_c, drops will not form. It is not as clear whether drops will form if h_0 is above h_c. The value of h_f is dependent on the pulse density at formation and hence sensitive to the initial conditions. Simulations by Kerchman and Frenkel indicate that, for h_0 slightly in excess of h_c, growing supercritical pulses or saturated subcritical pulses are equally likely as the final outcome. It is not until h_0 exceeds $3h_c$ that coalescence cascades driven by growing pulses always form from small-amplitude random initial conditions. This is also seen in Figure1 where h_f in Figure1b is about 1/3 the initial thickness. There is hence a band of initial wave thickness, $h_c < h_0 < 3h_c$, from which drop formation is possible but can only be induced with large and localized perturbations which produce a low pulse density at the formation stage. We suspect that Quere's experimental condition introduces such perturbations rather than small-amplitude random noise. This would explain why his measured critical h_0 is so close to h_c.

References

[1] Alekseenko, S. V., Nakoryakov, V. E. and Pokusaev, B. G., "Wave formation on a vertical falling Film", [in Rassian], Zhurn. Prikl. Mekh. Tekhn. Fiz., 6, 77-87 (1979).

[2] Alekseenko, S. V., Nakoryakov, V. E. and Pokusaev, B. G., "Wave formation on a vertical falling Film", AIChE J, 31, 1446-1460 (1985).

[3] Alekseenko, S.V., Nakoryakov, V. E. and Pokusaev, B.G. , " Wave flow of liquid films", Begel House, Inc (1994).

[4] Arneodo, A., Collet, P., Speigel, E.A., and Tresser, C., " Asymptotic Chaos", Preprint, Universite de Nice (1982).

[5] Arnold, V.I., Mathematical methods of classical mechanics. New York: Springer-Verlag, (1978).

[6] Arnold, V.I., " Reversible systems, - Nonlinear and turbulent processes in physics", Proc. Second Intern. Conf., Kiev, 1983.- Ed. R.Z.Sagdeev, New York: Harwood Acad. Publishers, 1161-1174 (1984).

[7] Arnold, V.I., and Sevryuk, M.B., "Oscillations and bifurcations in reversible systems", In: "Nonlinear Phenomena in plasma physics and hydrodynamics", Ed. R. Z. Sagdeev, Moscow, Mir Publishers, 31-64 (1986).

[8] Arnold, V.I., Ordinary differential equations. New York: Springer-Verlag (1992).

[9] Armbruster, D., Guckenheimer, J. and Holmes, P. "Heteroclinic cycles and modulated travelling waves in systems with $O(2)$ symmetry", Phis. D, 29, 257-282 (1988).

[10] Atherton, R.W. and Homsy, G.M., "On the derivation of evolution equations for interfacial waves",Chem. Eng. Comm. 2, 57 (1976).

[11] Bach, P. and Villadsen, J., "Simulation of the vertical flow of a thin, wavy film using a finite element method", J. of Heat Mass Transfer, 27, N 6. 815-827 (1984).

[12] Balmforth, N. J., Ierely, G. R. and Spiegel, E. A., "Chaotic pulse trains", SIAM J. Applied Math, 54, 1291-1334 (1994).

[13] Balmforth, N. J., "Solitary waves and homoclinic orbits", Ann. Rev. Fluid Mech., 27, 335-373 (1995).

[14] Balmforth, N. J., Ierely, G. R. and Worthing, R., "Pulse dynamics in a unstable medium", SIAM J. Applied Math, 57, 205-251 (1997).

[15] Benjamin, T.B., "Wave formation in laminar flow down an inclined plane", J. of Fluid Mech., 2, 554-574 (1957).

[16] Benney, B. J., "Long Waves in Liquid Films", J. Math. Phys., 45, 150-155 (1966).

[17] Bers, A., "Linear waves and instabilities, in 'Physique de Plasmas' ", edited by C. De Witt and J. Peyraud C Gordon and Breach, New York (1975).

[18] Betchov, R. and Criminale, W.O., "Stability of parallel flows", Academic Press, New York (1967).

[19] Binnie, A.M., "Experiments on the onset of wave formation on a film of water flowing down a vertical plate", J. Fluid Mech., 2, 551-553 (1957).

[20] Binnie A.M. "Instability in a slightly inclined water channel", J. Fluid Mech., 5, 561-570 (1959).

[21] Brauner, H. and Maron, D.M., "Modeling of wavy flow inclined thin films", Chem. Eng. Sci., 38. N 5, 775-788 (1983).

[22] Bretherton, F. P., "The motion of long bubbles in tubes", J. Fluid Mech., 10, 166-188 (1961).

[23] Brock, R. R., "Development of roll-wave trains in open channels", J.of Hydraulics Division, Proc. Am. Soc. Siv. Eng., July, HY 4, 1401-1427, (1969).

[24] Brock, R. R., "Periodic permanent roll waves", J. of Hydraulics Division, Proc. Am. Soc. Civ. Eng., Dec., HY 12, 2565-2580 (1970).

[25] Bruin, G. J., "Stability of layer of liquid flowing down an inclined plane", J. Eng. Math., 78, N 3, 1186-1201 (1974).

[26] Bunov, A.V., Demekhin, E.A. and Shkadov, V.Ya., "Stability of flows with free surface", Report N 2745, Institute of Mechanics (Moscow University) (1982).

[27] Bunov, A.V., Demekhin, E.A. and Shkadov, V.Ya., "On the non-uniqueness of non-linear wave solutions in a viscous layer", Prikl. Matem. Mekhan., 48, 4, 691-696 (1984).

[28] Bunov, A.V., Demekhin, E.A. and Shkadov, V.Ya., "Bifurcations of solitary waves in a flowing liquid film", Vestnik Moskovskogo Universiteta, Mekhanika, 41, N 2, 37-38 (1986).

[29] Bykov, V.V., "On the appearance of periodic motions from a separatrix contour of a three-dimensional system", Advantages of Mathematical Sciences, 32, N.6, 213-214 (1977).

[30] Bykov, V.V., "Bifurcation of dynamical system close to system with separatrix contour containing a saddle-focus. In "Methods of qualitative theory of differential equations", Gorkii University, Gorkii (in Russian), 44-72 (1980).

[31] Chang, H.-C., "Traveling waves on fluid interfaces: Normal form analysis of the KS equation", Phys. Fluids, 29(10), 3142-3147 (1986).

[32] Chang, H.-C., "Nonlinear waves on liquid film surfaces, I. Flooding in vertical tubes", Chem. Eng. Sci., 41, 2463-2476 (1986).

[33] Chang, H.-C., "Evolution of nonlinear waves on vertically falling films – a normal form analysis", Chem. Eng. Sci., 42, 515-533 (1987).

[34] Chang, H.-C., "Onset of nonlinear waves on falling Films", Phys. Fluids, A1, 1314-1327 (1989).

[35] Chang, H.-C., "Wave evolution on a falling film", Ann. Rev. of Fluid Mech., 26, 103-136 (1994).

[36] Chang, H.-C. and Demekhin, E. A., "Solitary wave formation and dynamics on a falling film", Adv. in Applied Mech., 32, 1-58 (1995).

[37] Chang, H.-C. and Demekhin, E.A., "Coalescence cascade towards drop formation", J. Fluid Mech., 380, 233-255 (1999).

[38] Chang, H.-C., Demekhin, E. A. and Kalaidin, E. N., "Interaction dynamics of solitary waves on a falling film", J. Fluid Mech., 294, 123-154 (1995).

[39] Chang, H.-C., Demekhin, E. A. and Kalaidin, E. N., "A simulation of noise-driven wave dynamics on a falling film", AIChE J., 42(6), 1553-1568 (1996a).

[40] Chang, H.-C., Demekhin, E. A., Kalaidin, E. N. and Y. Ye., "Coarsening dynamics of falling-film solitary waves", Phys. Rev. E., 54, 1467-1471 (1996b).

[41] Chang, H.-C., Demekhin, E. A. and Kalaidin, E.," Generation and suppression of radiation by solitary pulses", SIAM J. App Math., 58, 1246-1277 (1998).

[42] Chang, H.-C., Demekhin, E. A. and Kopelevich, D. I., "Nonlinear evolution of waves on a vertically falling film", J. Fluid Mech., 250, 443-480 ,(1993).

[43] Chang, H.-C., Demekhin, E. A. and Kopelevich, D. I., "Laminarizing effects of dispersion in an active-dissipative nonlinear medium", Physica D, 63, 299-320 (1993).

[44] Chang, H.-C., Cheng, M., Demekhin, E. A. and Kopelevich, D. I., "Secondary and tertiary excitation of three-dimensional patterns on a falling film", J. Fluid Mech., 270, 251-275 (1994).

[45] Chang,H.-C., Demekhin, E.A. and Kopelevich, D.I., "Stability of a solitary pulse against wave packet disturbance in an active medium", Phys. Rev. Lett, 75, 1747-1750 (1995).

[46] Chang, H.-C., Demekhin, E.A. and Kopelevich, D.I., "Local stability theory of solitary pulses in an active medium", Physica D, 97, 353-375 (1996).

[47] Chang, H.-C., Demekhin, E.A., Kopelevich, D.I. and Ye, Y., "Nonlinear wavenumber selection in gradient-flow systems", Phys. Rev. E, 55, 2818-2828 (1997).

[48] Chen, L.-H. and Chang, H.-C., "Nonlinear waves on liquid film surfaces, II. Bifurcation analyses of the long-wave equations", Chem. Eng. Sci., 41, 2477-2486, (1986).

[49] Cheng, M. and Chang, H. - C., "A generalized sideband stability theory via center manifold projection", Phys.Fluids, 2, 1364-1379 (1990).

[50] Cheng, M. and Chang, H.-C., "Subharmonic instabilities of finite-amplitude monochromatic waves", Phys. Fluids, 4, 505-523 (1992).

[51] Cheng, M. and Chang, H.-C., "Stability of axisymmetric waves on liquid films flowing down a vertical column to azimuthal and streamwise disturbance", Chem. Eng. Comm., (Special Issue in honor of S. G. Bankoff), 118, 327-340, (1992).

[52] Cheng, M. and Chang, H.-C., "Competition between Sideband and Subharmonic Secondary Instability on a Falling Film", Phys. Fluids, 6, 34-54 (1995).

[53] Chu, K.J. and Dukler, A.E., "Studies of the substrate and its wave structure",AIChE J., 20, 4, 695-706 (1974).

[54] Chu, K.J. and Dukler, A.E., "Structure of large waves and their resistance to gas films",AIChE J., 21, 3, 583-593 (1975).

[55] Collet, P. and Elphick C.," Topological defects dynamics and Melnikov's theory", Physics Letters A, 121, N 5, 233-236 (1987).

[56] Conrado, C. V. and Bohr, T., "Singular gwouth shapes in turbulent field theory", Phys. Rev. Lett., 72, 3522-3525 (1994).

[57] Cross, M.C. and Hogenberg, P.C., "Pattern formation outside of equilibrium, Reviews of Modern Physics", 65, N.3, July, 851-1112 (1993).

[58] Demekhin, E.A. and Kaplan, M.A., "Stability of stationary travelling waves on the surface of a vertical film of viscous fluid", Izv. Akad. NAUK SSSR, Mekh. Zhidk. i Gaza, 3, 23-41 (1989).

[59] Demekhin, E.A. and Kaplan, M.A., "Construction of exact numerical solutions of the stationary traveling wave type for viscouse thin films", Izv. Akad. NAUK SSSR, Mekh. Zhidk. i Gaza, 3, 23-41 (1989).

[60] Demekhin, E.A., Kaplan, M.A. and Shkadov, V.Ya., "Mathematical models of the theory of viscous liquid films", Izv. Akad. Nauk SSSR, Mekh. Zhidk. Gaza., 6, 73-81 (1987).

[61] Demekhin, E.A. and Shkadov, V. Ya., "On non-stationary waves in a viscous fluid layer", Izv.Akad. Nauk SSSR, Mekh. Zhidk., i Gaza 3, 151-154 (1981).

[62] Demekhin, E.A. and Shkadov, V.Ya., "Three-dimensional waves in a liquid flowing down a wall", Izv.Akad. Nauk. SSSR, Mekh. Zhidk., I Gaza 5, 21-27 (1984).

[63] Demekhin, E.A. and Shkadov, V.Ya., "Two-dimensional waves in a liquid film", Izv.Akad.Nauk. SSSR, Mekh. Zhidk., I Gaza 3, 63-67 (1985).

[64] Demekhin, E.A. and Shkadov, V.Ya., "Solitons in a dissipative media", *Hydrodynamics and Heat and Mass Transfer of free-surface Flows*, Institute of Thermophysics, Siberian Branch of the USSSR Academy of Science, Novosibirsk, 34-48 (1985).

[65] Demekhin, E.A. and Shkadov, V.Ya., "Two-dimensional wave regimes of a thin liquid film", Izv. Akad. Nauk SSSR, Mekh. Zhidk. Gaza., 3, 63-67 (1985).

[66] Demekhin, E.A. and Shkadov, V.Ya., "Theory of solitons in systems with dissipation", Izv.Akad.Nauk SSSR, Mekh. Zhidk, i Gaza, 3, 91-97 (1986).

[67] Demekhin, E.A., Tokarev, G.Yu. and Dyatlova, G.A, "Numerical simulation of nonstationary two-dimensional waves of freely falliny films, In book: Current Topics of Thermophysics", 105-112 (1984).

[68] Demekhin, E.A., Tokarev, G.Yu. and Shkadov, V.Ya., "Two-dimensional unsteady waves on a vertical liquid film", Teor.Osn.Khim.Tekhnol., 21, 177-183 (1987).

[69] Demekhin, E.A., Tokarev, G.Yu. and Shkadov, V.Ya., "On the existence of critical Reynolds number for the falling by gravity liquid film", Teor. Osn. Khim. Tekhnol., 21 , N 4, 555-559 (1987).

[70] Demekhin, E.A., Tokarev, G.Yu. and Shkadov, V.Ya., "Numerical simulation of evolution of three-dimensional waves in falling viscous layers", Vestnik MGU, ser. mat.-meh., 2, 50-54 (1988).

[71] Demekhin, E.A., Tokarev, G.Yu., and Shkadov, V.Ya., "Hierarchy of bifurcations of space-periodic structures in a nonlinear model of active dissipative media", Physica D, 52, 338-350 (1991).

[72] Dressler, P.S., "Mathematical solution of the problem of roll-waves in inclined open channels", Pure Appl. Math., 2, 149-194 (1949).

[73] Dukler, A.E., "The role of waves in two phase flow: some new understanding", Chem. Eng. Educ., 108-138 (summer 1976).

[74] Elphick, C., Iereley, G. R., Regev, O. and Spiegel, E. A., "Interacting Localized Structures with Galilean Invariance", Phys. Rev. A, 44, 1110-1122 (1991).

[75] Elphick, C., Meron, E. and Spiegel, E.A., "Spatiotemporal complexity in travelling patterns", Phys. Rev. Letters, 61, N 5, 496-499 (1988).

[76] Elphick, C., Meron, E. and Spiegel, E.A., "Patterns of propogating pulses", SIAM Appl. Math., 50, N 2, 490-503 (1990).

[77] Evans, J.W., "Nerve axon equations.III.Stability of the nerve impulse", Indiana Univ. Math.J., 22, 577-594 (1972).

[78] Fasel, H.F., Rist, U. and Konzelmann, U., "Numerical Investigation of the Three-Dimensional Development in Boundary Layer Transition", AIAA, 87, 1203-1234 (1987).

[79] Floryan, J.M., Davis, S.H. and Kelly, R.E., " Instabilities of a liquid film down a slightly inclined plane", Phys. Fluid, 30, 4, 983-989 (1987).

[80] Frenkel, A.L., "Nonlinear theory of strongly undulating thin films flowing down vertical cylinders", Europhys. Lett, 18, 583-588 (1992).

[81] Frenkel, A.L and Indireshkumar, K. "Derivation and simulations of evolution equations of wavy film flows", Math. Mod. and Sim. in Hydr. Stab., 35-81 (1996).

[82] Frenkel, A.L., Indereshkumar. "Wavy film down an inclined plane: Perturbation theory and general evolution equation for the film thickness", Phys. Rev. E, 60, N4, 4143-4157 (1997).

[83] Frish, U., Zhen, S. S. and Thual, O., "Viscoelastic behavior of solutions to the KS equation", J. Fluid Mech., 168, 221-240 (1986).

[84] Fulford, G.D.," The flow of liquids in thin films", Advan. Chem. Eng., 5, 151-236 (1964).

[85] Gardner, R. and Jones, C.K.R.T., "Stability of travelling wave solutions of diffusive predator-prey systems", Transactions of the Am. Math. Soc., 327, 2, 456-524 October (1991).

[86] Gaspard, P. and Nicolis, G., "What can we learn from homoclinic orbits in haotic dynamics?", Stat. Phys., 31, 3, 499-518 (1983).

[87] Gaster, M., "A note on a relation between temporally increasing and spatially increasing disturbances in hydrodynamic stability", J. Fluid Mech., 14, 222-224 (1963).

[88] Geshev, P.I., Ezdin, B.S., "Calculation of velocity profile and wave from in fallin liquid film, in; Hydrodynamics and Heat and Mass Transfer of Free-Surface Flows", [in Russian], Institute of Heat Physics, Siberian Branch of the USSR Academy of Science, Novosibirsk, 49-58 (1985).

[89] Gjevik, B., "Occurrence of finite-amplitude surface waves on falling liquid films", Phys. Fluids, 13, N 8, 1918-1925 (1970).

[90] Glendinning, P. and Sparrow, C., "Local and global behavior near homoclinic orbits", Stat. Phys., 35, N. 56, 645-696 (1984).

[91] Goldshtik, M.A., Saporozhnikov, V.A. and Shter, V. N., "Differential factorization method in problems of hydrodinamic stability",Proc. GAMM Conf. Numer. Method in Fluid Mech., Kölh, DFVLR, 52-59 (1975).

[92] Goncharenko, B.N. and Urintsev A.L., "On the stability of viscous liquid flow on an inclined plane", PMTF, 2, 172-175 (1975).

[93] Gorshkov, K.A., Ostrovsky, L.A. and Papko, V.V., "Interactions and bound states of solitons as classical particles", Sov. Phys. JETP, 44, N 2, 306-311 (1976).

[94] Guckenheimer, J. and Holmes, P. "Nonlinear oscillations, dynamical system and bifurcations of vector fields", Springer-Verlag (1983).

[95] Hammond, P.S., "Nonlinear adjustment of a thin annular film of viscous fluid surrounding a thread of another within a circular cylindrical pipe", J. Fluid Mech, 137, 363-384 (1983).

[96] Hanratty, T.J., "Waves on Fluid interfaces", edited by Meyer R.E., 221-245 (1983).

[97] Hecke, M., Wit, E., and Saarloos, W., "Coherent and Incoherent Drifting Pulse Dynamics in a Complex Ginzburg-Landau", Phys. Rev. Lett, 75, 3830-3833 (1995).

[98] Henstock, W.H. and Hanratty, T.J., "Gas Absorption by a Liquid Layer Flowing on the Wall of a Pipe", AIChE J., 25, 122-148 (1979).

[99] Hislop, P.D. and Crawford, J.D., "Application of spectral deformation to the Vlasov-Poisson system", J. Math. Phys., 30, 12, 2819-2837 December (1989).

[100] Ho, L.W. and Patera, A.T., "A Legendre spectral element method for simulation of unsteady incompressible viscous surface flow", Comp. Methods Appl. Mech. Eng., 80, 355-366 (1990).

[101] Huerre, P. and Monkewitz, P. A., "Local and Global Instabilities in Spatially Developing Flows", Annual Rev. of Fluid Mech., 22, 473-537 (1990).

[102] Hwang, S.-H. and Chang, H.-C., "Turbulent and inertial roll waves in inclined film flow", Phys. Fluids, 30(5), 1259-1268 (1987).

[103] Hyman, J.M. and Nicolaenko, B., "The Kuramoto-Sivashinsky equation: a bridg between PDE's and dynamical systems", Physica 18D, 113-126 (1988).

[104] Hyman, J.M. Nicolaenko, B. and Zaleski, S., "Order and complexity in the Kuramoto-Sivashinsky model of weakly turbulent interfaces", Physica 23D, 265-292 (1988).

[105] Ioss, G. and Joseph, D.D., "Elementary stability and bifurcation theory." Springer - Verlag, Berlin - Heidelberg - New York, 1980.

[106] Jones, L.O. and Whitaker S., "Experimantal study of falling liquid films", AIChE Journal, 12, 525-529 (1966).

[107] Joo, S.W., Davis, S.H. and Bankoff, S.G., "On falling film instabilities and wave breaking", Phys. Fluids, A3, 231-242 (1991).

[108] Joo, S.W. and Davis, S.H., "Instabilities of three-dimensional viscous falling films", J. Fluid Mech., 242, 529-547 (1992).

[109] Jurman, L.A., and McCready, M.J., "Study of waves on thin liquid films sheared by turbulent gas flows", Phys.Fluids A 1, 3, 522-536 (1989).

[110] Kadomstev, B.B. and Karpman, V.I., "Nonlinear waves", Uspekhi Fiz. Nauk, 103, N 2, 199-232 (1971).

[111] Kalaidin, E., "Spatial evolution of perturbations and soliton dynamics in falling films", PhD thesis, Novosibirsk (1996).

[112] Kalliadasis, S. and Chang, H.-C., "Drop Formation during Coating of Vertical Fibers", J. Fluid Mech., 261, 135-168 (1994a).

[113] Kalliadasis, S. and Chang, H.-C. "Apparent dynamic contact angle of an advancing gas-liquid meniscus", Phys. Fluids, 6, 12-23 (1994b).

[114] Kalliadasis, S. and Chang, H.-C., "Effects of wettability on spreading dynamics", Ind. Eng. and Chem. Fund, 35, 2860- 2874 (1996).

[115] Kamke, E., "Differentialgleichungen losungsmethoden und losungen." Verbesserte Auflage, New York (1948).

[116] Kapitza, P.L., "Wave flow of a thin viscous fluid layers", Zh. Eksp. Teor. Fiz., 18, 1 (1948).

[117] Kapitsa, P.L. and Kapitsa, S.P., "Wave flow of thin liquid layers", Zh.Eksp. Teor. Fiz., 19, 105-120 (1949).

[118] Kaplan, R.E.," The stability of laminar incompressible boundary layer in the presence of compliant boundaries", Aeroelastic and Structures Research Lab., ASRL-TR 116-1, Massachusetts Inst. of Technology (1964).

[119] Karpman, V.I., "Non-linear waves in dispersive media", Oxford, New York, Pergamon Press, 1975.

[120] Kath, W.L. and Smyth, N. F., "Soliton Evolution and Radiation Loss for the KdV Equation", Phys. Rev. E, 51, 661-670 (1995).

[121] Kawahara, T.,"Formation of saturated solution in a nonlinear dispersive system with instability and dissipation", Phys. Rev. Lett., 51, N 5, 381-383 (1983).

[122] Kawahara, T. and Toh, S., "Pulse interactions in an unstable dissipative-dispersive nonlinear system", Phys. Fluids, 31, 2103-2111 (1987).

[123] Kawasaki, K., and Ohta, T., "Kink dynamics in one-dimensional nonlinear systems", Physica 116 A, 573-593 (1982).

[124] Kerchman, V.I. and Frenkel, A.L. "Interactions of coherent structures in a film flow: simulations of a highly nonlinear evolution equation", Theoret. Comp. Fluid Dynamics, 6, 235-254 (1994).

[125] Kevrekides, I., Nicolaenko, B. and Scovel, J.C., "Back in the saddle again a computer-assisted study of the Kuramoto-Sivashinsky equation", SIAM J. Appl. Math., 50, 760-790 (1990).

[126] Kliakhandar, I.L., "Inverse cascade in film flows", J. Fluid Mech., 423, 205-255 (2000).

[127] Khesghi, H.A. and Scriven, L.E., "Disturbed Film Flow on a Vertical Plane", Phys. Fluids., 35, 990-997 (1987).

[128] Krantz, W.B., Goren, S.L., "Finite-amplitude, long waves on liquid films flowing down a plane", Ind. Engng. Chem. Fund., 9, N 1, 107-113 (1970).

[129] Krantz, W.B. and Goren, S.L., "Stability of thin liquid films flowing down a plane", Ind. Eng. Chem., 10, 91-101 (1971).

[130] Kuramoto, T. and Tsuzuki T., "Persistent propagation of concentration waves in dissipative media far from thermal equilibrium", Prog. Teor. Phys., 55, N 2, 356-369 (1976).

[131] Lamb H., Hydrodinamics, 4th ed., Cambridge Univ. Press, Cambridge (1947).

[132] Landau, L.D. and Lifshitz E.M., Fluid Mechanics, Pergamon (1959).

[133] Lee, J., "Kapitza's method of film flow destription", Chem. Eng. Sci., 24, 309-1313 (1969).

[134] Leslie, M. Mack,"A numerical study of the temporal eigenvalue spectrum of the Blasius boundary layer", J. Fluid Mech., 73, 3, 497-520 (1976).

[135] Levich, B., "Phisiochimical Hydrodynamics." Prentice Hall (1962).

[136] Lighthill, J.," Waves in fluids", Cambridge Univ. Press, Cambridge (1978).

[137] Lin, C.C., "The theory of hydrodynamic stability", Cambridge Univ. Press. Cambridge (1967).

[138] Lin, S.P., "Finite amplitude stability of a parallel flow with a free surface", J. Fluid Mech., 36, 113-126 (1969).

[139] Lin, S.P., "Instability of liquid film down an inclined plane", Phys. Fluids, 10, N 2, 308-313 (1967).

[140] Lin, S.P., "Finite amplitude side-band stability of viscous film", J. Fluid Mecj., 63, pt. 3, 417-429 (1974).

[141] Lin, S.P. and Krishna, M.V.G., "Nonlinear stability of viscous film with respect to three-dimensional side-band disturbances", Phys. Fluids, 20, N 7, 1039-1044 (1977).

[142] Lin, S.P. and Krishna, M.V.G., "Stability of liquid film with respect to initially finite three-dimensional disturbances", Phys. Fluids, 20, N 13, 2005-2011 (1977).

[143] Lin, S.P., "Film Waves in Waves on Fluid Interfaces", ed. R. E. Meyer, 261-290 (1983).

[144] Liu, J., Paul J.P. and Gollub, J.P., "Measurement of the primary instabilites of film flows", J. Fluid Mech., 220, 69-101 (1993).

[145] Liu, J. and Gollub, J.P., "Solitary wave dynamics of film flows", Phys. Fluids, 6, 1702-1712 (1994).

[146] Liu, J, Schneider, J.B. and Gollub, P.J., "Three-dimensional instabilitites on film flows", Phys. Fluids, 7(1), 55-67 (1995).

[147] Mack, L. M., "A numerical study of the temporal eigenvalue spectrum of the Blasius boundary layer", J. Fluid Mech., 73, 497-520 (1976).

[148] Malamataris, N.T. and Papanastasiou, T.C., "Unsteady free surface flows on truncated domains", Ind. Eng. Chem. Res., 30, 2211-2219 (1991).

[149] Mei, C.C., "Nonlinear gravity waves in thin sheet of viscous fluids", J. App. Math. and Phys., 45, N 2, 266-288 (1966).

[150] Michelson, D.M. and Sivashinsky, G.I., "Nonlinear analysis of hydrodynamic instability in laminar flames II", Numerical experiments. - Acta Astranautica, 4, 1207-1221 (1977).

[151] Michelson, D., "Steady solutions of Kuramoto-Sivashinsky equation", Physica 19D, 89-111 (1986).

[152] Nakaya, C., "Long waves on thin fluid layer flowing down an incline plane", Phys. Fluids, 18, 1407-1412 (1975).

[153] Nakaya, C., "Waves on a viscous fluid film down a vertical wall", Phys. Fluids, A1, 1143-1154 (1989).

[154] Nakoryakov, V.E. and Alekseenko, S.V., "Waves on liquid falling film falling down an inclined surface, in Wave processes in two-phase media", [in Russian], IT SO AN SSSR, Novosibirsk, 66-79 (1980).

[155] Nakoryakov, V.E., Pokusaev, B.G., Alekseenko, S.V. and Orlov, V.V., "Instantaneous velocity profile in a wavy liquid film", Inzh.-Fiz.Zh., 33, 3, 399-405 (1976).

[156] Nakoryakov, V.E., Pokusaev, B.G. and Alekseenko, S.V., "Stationary two-dimensional rolling waves on a vertical film of fluid", Inzh.-Fiz.Zh., 30, 780-785 (1976).

[157] Nakoryakov, V.E., Pokusaev, B.G., Khristoforov, V.V. and Alekseenko, S.V., "Experimental investigation of film on vertical wall", Inzh.-Fiz. Zh., 27, N 3, 397-401 (1974).

[158] Nakoryakov, V.E., Pokusaev, B.G. and Radev, K.B., "Waves and their effect on convective gas diffusion in falling films", [in Russian], Zhurn. Prikl. Mekh. Tekh. Fiz., 3, 95-104 (1987).

[159] Nayfeh, A.H., "Perturbation Methods", Wiley-Interscience (1973).

[160] Needham, D.J. and Merkin, J. H., "On roll waves down an inclined channel", Proc. R. Soc. London Ser. A, 394, 259-268 (1984).

[161] Nemet, I., and Sher, V., "Hydrodynamics of thin liquid films flowing down a vertical plane", Acta. Chin. Acad. Sci. Hung., 60, 103-121 (1969).

[162] Nepomnyashchy, A.A., "Stability of wave regimes in a film flowing down an incline plane", Izv.Akad. Nauk. SSSR Mekh. Zhidk. I Gaza, 3, 28-34 (1974).

[163] Nepomnyashchy, A.A.," Three-dimensional spatial-periodic motions in liquid films flowing down a vertical plane", In: Hydrodynamics, 7 [in Russian], Perm, 43-52 (1974).

[164] Nepomnhyashachy, A.A., "Wave motion in layer of viscous liquid flowing down an inclined plane", In: Hydrodynamics, [in Russian], Perm, 114-127 (1976).

[165] Nquen, L.T. and Balakotaian, V., "Modeling and experimental studies of wave evolution on free falling viscous films", Phys. Fluids, 12, 9, 2236-2256 (2000).

[166] Ooshida, T., "Surface equation of falling film flows which is valid even far beyound the criticality", Phys. Fluids, 11, 3247-3269 (1999).

[167] Orszag, S.A., "Accurate solution of the Orr-Sommerfeld stability equation", J. Fluid Mech., 50, 4, 689-704 (1971).

[168] Pego, R.L. and Weinstein, M.I., "Eigenvalues, and instabilities of solitary waves", Phil. Trans. R. Soc. London. A, 340, 47-94 (1992).

[169] Pego, R.L. and Weinstein, M.I., "Asymptotic stability of solitary waves", Commun. Math. Phys., 164, 305-349 (1994).

[170] Petviashvili, V.I. and Tsvelodub, O,Yu., "Horseshoe-shaped solitons on an inclined viscous liquid film", Dokl.Akad. Nauk SSSR, 238, 6, 1321-1323 (1978).

[171] Pierson, F.W. and Whitaker, S., "Some theoretical and experimental observations of the wave structure of falling liquid films", Ind. Eng. Chem. Fundam., 16, 401-408 (1977).

[172] Pokusaev, B.G. and Alekseenko, S.V., "Two-dimensional waves on a vertical liquid film", In: Nonlinear Wave Processes in Two-Phase Media, [in Russian], Novosibirsk, 158-172 (1977).

[173] Portalsky, S., "Velocities in film flow of liquids on vertical plates", Chem. Eng. Sci., 19, 575-582 (1964).

[174] Portalsky, S. and Clegg, A.J., "An experimental study of falling liquid films", Chem. Eng. Sci., 27, 1257-1265 (1972).

[175] Prokopiou, T., Cheng M. and Chang, H.-C.,"Long waves on inclined films at high Reynolds number", J. Fluid Mech., 222, 665-691 (1991).

[176] Pukhnachev, V.V., "Two-dimensional stationary problem with free boundary for Navier-Stocks equation", Zh. Prikl. Mekhan. Tekh. Fiz., 3, 91-102 (1972).

[177] Pumir, A., Manneville, P. and Pomeau, Y., "On solitary waves running down an inclined plane", J. Fluid Mech., 135, 2-50 (1983).

[178] Quere, D., "Thin films flowing on vertical fibers", , Europhys. Lett, 13, 721-726 (1990).

[179] Rabinovich, M.I. and Trubetzkov, D.I., "Introduction in theory of oscillations and waves", [in Russian], Nauka, Moscow, (1984).

[180] Radev, K.B., "Wave effects on heat and mass transfer at film condensation", [in Russian], Preprint No 9, ITMO AN BSSR, Minsk (1989).

[181] Ramaswamy, B., Chippada, S. and Joo, S.W., J. Fluid Mech., 325, 163-194 (1996).

[182] Reed, M. and Simon, B., "Methods of Modern Mathematical Physics. IV Analysis of Operators", New York, Academic Press (1978).

[183] Roberts, A.J., "Boundary Conditions for Approximate Differential Equations", J. Austral. Math. Soc., B34, 54 (1992).

[184] Roskes, G.J., "Three-dimensional long waves on liquid film", Phys. Fluids, 13, N 6, 1440-1445 (1970).

[185] Rushton, E. and Davis, G., "Linear analysis of liquid film flow", AIChE Journal, 17, N 3, 6701-676 (1971).

[186] Ruyer-Quil, C. and Mannevile, P.,"Modeling film flows down inclines planes", Eur. Phys. J., B 6, 277-292 (1998).

[187] Ruyer-Quil, C. and Mannevile, P., "On the modeling of flows down inclines: Convergence of weighted-residual approximations", Phys. Fluids, (to be published) (2001).

[188] Salamon, T.R., Armstrong, R.C. and Brown, R. A., "Travelling waves on inclined films: numerical analysis by the finite-element method", Phys. Fluid, 6, 2202-2220 (1994).

[189] Sangalli, M., Prokopiou, T., McCready, M. J. and Chang, H.-C., "Observed transitions in two-phase stratified gas-liquid flow", Chem. Eng. Sci., 47, 3289-3291 (1992).

[190] Sangalli, M., McCready, M. J. and Chang, H.-C., "Stabilization mechanism of short waves in stratified gas-liquid flow", Phys. Fluids, 9, 919-939 (1997).

[191] Schlichting, H., "Boundary layer theory", McGraw-Hill, New York (1968).

[192] Sevryuk, M.B., "Reversible systems", Lect.Notes in Math. Berlin: Springer- Verlag, 1211, (1986).

[193] Shilnikov, L.P., "A case of the existence of a countable set of periodic solutions", Sov. Math. Dokl., 6, 163-166 (1965).

[194] Shilnikov, L.P., "On a Poincare-Birkhoff problem", Math. USSR Sb. 3, 353-371 (1976).

[195] Shkadov, V.Ya., Belikov, V.A. and Epikhin, V.E., "Stability of flows with free surface", Report N2450, Institute of Mechanics (Moscow Unerversity) (1980).

[196] Shkadov, V.Ya., "Wave modes in the gravity flow of a thin layer of a viscious fluid", Izv. Akad. Nauk. SSSR, Mekh. Zhid.i Gaza, 3, 43-51 (1967).

[197] Shkadov, V.Ya., "Theory of wave flows of a thin layer of a viscous liquid", Izv. Akad. Nauk. SSSR Mekh. Zhidk I Gaza 2, 20-25 (1968).

[198] Sivashinsky, G.I., "Nonlinear analysis of hydrodynamic instability in laminar flame", I. -Derivation of basic equation, Acta Astronautica, 4, 1177-1206 (1977).

[199] Squire, H.B., "On the stability of three-dimensional disturbances of viscous flow between parallel walls", Proc. Roy. Soc., A142, 621-628 (1933).

[200] Stainthorp, F.P. and Allen, J.M., "The Development of Ripples on the Surface of Liquid Film Flowing Inside a Vertical Tube", Trans. Inst. Chem. Eng., 43, T85-T91 (1965).

[201] Stoker, J.J., "Water waves." Interscience, (1957).

[202] Strobel, W.J. and Whitaker, S., "The effects of surfactants on the flow characteristics of falling liquid films",AIChE J., 15, 4, 471-476 (1977).

[203] Swinton, J., "The stability of homoclinic pulses: a generalization of Evans' method", Phys. Lett.A, 163, 57-62 (1992).

[204] Tailby, S.R. and Portalski, S., "The hydrodynamics of liquids films flowing on vertical surface", Trans. Instn. Chem. Engrs., 38, 324-330 (1960).

[205] Takaki, R., "Wave motions on liquids layer falling along a vertical wall", J. Phys. Soc. Japan, 27, N 6, 1648-1654 (1969).

[206] Telles, A.S. and Dukler, A.E., "Statistical characteristics thin vertical wavy liquid films", Ind. Engng. Chem. Fund., 9, N 3, 412-421 (1970).

[207] Thomas, L.H., "The stability of plane Poiseuille flow", Phys. Rev., 91, 780-783 (1953).

[208] Toh, S., Iwasaki, H. and Kawahara, T., "Two-dimensionally localized pulses of a nonlinear equation with dissipation and dispersion", J. Phis. Rev., 40, 9, 5472-5475 (1989).

[209] Topper, J. and Kawahara, T., "Approximate equation for long nonlinear waves on viscous fluid", J. Phys. Soc. Japan, 44, N 2, 663-666 (1978).

[210] Trifonov, Yu.Ya., "Steady-state traveling on the surface of a viscous liquid film falling down on vertical wires and tubes", AIChE J, 38, 821-834 (1992).

[211] Trifonov, Yu.Ya. and Tsvelodub, O.Yu., "Nonlinear waves on the surface of liquid film flowing down vertical wall", Zh. Prikl. Mekhan. Tekhn. Fiz., N 5, 15-19 (1985).

[212] Trifonov, Yu.Ya. and Tsvelodub, O.Yu., "Nonlinear waves on the surface of a falling liquid film. Part 1.", J. Fluid Mech., 299, 531-554 (1991).

[213] Tsvelodub, O.Yu., "Steady Travelling Waves on a Vertical Film of Fluid", Izv. Akad. Nauk SSSR, Mekh. Zhidk. i Gaza., 4, 142-146 (1980).

[214] Tsvelodub, O.Yu., "Stationary running waves on a film falling down inclined surfaces", Izv. Akad. Nauk SSSR, Mekh. Zhidk. i Gaza, 4, 142-146 (1980).

[215] Tsvelosub, O.Yu., and Kotychenko, L.N., "Spatial wave regimes on a surface of thin viscous liquid film", Phis. D, 63, 361-377 (1993).

[216] Tsvelodub, O.Yu. and Trifonov, Yu.Ya., "Nonlinear waves on the surface of a falling liquid film. Part 2.", J. Fluid Mech., 244, 149-169 (1992).

[217] Tuck, E.O., "Continuous coating with gravity and jet tripping ", Phys. Fluids, 26, 2352-2358 (1983).

[218] Van Dyke, M., "Perturbation methods in fluid mechanics", The Parabolic Press, Stanford, California (1975).

[219] Wasden, F.K. and Dukler, A.E., "Insights into the Hydrodymanics of Free Falling Wavy Films", AIChE J., 35, 187-192 (1989).

[220] Whitaker, S. and Jones, L.O., " Stability of falling liquid films. Effects of interface and interfacial mass transport", AIChE J., 12, 3, 421-431 (1966).

[221] Whitham, G.B., "Linear and nonlinear waves", Wiley Chichester (1974).

[222] Wilkes, J.O. and Nedderman, P., "The measurement of velocities in thin films of liquid", Chem. Eng. Sci., 17, 177-187 (1962).

[223] Wilson, S.D.R., "The drag-out problem in film coating theory", J. Engng Maths, 16, 209-221 (1982).

[224] Wilson, S.D.R. and Jones, A.F., "The entry of a falling film into a pool and the air entrainment problem", J. Fluid Mech, 128, 219-230 (1983).

[225] Yamada, T. and Kuramoto, Y., "A reduced model showing chemical turbulence", Prog. Theor. Phys., 56, N 2, 681-683 (1976).

[226] Ye. Y. and Chang, H.-C., "A spectral theory for fingering on a prewetted plane", Phys. Fluids, 11, 2494-2515 (1999).

[227] Yih, C.-S., "Stability of parallel laminar flow with a free surface", Proc. Of the Second U.S. National Congress of Applied Mech., N.Y., 623-631 (1955).

[228] Yih, C.-S., "Stability of liquid flow down an inclined plane", Phys. Fluid, 6, N 3, 321-334 (1963).

Index

absolute instability, 87, 345
active medium, 111, 208, 273
averaging, 2, 3, 51, 54, 84, 162

Benney's equation, 49, 58, 60, 61
bifurcation, 3, 33, 85, 112, 115, 116, 130, 132, 134–136, 139, 141, 145–147, 149, 150, 154, 161–167, 175–178, 207, 208, 217, 220, 320, 321, 325, 326, 338
blow-up, 60, 61, 337, 338, 377, 379, 381, 382
boundary layer equation, 11, 160, 161, 166

capillary length, 363
chaos, 98, 101, 213, 271, 330
coalescence dynamics, 108, 245, 271, 342, 364
coherent structure, 3, 107, 108, 187, 191, 192, 207, 214, 215, 218, 271, 272, 289, 313, 330
continuation, 65–67, 327
continuous spectrum, 218, 248, 273
convective instability, 92, 209, 212, 330, 345

dimple, 152, 229, 288, 366, 369, 383
discrete spectrum, 200, 201, 207, 208, 211, 214, 217, 218, 222–224, 230, 263, 273, 277, 278, 327, 371, 372
dispersion, 18, 20, 23, 33, 35, 36, 38, 42–44, 46, 59, 63–66, 112, 147, 149, 150, 166, 184, 193, 201, 210, 213, 224, 225, 239, 263, 266, 275, 282, 285, 294, 295, 318, 328, 330, 344, 370, 372, 377
dispersive medium, 201, 223, 280
dissipative medium, 111
drainage, 87, 199, 218, 219, 225, 226, 228, 230, 232, 236, 239, 243, 245, 248, 253, 255, 257, 258, 269, 271, 287, 288, 293, 310, 311, 368, 371, 372, 378, 383

essential spectrum, 201, 202, 204–208, 211, 216, 218–220, 222–226, 230, 231, 235–237, 242, 262, 264, 267, 268, 323, 370–372
Evans' function, 203, 205

filtering, 100, 101, 294, 296, 298, 313, 316, 344, 365
fixed point, 116, 117, 119–123, 127, 129, 132, 135, 145, 147, 150, 164, 166, 179
Floquet multiplier, 140, 141
Floquet theory, 179, 198, 240, 306, 314
friction factor, 343
front instability, 4, 198, 338
Froude number, 8, 31, 342, 343, 352, 353

Galilean symmetry, 218, 251
generalized Kuramoto-Sivashinsky (gKS) equation, 38, 97

generalized Kuramoto-Sivashinsky equations, 4
gravity, 5, 7, 66, 112, 288, 318, 320, 364, 375

Hamiltonian system, 116, 141
heteroclinic orbit, 3, 179, 258
homoclinic orbit, 3, 271, 275, 278, 288, 289
homoclinic tangle, 145
hydraulic shocks, 4

inception, 4, 63, 65, 73, 75–79, 81, 82, 87, 91, 97–99, 101–103, 152, 167, 168, 180, 183, 184, 189, 192, 193, 215, 294, 307, 309, 311, 313, 344, 345, 350, 362–364
inertia, 1, 16, 66, 181, 187, 274
invariance, 200, 207, 217, 253, 254, 273, 278, 303, 306, 322, 348, 350, 351, 353, 356, 360, 362, 370

KAM theory, 116
Kapitza number, 5, 8, 19, 25, 26, 29–31, 35, 60, 151, 174
Kawahara equation, 38, 145, 264, 282, 285, 317
Kuramoto-Sivashinsky (KS) equation, 35, 36, 49, 97, 121, 133, 146, 160
Kuramoto-Sivashinsky equation, 2

laminar, 79
limit-cycles, 3, 115, 129, 147, 150
longwave instability, 5, 17, 19, 33, 34, 62, 112, 300
lubrication theory, 46

modulation instability, 300, 302, 306, 308, 313, 314, 321

Navier-stocks equation, 340
Navier-Stokes equation, 3–5, 11, 32, 60, 151, 160, 174, 317
Newton scheme, 95

noise, 4, 69, 87, 91, 92, 96, 97, 99, 100, 102, 103, 105, 107, 108, 179, 184, 192, 194, 195, 198, 211, 213–215, 293–295, 299–301, 309, 313–316, 331, 340, 345, 348, 362, 383
normal form, 3, 116, 145, 146, 148–150, 275–277

Orr-Sommerfeld, 2–5, 12, 13, 23–26, 46, 63, 66, 183
overtone, 103

Padé approximation, 49, 50
periodic solutions, 38, 40, 116, 139, 141, 149, 150, 175, 300, 351
phase velocity, 16, 22, 23, 27, 32, 33, 47, 55, 78, 85, 112, 152, 313, 319
pitchfork bifurcation, 139, 152, 162, 163
primary instability, 2, 3, 61, 191
pulses, 60, 61, 71, 77, 87, 88, 97–109, 115, 120, 121, 125, 128, 129, 131, 132, 138, 150, 154, 183, 194, 195, 198–200, 205, 208, 211, 213–219, 223, 224, 226, 227, 229, 235, 236, 239, 241–245, 247–249, 251, 252, 254, 255, 259–262, 264, 267, 269–275, 278, 280, 282, 284, 285, 287–290, 293, 303, 307, 309–314, 316, 321–324, 327, 330, 331, 333, 336, 338, 340, 363, 364, 366, 368–370, 372–375, 380–383

radiation, 206–208, 210, 211, 213–219, 223, 225, 236, 239, 241, 242, 248, 262, 316, 327, 370
Rayleigh instability, 342, 363, 364

reflection symmetry, 146, 147, 150, 210
resonance pole, 3, 201, 208, 216–220, 222–228, 230, 231, 235–238, 243, 245, 253, 254, 259, 262, 267–270, 273, 313, 371–373
resonant triad, 319, 320
reversible system, 116, 140
Reynolds number, 1, 5, 8, 9, 11, 13, 14, 18–21, 23–25, 27, 29, 31, 36, 46, 47, 49, 51, 52, 56, 60, 61, 66, 69, 70, 72–75, 77–84, 97, 101, 135, 139, 151, 160, 162, 163, 168, 174, 175, 264, 293, 310, 317, 340, 342
Riemann surface, 224, 372

saddle points, 63–66, 116, 208, 209, 220, 224, 226, 267, 291, 292, 328
scallop waves, 91, 102, 316, 322, 325, 330, 338
secondary instability, 13, 87, 180, 183, 316
shear instability, 22, 27, 29
Shkadov model, 4, 53, 54, 57, 58, 60, 61, 112, 151, 152, 154, 160, 162, 164, 167, 174, 198, 243, 245, 262, 268, 287, 293, 294, 300, 302, 338
shocks, 115, 117, 258, 285, 340, 345, 348
shooting method, 125, 204, 254, 278
shortwave instability, 5, 19, 22, 36, 40
solitary waves, 1, 3, 49, 60, 72, 75–79, 83, 86–88, 91, 111, 114, 115, 117, 120, 121, 126–128, 135–137, 147, 148, 150, 154, 161, 164–167, 169–172, 175, 176, 180, 182, 187, 189–194, 196, 198, 289, 290, 293, 314, 316, 325, 340, 342, 354

soliton, 3, 74, 77, 91, 120, 139, 146, 157, 162, 227, 277, 325
steepest descent, 63, 208, 223, 266
subharmonic instability, 87, 111, 180, 181, 192, 240
surface tension, 2, 13, 16, 18–20, 22, 27, 29, 30, 35, 43, 44, 49, 174, 175, 178, 187, 344
symmetry, 92, 112, 115, 116, 118, 134, 136, 139, 142, 146, 152, 162–164, 191, 199, 200, 207, 227, 251, 253, 254, 256, 258, 285, 303, 330, 342, 350, 371

torus, 145
translational symmetry, 302
travelling waves, 3, 38, 40, 55, 93, 97, 108, 115, 116, 132, 135, 136, 139, 141, 142, 163, 198, 300, 325, 331
turbulence, 3, 271, 338
turbulent spot, 199, 330, 331, 338
turning point, 61

viscosity, 8, 16, 19, 344, 362

wave coarsening, 105, 342, 344, 350
wave number, 319, 325, 327
wave packet, 64, 193, 194, 196, 198, 209–211, 213–216, 239, 262, 264, 285, 357, 374
wave spectrum, 81, 107
Weber number, 5, 8, 27, 28, 340, 342

zero mode, 200, 224, 227, 230, 253–256, 258, 271, 274, 303, 305, 372